ENCYCLOPAEDIC DICTIONARY
OF PHYSICS

SUPPLEMENTARY VOLUME 5

Pergamon Press Ltd., Headington Hill Hall, Oxford
Pergamon Press Inc., Maxwell House, Fairview Park, Elmsford, New York 10523
Pergamon of Canada Ltd., 207 Queen's Quay West, Toronto 1
Pergamon Press (Aust.) Pty. Ltd., 19a Boundary Street, Rushcutters Bay, N.S.W. 2011, Australia

Copyright © 1975 Pergamon Press Ltd.
All Rights Reserved. No part of this publication may be reproduced, stored in a retrieval system, or transmitted, in any form or by any means, electronic, mechanical, photocopying, recording or otherwise, without the prior permission of Pergamon Press Ltd.

First edition 1975

Library of congress catalog card number: 74-5600

Printed in Great Britain by A. Wheaton & Co. Exeter.

0 08 017056 0

ARTICLES CONTAINED IN THIS VOLUME

Acoustics in the fishing industry . . . . R. E. Craig
Advances in oxide materials . . . . B. Cockayne
Applications of reverse osmosis and ultrafiltration . . . . J. R. Grover
Archaeological dating by physical methods . . . . M. J. Aitken

Barium clouds for experiments in space . . . . E. Rieger
Beam-foil spectroscopy . . . . I. Martinson
Biochemical fuel cells . . . . R. F. Acker
Black holes in space . . . . H. I. Ellington
Blocking patterns from crystals . . . . R. S. Nelson

Calculation of thermodynamic properties from spectroscopic data . . J. C. Lockhart
Carbon fibre . . . . W. G. Harland
Charge-transfer complexes . . . . R. Foster
Circular dichroism . . . . S. F. Mason
Colour transparencies, copying of . . . . R. G. W. Hunt
Counting technology, recent advances in . . . . H. W. Wilson
Critical point . . . . D. Ambrose
Crystal structure determination, direct methods of . . . . J. Karle

Detection and identification of individual atoms and ions in mass
  spectrometry . . . . G. D. W. Smith
Digital learning computers . . . . I. Aleksander

Electret transducers . . . . C. W. Reedyk
Electric motors using transverse flux . . . . E. R. Laithwaite
Electrogravimetric analysis . . . . J. V. Westwood
Electromagnetic gun . . . . E. R. Laithwaite
Enthalpimetry . . . . H. J. V. Tyrell

Ferroelectrics—recent applications of . . . . A. M. Glazer
Fibre reinforcement of cement and concrete . . . . R. W. Nurse
Field-ion microscopy and the atom-probe FIM . . . . E. W. Müller
Four-colour problem . . . . J. H. Thewlis
Fractional horsepower motors . . . . E. R. Laithwaite

Geometrodynamics . . . . D. Brill

Helium, negative ion of . . . . I. G. Main
Helium dilution refrigeration . . . . D. S. Betts
High-energy heavy ions . . . . R. H. Thomas
Holography, engineering applications of . . . . E. Archbold

Infrared galaxies . . . . N. C. Wickramasinghe
Insulation of electric power cables . . . . B. M. Weedy
Ion implantation . . . . G. Carter and W. A. Grant

Lasers in biology and medicine	D. Smart
Location and reduction of noise	B. J. Fielding
Mach effect (visual)	W. T. Welford
Magnetism, medical applications of	E. H. Frei
Medical ultrasonics	P. N. T. Wells
Mesonic atoms	Y. N. Kim
Meteorological observations from satellites	R. S. Harwood
Microwave phonons	K. W. H. Stevens
Mirror electron microscope	M. E. Barnett
Modular electronic equipment	H. Bisby
Modulation spectroscopy	D. E. Aspnes and J. E. Fischer
Monsoon meteorology	C. S. Ramage
Myoelectric control	A. K. Godden
Neutron radiography—the present position	H. Berger
Neutron stars	H. I. Ellington
Neutrons, chemical binding of	H. Bridge
Nondestructive testing, automation of	R. S. Sharpe
Optical communication systems	J. E. Midwinter
Optical data processing	A. C. Moore
Optical processing of radar and sonar signals	I. Firth
Palaeogeophysics	J. C. Briden
Pattern recognition	A. M. Andrew
Planets, magnetic fields of	I. Wilkinson
Pollution of the air	W. S. Moore
Power sources, isotopic	A. W. Penn
Pulsars	F. G. Smith
Quasars	R. W. Hunstead
Radiation and entry of space vehicles into planetary atmospheres	D. B. Olfe
Radiation and shielding in space	J. W. Haffner
Radiation dosimeter, colour	W. L. McLaughlin
Radiation, non-ionizing, hazards of	R. Oliver
Recent advances in radio and x-ray astronomy	S. Mitton
Reflection spectroscopy	J. Chamberlain
Regenerative flywheels	S. J. Peerless
Regge model of elementary particles	E. J. Squires
Resonance in the orbits of the earth and an asteroid	L. Danielsson
Signal detection, statistical theory of	A. D. Whalen
Spectroscopy, biological applications of	K. R. Naqvi
Spectroscopy, photoelectron	W. C. Price
Speech spectrometry	R. E. Bogner
Superconducting machines	A. D. Appleton and B. E. Mulhall
Superconductivity, applications of	M. W. Wilson
Super-heavy elements beyond the known periodic table	P. R. Fields and J. K. Unik
Sweep circuits	J. D. Weaver
Tachyons	T. P. Swetman
Technological forecasting	I. C. Hendry
Thermal desorption spectroscopy	J. Yarwood
Thermoluminescent dating	M. J. Aitken
Thermonuclear power—the present position	I. Cook

Thick film technology . . . . R. G. LOASBY
Time, atomic. . . . . B. W. PETLEY

Ultrasonics in medical diagnosis . . . . W. N. MCDICKEN

Void formation in fast-reactor materials . . . . R. S. NELSON

Wankel engine . . . . J. A. BARNES and N. WATSON

X-ray interferometry . . . . M. HART

LIST OF CONTRIBUTORS TO THIS VOLUME

ACKER, R. F. (*Evanston, Illinois*)
AITKEN, M. J. (*Oxford*)
ALEKSANDER, I. (*Canterbury, Kent*)
AMBROSE, D. (*Teddington, Middlesex*)
ANDREW, A. M. (*Reading*)
APPLETON, A. D. (*Newcastle*)
ARCHBOLD, E. (*Teddington, Middlesex*)
ASPNES, D. E. (*Murray Hill, N.J.*)

BARNETT, M. E. (*London*)
BARNES, J. A. (*London*)
BERGER, H. (*Argonne, Illinois*)
BETTS, D. S. (*Brighton*)
BISBY, H. (*Harwell*)
BOGNER, R. E. (*Adelaide*)
BRIDEN, J. C. (*Leeds*)
BRIDGE, H. (*Warrington*)
BRILL, D. (*München*)

CARTER, G. (*Salford*)
CHAMBERLAIN, J. (*Teddington, Middlesex*)
COCKAYNE, B. (*Great Malvern, Worcs.*)
COOK, I. (*Culham, Abingdon*)
CRAIG, R. E. (*Aberdeen*)

DANIELSSON, L. (*Stockholm, Sweden*)

ELLINGTON, H. I. (*Aberdeen*)

FIELDING, B. J. (*Manchester*)
FIELDS, P. R. (*Argonne, Illinois*)
FIRTH, I. (*Fife, Scotland*)
FISCHER, J. E. (*New Jersey, U.S.A.*)
FOSTER, R. (*Dundee*)
FREI, E. H. (*Rehovet, Israel*)

GLAZER, A. M. (*Cambridge*)
GODDEN, A. K. (*Oxford*)
GRANT, W. A. (*Salford*)
GROVER, J. R. (*Harwell*)

HAFFNER, J. W. (*California*)
HARLAND, W. G. (*Manchester*)
HART, M. (*Bristol*)
HARWOOD, R. S. (*Oxford*)
HENDRY, I. C. (*Aberdeen*)
HUNSTEAD, R. W. (*Sydney*)
HUNT, R. G. W. (*Harrow, Middlesex*)

KARLE, J. (*Washington*)
KIM, Y. N. (*Texas*)

LAITHWAITE, E. R. (*London*)
LOASBY, R. G. (*Reading*)
LOCKHART, J. C. (*Newcastle*)

MAIN, I. G. (*Liverpool*)
MARTINSON, I. (*Stockholm, Sweden*)
MASON, S. F. (*London*)
McDICKEN, W. N. (*Glasgow*)
McLAUGHLIN, W. L. (*Washington*)
MIDWINTER, J. E. (*Ipswich*)
MITTON, S. (*Cambridge*)
MOORE, A. C. (*Malvern, Worcs.*)
MOORE, W. S. (*Nottingham*)
MÜLLER, E. W. (*Pennsylvania*)
MULHALL, B. E. (*Newcastle*)

NAQVI, K. R. (*Sheffield*)
NELSON, R. S. (*Harwell*)
NURSE, R. W. (*Hertfordshire*)

OLFE, D. B. (*California*)
OLIVER, R. (*London*)

PEERLESS, S. J. (*London*)
PENN, A. W. (*Harwell*)
PETLEY, B. W. (*Teddington, Middlesex*)
PRICE, W. C. (*London*)

RAMAGE, C. S. (*Honolulu, Hawaii*)
REEDYK, C. W. (*Ottawa*)
RIEGER, E. (*München*)

SHARPE, R. S. (*Harwell*)
SMART, D. (*Newcastle*)
SMITH, F. G. (*Manchester*)
SMITH, G. D. W. (*Oxford*)
SQUIRES, E. J. (*Durham*)
STEVENS, K. W. H. (*Nottingham*)
SWETMAN, T. P. (*Cambridge*)

THEWLIS, J. H. (*Glasgow*)
THOMAS, DR. R. H. (*California*)
TYRELL, H. J. V. (*London*)

UNIK, J. K. (*Argonne, Illinois*)

WATSON, N. (*London*)
WEAVER, J. D. (*London*)
WEEDY, B. M. (*Southampton*)
WELFORD, W. T. (*London*)
WELLS, P. N. T. (*Bristol*)
WESTWOOD, J. V. (*London*)
WHALEN, A. D. (*New Jersey*)
WICKRAMASINGHE, N. C. (*Cardiff*)
WILKINSON, I. (*Newcastle*)
WILSON, H. W. (*Glasgow*)
WILSON, M. W. (*Didcot, Berks.*)

YARWOOD, J. (*London*)

A

ACOUSTICS IN THE FISHING INDUSTRY.
Since the possibilities were first recognized in the early 1930s, the applications of acoustics in fisheries have developed steadily, and appear to be extending with each year that passes. Among the aspects of current or potential importance may be listed:

sonar, including echo-sounding;
passive listening, to locate or identify fish;
Telemetry, to convey information from gear to ships;
Silencing of ships and gear to improve catching efficiency;
Artificial sound sources to attract fish, or drive them in a desired direction.

Since only the first of these is a well-established technique, with a considerable literature and commercial importance, it will be dealt with last, allowing only a few remarks to the other items. It is first necessary to introduce a few definitions and general principles.

Definitions and Principles

The *intensity I* of sound in the sea depends on the square of the sound pressure. The standard unit of intensity is that of a sound field with an r.m.s. pressure of 1 pascal.† (This corresponds to 6.5×10^{-7} W m^{-2}.) Observed sound intensities are described as plus or minus so many decibels with respect to one pascal.

Standard *source level* is such that it gives rise to a sound intensity of 1 pascal r.m.s. at a distance of 1 m, in the direction of greatest intensity. Thus we can describe a source level S in decibels with respect to a standard source. In dealing with sonar we may be effectively radiating energy at a single frequency, and no ambiguity arises. When dealing with a natural source, it will be necessary to specify the frequency range under consideration when describing the source level. In a large homogeneous body of water, the energy spreads uniformly so the intensity at a range R from a source level S would be, in the absence of losses,

$$I = S - 20 \log R \text{ (logarithms to the base 10).}$$

† 1 pascal = 1 newton per square metre.

There is, however, some attenuation due to the conversion of sound energy to heat within the medium. Let this be A decibels per metre, then the more adequate expression becomes

$$I = S - 20 \log R - AR.\dagger$$

This is the most used expression, but it may be in error for signal sin the horizontal direction; (a) where the depth is small compared with the range, so that spreading is limited by the boundaries. In such cases empirical formulae are required, taking account of the nature of the substrate; (b) when refraction is significant, causing the sound to be channelled, and so giving excess energy in some directions, and a deficiency in others. The value of A is frequency dependent (Fig. 1) and the magnitude of the attenuation term is the main factor controlling the usable range of high-frequency equipments.

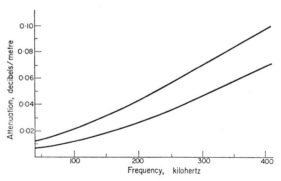

Fig. 1. Attenuation of acoustic energy in sea water. Authorities vary somewhat, and values probably lie between the limits shown.

The velocity of sound in the sea is dependent on the temperature, salinity and hydrostatic pressure (Fig. 2). For a wide range of purposes, however,

---

† For directness of presentation it is best to adhere to the decibel notation consistently. The reader should note that the above relationship would in arithmetic units merely express an inverse square law of intensity, modified by an exponential term representing the energy conversion losses.

the round figure of 1500 m/sec may be used with confidence. When very accurate soundings are required, some correction is needed, but the most important effect of the small changes that occur is to cause refraction, which is of great importance in echo-ranging over distances in excess of about a kilometre.

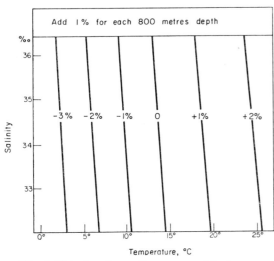

Fig. 2. Velocity of sound in sea water. The figure shows the variation with temperature, salinity, and depth, from the notional value of 1500 m/sec.

The final term to be defined is *Target Level* which again is conveniently expressed in decibels relative to a standard. The incidence of acoustic energy on a target causes the target to act as a secondary source. If the incidence of standard intensity causes the target to act as a standard source, then that target is said to be of standard level, or a "0 db" target.

If the range of the target from the equipment be R, the target strength T, and the intensity incident upon the target I_T, then the intensity of the echo signal at the receiver will be, as discussed,

$$I_R = I_T + T - 20 \log R - AR,$$

we have already noted, however, that

$$I_T = S - 20 \log R - AR$$

and combining these pieces of information we see that

$$I_R = S - 40 \log R - 2AR + T$$

and this equation is in fact the clearest definition of target strength. In ordinary usage the term is applied to small objects, such as might exist wholly within the beamwidth of the search equipment. Where an extended target exists, larger than the equipment beamwidth, a larger and larger section of the target becomes involved in reflection as the range increase. A useful example is that of a perfectly reflecting plane mirror placed at a range R so as to reflect energy directly back to the equipment. So far as the receiver is concerned, it looks at an image of the transmitter sited at a distance of $2R$. Thus the received intensity is

$$I_R = S - 20 \log 2R - 2AR$$
$$= S - 20 \log R - 20 \log 2 - 2AR$$
$$= S - 40 \log R - 2AR + (20 \log R - 6).$$

We recognize the term in brackets as corresponding to the target strength of a small object, and may conclude that the target strength of an extensive plane mirror is represented by $(20 \log R - 6)$ decibels. This expression is useful in the calibration of equipment since a water surface regarded from below is close to being a plane mirror for acoustic energy.

Passive Listening

While for experimental purposes it is perfectly feasible to listen to sounds made by fish in aquaria or in selected situations in nature, and although a considerable body of information exists, no practical application has yet proved possible in advanced fisheries. This is because the source level of most individual fish is of the order of 10 decibels only, and typical background intensities of sound at the low frequencies concerned are about -30 decibels, in the range 200 Hz to 1 kHz, depending of course on sea state and other conditions. Thus the fish noise is lost in the background at a range of about 100 m. In practice the presence of machinery and propeller noise from ships worsens the situation considerably. Future development may exploit listening hydrophones, linked by radio with the skipper or with some central control station, and there are possibilities of exploiting directional hydrophones, listening on selected frequency bands.

Telemetry

Acoustic links to convey information from the net to the ship have found application in research, and are beginning to be accepted commercially. While an acoustic link is in general inferior in reliability to a cable link, the greater convenience of the former tends to make it more acceptable. Multichannel F.M. systems have been most generally used, the information being modulated to a depth of a few kHz on a carrier of around 50 kHz. Pulse-code systems have also been developed, which are likely to predominate in the future.

A special aspect of telemetry is the use of an F.M. link on one frequency to relay information from a self-contained echo-sounder on the net to a recorder onboard ship.

Silencing of Ships and Gear

Practical experience has shown that ship noises, in particular propeller noise, can be a factor influencing the capture of herring and similar fish by purse seine. Since the low-frequency hearing of fish is acute (being usually limited by the level of background noise), this is not very surprising. Research continues on the identification of the most significant components of ship noise, and on means of reducing these to a minimum.

Use of Artificial Sound Sources

Only the most tentative experiments have so far been carried out on this potentially important subject, and there is good reason to expect progress in the very near future.

Sonar

*So*und *Na*vigation and *R*anging is a subject on which extensive research and development has been carried out for both civil and military purposes. Echo-sounders for navigation presented fishermen with one type of ready-made tool which found application in finding fish and assessing the nature of the seabed. Similarly, equipment already highly developed for finding submarines was found to be a powerful means of seeking out shoals of pelagic fish, and has made possible the development of new methods of fishing. From these bases of established hardware, interested firms and fishery researchers have sought to develop equipment for the needs of fishermen, and suited to the type of vessels in use by them. The target levels of important fish range from about −20 decibels for a large cod to about −40 decibels for an adult herring. The smallest fish of commercial interest have target levels of perhaps −50 decibels. To detect even herring as individuals in depths of say 200 m, requires more sensitive equipment than is needed for most navigational purposes. The target strength of the sea-bed is about −10 decibels per square metre, and so for a typical beamshape the sea-bed at 200 m has a target level of about +20 decibels. (It has been assumed that effectively 1000 square metres fall within the acoustic beam.) Thus the overall sensitivity needed to detect a single herring at this depth is some 60 decibels greater (a voltage amplification 1000 times greater) than is required simply to detect the seabed. Detection of fish by the earlier low-powered sounders depended on the presence of large shoals of fish, for example 100,000 herring in a mass, which might give an effective target level of −10 decibels or thereabouts. This is only a rough estimate, as the level of such a target depends on the pulse length and beamwidth of the equipment in use.

A typical fishing echo-sounder or sonar in 1974 would consist of a pulse transmitter, giving a 1-msec pulse with a source level around 110 decibels, and having a beamwidth of the order of 15° measured between the half-power points. (Beamwidth is controlled by the size of the transducer face relative to the acoustic wavelength, in the identical way as an optical beam depends on the size of the pinhole or slit forming it.) The carrier frequency would probably be 30 or 50 kHz and the transducers constructed either from nickel laminations (magnetostrictive) or from ceramics (piezo-electric or electrostrictive). The receive transducer would be similar, or commonly would be the same transducer. A target of −40 decibels at 200-m range would give an echo-intensity at the receiver of

$$I_R = S - 40 \log R - 2AR + T$$
$$= 110 - 92 - 6 - 40$$
$$= -28 \text{ decibels.}$$

This signal would give an input of 1 to 2×10^{-11} W to the receiving amplifier input, depending on the efficiency and size of the transducer and the adequacy of matching between the transducer and the input circuits. Thus to give an output of 1 W to the display, a gain of about 110 decibels is required in the receiving amplifier. The same target at a range of only 100 m would give an echo signal higher by 15 decibels, and the gain required to present this to the display as a 1-W signal would be only 95 decibels.

Since it is obviously desirable that the signal at the display should depend on the target level of the fish, and not be influenced by the range, it should be arranged that the gain of the receiving amplifier is low at the instant of transmission, and rises uniformly at a rate which will keep the amplified echo from a given target at the same level regardless of range. Clearly the appropriate law is

$$G = \text{constant} + 40 \log R + 2AR.$$

The value of the constant will depend on the target level which it is required to detect. Two remarks may be made here:

(a) The time-varied gain must necessarily be limited at a point where noise or volume reverberation become excessive. Beyond this range it is necessary to keep the gain constant, and small targets will not be detected.

(b) Where fish are present in large numbers, so that one's concern is with measurement of volume reverberation, or if it is for some reason desired to keep the *seabed* echo approximately constant irrespective of range, the appropriate T.V.G. law becomes

$$G = \text{constant} + 20 \log R + 2AR.$$

Sonar Equipments

The principles described above apply to echo-sounding, and with the qualifications made, to all echo methods. The normal side-search or "searchlight sonar" is simply an echo-sounder mounted so as to be trainable in any direction, and usually so that the beam can be tilted downward as required. These functions are controlled from a console in the bridge, and automatic search programmes are frequently incorporated. The skilled operator makes use of an audio presentation of the echoes, as well as of some form of paper recorder. The audio presentation is valuable in using a sense not distracted by other duties, and is helpful in interpretation carrying as it does a certain amount of Doppler information. The direction of the target is found by noting the direction of the sound beam for best signal strength. In a simple searchlight sonar, this is of course the direction from which the signal is *in phase* at all points on the transducer face. This is the only way in which phase relationships are exploited by the simple system. However, by dividing the transducer into sections it is possible to sample the phase of the incoming signal at various points. By suitable processing of the signals it is possible to determine, along with the range, the *direction* of each target relative to the transducer. Equipments using signal processing of this kind are already in use for specialized purposes and, with the advent of integrated circuits, signal processing is made cheaper and more reliable so it is likely that fishing equipments will shortly be available, based on these principles, and giving much enhanced performance.

R. E. CRAIG

ADVANCES IN OXIDE MATERIALS

1. Introduction

Traditionally, oxides have been recognized mainly as refractories, ceramics, glass-forming materials and phosphors. Oxides are also widely distributed in Nature and form the basis of many ores for metal extraction; in somewhat rarer natural deposits, oxides occur as single crystals which find application as gemstones.

During the past decade, important developments in solid state physics, such as the discovery of the crystalline laser, have opened entirely new fields for the use of oxides, particularly in optics and electronics.

Synthetic single crystals possessing a high degree of structural and chemical homogeneity have been required for these new fields of interest, hence, the development of techniques for producing single crystals with properties which can meet stringent device requirements has been one of the most important aspects of the progress made in using oxides and is reviewed in Section 2.

The provision of single-crystal materials and the knowledge gained by measuring the relevant properties has stimulated the demand for improved materials so, during the past decade, a vast number of existing and previously unknown chemical compounds has been investigated. For a full list of compounds, the reader is referred to the bibliography, but in Section 3 the more important materials are discussed in terms of the properties which make them useful for particular device applications.

Substantial developments in the traditional uses of oxides have also taken place during the past 10 years and these are reviewed in Section 4. Traditional uses often tend to be ignored amongst the excitement of innovating new physical devices and effects but it is as well to realize that, in terms of tonnage and profitability, their importance frequently outweighs that of the newer developments in oxide materials.

The basic feature common to all oxides is that they are compounds of oxygen containing one or more metallic elements. Most oxides possess a substantial degree of ionic bonding and therefore can exhibit properties characteristic of ionic crystals such as optical transparency, high electrical resistivity and low thermal conductivity. They are also refractory, chemically inert, hard, and fracture in a brittle manner. These properties have always formed the framework within which oxide materials can be used. At first sight this framework appears restrictive but, during the past decade, new ways of exploiting the various combinations of these properties have been discovered, thereby accounting for the advances in oxide materials described herein.

Inevitably, exceptions exist to this general framework of properties and these have also promoted the use of oxides. For instance, beryllia (BeO) is a good thermal conductor making it an attractive substrate material whilst β-alumina ($NaAl_{11}O_{17}$) is a useful ionic conductor.

2. Single-crystal Growth

The single-crystal growth of oxides has been carried out commercially for many years. The best known examples are ruby and other alumina-based crystals, grown by the Verneuil flame fusion process (Verneuil, 1902; Bauer and Field, 1963) for either watch bearings or synthetic gemstones, and quartz, produced by the hydrothermal technique (Ballman and Laudise, 1963) for piezoelectric applications. On a research scale, a wide range of oxide materials has been produced by crystallization from solution at high temperatures (Laudise, 1963), a technique often referred to as flux growth.

2.1. Czochralski developments

The most significant of the recent developments in oxide single-crystal growth has been the adaptation of the Czochralski technique (Nassau and Broyer, 1962; Cockayne *et al.*, 1967), which crystallizes materials directly from the melt, to the high temperatures at which oxide materials fuse. The principal features of the process are that a seed crystal is dipped into a crucible containing the melt and then slowly withdrawn. Melt solidifies on to the seed during the withdrawal process and the seed is grown into a crystal of the requisite shape by changing the heat input into the melt in order to control the crystal diameter.

The Czochralski technique has proved capable of producing oxide crystals of a larger size and with a higher degree of chemical and structural uniformity than other processes can provide. There are many reasons for this success. Firstly, the inherent characteristics of the technique which allow the crystal to be extracted conveniently from both crucible and melt and the relatively large thermal masses involved which confer stability upon the growth process. Secondly, the development of suitably shaped crucible and refractory materials which are chemically and mechanically compatible with molten oxides and the inert or slightly oxidizing growth atmosphere; crucibles are usually manufactured from the precious metals platinum (maximum operational temperature 1550°C: m.p. 1773°C) and iridium (maximum operational temperature 2200°C: m.p. 2443°C) whilst alumina, magnesia and zirconia are the commonly used refractories (Cockayne, 1968). Thirdly, the scientific understanding which has been obtained about the nature and formation of crystal defects and thereby has allowed development of methods for controlling defect formation (Cockayne, 1968); the recognition that phenomena occurring in the melt (e.g. convection), at the solid/liquid interface (e.g. segregation of gases and impurities) and in the crystal subsequent to growth (e.g. dislocation generation and propagation, solid state precipitation) are all important, has contributed substantially to the improvements made in the chemically and structurally sensitive properties of oxide single crystals.

The extensive application of Czochralski growth is illustrated by the materials and melting points encompassed, ranging from lead molybdate (Pb_2MoO_5: m.p. 950°C) to magnesium aluminate spinel ($MgAl_2O_4$: m.p. 2150°C) and including important device materials such as lithium niobate ($LiNbO_3$: m.p. 1260°C), yttrium aluminium garnet ($Y_3Al_5O_{12}$: m.p. 1970°C) and sapphire (Al_2O_3: m.p. 2050°C).

A very recent ingenious development of the Czochralski technique has been the production of continuous lengths of sapphire filament and a variety of other cross-sectional shapes by making use of a shaped die (Chalmers et al., 1971). The liquid pool, from which the crystal is withdrawn, is formed on the top surface of the die and is fed by capillaries in the die which lead down into the bulk melt. Shaping of the crystal is achieved by controlling the geometry of the top surface of the die and maintaining the correct contact angle between the liquid and the die material.

2.2. Crucible-less techniques

The main restrictions on the Czochralski technique at the current stage of development are the limited number and the expense of compatible crucible materials, particularly for materials which melt at temperatures above 2200°C. Hence, some attention has been paid to crucible-less techniques. Direct radio-frequency heating (Montforte et al., 1961), direct arc heating (Abraham et al., 1971) electronbeam heating (Class, 1968) and indirect optical heating (Field and Wagner, 1968, Gasson and Cockayne, 1970) of oxide materials have been attempted.

Direct coupling suffers from the disadvantage of having to preheat the oxide by ancillary means until it becomes sufficiently conducting to couple into the radio-frequency field whilst the radiation flux available for current optical heating, using either arc lamps or Co_2—N_2—He gas lasers, restricts the volume of material which can be melted at any one time. Direct-arc heating is normally used to fuse a large mass of material which is slowly cooled to yield large polycrystals.

In principle, heating methods of these types can be applied to Czochralski growth but, in practice, development has been slow, partly because of the difficult high-temperature technology and partly because of the absence of an important device material amongst the higher melting-point compounds. Even so, the floating-zone recrystallization technique of crystal growth, used in conjunction with optical heating, has been developed to produce small single crystals of yttria (Y_2O_3: m.p. 2450°C) and other rare-earth oxides whilst direct arc heating has yielded crystals of calcia (CaO: m.p. 2580°C) zirconia (ZrO_2: m.p. 2700°C) and magnesia (MgO: m.p. 2800°C).

Much more development is needed before these technologies can compete commercially with the current Czochralski technique. The relative stages of development are well illustrated by ruby crystal growth where Czochralski growth from iridium crucibles can yield 60 cm long × 6·5 cm diameter crystals compared to 5 cm long × 0·5 cm diameter for material produced by floating-zone recrystallization.

2.3. Epitaxial techniques

A requirement for thin single-crystal films of oxide materials based on the magnetic garnets (Bobeck et al., 1969) has recently stimulated the development of epitaxial growth processes for oxides. These processes can be divided into three basic groups, namely, liquid phase epitaxy (LPE), hydrothermal epitaxy (HE) and chemical vapour deposition (CVD). Growth is produced in LPE and HE by crystallization on to a substrate from a supersaturated solution and in CVD by similar crystallization from a supersaturated vapour (Laudise, 1972). These processes require a chemically and structurally homogeneous substrate of similar general structure to the material forming the epitaxial layer. Substrates are normally prepared as thin slices from Czochralski-grown materials, an important requirement being the production of a damage-free surface in order to inhibit defect generation at the substrate/thin-film interface and prevent consequent impairment of the thin film properties.

3. Single-crystal Applications

3.1. Crystalline lasers

The expansion of interest in single crystal oxides has been stimulated mainly by the discovery of the solid state laser. Pulsed laser action in the solid state was first observed in ruby, Al_2O_3/Cr^{3+} (Maiman, 1960), and the first continuously emitting room temperature laser was the mixed oxide calcium tungstate, $CaWO_4/Nd^{3+}$ (Johnson et al., 1962). Currently, another mixed oxide, yttrium aluminium garnet, $Y_3Al_5O_{12}/Nd^{3+}$, is the established material for a wide range of laser devices (Geusic et al., 1964).

The key properties of oxides exploited in laser use are optical transparency, hardness, mechanical stability and chemical inertness.

In order to stimulate laser action efficiently it is essential for the host lattice (e.g. Al_2O_3) to be completely transparent at both the absorption and emission wavelengths of the lasing ion (e.g. Cr^{3+}); the wavelength range 0·5–3·0 μm meets the demands of most laser ions and corresponds with the maximum transmissivity of most oxides. Other optical requirements are freedom from scattering particles, absorption centres and changes in refractive index which inhibit laser action; these properties are most readily obtained in Czochralski grown single crystals of the hard stable lattices provided by the strongly-bonded high melting point oxides (Cockayne, 1972).

A laser medium is normally used in the form of a cylindrical rod, typically 0·5 cm diameter × 7·5 cm long and with accurately polished plane and parallel ends, which has to be machined from a single crystal. Thus, mechanical stability and hardness are important properties during the respective shaping and polishing processes. Good mechanical properties are also required during the broad band optical pumping process used to stimulate laser action, when stresses develop due to the non-uniform absorption of heat energy within the rod; freedom from cleavage for fracture planes is a particularly important characteristic, exhibited by both $Y_3Al_5O_{12}$ and Al_2O_3.

Chemical stability is a property pertinent to the processes used to deposit reflecting ends on the laser rod and to the liquid coolants circulated within a laser cavity to assist the removal of excess heat.

Within this key framework of properties, a search for materials with optimized laser properties has occurred. Solid solubilities for the lasing ion >1 at % are preferred in order to provide reasonable gain. Crystal lattice sites consistent with broad absorption bands, a high quantum efficiency and a narrow output linewidth are also desirable in continuous wave operation; these conditions can be relaxed to a certain extent in pulsed operation where energy storage capability is more important. Crystal structure characteristics related to crystal growth, such as an absence of phase transformation, are of paramount importance.

Inevitably, no material has all the requisite properties and a compromise has to be effected. The materials in current use, namely Al_2O_3/Cr^{3+}, $Y_3Al_5O_{12}/Nd^{3+}$ and some Nd^{3+} doped glass oxides, are the most effective compromises so far obtained in relation to device use.

The predominant use of oxide lasers has been distance measurement but other applications include specialized welding, drilling, material shaping, material removal, plasma research and eye therapy.

3.2. Non-linear optics

The basic properties demanded from non-linear optical materials are very similar to those for laser materials. Thus, single crystals of good optical quality which are hard and stable are required. This is not surprising because readily detectable non-linear-effects are only induced at optical frequencies by laser radiation. Non-linear optical effects can be used in two main ways, either to generate harmonics of the fundamental laser radiation impinging upon a crystal (e.g. second harmonic generation, SHG) or to produce a change of refractive index within the crystal by the application of an electric field (the electro-optic effect). Thus, non-linear optical crystals can be used either to switch or change the frequency of a laser beam (Hulme, 1972).

In applications of this type, the subsidiary properties required strictly limit the number of available materials. Acentricity of the crystal structure is obligatory for SHG and preferred for electro-optic switching materials. In both applications, the ability to withstand high optical power densities without surface or bulk damage is essential and for SHG, adequate birefringence for phase-matching is an additional criterion. Consequently, only two oxides have proved of significant use, lithium niobate $LiNbO_3$ for SHG and electro-optic Q-switches (Boyd et al., 1964), and barium sodium niobate, $Ba_2NaNb_5O_{15}$, also for SHG (Geusic et al., 1968); SHG has been applied mainly to the generation of 0·53 μm green light from 1·06 μm Nd-based lasers. Both these materials are used in relatively small sections, typically 1 cm cubes, accurately orientated and polished and cut from Czochralski grown crystals.

Chemical inertness is even more important here than in laser use, compatibility with electrodes being necessary in the Q-switches and stability at the phase matching temperatures (80–200°C) being essential for SHG.

3.3. Acousto-optics

Acousto-optic interactions within solid state materials also offer ways of controlling laser beams for a diversity of applications which include optical stores and laser Q-switches (Hobden, 1972). Stable, durable, optically perfect materials are again demanded. Devices based on acousto-optic interactions use an acoustic strain wave generated within the material to diffract light beams passing through it. Hence, properties which govern the diffracting efficiency and acoustic propagation are also important. Chemical compatibility with the transducer material and stability during the bonding of the transducer are essential.

Suitable properties are often expressed as a figure of merit, which can be confusing as the definition of the merit figure varies according to the application. For modulator applications, the merit figure M_2, is defined as $n^6 p^2 \varrho^{1/2} c^{-3/2}$ (n is refractive index, p is the relevant photo-elastic coefficient, ϱ is density and c is the relevant elastic modulus). Thus dense materials with a high refractive index are preferred. Low melting point mixed oxides of the lead molybdate type, Pb_2MoO_5 and $PbMoO_4$, meet these two requirements but, even in single-crystal form, the low acoustic and optical absorptions required for efficient use are not consistently obtained. Hence, more optically perfect materials with lower figures of merit, such as SF4 glass, are used in some circumstances.

3.4. Delay lines

Oxides have recently found considerable application in ultrasonic delay lines both as transducers and delay media (Lewis, 1972). Typical transducer materials are barium titanate ($BaTiO_3$) and zinc oxide (ZnO). The delay lines are of two basic types, those employing bulk waves propagated in either shear or longitudinal modes and those employing surface waves.

For bulk media, the most important properties are compatibility of the acoustic impedance with that of the input and output transducers, mechanical hardness for accurate machining and polishing, and low ultrasonic propagation losses. The acoustic wavelength at the frequencies used is similar to the wavelength of light so many of the properties and tolerances demanded are identical to those for lasers. Large Czochralski grown crystals (typically 5–10 cm long × 0·5–2·0 cm diameter) of sapphire, spinel and yttrium aluminium garnet have all been used as bulk delay media. Thus far, spinel in shear wave propagation has proved to have the lowest attenuation.

Surface acoustic wave devices consist of an interdigital metal finger array mounted on a piezoelectric medium. The media in current use are single crystals of the oxide materials lithium niobate and quartz, the former being produced by the Czochralski technique and the latter by the hydrothermal process. The properties required for surface wave media are similar to those for bulk media but because the surface devices generally operate at higher frequencies, it is especially important that acoustic losses are low and surface-wave velocities are high. A zero temperature coefficient for the delay time is also an important operational consideration and quartz is the only known material to have this property.

Magnetoelastic delay lines can be produced in ferrimagnetic insulators by using spin waves to produce delay in a manner analogous to the use of sound waves. Only one suitable material is presently known. This is the mixed oxide, yttrium iron garnet ($Y_3Fe_5O_{12}$) which can be produced as single crystals by flux growth techniques.

3.5. Substrates for semiconductors

Single crystal oxides are useful substrate materials for the epitaxial deposition of some single crystal thin films of semiconductor materials (Filby, 1972). In the field of microelectronics, the insulating properties of oxide substrates allow complete electrical isolation of adjacent regions of semiconducting material to be obtained. This eliminates the parasitic capacitance associated with the back-biasing techniques devised to produce electrical isolation when a semiconducting substrate is used and permits the operational speed of the circuit to be increased. The principal oxide substrates for semiconductors are sapphire and magnesium aluminate spinel; both have been used in conjunction with thin films of silicon but spinel has also been employed in the production of gallium arsenide layers. The substrate materials must exhibit other properties in addition to low electrical conductivity. For instance, a reasonable atomic fit at the substrate/film interface is necessary for epitaxy, so similar lattice spacings for the semiconductor and substrate are essential. As the deposition temperature is approximately 1000°C, similar expansion coefficients are also an advantage, if films are to be obtained stress-free at room temperature. Other important features are chemical stability during deposition and the ability to provide a smooth damage-free surface for deposition.

3.6. Magnetic devices

Certain single-crystal magnetic oxides such as the rare-earth orthoferrites ($RFeO_3$) and gallium-doped yttrium iron garnet, often containing additional rare earth substitutions, can be used to support bubble domains when prepared as thin films (Bobeck et al., 1969). These domains can be manipulated under the influence of a magnetic field and can thus be applied to a number of logic and storage devices. The ability of a material to form magnetic bubbles depends upon the presence of magnetic anisotropy. This is intrinsic in the anisotropic crystal structure of the orthoferrites but is normally strain-induced in the garnets by depositing the magnetic film on to a substrate with a slight lattice mismatch. Currently, the use of garnets is favoured because of the smaller bubble diameters (5–10 µm, cf. 25–100 µm for the orthoferrites) which facilitate higher packing densities. The substrates used must be non-magnetic, have the necessary mismatch and be available as structurally and chemically homogeneous single crystals. Some of the Czochralski-grown rare-earth garnets meet these requirements. Thus far, gadolinium gallium garnet ($Gd_3Ga_5O_{12}$) has been the substrate material used most extensively.

3.7. Pyroelectric detection

A number of oxide single crystals are of current interest for the detection of infra-red radiation using the pyroelectric effect (Putley, 1971). Important property requirements are a high pyroelectric coefficient, and a low relative permittivity and dielectric loss. Strontium barium niobate, $Sr_{1-x}Ba_xNb_2O_6$, (Glass, 1969), and lead germanate, $Pb_5Ge_3O_{11}$, (Jones et al., 1972), both exhibit useful properties in single crystal and ceramic form. The chemical and thermal stability of these materials combined with reasonable mechanical properties are of practical importance for device manufacture, which can involve high-temperature vacuum treatment, and stability under a range of operational conditions.

3.8. Gemstones

Several of the higher-melting-point oxide single crystals are sufficiently hard, transparent and dispersive to be attractive as gemstones. Ruby has been available as a synthetic gemstone since the early part of this century, but other materials such as strontium titanate ($SrTiO_3$) and yttrium aluminium garnet have become fashionable more recently, the latter only because of its primary development as a laser material. The use of the garnet structure opens a wide selection of possible colours for synthetic gemstones because of the substitutions which can be made on both the Y^{3+} and Al^{3+} lattice sites by deliberately added impurity ions.

3.9. Composites

Oxides have a number of useful mechanical properties such as high strength and high rigidity in addition to thermal and chemical stability but from an engineering viewpoint, their use is limited by marked brittleness. However, by combining brittle oxides with more ductile materials in the form of composites, more favourable engineering properties can be realized (Harris, 1972). The most widely investigated oxide single-crystal system used in this manner has been sapphire whiskers embedded in an aluminium matrix.

4. Developments in Traditional Uses

4.1. Polycrystalline oxide materials

The developments are almost limitless and all the basic properties such as hardness, optical transparency, chemical stability, electrical insulation and refractoriness have been exploited in addition to ferroelectric, piezoelectric, pyroelectric and magnetic characteristics.

The traditional method for making ceramics has been to cold form the material as a powder with subsequent high-temperature sintering to yield a high density product. In recent years, this process has been extended to larger sized articles, for instance $35 \times 45 \times 250$ cm alumina blocks are now available. The use of plastic additives in conjunction with extrusion techniques has permitted complex shapes to be produced. Hot-forming and melt-fusion processes have also been developed; the former is a combined pressing and sintering process whilst the latter involves deposition of oxide coatings from molten oxides produced in a flame or plasma torch. Many of these processes are commercially confidential because of the marked relationship between the properties of the final product and the fine detail of the manufacturing method. Hence, properties are more in evidence in the literature than processes. However, the evolution of ceramic technology has undoubtedly continued with better understanding of the relationship between microstructure and properties, particularly with respect to chemical purity, grain size, bonding across grain boundaries and porosity (Matkin, 1972).

It is now realized that translucent or even transparent ceramics can be obtained when the porosity approaches zero and, as a consequence, ceramics with these optical properties based on magnesia, yttria and alumina (lucalox) are now available. Transparent ferroelectric materials, with useful optical switching properties, based on compounds such as lead zirconate–titanate ($PbTiO_3$–$PbZrO_3$) have also been developed (Ainger, 1972). The piezoelectric and pyroelectric properties of this material are also respectively exploited in electromechanical transducer and radiation detector applications.

Polycrystalline ceramics based on alumina and quartz have been used as substrates for some thin- and thick-film semiconductor circuits where the electrical insulation properties are exploited in a manner similar to that described for single crystal substrates (Filby, 1972). Some polycrystalline oxides are electrically conducting due to a high degree of ionic mobility (e.g. β-alumina) and are beginning to find application as electrolytes in batteries and fuel cells (Greene, 1972).

Pressed magnetic oxides based on ferrites can be used in permanent magnets in addition to their uses in communication components, as toroids for computer main stores and magnetic recording tape (Marshall, 1972).

The largest application of oxide ceramics is still for refractories in the iron and steel and glass-making industries, where new production processes have stimulated the development of high-performance materials with increased refractories and better mechanical and chemical stability. Purer materials and improved bonding techniques have been developed to meet the demanded properties (Matkin, 1972).

The high-temperature mechanical stability of oxides has been further utilized in the provision of ceramic cutting tools, based on alumina, and in ceramic dies and cones used in the wire drawing of metals. Room temperature uses in items such as textile thread guides and ceramic bearings also make use of the wear-resistant properties of oxides.

The low density of oxides such as alumina, silica and beryllia has been coupled with good mechanical strength in a number of applications such as radomes, missile noses and transparent window armour.

4.2. Amorphous oxide materials

The most important amorphous materials are glasses and those made in greatest quantities, namely soda–lime–silica, pyrex and lead silicate, are all made from oxide components and are used to manufacture most of the commonplace glass articles. These materials have been further developed to produce a whole range of glasses for technical purposes such as the low dielectric loss materials used in the electrical industry and sealing glasses with specific expansion properties for electronic applications (Savage, 1972).

A number of entirely new oxide glasses have also been developed. Some, based on boric oxide, cerous oxide and ferrous oxide, have semiconducting properties and can be used in channel plate multipliers whilst others, based on alkali/alkaline earth silicate compositions and doped with neodymium, have found application in laser systems. Fibre optical applications are a further new use for glasses. Glass fibres are also potentially useful in laser communication systems and for reinforcing composite materials.

Other innovations for oxide glasses include the production of phase-separated materials, photosensitive compositions and glass ceramics; the latter exhibit a wide range of properties including low and high expansion characteristics, low dielectric loss and high strength.

Other amorphous oxide materials with novel uses are boric oxide and the oxides of silicon. Boric oxide is used as a liquid encapsulant in the Czochralski growth of some compound semiconductors, such as gallium arsenide and indium phosphide, in order to restrict loss of the volatile species at the melting point of the compound (Mullin et al., 1964). Extensive use is made of silicon oxides as masks in the fabrication of microelectric devices (Filby, 1972).

4.3. Crystalline oxide powders

The principal uses of oxides in the form of crystalline powders are as phosphors and basic chemicals.

Phosphor technology is very empirical and is not confined to oxides. Both the literature and the number of recipes are vast. Recent developments where oxides have played an important part (Taylor, 1972) have been concerned with improving efficiency, colour and temperature stability. The main applications have remained cathode-ray tubes and lamps. The most significant development involving an oxide is probably the production of the europium-doped yttrium orthovanadate phosphor for emitting red light in colour-television applications.

The role of oxides as basic chemicals is an area which is frequently overlooked. Very few oxides occur in Nature with either a purity or form directly suitable for device application but, until recently, many industrial chemicals were also insufficiently pure to meet the demands of many of the new applications described here. However, as devices based on oxides developed, the dependence of properties upon chemical purity became much more evident. Chemical manufacturers have responded well to this challenge and by employing solvent extraction, ion-exchange, zone-refining and other fractional crystallization procedures, have produced much purer grades of material, in some cases with a 99·9999% specification. As a consequence, a wide range of high purity oxide chemicals are now available for the single crystal and polycrystalline materials used in both new and traditional applications.

Bibliography

ABRAHAM, M. M., BUTLER, C. T. and CHEN, Y. (1971) *J. Chem. Phys.* **55**, 3752.

AINGER, F. W. (1972) *Modern Oxide Materials*, p. 147, Academic Press.

BALLMAN, A. A. and LAUDISE, R. A. (1963) *The Art and Science of Growing Crystals*, p. 239, John Wiley and Sons.

BAUER, W. H. and FIELD, W. G. (1963) *The Art and Science of Growing Crystals*, p. 399, John Wiley and Sons.

BOBECK, A. H., FISCHER, R. F., PERNESKI, A. J., REMEIKA, J. P. and VAN UITERT, L. G. (1969) *IEEE Trans. Mag.*, *MAG* **5**, 544.

BOYD, G. D., MILLER, R. C., NASSAU, K., BOND, W. L. and SAVAGE, A. (1964) *Appl. Phys. Lett.* **5**, 234.

CHALMERS, B., LABELLE, H. E. and MLAVSKI, A. I. (1971) *J. Cryst. Growth* **13/14**, 84.
CLASS, W. (1968) *J. Cryst. Growth* **3/4**, 241.
COCKAYNE, B. (1968) *Plat. Met. Rev.* **12**, 16.
COCKAYNE, B. (1968) *J. Cryst. Growth* **3/4**, 60.
COCKAYNE, B. (1972) *Modern Oxide Materials*, p. 1, Academic Press.
COCKAYNE, B., CHESSWAS, M. and GASSON, D. B. (1967) *J. Mat. Sci.* **2**, 7.
FIELD, W. G. and WAGNER, R. W. (1968) *J. Cryst. Growth* **3/4**, 799.
FILBY, J. D. (1972) *Modern Oxide Materials*, p. 203 Academic Press.
GASSON, D. B. and COCKAYNE, B. (1970) *J. Mat. Sci.* **5**, 100.
GEUSIC, J. E., LEVINSTEIN, H. J., SINGH, S., SMITH, R. G. and VAN UITERT, L. G. (1968) *Appl. Phys. Lett.* **12**, 306.
GEUSIC, J. E., MARCOS, H. M. and VAN UITERT, L. G. (1964) *Appl. Phys. Lett.* **4**, 182.
GLASS, A. M. (1969) *J. Appl. Phys.* **40**, 4699.
GREENE, G. (1972) *New Scientist* **54**, 321.
HARRIS, B. (1972) *Modern Oxide Materials*, p. 258, Academic Press.
HOBDEN, M. V. (1972) *Modern Oxide Materials*, p. 29, Academic Press.
HULME, K. F. (1972) *Modern Oxide Materials*, p. 53, Academic Press.
JOHNSON, L. F., BOYD, G. D., NASSAU, K. and SODEN, R. R. (1962) *Phys. Rev.* **126**, 1406.
JONES, G. R., SHAW, N. and VERE, A. W. (1972) *Elec. Lett.* **8**, No. 14.
LAUDISE, R. A. (1963) *The Art and Science of Growing Crystals*, p. 252, John Wiley.
LAUDISE, R. A. (1972) *J. Cryst. Growth* **13/14**, 27.
LEWIS, M. F. (1972) *Modern Oxide Materials*, p. 74, Academic Press.
MAIMAN, T. H. (1960) *Nature*, **187**, 493.
MARSHALL, D. J. (1972) *Modern Oxide Materials*, p. 176, Academic Press.
MATKIN, D. I. (1972) *Modern Oxide Materials*, p. 235, Academic Press.
MONTFORTE, F. R., SWANEKAMP, F. W. and VAN UITERT, L. G. (1961) *J. Appl. Phys.* **32**, 959.
MULLIN, J. B., STRAUGHAN, B. W. and BRICKELL, W. S. (1964) *J. Phys. Chem. Sol.* **26**, 782.
NASSAU, K. and BROYER, M. (1962) *J. Appl. Phys.* **33**, 3064.
PUTLEY, E. H. (1971) *Optics and Laser Technology*, p. 150.
SAVAGE, J. A. (1972) *Modern Oxide Materials*, p. 99, Academic Press.
TAYLOR, M. J. (1972) *Modern Oxide Materials*, p. 120, Academic Press.
VERNEUIL, A. (1902) *Comptes Rendus*, **135**, 791.

B. COCKAYNE

APPLICATIONS OF REVERSE OSMOSIS AND ULTRAFILTRATION.

Reverse osmosis and ultrafiltration are membrane separation processes which can be applied to both aqueous and non-aqueous systems. A specific membrane is normally characterized by the molecular weight and size of the molecules or the percentage of salts which are retained by the membrane. The name reverse osmosis or hyperfiltration is given to the technique when molecular weights up to 1000 are involved, whereas for the separation of larger organic molecules with molecular weights above 1000, the name ultrafiltration is normally used. However, there are many applications which come at this arbitrary boundary. Reverse osmosis membranes are normally characterized by the percentage salt rejection under standard conditions.

From a given feed, these membrane processes produce two streams:

(1) Normally the larger fraction that permeates the membrane. In aqueous solutions this is substantially purified water, practically free of colloidal, particulate and microbiological species and relatively low in dissolved organic and inorganic species.

(2) A stream of reduced volume in which are concentrated all the various species retained by the filtration properties of the membrane.

In many applications it is the purified water stream which is of prime importance and the concentrate is a waste stream. Where one is interested in concentration as a means of recovering valuable materials, or to dewater heat sensitive materials, or to reduce waste disposal costs by volume reduction, it is the second stream that is of importance. In some applications, the ability to reclaim relatively pure water is accompanied by the simultaneous concentration of a useful product or waste. In general membrane properties can be chosen to satisfy these widely differing requirements. The main areas of application of reverse osmosis will be considered in turn, with some examples.

(a) Brackish Water Demineralization

Reverse osmosis has been very successfully applied to the demineralization of brackish water containing up to, but not necessarily limited to, 5000 mg/l total dissolved solids (TDS). This is not a fixed upper limit as there are many cases where high water recovery is not essential and the composition of the water is such that acceptable water quality can be obtained. This is generally dependent upon the mono- and divalent ion concentrations and ion inter-relationships. Three examples are given in Tables 1–3 to illustrate typical results for three widely differing feed streams. The applications are generally purified water for industrial, commercial or potable purposes.

Tables 1 and 2 show feed, product and concentrate analyses from tests on water from two boreholes at Appleby Parva, Leicestershire. Table 1 is for a carbonate rich water, whereas Table 2 is for

Table 1. Water Analyses From Reverse Osmosis Test on Bunter Borehole Appleby Parva (concentrations in mg/l)

Analysis	Raw feed	Overall product	Concentrate
pH	7·1	7·1	7.3
Electrical conductivity (μmho)	995	100	1600
Total dissolved solids	610	50	1140
Hardness—$CaCO_3$	386	22	700
Alkalinity—$CaCO_3$	276	19	503
Calcium	91	6·4	179
Magnesium	38	1·5	61
Sodium	76	13·1	113
Potassium	11·4	1·3	22
Carbonate	166	11·4	302
Sulphate	171	4·8	271
Chloride	76	20	134
Nitrate	1·1	<1	<1
Silica (SiO_2)	11	4	18
Phosphate	<0·1	<0·1	1·8

46·5% recovery at 600 psi.

Table 2. Water Analyses from Reverse Osmosis Test on Keuper Borehole at Appleby Parva (concentrations in mg/l)

Analysis	Feed	Product	Concentrate
pH	7.0	6.3	7.5
Electrical conductivity (μmho)	1950	110	4400
Total dissolved solids	1682	70	4893
Hardness—$CaCO_3$	1215	36	3440
Alkalinity—$CaCO_3$	259	25	663
Calcium	334	8	976
Magnesium	92·3	3·9	243
Sodium	38·2	8·9	80·3
Potassium	5	1·3	13·2
Carbonate	155	15	398
Sulphate	950	12.9	2720
Chloride	28	8	77
Nitrate	14·6	8·7	19
Silica (SiO_2)	14	4	32

% water recovery based on TDS—66·6%.
TDS desalination factor based on feed—24.
TDS desalination factor based on reject—70.
600 psi.

a high sulphate water. Table 3 shows results from a United States Office of Saline Water (OSW) test on a well at San Diego in which the feed contained a high sodium chloride content and was over 4000 mg/l TDS.

Table 3. Water Analyses from Reverse Osmosis Test on Brackish Well Water, San Diego (concentrations in mg/l)

	Well	Feed*	Product	Concentrate
pH	7·0	6·7	5·3†	6·8
Conductivity at 25°C (μhos)	7400	7560	610	24,000
Dissolved solids (180°C)	4460	4580	310	17,720
Hardness—$CaCO_3$	1830	1880	20	7350
Alkalinity—$CaCO_3$	380	280	10	940
Calcium	360	360	7·6	1400
Magnesium	230	240	0·5	940
Sodium	900	900	110	3400
Potassium	26	26	3·8	91
Iron (Fe^{2+})	6·5	6·5	0·12	24
Manganese (Mn^{2+})	3·8	3·8	0·0	16
Bicarbonate	460	340	12	1150
Sulfate	620	630	0·0	2580
Chloride	1960	2020	170	7850
Nitrate	0·53	0·53	0·26	3·3
Fluoride	0·70	0·70	0·12	1·3
Boron (BO_3)	0·3	0·3	0·14	0·4
Silica (SiO_2)	33	34	7·6	120
Phosphate (Total)	0·36	10.0	0·18	40

* Reflects pretreatment—acid addition and sodium hexametaphosphate.
† Mostly CO_2.
75% recovery at 600 psi.

(b) Production of High-purity Water

Reverse osmosis has become established as an economical precursor to final ion exchange polishing to produce ultra-pure water for semiconductor manufacture or as feed water for high-pressure boilers. The main advantage of reverse osmosis is the virtually complete elimination of particulate matter and the high reduction in silica and organic content, all of which lead to significantly reduced regeneration of the ion exchange units.

(c) Acid Mine Drainage Liquor

A specialized application of brackish water demineralization is the treatment of acid mine drainage liquor to produce a large fraction of purified water and a small volume of concentrate which can be more economically neutralized for final disposal.

Many of these applications occur in the U.S.A. and a typical example is given in Table 4.

Table 4. Water Analyses from Reverse Osmosis Test on Acid Mine Drainage Liquor
(concentration in mg/l)

	Raw feed	Product	Concentrate
pH	2·7	4·4	2·4
Total dissolved solids	1280	10	4075
Acidity—$CaCO_3$	644	6	2330
Calcium	115	2	400
Magnesium	38	0·9	146
Aluminium	38·5	3·1	153
Total iron (Fe)	153	0	566
Sulfate	936	4·2	2810

612 psi, 75% recovery.

(d) Treatment of Sea Water

The ultimate goal of reverse osmosis in the field of desalination is the production of potable water from sea water, a field covered almost exclusively by distillation to date. The sea water salinity of 35,000 mg/l has to be reduced to less than 500 mg/l, preferably to 100 mg/l if it is to be fed into an existing water supply system. Many small pilot plants have been in operation over the past few years, in most cases with limited success. Current cellulose acetate membranes are not stable enough in the high salt concentrations and at the high operating pressures. Few people have achieved adequate desalination factors to reach the required product water salinity in a one-stage process, but many pilot plants have been operating with a two-stage process which does not put such severe restraints on the membrane properties.

(e) The Dairy Industry

Dairy industry applications are part of the more general area of reverse osmosis applications in the food industry considered in the next section. However, as whey treatment has received detailed attention it will be discussed separately.

Whey is a by-product from cheese manufacture which contains about half of the original milk solids including most of the original lactose, some protein, amino-acids and salts, together with over 90% water. The material has a high B.O.D. (Biological Oxygen Demand) which presents an effluent disposal problem. Large quantities are used as an animal feed, but it is not economic to transport it long distances. If an open membrane (ultrafiltration) is used, all of the protein can be removed, together with some lactose, in the concentrate stream. If a reverse osmosis membrane is used, both the protein and lactose can be separated almost completely from the salts. In this way the nutritive constituents can be more efficiently utilized, and the pollution problem is eased as the B.O.D. level is considerably reduced in the effluent.

Reverse osmosis has also been used for the treatment of skim milk. The concentrate has been used for cheese of the Camembert type. The concentrate can also be evaporated further and spray dried to produce a dried whey or milk product.

(f) The Food Industry

This is a large, but diffuse, area of applications in which it is the concentrate stream that is generally of interest.

Apple, orange, lemon, grape and plum juice concentration are examples where the aroma constituents are often lost by evaporation methods. Some technique of concentration of these fruit juices is required to reduce the bulk and hence the costs of packaging, storage and transport. Individual tests are required for each juice to establish the behaviour of the aroma constituents with different types of membranes.

Reverse osmosis has been proposed as a means of preliminary concentration prior to spray drying for instant coffee and tea. It can also be applied to jams, jellies, essences, soups, purées and meat stocks.

The refining of raw beet or cane sugar should also be possible be reverse osmosis as an alternative to sedimentation and filtration. Maple syrup has also been treated.

Egg white, which is used extensively in the baking industry, can also be concentrated by reverse osmosis although high shearing forces must be avoided on the concentrate.

Apart from the specific applications listed above, the general problem of treatment of effluents from the food industry, which have a high B.O.D., can also be solved by the use of reverse osmosis.

(g) Biochemical Applications

These are also as wide and diffuse as the food industry applications considered above. The production of enzymes, proteins, nucleic acids, vaccines and viruses is possible by reverse osmosis with advantages of high yields and maintenance of high biological activity because the structure of the solutes is not appreciably altered.

Separation of blood proteins (and viruses) should be possible leading to the production of virus free blood plasma.

Applications in the fermentation industry are also being developed.

(h) Electrophoretic Paint Recovery

This is a fairly recent and novel application using very open membranes (ultrafiltration) in which large volumes of wash water used to rinse car bodies after electrophoretic painting can be treated to recover the paint and ease the effluent problem.

(i) Industrial Effluent Treatment

Worldwide attempts are being made to improve the quality of rivers. To achieve this, restrictions on the discharge of effluents are being tightened. At the same time, raw water costs are steadily rising and serious attention is being given to the re-use of water. Reverse osmosis is an attractive process which is receiving more and more attention in such situations, although it is not possible to discuss them in detail as individual circumstances vary for each case.

(j) Sewage Treatment

The discharges from sewage treatment works represent very large sources of water, the majority of which is discharged to rivers and estuaries. Much of this water is of good quality and, with simple treatment, is reuseable. However, as general salinity levels increase it may be necessary to consider some reduction in solids content for potable reuse. At the present time, reverse osmosis is too expensive compared with other conventional techniques, for widespread use. However, many pilot plants have been operated on sewage effluents and the product water has been potable after threshold chlorination in case of possible bacterial contamination.

Bibliography

LACEY, R. E. and LOEB, S. (Eds.) (1972) *Industrial Processing with Membranes.* New York: Wiley-Interscience.

SOURIRAJAN, S. (1970) *Reverse Osmosis.* London: Logos Press.

J. R. GROVER

ARCHAEOLOGICAL DATING BY PHYSICAL METHODS

Introduction

Within the material remains associated with ancient man continuing physical and chemical changes are taking place. For some characteristics the amount of change is proportional to the age, thereby giving a method of dating. The method of predominant importance is *radiocarbon dating*; this is the basis of nearly all prehistoric chronology back to about 50,000 years ago. The earliest historical chronology is the Egyptian calendar beginning in 3000 B.C. and based on recorded astronomical observation. For earlier millennia the only generally applicable means of absolute dating is by physics.

In addition to radiocarbon dating this article outlines *thermoluminescent dating, archaeomagnetism (magnetic dating), potassium-argon dating, uranium-series dating, fission-track dating*, and *dating by chemical change* (bone, obsidian, and glass).

Radiocarbon

In the carbon dioxide of the atmosphere, in living trees, plants and animals, and in the dissolved carbonates of the ocean, there is a minute amount of the weakly radioactive isotope *carbon-14*. The concentration is about 1 part per million million parts of natural carbon (^{12}C) and it is effectively uniform; the uniformity obtains because of comparatively rapid mixing throughout the *carbon exchange reservoir* which consists of the atmosphere, the biosphere and the ocean. The concentration stays approximately constant because it represents the equilibrium level established, on a global scale, between loss by radioactive decay and production by cosmic rays; the latter generate neutrons in the upper atmosphere and these transmute ^{14}N into ^{14}C. On the other hand, in animals and plants that are preserved after death (and thereby removed from the exchange reservoir) the concentration decays exponentially with the 5730-year half-life of ^{14}C, and by comparing the concentration found in such material with the concentration for living material the time that has elapsed since death can be determined. The level of ^{14}C is determined by measuring the beta-particle activity of the sample. For living material this is approximately 15 disintegrations per minute (that is about 10^{-11} curies per gram of natural carbon).

The maximum age that can be reached is about 50,000 years; the limit is set by the difficulty of measuring the very low level of activity—about 0.2% of the level for living materials.

The essential requirement in sample material is that at death the material becomes isolated from the exchange reservoir so that no ^{14}C atoms enter the material subsequently. Exchange with the reservoir after death is least likely when the molecular structure is large; in addition, the sample should be in a well-preserved condition since decomposition may be associated with the assimilation of fresh carbon. Charcoal is one of the best types of material and this is fortunate since it is widely associated with human occupation. Other suitable materials are wood, cloth, peat, leaves, hair, skin, antlers, leather and paper. Bone and shell can also be dated but special pre-treatments are necessary; another difficulty with bone is that there is not much carbon present and inconveniently large samples are necessary—hundreds of grams—whereas for most other materials some tens of grams are enough, depending on the design of the measuring apparatus. With charcoal and wood there is the possibility that the sample represents inner rings of a slow-growing tree and in such a case the radiocarbon date can be several hundred years earlier than the archaeological event of felling the tree; this is because when a ring is formed the carbon atoms are fixed in it and the ^{14}C activity found in it is determined by the decay that has taken place since the date of formation despite the presence of higher and lower activities on either side. This puts a premium on short-lived materials such as straw, seeds, nuts, grasses, twigs, etc.

The idea of age determination by ^{14}C arose in the course of a study of the effects of cosmic rays on the Earth's atmosphere by W. F. Libby and his group at the University of Chicago, in the late 1940s. Widespread archaeological application of the method took place during the 1950s, establishing its overwhelming importance for prehistoric research; today there are 100 radiocarbon dating laboratories in operation though these serve too the equally important application fields of paleobotany and geology. With improvements in measurement accuracy it became evident from tests with known-age samples during the 1960s that one of the basic assumptions was not precisely valid. This was the assumption that the concentration of ^{14}C in the exchange reservoir has remained constant, a condition that could be upset by several causes of which the most obvious is long-term fluctuation in cosmic-ray intensity. The main source of known-age samples for these tests has been wood from long-lived trees dated by the techniques of tree-ring counting (*dendrochronology*); in particular Bristlecone pines of the White Mountains of California have provided dated samples covering the past 7000 years and these indicate that prior to 1000 B.C. the ^{14}C concentration was significantly higher than subsequently. The error in radiocarbon age reaches 700 or 800 years in the fourth and fifth millennia B.C., in the sense that the radiocarbon date is too recent. The corrections determined from the Bristlecone pine measurements can be used to "calibrate" radiocarbon ages and so obtain calendar ages; however, there is loss of precision in this process due to short-term fluctuations in the ^{14}C concentration and this makes it unrealistic in general to assign an error limit (at the 68% level of probability) of less than ± 150 years. The error limit usually quoted for a radiocarbon date includes only the effect of counting statistics and for a good measurement this limit is around ± 40 years, corresponding to $\pm 0.5\%$ in the count. Published dates are usually based on the "old" or "Libby" half-life of 5568 years which is some 3% lower than the currently accepted best value of 5730 years.

The changes in cosmic-ray intensity responsible for the fluctuations in ^{14}C concentration are probably associated with changes in the earth's magnetic moment (for long-term fluctuations) and with sunspot activity (for short-term fluctuations). It seems likely that climate plays a part too. Despite these complicating effects radiocarbon dating remains of predominant importance in archaeology and forms the basis of nearly all prehistoric chronology back to about 50,000 years ago.

Thermoluminescence

If a ground-up sample of ancient pottery is heated rapidly to 500°C there is a weak, but measurable, emission of light. For a second heating of the same sample the emission consists only of "red-hot glow", or incandescence. The extra light in the first heating is *thermoluminescence* (TL) and it results from the release, by thermal stimulation, of electrons that were trapped at defects in the crystal structure of constituent minerals in the pottery. The number of trapped electrons is proportional to the amount of ionizing radiation to which the minerals have been exposed since the pottery was fired. This radiation comes mainly from radioactive impurities in the pottery and the surrounding burial soil—a few parts per million of ^{238}U, ^{232}Th and ^{40}K. The half-lives of these are very long (10^9 years or more) so that the radiation dose-rate is constant and the number of trapped electrons grows uniformly with time. The act of firing "drains" the clay of all previously acquired TL thereby setting the TL "clock" to zero.

The amount of TL depends not only on the age of the pottery but also on its sensitivity to acquiring TL and to the concentration of the radioactive impurities in the pottery and the soil. The sensitivity is determined by measuring the TL induced by exposure to a known amount of nuclear radiation from an artificial radioisotope source, and the impurity concentration by radioactive counting and chemical analysis.

At present (1972), in good circumstances an absolute accuracy of between 10% and 5% of the age can be achieved; for various reasons 5% is likely to be the limit. This is somewhat poorer than is usually quoted for radiocarbon dating though probably no worse than indicated by a realistic assessment of accuracy in the light of the short-term ^{14}C fluctuations. However, the TL method has the advantage that the event dated is the actual firing of the pottery, whereas in the case of radiocarbon dating of wood or charcoal the event dated may precede involvement with ancient man by several hundred years. A further advantage is that the changing technique and style of pottery often forms the basis of archaeological dating so that TL dates are directly related to chronologically significant objects.

Besides dating pottery found in excavations an important application is in detecting forgeries of works of art. The technique is more fully discussed in a separate entry under THERMOLUMINESCENT DATING.

Archaeomagnetism (*Magnetic Dating*)

The time-dependent quantity in this case is the Earth's magnetic field. When baked clay cools down from firing it acquires a weak permanent magnetization in the same direction as the field and of a strength proportional to the intensity of the field. This is due to the *Thermoremanent Magnetism* of iron oxide present in the clay as haematite (α-Fe_2O_3) or magnetite (Fe_3O_4). To be a useful record of direction the clay must be part of a kiln, hearth or oven so that it has exactly the same orientation today as when it cooled down. On the other hand, intensity information can be obtained from bricks and pottery fragments as well as from kilns, hearths and ovens.

The direction changes appreciably from decade to decade (see Fig. 1) and it was hoped initially that "magnetic dating" would be a powerful technique. However, its useful applications have been rather restricted. One difficulty is that the variation of direction does not follow any predictable law and calibration from samples of known date is first necessary; this calibration holds for only a region 500–1000 miles across. Another difficulty is that a given direction may recur after an interval of a few centuries. A practical difficulty is that in many structures the recorded direction is slightly distorted by the magnetism of the baked clay itself. The foregoing is in respect of the small-scale *secular variation* of direction. From time to time in the geological past there has been a worldwide *reversal* of direction and there is evidence that this has occurred two or three times in the last 200,000 years; these reversals will be recorded in the hearths of some Mesolithic and Palaeolithic levels and may provide useful time-marks.

The primary interest in determining past variations in intensity is in connection with radiocarbon dating. The Earth's magnetic field acts as a shield against cosmic rays and during millennia when the intensity has been low the rate of radiocarbon production will have been high. Such study goes some way to explaining the known long-term systematic error in radiocarbon dating and may eventually allow prediction of the systematic error in earlier periods. Apart from this direct feedback into archaeology, archaeomagnetism has provided geophysical data, in respect of both intensity and direction, that are unobtainable except through archaeology. These data are of considerable interest for the

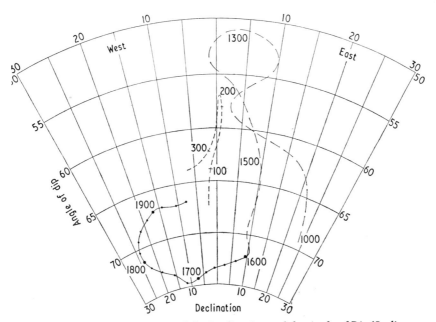

Fig.1. The secular variation of the Declination and the Angle of Dip (Inclination), for London. The dates shown are years A.D. *Subsequent to* A.D. *1580 the data are from historically recorded observations; before* A.D. *1580 the data are from archaeomagnetic measurements.*

Potassium–argon Dating

The basis is the build-up of argon formed by the radioactive decay of ^{40}K in volcanic rocks, the amount present being a measure of the time that has elapsed since the rocks last cooled down. Natural potassium is present in rocks in the range 0–10% and, in addition to the stable isotopes ^{39}K and ^{41}K, contains 0·012% of the weakly radioactive isotope ^{40}K. This has a half-life of $1·3 \times 10^9$ years and decays either to ^{40}Ar or ^{40}Ca. The age determination is based on argon rather than calcium because the former can be detected with very high sensitivity in a mass spectrometer and because ^{40}Ca is already abundant in most rocks anyway. Also, because argon is a gas, it is less likely to exist at the time of the event being dated—namely, the volcanic eruption. Thereafter the argon builds up according to

$$N_2 = N_1 \frac{1 - \exp(-\lambda t)}{\exp(-\lambda t)}$$

where N_1 and N_2 are the numbers of ^{40}K and ^{40}Ar atoms in a given amount of sample and λ is the disintegration rate. The ^{40}K content is determined either by measuring the total potassium content chemically (by flame photometry) or by measuring the ^{39}K content using neutron activation. In either case knowledge of the abundance ratio for the potassium isotopes allows derivation of the ^{40}K content itself.

The difficulty of using the method on rocks that are young enough to be of archaeological relevance is due to the very slow rate of formation of argon. For example, rocks with a potassium content of 2%, formed 10,000 years ago, have a fractional radiogenic ^{40}Ar content of only about 10^{-12}. Beside the need for high instrumental sensitivity there is also the difficulty of contamination by ^{40}Ar from the atmosphere; this occurs by absorption on the sample material and on the walls of the measuring apparatus. The minimum age that can be measured depends on the condition of the minerals present as well as on the potassium content. The limit to which the method is considered generally useful is around 100,000 years though in special cases meaningful ages have been obtained for much younger rocks.

Uranium-series Dating (including ionium dating)

In a sample that is old enough the uranium radioactive decay series is in equilibrium; that is, the amount present of any member of the series is proportional both to the amount of uranium present and to the half-life of the member. In a young sample the series may not be in equilibrium because at formation there may have been chemical discrimation for or against some members of the series. The degree to which equilibrium has been re-established can be used to determine the time that has elapsed since formation, as long as the details of the initial disequilibrium condition can be presumed.

One example is the ionium dating of ocean sediments. This isotope (^{230}Th) tends to be precipitated from sea water whereas the parent uranium remains in solution. Hence in deposited sediment the ^{230}Th is "unsupported" and its concentration decays to a negligible level at a rate determined by its half-life of 75,200 years. Consequently in sediment that has been deposited at a uniform rate the ^{230}Th decreases exponentially with depth and the time that has elapsed since the sediment at a given depth was deposited can be determined.

Of more direct archaeological relevance is application to shell, stalagmite and bone. Here it is assumed that uranium is incorporated at formation (or shortly after) without any accompanying ^{230}Th. The degree to which the ^{230}Th has built up towards its equilibrium value indicates the time that has elapsed since formation. Similar considerations apply in respect of protactinium-231 (half-life 32,500 years). Obviously the validity of the method rests critically on the assumption that shortly after formation there is no further uptake of uranium, nor any loss. It is well known that in bone uptake does go on for some time after burial since the uranium concentration is one way of distinguishing old bones from later intrusions, in the same deposit. It is as yet (1972) too early to judge how reliable the method may prove to be in respect of bone but such application would be of the highest potential importance for the Palaeolithic period since this is mostly beyond the range of radiocarbon dating.

Fission-track Dating

In minerals and glasses that contain traces of uranium, substantial damage to the lattice is caused by recoiling fragments from the spontaneous fission of the ^{238}U nucleus. These damage tracks can be made visible under the microscope by etching with hydrofluoric acid, the damaged regions being less resistant to attack. The tracks so revealed are between 5 and 20 µm long and can be distinguished from etch pits due to dislocations. The tracks are sufficiently stable (a temperature of the order of at least 500°C typically being needed to anneal them) for the number present to be used as a measure of age— in a similar way to the build up of thermoluminescence. The uranium concentration is conveniently evaluated by measuring the increase in track density that results from exposing the sample to a known dose of thermal neutrons in a nuclear reactor. Thermal neutrons induce fission in ^{235}U and since the cross-section for this, the present-day abundance ratio of ^{235}U/^{238}U and the spontaneous fission decay constant are all known, the age may be determined.

The technique is primarily of use for geological dating, the event dated being the formation of the mineral concerned. Some archaeological application has been made but lack of sensitivity limits this technique to materials with a fairly high uranium content. One such example is the dating of pottery by means of the fission-tracks in uranium-rich inclusions in it—such as zircons.

A possible associated technique, which would be much more applicable to archaeology because of its potentially higher sensitivity, is the measurement of *recoil tracks* accompanying alpha particle emission in the decay of uranium and thorium. These tracks are very much shorter (0·01 micron) and consequently rather difficult to identify with reliability.

Chemical Change

Bone. The uptake of uranium by buried bone has already been mentioned. Uptake of fluorine also occurs, both it and uranium being incorporated in the phosphatic mineral (hydroxyapatite) of which bone is mainly composed. On the other hand, the nitrogen content decreases due to the gradual disappearance of protein (collagen). Because of the dependency on external conditions these methods are restricted to relative dating of bones found in the same deposit.

Obsidian. Obsidian, a variety of volcanic glass, was widely used for prehistoric tools. A fresh surface (as made by ancient man in chipping the material to form the tool) slowly absorbs water and in the course of time there is a measurable hydration layer. The rate of growth depends on temperature, and consequently, in using the thickness of the layer as a measure of age, it is necessary to take this into account. Typical growth rates for different climatic regions have been established; rather surprisingly the rate of growth is independent of the wetness of the burial circumstances. The chemical composition influences the hydration rate but fortunately there are unlikely to be more than one or two effective types within a given region. The actual thickness of the layer for a 4000-year-old specimen (as an example) is about 7 μm for a hot region such as Egypt but only about 1 μm for the Arctic. On theoretical grounds the thickness is proportional to the square root of the age. The hydrated layer contains about 3% water which is about ten times the water content of fresh obsidian.

Glass layer counting. Some ancient glass exhibits iridescence and studies of glasses from Nineveh by Brewster (1863) showed this to arise from diffraction effects associated with weathering layers on the surface. For glasses which have a poor corrosion resistance and which have been buried in moist conditions, the weathering may be sufficient to give rise to a crust which is sometimes several millimetres in overall thickness. Study of some well-dated specimens by Brill and Hood (1961) revealed the remarkable fact that the number of layers within the crust was equal to the age, suggesting the possibility of "layer counting" as a powerful method of dating.

The validity of the technique has been questioned by Newton (1971) who points out that layered crusts can also be produced in accelerated weathering experiments in which the conditions (temperature, humidity) are kept constant. Also, layers are exhibited by ancient glasses which have been subject to negligible annual variation in environment, such as obtains on the bed of the sea. The mechanism that gives rise to the layers is not at all understood, but it seems possible that they result from internal strains set up by chemical change in which there is an associated expansion or contraction. In the weathering crusts the alkali of the glass has been replaced by water, the degree of hydration being about 20%. The layer thicknesses lie in the range 0·5 to 20 microns.

See also: Thermoluminescent dating.

Bibliography

Books:

AITKEN, M. J. (1974) *Physics and Archaeology*, Oxford University Press.

ALLIBONE, T. E. (ed.) (1970) The Impact of the Natural Sciences and Archaeology, A Joint Symposium of the Royal Society and the British Academy (Oxford University Press, and *Phil. Trans. Roy. Soc. Lond.* A, **269**).

BROTHWELL, D. and HIGGS, E. (eds.) (1969) *Science in Archaeology*, 2nd. Ed., London: Thames & Hudson.

LIBBY, W. F. (1955) *Radiocarbon Dating*, University of Chicago Press.

OLSSON, I. U. (ed.) (1970) Radiocarbon Variations and Absolute Chronology (Stockholm: Almqvist and Wiksell), *Proceeding of the 12th Nobel Symposium at Uppsala, August 1969*.

PYDDOKE, E. (ed.) (1963) *The Scientist and Archaeology*, Phoenix.

Review Articles:

AITKEN, M. J. (1970) Dating by archaeomagnetic and thermoluminescent methods, *Phil. Trans. Roy. Soc. Lond.* A, **269** (1193), 77–88.

AITKEN, M. J. (1970) Physics applied to archaeology. I. Dating, *Reports on Progress in Physics*, **33**, 941–1000.

THELLIER, E. (1966) Le champ magnétique terrestre fossilé, *Nucleus*, **7**, 1–35.

TITE, M. S. (1970) The impact of the natural sciences in archaeology, *Contemporary Physics*, **11**, 523–39.

Journals:

Archaeometry (Cambridge University Press, Bentley House, 200 Euston Road, London, N.W. 1).

Radiocarbon (American Journal of Science, Yale University, New Haven, Connecticut, U.S.A.).

M. J./AITKEN

B

BARIUM CLOUDS FOR EXPERIMENTS IN SPACE

Introduction

In 1951 L. Biermann suggested that the observed acceleration of ionized molecules in comet tails was due to a continuous outpouring of plasma from the Sun, the solar wind. But the physics of the interaction of the cometary plasma with the plasma of the solar wind is not yet understood in detail. Furthermore, the chemical composition of a comet is unknown *a priori* and comets with easily observable ionized tails are rare cosmic phenomena. Therefore L. Biermann *et al.* (1961) proposed to conduct an experiment involving the release of an ion cloud in the interplanetary medium, in order to study the detailed interaction of the artificially injected plasma with the solar-wind plasma and to learn more about the comets. In pursuance of this objective the Max-Planck-Institut für extraterrestrische Physik in Garching near Munich has directed efforts in the last 10 years towards producing metal vapour clouds in the ionosphere and magnetosphere.

Choice of the Elements

To make the experiments as cheap as possible the observation of the clouds must be made from the ground and the vaporization should be achieved by a chemical reaction. The ionization of the metal vapour cloud is caused by the solar ultraviolet radiation. Therefore the suitable elements must fulfil the following conditions:
(1) the resonance lines of the ions and neutrals must lie within the optical window of the Earth's atmosphere;
(2) the chemical reaction must produce a sufficient amount of metal vapour and
(3) the excitation times of the lines and the photoionization of the neutral atoms must be sufficiently short. In view of these constraints the alkaline earth metal Ba has proved to be the most favourable element. Experiments with the rare earth element Eu have been carried out recently, which showed that Eu is also a suitable element.

Experimental Technique

Method of evaporation

The Ba is vaporized by the reaction of a mixture of copper oxide and barium

$$(n + 1)\,Ba + CuO \rightarrow BaO + Cu + n\,Ba\,(vapour)$$

The barium is partially burned and provides the heat for the evaporation of the barium which is in stoichiometric surplus (nBa). Experiments have shown that $n = 1 \cdot 5$ gives a good efficiency of evaporated metal. Referred to the total weight of the mixture about 5–10% of Ba can be vaporized.

The ionization

The time for the photoionization of barium turned out to be very short (~ 20 sec). The reason is that the barium is not ionized directly from the ground state ($6s^2\ ^1S6$) but from two metastable states ($6s\ 5d\ ^1D_2$ and $6s\ 5d\ ^3D_{1\,2\,3}$) which are occupied from the ground state over higher levels. To ionize the barium from the ground state we need photons with a wavelength of less than 2380 Å. This would give us a time of ionization of about 15 min. To ionize the barium from the two metastable states we need photons with a wavelength of less than 3266 Å. In this part of the spectrum the solar radiation is about 30 times more intense than in the shorter wavelength range around 2380 Å. (See, for instance, L. Haser, 1967.)

Observation

The observation is carried out from the ground by photographic and photoelectric devices. The cloud emits the light by resonance fluorescence, so it has to be illuminated by the Sun, whereas the observer must be in the darkness (Fig. 1). This is one of the greatest disadvantages of the ion cloud technique because one is bound to twilight conditions. Thus the observation time in equatorial and mid-latitudes is limited to about 30 min. Longer observation times (1–2 hr) can only be achieved when we evaporate a sufficient quantity of barium at large heights (a few 1000 km) or in experiments carried out in high geographic latitudes.

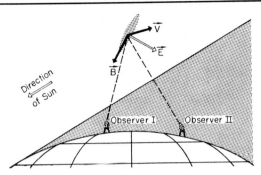

Fig. 1. Triangulation of the clouds. **B** $=$ *direction of the magnetic field.* **V** $=$ *direction of the drift motion.* **E** $=$ *electric field, which is calculated from the equation.*

The General Behaviour of the Barium Clouds

The neutral cloud

To investigate the atmosphere we observe the neutral cloud and the ion cloud. The neutral cloud interacts only with the neutral part of the atmosphere, so we can derive some of the physical parameters of the neutral atmosphere by the study of its behaviour. After the evaporation the neutral cloud expands in a spherical shell with an expansion velocity of about 1 km/sec (the linear phase). Collisions of the atoms with the neutral atmosphere lower this expansion velocity until after some time the cloud expands merely by diffusion. The final radius of the cloud in the linear phase is a measure of the mean free path of the barium atoms in the ambient atmosphere from which the density can be derived. The radial growth of the cloud in the diffusional phase gives us the diffusion coefficient which depends on the density and temperature of the atmosphere. Therefore an estimate of the temperature can also be obtained. Owing to the short ionization time of 20 sec the neutral barium cloud would be very faint and hard to observe at times greater than 20 sec. But in the CuO/Ba mixture there is always some strontium as an impurity which is easily evaporated. The strontium has a much longer ionization time than the barium ($\sim 1 \cdot 5$ hr). This and the fact that barium is oxidized at heights around and below 200 km to some extent makes it possible to extend the observation of the neutral cloud considerably.

By measuring the motion of the neutral cloud as a whole we get the streaming velocity of the neutral atmosphere. The horizontal wind velocities obtained in this way are of the order of 30–250 m sec^{-1} at altitudes between 130 and 250 km in equatorial and mid-latitudes. In high geomagnetic latitudes we sometimes get wind velocities as high as 500 m sec^{-1}, while vertical wind velocities are between $+50$ and -50 m sec^{-1}. In most cases the vertical air motion is very small.

The ion cloud

The neutral and the ion cloud differ in their appearance. The neutral cloud expands in a spherical shell, whereas the ion cloud takes the shape of a cigar. This is caused by the action of the Earth's magnetic field upon the ions and electrons. They are forced to spiral around the field (Fig. 2). The longer axis is directed along the magnetic field which becomes visible in this way (Fig. 3). Also

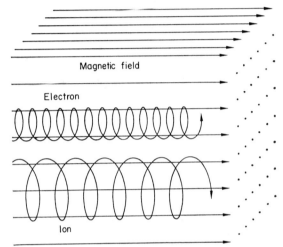

Fig. 2. Spiral motion of the electrons and ions around magnetic field lines. This is a combination of a circular motion around the field and a linear motion along the field (taken from Haerendel and Lüst, 1968).

owing to the collisions of the ions with the atmospheric particles, the diameter perpendicular to the magnetic field will grow, but at a lower rate than that of the long axis of the cloud. In an experiment at 2000 km where collisions with the atmosphere are negligible and the diameter of the cloud is determined by the gyration of the barium ions around the field lines, we observed very long needle shaped ion clouds (Fig. 4).

The magnetic field determines the shape of the ion cloud. If there is also an electric field present, the ion cloud drifts in a direction perpendicular to both, as is shown in Fig. 5 for a single ion and electron. The drift is in the same direction for the ion and the electron. It is given by the cross-product of the electric and magnetic field vectors **E** and **B** respectively,

$$\mathbf{v}_D = c \frac{\mathbf{E} \times \mathbf{B}}{B^2}$$

where c is the velocity of light.

Since the magnetic field is known in most cases, we can derive in a very simple way the electric field at the height of our experiment by measuring the drift motion of the cloud perpendicular to the magne-

Fig. 3. Barium ion cloud (elongated) and neutral strontium cloud (spherical) at a height of 200 km over the Algerian Sahara. Both clouds have been made in one evaporation. They have separated because the ion cloud moves with the ionized part of the upper atmosphere and the neutral cloud with the neutral part.

Fig. 4. Ion clouds at a height of 2000 km over the Algerian Sahara. The long ion cloud has a length of about 900 km 4 min 50 sec after evaporation. The short ion cloud is shown 50 sec after evaporation. Neutral strontium (spherical cloud) is still visible. The clouds were also observed from Europe (Rieger et al., 1970).

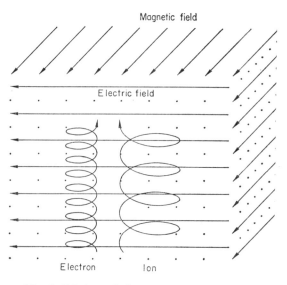

Fig. 5. Motion of electrons and ions in a perpendicular magnetic and electric field. Positive and negative charged particles move in the same direction. The motion is not helical as in Fig. 2, but proceeds in a plain. The component along the magnetic field was set equal to zero (taken from Haerendel and Lüst, 1968).

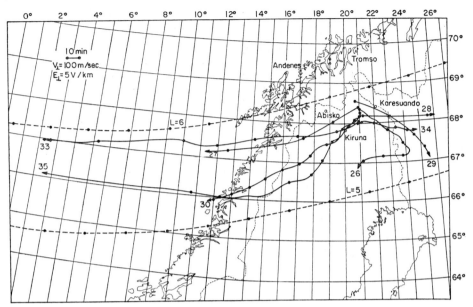

Fig. 6. Drift of ion clouds released over northern Scandinavia at heights around 230 km. The numbers at the paths refer to experiment numbers.

tic field. The above equation is applicable for single particles; for an ion cloud it is a simplification because the motion of the cloud is influenced by the neutral wind and the ion cloud itself disturbs the surrounding medium by changing the electrical conductivity. The first effect can be kept small if we make our experiments at heights where the collision frequency of the ions and electrons with the neutral atmosphere is small compared to the gyration frequency of the ions and electrons around the lines of force. This condition is fulfilled at heights above 180 to 200 km. The effect of the ion cloud upon the surrounding medium can be kept small if we release small amounts of barium. At heights around 200 km the total mass of barium ions should not be much greater than about 15 g. Even this small amount gives clouds which are clearly observable for a sufficiently long time (Haerendel et al., 1967).

Measurement of the Electric Field

Ionospheric experiments

Since November 1964, when the first barium ion cloud was made over the Algerian Sahara by the Max-Planck-Institut für extraterrestrische Physik, many experiments have been carried out in the ionosphere at different geomagnetic latitudes. Experiments carried out at the geomagnetic equator helped us to understand some of the phenomena of the *equatorial electrojet*. Experiments made in geomagnetic mid-latitudes have given support to the *dynamo theory of Stewart*† as the cause for the ionospheric current system during quiet solar periods, the so-called *Sq-current system*.† The electric field is of the order of a few V/km.

It is of great interest to make these experiments at high geomagnetic latitudes, namely in the auroral zone. In this region the magnetic lines of force that are linked to the distant part of the magneto-

---

† Due to the heating of the atmosphere by the Sun during the day a world-wide circulation is set into motion. At heights around 100 km, charged particles, atmospheric ions, participate in this motion via collisions with the neutrals, which in the presence of the geomagnetic field leads to an electromotive force. This is called the *atmospheric dynamo*. Above about 90 km the electrons on the other hand are much less influenced by the motion of the atmosphere. We therefore get an accumulation of space charges around the globe, which causes an electrostatic field. This is the field we measure with the barium ion cloud technique if we observe the constraints which were mentioned above. The electromotive force and the electrostatic field set up a current, the *Sq current system*. The Sq means the current system during quiet solar periods. The *equatorial electrojet* is a part of the Sq current system. It is confined to the geomagnetic equator and has its maximum intensity around noon.

sphere meet the surface of the Earth. Under magnetically not too much disturbed conditions the electrical conductivity along the magnetic field is much greater than the conductivity perpendicular to it. That means that potential differences along the magnetic field are negligible and that magnetic field lines are practically equipotential lines. So the electric field which determines the motion of the distant magnetospheric plasma can be measured at ionospheric heights. Of course there may also be electric fields of ionospheric origin, which are caused by the atmospheric dynamo. But it turns out that their contribution to the fields of magnetospheric origin is small. Figure 6 shows the drift motion of ion clouds released over Northern Scandinavia. We see that the east–west motion of the clouds dominates the north–south motion, showing that the electric field is mostly north-south. The field is normally around 10–20 V/km and may be as high as 100 V/km during magnetically disturbed conditions. Because of the high parallel conductivity the electric field measured at ionospheric heights can be projected along the magnetic field lines into the magnetospheric equatorial plane.

In Fig. 7 we have projected the drift along the magnetic field lines into the magnetic equatorial plane. For simplicity we have taken a dipole configuration of the magnetic field. We see that in the early evening hours the drift is clockwise around the earth if we look from above to the north pole. In the late evening hours the motion is more random. During midnight and in the morning the motion is counter-clockwise. This picture of the magnetospheric convection during not too much disturbed periods fits well into the general ideas which one has about the motion of the magnetospheric plasma (Axford, 1969).

Magnetospheric experiments

Although it is possible with certain assumptions to extrapolate our high latitude measurements of the electric field into the distant magnetosphere it is necessary to carry out barium cloud experiments in the magnetosphere itself. This gives us the possibility not only of measuring electric fields but also of studying the interaction of an artificial plasma with the dilute magnetospheric plasma where collisions are completely absent. In contrast to ionospheric experiments the injected plasma is a very heavy load on the weak magnetic field lines and we cannot expect the velocity of the barium cloud to overtake that of the ambient plasma in a time of 1 sec, as is the case in our ionospheric experiments. The cloud will more or less continue to move with the velocity of the space vehicle. But the magnetic field tries to penetrate from the dim outskirts into the cloud thereby peeling it off like an onion. By studying the motion of this peeled off thin artificial plasma we can hope to measure electric fields.

Until now two experiments have been carried out in the distant magnetosphere. On March 18, 1969 at 7.20 universal time about 150 g of barium were evaporated at a distance of 74,300 km from the Earth's surface on the morning side of the magnetsphere. The cloud was observed in North and South America for about 25 min. Figure 8 is a 3-min

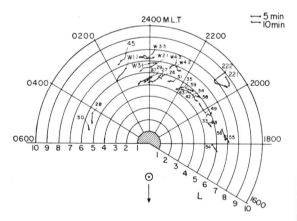

Fig. 7. Drift of ion clouds released in the auroral ionosphere and projected along the magnetic field lines into the geomagnetic equatorial plane, by taking a dipole field. The Earth is indicated by the hatched area. The direction to the Sun is shown by a "☉". The numbers at the paths are experiment numbers.

Fig. 8. Ion cloud released in a distance of 74,300 km, 7 min after evaporation; 3-min exposure of the 48-in. Schmidt-telescope on Mount Palomar.

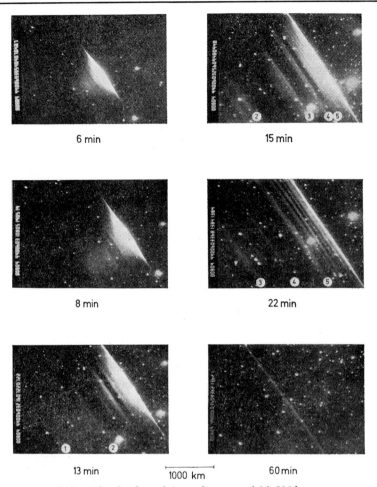

Fig. 9. Ion cloud released in a distance of 31,600 km as seen from Kitt Peak, Arizona with a Super-Schmidt-lens and a TV camera. The most prominent structures are numbered in order to show the separation from the main part of the cloud (the core).

exposure of the cloud taken with the 48-in. Schmidt-telescope on Mount Palomar 7 min after evaporation. The cloud is divided into striations which are formed by plasma instabilities. For the longer part of the observation time the cloud was not markedly affected by the ambient plasma flow. However, the third weak feature on the right-hand side of the cloud showed a separation from the main body. The motion of this feature allowed us to set a lower limit of 0·085 V/km to the ambient electric field magnitude. Projecting the direction of the electric field downward into the ionosphere we obtained evidence for the existence of a current along the magnetic field into the ionosphere during a magnetic substorm.

The experiment carried out on September 21, 1971 at 3.4.52 universal time was much more spectacular. At a height of 31,600 km about 1·5 kg of barium were evaporated near magnetic midnight. The observation time was about 80 min. In Fig. 9 the cloud is shown at six successive times as seen from Kitt Peak, Arizona. The main part of the cloud is elongated along the magnetic field lines and moves, more or less unaffected by the ambient plasma, with the rocket's velocity from left to right, leaving a tail behind which consists of barium plasma already interacting with the magnetic field. Fortunately this plasma forms striations so that an identification is possible for observers at different places. To visualize the separation from the main part of the cloud the most prominent features have been numbered. After an acceleration phase of a few hundred seconds the features showed a constant velocity of separation (Fig. 10). In this phase we believe that they have attained the velocity of the ambient plasma flow. Remarkably this velocity was not the same for all the features, showing that local gradients existed in the magnetospheric plasma motion. Such local gradients have been detected already in a few of our ionospheric experiments carried out in the auroral zone. The electric field derived from the motion of the different structures was between 0·1 and 0·64 V/km in the magnetosphere.

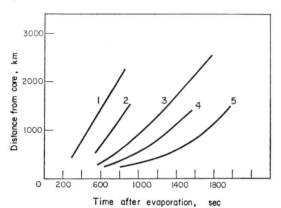

Fig. 10. *Separation of the structures from the main part of the cloud over time.*

As it is of interest to compare magnetospheric electric field measurements with ionospheric because we learn something about the coupling of the magnetosphere to the ionosphere via the magnetic field, electric-field measurements have been carried out in the area where the magnetic field line of the barium cloud experiment enters the ionosphere. This was near Great Whale River at the east coast of the Hudson Bay. Unfortunately we missed the field line which comes down over Great Whale because the configuration of the distant magnetic field is hard to predict. The field line where the barium experiment took place met the ionosphere about 200 km south of the station. The electric fields measured in the magnetosphere and in the ionosphere were in opposite directions to each other, which is not very surprising because of the displacement of 200 km in the ionosphere for the two measurements and because both measurements showed local gradients of the electric field.

To connect electric field measurements in the ionosphere with those in the magnetosphere new developments have been started with the aim of ejecting barium clouds with rather high velocities up to 14 km/sec along a magnetic field line using shaped charges. By this method a barium cloud could be ejected from comparatively small rockets at altitudes of around 500 km so as to travel along the magnetic field line through the distant magnetosphere to the opposite hemisphere. This would enable us to investigate magnetic conjugate-point phenomena and to study electric fields with a more dilute artificial plasma on the magnetic field lines.

Conclusion

The two experiments in the distant mangetosphere can already be considered as a step towards an artificial comet. The essential difference is the high streaming velocity of the solar wind with respect to the comet. It is large compared to the Alfvèn-velocity. In our magnetospheric experiments the cloud's velocity with respect to the magnetospheric plasma is small compared to the Alfvèn-velocity. Information along the magnetic field lines is passed with Alfvèn-velocity. Therefore the magnetic field lines on which the barium clouds are made are practically undeformed as is evident in the structure of the cloud and of the striations (Figs. 8 and 9). In contrast to the this magnetic field is bent around the head of the comet and is bundled to an axis as is indicated in the structure of the rays of the comet's plasma tail (Fig. 11).

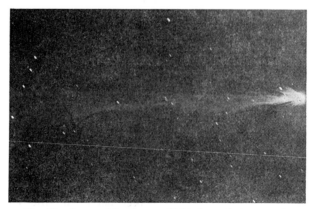

Fig. 11. *Comet Morehouse 1908 (Royal Greenwich Observatory, Nov. 25, 1908).*

Bibliography

AXFORD, W. I. (1969) Magnetospheric convection, *Rev. Geophys.* **7**, 421.

BIERMANN, L. (1951) Kometenschweife und solare Korpuskularstrahlung, *Z. Astrophys.* **29**, 274.

BIERMANN, L., LÜST, R., LÜST, RH. and SCHMIDT, H. U. (1961) Zur Untersuchung des interplanetaren Mediums mit Hilfe künstlich eingebrachter Ionenwolken, *Z. Astrophys.* **53**, 226.

HAERENDEL, G., LÜST, R. and RIEGER, E. (1967) Motion of artificial ion clouds in the upper atmosphere, *Planet. Space Sci.* **15**, 1.

HAERENDEL, G. and LÜST, R. (Nov. 1968) Artificial plasma clouds in space, *Scientific American* **219**, No. 5, p. 81.

HASER, L. (1967) Use of artificial barium clouds to study magnetic and electric fields in the atmosphere, *Aurora and Airglow* (B.M. McCormac, ed.), p. 391.

RIEGER, E., NEUSS, H. LÜST, R., MEYER, B., HASER, L., LOIDL, H., STÖCKER, J. and HAERENDEL, G. (1970) High altitude releases of barium vapor using a rubis rocket, *Ann. Geophys.* **26**, 44.

E. RIEGER

BEAM-FOIL SPECTROSCOPY

1. Introduction

Beam-foil spectroscopy (BFS) is an experimental technique for investigations of spectra and energy levels of atoms and ions, mean lifetimes of excited states, and atomic fine-structure effects, including the Lamb shift. This technique, which is based on the excitation of fast heavy ions in a thin foil, has been in frequent use since the early 1960s (Kay, 1963; Bashkin, 1964).

The principles of BFS are shown in Fig. 1. The ions are produced in the ion source of an accelerator (e.g. a Van de Graaff generator). After acceleration and mass analysis the collimated, isotopically pure beam of positive ions is directed through a very thin exciter foil, located in an evacuated target chamber. Charge exchange processes (further ionization or electron capture) thereby affect the accelerated particles, which often also emerge from the foil in excited electronic states. On the downstream side of the foil these states decay spontaneously (collisions can be neglected if the pressure in the target chamber is below 10^{-5} torr) with the emission of photons (or electrons, autoionization). The electromagnetic radiation can be analysed with a monochromator. The foil-excited particle beam thus forms a spectroscopic light source which has several advantages: (1) high chemical and isotopic purity, (2) possibilities of obtaining atomic spectra from many ionization degrees, including very high ones, by varying the accelerator energy, (3) time resolution (the velocity of the ions is known) which permits measurements of decay constants for spectral lines, as well as intensity-decay studies in perturbing external fields.

The wavelength accuracy of beam-foil measurements has some inherent limitations, caused by doppler effects of the moving ions, and it is very

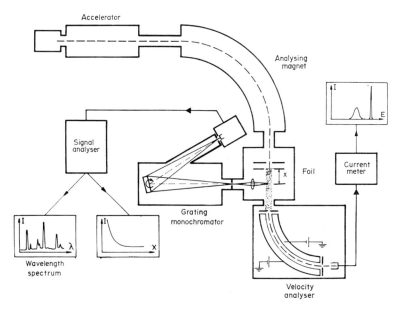

Fig. 1. Experimental arrangement for studies of atomic spectra and lifetimes with the beam-foil technique.

difficult to obtain smaller linewidths than 0·5 Å at 5000 Å (Stoner and Leavitt, 1973).

2. Experimental Equipment

Most initial beam-foil studies were performed with single-stage Van de Graaff accelerators with 400 kV–5·5 MV terminal voltages. More recently tandem Van de Graaffs have also been employed in BFS, and by accelerating multiply ionized atoms, energies up to 60 MeV have been obtained. In addition to gases or elements forming gaseous compounds, alkali metals as well as Fe and Ni have been successfully accelerated in Van de Graaffs for beam-foil purposes. Electromagnetic isotope separators and heavy ion accelerators which are equipped with universal ion sources have been employed in beam-foil studies of many elements, including Be, Al, Si, P, transition metals, rare earths, Tl, and Pb. The energy range has been 50 keV–1·2 MeV. Very high energies have been obtained with heavy ion linear accelerators. Using the Berkeley HILAC, Marrus and Schmieder (1972) accelerated Ar ions to 412 MeV.

In most cases self-supporting carbon foils (2–15 μg/cm^2) are used to excite the beam. When the incoming ion velocity is low, very thin foils of low atomic number material (Be or C) are necessary in order to minimize the effects of angular divergence and velocity straggling after the foil. The foils have a limited lifetime against damage (typically 5–30 min), depending on mass, velocity and intensity of the incident ions.

When the ion current is 0·5 μA or more, the light emitted by the ions can easily be viewed by the unaided eye. Wavelength studies are often made with relatively fast grating monochromators (focal length 0·3–3 m), equipped with photomultipliers or channel electron multipliers. In this way beam-foil spectra between 15 and 9000 Å have been observed. (For wavelengths below 400 Å grazing-incidence spectrometers are needed.) The use of solid state X-ray detectors and electron spectrometers has permitted studies of X-rays and Auger electrons, emitted by highly stripped beams.

The particle velocity after the foil, which is necessary for lifetime determinations, can be measured with an electrostatic analyser (Fig. 1). Alternatively, it can be obtained by correcting for the energy loss in the foil, using known range–energy relationships.

3. Results

Atomic spectra

Beam-foil experiments have provided much new information about atomic structure, above all for highly ionized atoms. By accelerating O and Cl ions in a tandem Van de Graaff to 6–42 MeV, Hallin *et al.* (1973) observed many previously unreported transitions in O VI–O VIII and Cl VII–Cl XIII (O VI indicates five times ionized oxygen, etc.). Figure 2 shows examples of their spectra. The indicated transitions, which have been observed only in beam-foil spectra, occur between high n-states (e.g. 7–8) in hydrogen-, helium-, and lithium-like oxygen. Similar results have also been reported by Denis *et al.* (1971).

While all these transitions are allowed by the selection rules (electric dipole or E1 transitions), forbidden transitions, e.g. magnetic dipole (M1) and quadrupole (M2), electric quadrupole (E2) as well as two-photon decay (2 E1) of metastable states, have also been observed with BFS at very high particle energies. In the spectra of Ar XVII and Ar XVIII (helium- and hydrogen-like argon) Marrus and Schmieder (1972) measured the absolute rates of these processes. Their results provide important checks of relativistic quantum mechanics and quantum electrodynamics.

However, even at relatively low energies much spectroscopic information can be obtained using BFS. Several new transitions in elements such as Be, B, C, N, and O have been located with low-energy isotope separators or 2 MV Van de Graaffs.

As a result of the interaction with the foil atoms the emerging particles are often in multiply excited states, such as $2p^2\ ^3P$ in He I, $1s2s3d\ ^4D$ in Be II, and $1s2p^2\ ^4P$ in O VI. The multiply-excited states often lie energetically above the ionization potential of the corresponding atom or ion and hence often autoionize by electrostatic interaction. When this process is forbidden by the selection rules, radiative transitions and magnetic interaction (in highly-ionized atoms) become prominent. Usually the multiply excited states are difficult to observe in emission with other light sources (arcs, hollow cathodes, sparks, etc.) but they are abundantly populated in BFS (c.f. Berry *et al.*, 1971, and Sellin *et al.*, 1971).

In the simplest case the intensity $I(x)$ of a spectral line varies with the distance x from the foil according to

$$I(x) = I(0) \exp(-x/v\tau) \qquad (1)$$

where $I(0)$ is the initial intensity, v the velocity of the ions after the foil, and τ the radiative lifetime of the level under study. The lifetimes of atomic levels are typically of the order of 10^{-7} to 10^{-11} sec (allowed transitions), and such times are easily measurable with BFS. (The velocity of, for example, 3 MeV $^{12}C^+$ ions is 6·95 mm/nsec.)

Figure 3 shows the decay curves for two beryllium lines, measured with 56 keV Be$^+$ ions. The decay of the 3130 Å line ($2s\ ^2S - 2p\ ^2P$ in Be II) is represented by a single exponential, whereas the curve for the 3321 Å line ($2p\ ^3P - 3s\ ^3S$ in Be I) shows a more complex structure, caused by cascading transitions which repopulate the decaying level, $3s\ ^3S$. Due to the beam-foil excitation mechanism many levels are likely to be populated, and cascade effects

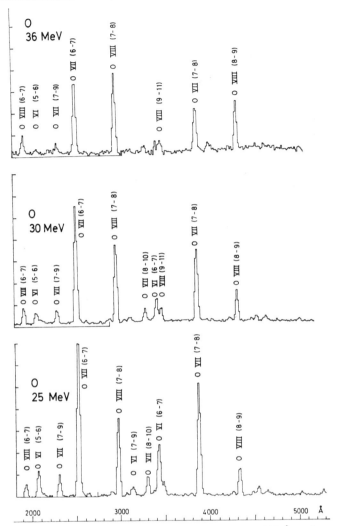

Fig. 2. Beam-foil spectra of oxygen between 2000 Å and 5000 Å. Oxygen ions of 25, 30, and 36 MeV were obtained from a tandem Van de Graaff (Hallin et al., 1973).

which limit the accuracy of lifetime measurements are quite frequent. It is usually difficult to reduce the uncertainties to below 5%.

From the measured lifetimes the radiative transition probabilities (or oscillator strengths) can be obtained. These quantities, which are often very sensitive tests of the accuracy of the wave functions, used in atomic structure calculations, have also important applications to astrophysics (studies of element abundances in the sun and the stars), plasma diagnostics, and laser physics.

Several experimental techniques are available for measurement of lifetimes and transition probabilities, e.g. the Hanle effect, anomalous dispersion, electron excitation, and spectral line intensity measurements (Foster, 1964; Corney, 1970). The main advantage of BFS lies in the applicability of the work to several times ionized atoms. Lifetime measurements with BFS have been reported for many elements from H to Tl. The experimental results are particularly valuable for atoms and ions with two or more electrons outside a closed shell, because here perturbations may cause substantial uncertainties in theoretical calculations. Regularities in atomic oscillator strengths have also been investigated with BFS, both for isoelectronic (the same number of electrons, e.g. B I, C II, N III, O IV, ...) and homologous (the same outer electron structure, e.g. B I, Al I,

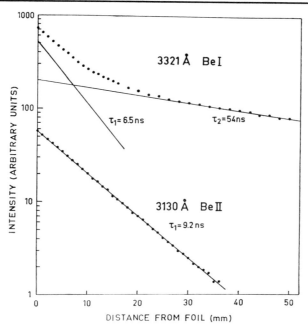

Fig. 3. *Two examples of decay curves.*

Ga I, In I, and Tl I) systems. In the Na I isoelectronic sequence, for example, the probability for the resonance transition $3s\,^2S - 3p\,^2P$ has been measured for Na I–Ar VIII.

Solar and stellar abundances of chemical elements are often derived from observed spectral line intensities which are proportional to the transition probabilities. Beam-foil measurements have provided decay constants for many spectral lines observed in solar and stellar atmospheres. Particularly valuable results have been obtained for the iron group elements. The measurements of Fe I lifetimes of Whaling *et al.* (1969) thus support an upward revision of the iron abundance in the solar photosphere by approximately a factor of 10, thereby eliminating previously reported discrepancies between the Fe abundances in the photosphere and the solar corona. Lifetime measurements with the beam-foil method have also been made for Sc, Ti, V, Cr, Mn, Co, and Ni, as well as for other astrophysically important elements, Be, Mg, Si, Ca, etc.

Fine structure

When the excited state of an atom (or ion) is an "eigenstate" (a solution to the time-independent Schrödinger equation), the intensity decay follows the simple exponential law (eq. 1). If the initial state is a superposition of two (or more) eigenstates, then interference effects between these states may cause periodic modulations (quantum beats) on the exponential decay curves.

Because of the sudden excitation ($<10^{-14}$ sec) in beam-foil spectroscopy, there exists an energy uncertainty which may be of the same order as the energy separation between close-lying states. An example of the coherent decay of such states (or the "Zero Field Quantum Beats") is showing in Fig. 4, which originates from a beam-foil study of the He I $2s\,^3S_1 - 3p\,^3P_{0,1,2}$ multiplet at 3889 Å. Oscillations are possible between all three J-states, $^3P_0$, $^3P_1$, and $^3P_2$, but only those between $^3P_1$ and $^3P_2$ are observable in Fig. 4. (The oscillations,

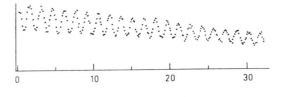

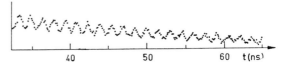

Fig. 4. *Zero-field quantum beats between 3p $^3P_1$ and 3p $^3P_2$ in He I, corresponding to a separation of 658 MHz (H. J. Andrä, unpublished result, 1972). The oscillations are superimposed on the exponential decay of the 3p ^3P term (mean lifetime 95 nsec).*

between e.g. $^3P_0$ and $^3P_1$ are of higher frequency c.f. Wittmann et al. (1972).) The observed frequency of 658 MHz is a measure of the energy separation between $^3P_1$ and $^3P_2$ states. Whereas such fine-structure effects in neutral atoms can also be precisely measured with level-crossing or radiofrequency (rf) techniques, BFS has made the extension of the work to ions possible.

Quantum beats may also be induced through external perturbations, such as electric or magnetic fields. An electric field can couple adjacent states of opposite parity, e.g. $2s\ ^2S_{1/2}$ and $2p\ ^2P_{1/2}$ in the hydrogen atom. Measurements of the intensity decay of the $1s\ ^2S$–$2p\ ^2P$ resonance line (1216 Å) in static fields of 50 V/cm or higher show superimposed sinusoidal oscillations (Stark beats) on the exponential decay of the $2p\ ^2P$ state (lifetime 1·60 nsec). The frequency and amplitude of the oscillations increases with the field strength. Similar field-induced oscillations have also been observed for higher levels in hydrogen and excited levels in singly-ionized helium.

The oscillations can also be observed if the foil-excited beam is subjected to a linearly varying, perpendicular magnetic field. The intensity of a given spectral line, observed at a fixed distance from the foil, thereby shows sinusoidal variations, the peroid of which is proportional to the Landé g-factor of the corresponding upper level (Liu et al., 1971). Such g-factor determinations are particularly valuable for excited levels in ions, since here other techniques (e.g. the Zeeman effect) may be difficult to apply.

The energy separation between $2s\ ^2S_{1/2}$ and $2p\ ^2P_{1/2}$, i.e. the $n = 2$ Lamb shift, has for H and He$^+$ been precisely measured with rf techniques, but these are difficult to apply to Li^{++} and other one-electron systems. Ion-beam techniques have been used for determinations of the Lamb shift in Li^{++}, C^{5+}, and O^{7+}. The experiments consist in accelerating and stripping the particles, which then are passed through a gas target, where some stripped ions capture an electron into the metastable $2s\ ^2S_{1/2}$ state. In a strong electric field the $2s\ ^2S_{1/2}$ and $2p\ ^2P_{1/2}$ states perturb each other, and the decay constant of the mixed level contains the $n = 2$ Lamb shift as a parameter.

Bibliography

ANDERSEN, T. and SØRENSEN, G. (1972) *Phys. Rev.* A5, 2447.
ANDRÄ, H. J. (1970) *Phys. Rev.* A2, 2200; *Phys. Rev. Letters* 25, 325.
BASHKIN, S. (1964) *Nucl. Instr. Methods* 28, 88.
BASHKIN, S. (1968) *Beam-foil Spectroscopy: Proceedings of the Conference on Beam-foil Spectroscopy, Tucson, Arizona, November 1967.* New York: Gordon & Breach.
BERRY, H. G., MARTINSON, I., CURTIS, L. J. and LUNDIN, L. (1971) *Phys. Rev.* A3, 1934.
BICKEL, W. S. (1967) *Applied Optics*, 6, 1309.
CORNEY, A. (1970) In *Advances in Electronics and Electron Physics*, 29, 116.
DENIS, A., DESESQUELLES, J. and DUFAY, M. (1971) *Compt. Rend. Acad. Sci.* 272B, 789.
FOSTER, E. W. (1964) *Rept. Prog. Phys.* 27, 469.
HALLIN, R., LINDSKOG, J., MARELIUS, A., PIHL, J. and SJÖDIN, R. (1973) *Physica Scripta* 8 209.
KAY, L. (1963) *Phys. Letters*, 5, 36.
LAWRENCE, G. P., FAN, C. Y. and BASHKIN, S. (1972) *Phys. Rev. Letters*, 28, 1612.
LIU, C. H., BASHKIN, S., BICKEL, W. S. and HADEISHI, T. (1971) *Phys. Rev. Letters*, 26, 222.
MARTINSON, I., BROMANDER, J. and BERRY, H. G. (1970) *Beam-foil Spectroscopy: Proceedings of the Second International Conference on Beam-foil Spectroscopy, Lysekil, Sweden, June 1970.* Amsterdam: North-Holland.
MARRUS, R. and SCHMIEDER, R. W. (1972) *Phys. Rev.* A5, 1160.
MURNICK, D. E., LEVENTHAL, M. and KUGEL, H. W. (1971, 1972) *Phys. Rev. Letters*, 27, 1625; ibid. 28, 1609.
SELLIN, I. A., PEGG, D. J., BROWN, M., SMITH, W. W. and DONNALLY, B. (1971) *Phys. Rev. Letters*, 27, 1108.
STONER, J. O. and LEAVITT, J. A. (1973) *Optica Acta* 20, 435.
WHALING, W., KING, R. B. and MARTINEZ-GARCIA, M. (1969) *Astrophys. J.* 158, 389.
WITTMANN, W., TILLMANN, K., ANDRÄ, H. J. and DOBBERSTEIN, P. (1972) *Z. Physik*, 257, 279.

I. MARTINSON

BIOCHEMICAL FUEL CELLS. Because of publicity given the American space programme, and particularly because of the news coverage given the Gemini and Apollo flights through the 1960s and 1970s, we have all come to know something about fuel cells. We know that they are a source of electric power, and most laymen probably consider them to be a kind of battery. That concept is not entirely erroneous, and yet there are fundamental differences between fuel cells and storage batteries which deserve to be explained.

The commonplace automobile battery, like the fuel cell, is an electrochemical cell, i.e. it depends upon the migration of ions through an electrolyte for the generation of electrical power. The positive and negative poles of an automobile battery are the electrodes. From one (the anode) electrons are released, and at the other (the cathode) electrons are absorbed. Sulphuric acid in the storage battery serves as an electrolyte because it breaks down, or dissociates, into ions. Such dissociation is essential to the operation of a battery, and an electrolyte is always defined in these terms. In operation, the ions migrate from anode to cathode, carrying electrons along with them. An electrical current is thereby created.

If we examine the typical lead–acid automobile battery a bit more closely, we find that it is a "storage" cell because it can be discharged as energy (electric current) is used or recharged as energy is supplied from an external source or generator. The anode is a multiple series of lead plates which are matched by an equal number of lead peroxide plates (the cathode). During the discharge cycle, lead sulphate and water are formed. The former coats the electrodes and the latter dilutes the electrolyte. During the charge cycle, with a generator supplying energy, the sequence is reversed, i.e. lead sulphate and water react to form pure lead and sulphuric acid.

Fuel cells are really much more simple in operation than storage batteries. For one thing, they do not store energy. Instead, energy in the form of ions is supplied directly from conventional fuel and oxygen. For Gemini and Apollo spaceships, the common fuel has been hydrogen—stored in liquid form under pressure. When released from its storage tank, hydrogen becomes gaseous, and can be directed into the anode of the fuel cell. At this point the hydrogen molecules become ions and electrons are released.

In a fuel cell used aboard a spacecraft, the anode and cathode are frequently separated by a special solid polymer electrolyte which functions as efficiently as any liquid. Electrons and hydrogen ions migrate through this membrane to be trapped at the cathode, where they combine with molecular oxygen to form water. The water is a valuable product of the fuel cell's electrochemical reaction because it may be used to quench the thirst of the space capsule's inhabitants.

Power for the Gemini spacecraft was supplied by General Electric fuel cells. Power for the Apollo series was supplied by fuel cells constructed by Pratt and Whitney, who manufactured several kinds of fuel cells for varied purposes. For example, the Apollo Command and Service Module powerplant was designed to provide electrical power for life support, guidance and communications systems. Its power output ranged from 563 W to 2.3 kW. The water produced as a by-product of this fuel cell's operation was used both for drinking by the crew and for cooling of certain of the spacecraft's components. Three of these powerplants were installed in early Apollo spacecraft to insure sufficient power reserve for an extended lunar mission. A powerplant of somewhat different design was used in the Apollo Lunar Excursion Module, which carried Apollo astronauts through their descent to the lunar surface and back again into lunar orbit. Each Lunar Excursion Module fuel cell provided electrical power over a range from 134 W to 1.1 kW for the several days of each moon exploration.

Closed-System Technology

Spacecraft, submarines and underwater habitats of various sorts share one feature in common—they function in environments which are hazardous to man. Man cannot survive in the low temperatures and reduced or elevated pressures unless the system around him (spacecraft or habitat) is equipped to provide an environment compatable with his human biological and psychological needs. The gaseous environment must be such that he can breathe; the temperature must be within a range that permits him to work in reasonable comfort; the food supply must nourish him adequately without displacing an undue amount of space; waste products of human metabolism must be contained or disposed of, so as not to be objectionable. Closed-system technology provides for these basic human requirements within the confines of a contrived artificial environment. Obviously, short ventures in closed systems require less sophistication than longer confinements. A relatively short trip of a few days' duration in a space capsule is one thing—an extended voyage or a lengthy confinement in a space station in permanent orbit is yet another. In the latter case, human needs are multiplied; the amount of stored food required is greater, the problem of disposal of solid and liquid wastes is amplified.

Closed-system technologists, when faced with the problems of longer confinements for man, turned their attention to cyclical systems. For example, if simple plants could be introduced to the sealed enclosure, perhaps they could utilize some of the waste products of man and yield products of value in flight. Simple plants, like the green alga, *Chlorella*, have been grown in quantity and under a variety of closed-space conditions, to test their usefulness as components of cyclical systems. *Chlorella* seemed to have promise for a number of reasons; for one, this alga is nourishing and has already proved feasible as an additive in animal feeds. Secondly, it can be grown readily in a variety of culture vessels. Thirdly, it produces oxygen as a product of its metabolism. Early investigators had hoped that they might feed modified wastes to the algal culture, and derive benefits in the form of oxygen and food production. Unfortunately this, and other kinds of experimental cyclical ventures, have not proved feasible. Always there has been a level of complexity which could not be reduced. Always there have been variations in performance which defied standardization, and which prevented the sought-after dependability.

Biochemical Fuel Cells

Biochemical fuel cells are simply those which utilize biochemical components—either in purified or natural form. The energy transport systems of yeasts or bacteria have, for example, been utilized. These microscopic forms of life are like tiny fuel cells in themselves, because they absorb a variety of different organic substrates (foodstuffs) which they decompose to yield energy. The steps in the breakdown of glucose, for illustration, are a series of simple oxidation-reduction reactions, leading to simpler and simpler products with each succeeding step. The specific chemical reaction which takes place in each step is mediated by an organic catalyst, or enzyme. The oxidation-reduction chain of events is like a microminiature "bucket-brigade" in which hydrogen atoms and electrons are passed along from one compound to the next. With transfer of hydrogen, the donor compound becomes oxidized; the recipient compound becomes reduced.

Barnet Cohen, a microbiologist, demonstrated in 1931 that electrodes immersed in a bacterial culture produced a measurable potential. His paper, entitled "A bacterial culture as a galvanic half cell", pointed out that an electromotive force or voltage could be measured when two electrodes connected to a voltmeter were immersed in a bacterial culture. When an oxygen electrode is used as the cathode, and the anode is immersed in a growing culture, the open-circuit voltage increases gradually, and when the circuit is closed, an electric current of a few milliamperes magnitude may be measured. It was this kind of observation, coupled with the rapid strides in fuel cell technology, which stimulated the imagination of biochemists and electrochemists in the late 1950s and early 1960s. It was during that period that substantial investments were made by government and industry in biochemical fuel-cell development.

Each bacterial cell receives fuel in the form of organic or inorganic compounds from the liquid environment in which it prospers (bacteria will not grow when deprived of moisture). It oxidizes the substrates which it can utilize as foodstuffs, and passes hydrogen and electrons to receptors inside and outside the cell. The extracellular environment, by this process, becomes somewhat reduced, particularly in relation to an oxygen electrode in an adjacent half-cell of sterile medium devoid of microorganisms. Knowledge of this phenomenon led in the 1950s and 1960s to an investigation of a myriad of combinations of different kinds of microorganisms in linkages with varying environments and varying electrochemical circuits.

Between 1960 and 1964 the U.S. Navy Department was hopeful that a low-power and long-life device could be developed for operation of beacons, transmitters and buoys for marking sea lanes. Likewise, the U.S. National Aeronautics and Space Administration were mindful of the need for new power sources in space, including those which utilized human wastes as sources of biochemical fuel.

In 1964 several investigators published the results of their work with selected bacterial species which utilized sewage in the production of ammonia. The bacteria selected for these experiments were those which possessed the enzyme urease in their cellular makeup, and were therefore capable of utilizing the urea in human wastes for the production of ammonia. The simplified chemical reaction for this process is as follows:

$$\underset{\text{urea}}{CO(NH_2)_2} + \underset{\text{water}}{H_2O} \xrightarrow{\text{urease}} \underset{\substack{\text{carbon}\\\text{dioxide}}}{CO_2} + \underset{\text{ammonia}}{2NH_3}$$

The best known urea-decomposing bacterial species is *Bacillus pasteurii*, characterized by its possession of relatively large intracellular amounts of urease, and by its capability for growth in the presence of urea and an alkaline environment. As this microorganism decomposes urea and produces ammonia, the environment becomes sufficiently alkaline to kill all organisms other than *Bacillus pasteurii*. In brief, it selects its own best conditions for survival and tends to effectively eliminate competitors.

Ammonia gas generated by bacterial cultures can be used directly as a fuel in a fuel cell. The ammonia molecules, like those of other fuels, become ionized carriers of electrons under suitable conditions.

Other microbiologists and biochemists, during the same period, investigated the production of hydrogen by bacteria. *Clostridium welchii* became the object of study because it possessed large amounts of the enzyme hydrogenase, which catalyses the release of hydrogen from suitable bacteriological foodstuffs. Efforts with *Clostridium welchii*, grown in numerous combinations of medium ingredients, and under a variety of environmental conditions, led to the production of impressive amounts of hydrogen. Some cultures, for example, produced as much as 8·2 litres of hydrogen per hour.

The disadvantage associated with bacterial production of either ammonia or hydrogen was the lack of adaptability to long-term use. No means could be found for sustaining the production on a continuous basis. No means could be found for supplying fresh nutrients and for removing spent medium without undue complexity in the production system.

Tests Employing Enzyme Systems

In the belief that enzyme-substrate systems free of living cells might be used to improve electrical output, several workers undertook to optimize such systems. A number of enzymes were studied, including D-amino acid oxidase, urease and glucose oxidase. Urease, in purified form, was found to be less effective than impure urease and no more effective than urease-producing bacteria in the production of ammonia. Of the many oxidative enzymes tested, only D-amino acid oxidase appeared promising in the oxidation of selected D-amino acids to yield a pyruvic acid derivative, ammonia, and hydrogen peroxide. Tryptophan was the most active substrate for enzyme action and ion–electron production; tyrosine and histidine were second in ability to yield electrons and measurable current.

The Seawater Battery

Of special interest to the U.S. Navy was the magnesium seawater cell, incorporating a wasting magnesium anode and an inert cathode in seawater as the electrolyte. The corrosion of the magnesium atom theoretically yields one magnesium ion and two electrons. At the cathode, water and electrons are pictured as reacting to yield hydrogen atoms and hydroxyl ions. These reactions should produce a calculated potential of 2·37 V, but only 1·5 were actually observed in seawater. When a mixed culture of bacteria, consisting mainly of the anaerobic species *Desulfovibrio*, was employed to coat the cathode, a marked improvement in the performance of the magnesium seawater cell was noted.

The advantage of bacterial colonized electrodes over control electrodes was apparent only at low current densities, and was difficult to analyse. The most reasonable explanation, however, was that bacteria played a role in removing hydrogen atoms from the cathode. If we visualize a sort of hydrogen "log jam" taking place, with hydrogen atoms accumulating at a greater rate than they can be removed or absorbed, then we can appreciate the contribution of a bacterial film which eliminates excess atoms and permits continuous absorption of electrons. Since it is known that hydrogenase-containing *Desulfovibrio* species consume hydrogen as well as sulphate, one can visualize them overcoming resistance at the cathode in this fashion. This kind of "depolarizing" activity of bacteria is effective only at low current densities in the range of less than one ampere per square foot.

The depolarizing role of bacteria is also substantiated by the knowledge that sulphate reducers lacking dehydrogenase are incapable of improving electrochemical performance of the seawater battery.

Yet another impressive test of the effect of bacteria was conducted by increasing the temperature at which magnesium seawater cells were operated. The optimal growth temperature for *Desulfovibrio estuarii* is in the 25°C to 35°C range with rapidly ensuing death above 45°C. Performance curves were constructed from a series of tests run at increasing temperatures. As temperature increased above 30°C, performance was seen to decline. After the biocell had been heated to 85°C and allowed to cool again at the termination of the tests, its performance again paralleled that of control cells devoid of bacteria.

These findings involving seawater cells suggest that a scientific break-through has been realized—that only the application of principle to human needs remains to be exploited. Unfortunately, the promise of the earlier findings has been negated by later results. The seawater cell with a colonized electrode performs well for up to a week. Beyond that time, in actual ocean tests, a number of variables come into play. Within a matter of hours after placing biocells and control cells beneath the ocean surface, the output voltage of the biocells drops to that of the control cells. A kind of periodic performance pattern develops thereafter, with biocells producing increases in power output in weekly pulses for 4 or 5 successive weeks. After about a month of such behaviour, the biocell performance falls to that of the controls. After a total submersion of 2 to 5 months, the power output of both types of cells decreases to zero.

The gradual decline in performance was attributable to two causes. First, white chalky deposits developed with time on the inner surfaces of the electrodes facing one another. The deposits, upon analysis, were found to contain large quantities of magnesium and calcium. Because of the presence of considerable amounts of dissolved carbon dioxide in seawater, it was concluded that the deposits on the electrodes consisted of calcium and magnesium carbonates. The second deterrent to sustained performance was the concomitant build-up on the outer surfaces of the electrodes, and elsewhere on the cells, of a zoogleal-type film. This film, which was found to contain magnesium, calcium and silicon, was presumably formed by a mixture of microorganisms unrelated to the sulphate-reducing species.

We are faced at this point with a hopelessness that is not easily dismissed. Many followers of fuel-cell technology are asking what is left in biochemical fuel-cell investigation swhich has not already been undertaken and found wanting. The answer lies in a more thorough understanding of electrode reactions and biological influences on electrochemical systems. Further, it seems possible that enzyme systems and electron transport mechanisms in biological cells may still be harnessed. The work on biochemical fuel cells, to date, has been largely empirical. The problems awaiting solution are highly sophisticated ones that will require the combined energies of experts in electrochemistry, biochemisty, microbiology, and enzymology. University laboratories and minds, as well as industrial talents, are needed for this effort. The results, when available, should point the way to low-level and long-term power generation from biochemical electron transport systems.

R. F. ACKER

BLACK HOLES IN SPACE. A black hole is a region where the gravitational collapse of a very large or highly-concentrated mass aggregate has caused the fabric of space-time to become so highly curved that no signal of any sort can escape from within its boundaries. The possibility of the existence of such objects has been known for some time and their properties have been extensively investigated by theoretical physicists during recent years. A number of objects which may turn out to be black holes are at present being investigated by astronomers, but, at the time of writing (July 1972), no positive identification of a black hole has been made.

The possible existence of black holes in space was first suggested by Laplace in 1798. Using Newtonian mechanics, he showed that if a body had sufficient mass the escape velocity at its surface would exceed the speed of light; thus, any light emitted from its surface would be pulled back and the body would be invisible. Since the escape velocity at the surface of a spherically-symmetrical body of mass M and radius r is $\sqrt{2MG/r}$, where G is the gravitational constant, Laplace pointed out that such an object would be invisible provided that r was less than $2MG/c^2$, where c is the velocity of light. In the case of a body of the same density as the Earth, this condition would be satisfied if its radius were 250 times that of the Sun, and since the existence of such massive bodies seemed highly improbable in Laplace's time, his work was largely ignored.

Modern interest in black holes dates from 1916, when Einstein published his general theory of relativity and Schwarzschild solved its field equations for the case of a single spherically-symmetrical gravitating mass. Schwarzschild showed that the *radius of curvature of space-time* at the surface of a body of mass M decreases as the radius of the body decreases, becoming equal to the radius of the body when the latter is equal to $2MG/c^2$—the so-called *Schwarzschild radius* of the body. He also showed that a body whose radius was less than its Schwarzschild radius would constitute a black hole, since no light could escape from its surface. Note that Schwarzschild's criterion for the production of a black hole, obtained using the general theory of relativity, is identical to that obtained by Laplace using classical mechanics; this agreement is purely coincidental however.

Until 1939, black holes were of purely hypothetical interest, since the existence of bodies of the enormous mass or high concentration needed to produce such objects appeared to be highly unlikely. (In the case of a body of solar mass, for example, the Schwarzschild radius could be shown to be roughly 3 km, corresponding to a seemingly impossible density in excess of 10^{19} kg m^{-3}.) In 1939, however, Oppenheimer and Snyder showed that a black hole could conceivably result from the *gravitational collapse* of a sufficiently massive star that had exhausted its thermonuclear fuel resources. During most of its active life, the tendency for a star to collapse under the influence of its own gravitational field is counteracted by the pressure that is produced by the thermonuclear reactions that take place in its interior, so that the mean density of the stellar material does not rise above about 10^4 kg m^{-3} at the very most. When the star has exhausted its fuel, however, this source of internal pressure is no longer present, and the remains of the star embark on a process of gravitational contraction that can end in one of three ways, producing a white dwarf, a *neutron star* or a black hole depending on the original mass of the star. In the case of a star of less than about 1·5 solar masses, Chandrasekhar has shown that the build-up of *degeneracy pressure* in the electron gas that permeates the stellar material causes the contraction to stop when the mean density has risen to roughly 10^{10} kg m^{-3}, thus producing a white dwarf with a radius comparable with that of the Earth. In the case of heavier stars, however, the gravitational forces are too powerful for the contraction to be halted in this way, and it is now believed that such stars, after undergoing one or more catastrophic explosions in which they lose a large part of their mass, end up as neutron stars or black holes. Oppenheimer and Volkoff have shown that in the case of a stellar remnant of less than a certain critical mass (originally given as 0·7 solar masses but now thought to be nearer 3 solar masses) the build-up of degeneracy pressure in the free neutron fluid of which the contracting star becomes increasingly composed causes the collapse to stop when the mean density has risen to roughly 10^{17} kg m^{-3}, thus producing a stable neutron star with a radius of roughly 10 km. If the mass exceeds the *Oppenheimer–Volkoff limit*, on the other hand, the gravitational forces are able to overcome even this new degeneracy pressure and the star continues to collapse, becoming a black hole when its radius falls below its Schwarzschild radius. Oppenheimer and Snyder showed that the collapse of such a body should continue indefinitely, since there are no known forces capable of bringing it to a halt, so that the star should eventually contract to a point mass of infinite density. The discovery that black holes could probably be produced during the normal evolution of massive stars gave a great stimulus to their study.

Since 1939, the properties of black holes have been thoroughly investigated by theoretical physicists, and a fairly complete picture of the final state of such objects has now been built up by Kerr, Israel, Penrose, Carter and Hawking. Firstly, it now appears to be fairly certain that any object which collapses to form a black hole must inevitably undergo complete gravitational collapse, its matter being crushed to a density that becomes infinite in a finite time. This causes the curvature of space-time at the centre of the black hole to become infinite, producing a so-called *space-time singularity*—a region in which space and time have become so distorted that the normal laws of physics are no longer valid. As the collapse proceeds towards its final stages, the enormous pressure that is produced in the contracting matter becomes in itself a source of extra gravitational attraction, and when the matter becomes sufficiently concentrated this extra gravitational pull becomes more than sufficient to overcome the outward forces that are produced directly by the pressure; thus a black hole eventually becomes independent of the body that produced it, becoming, in effect, a self-sustaining gravitational field.

Secondly, it has been shown that the final state of all black holes can be described in terms of the solution of the field equations of Einstein's general theory of relativity that was developed by Kerr in 1963—a solution that describes a rotating black hole and which reduces to the original Schwarzchild solution when no rotation is present. Kerr showed that the final state of such a black hole can be described in terms of only two parameters, one describing its mass and the other its angular momentum; it has since been shown that the hole may also possess an electric charge, but it is believed that all other properties of the body that produced it (shape, baryon and lepton number, electromagnetic multipole moments, etc.) are lost or radiated away during the process of collapse. Kerr's solution pictures a black hole as being bounded by a spherical surface known as the *absolute event horizon*, this being the boundary of the region from within which no signal can escape. The event horizon has a radius equal to the Schwarzchild radius corresponding to the mass of the black hole. Beyond the event horizon is a second surface known as the *stationary limit*, this being the surface at which a particle would have to travel at the local velocity of light in order to appear stationary to an observer at infinity. The stationary limit (an oblate spheroid) touches the event horizon only where the two surfaces intersect the axis of rotation of the hole, the region between the two being known as the *ergosphere*; in the case of a non-rotating black hole, the two surfaces coincide and the ergosphere vanishes.

The "size" of a black hole is determined by the radius of its event horizon, although it should be stressed that the event horizon does not constitute a physical barrier of any sort. Indeed, an observer who was unfortunate enough to fall into a black hole would be unaware that he had crossed the event horizon since signals from the outside world would continue to reach him as before; the observer would, however, become invisible to the outside world at the moment when he crossed the event horizon. It is, however, extremely doubtful whether our hypothetical observer would be in any fit state to make observations, since the enormous *tidal forces* produced by the highly non-uniform gravitational field that exists in and around a black hole would probably kill him even before he reached the event horizon (except in the case of an extremely massive black hole, when the tidal forces would not reach a lethal level until the observer was well inside the event horizon). As the observer approached the centre of the black hole, the strength of these tidal forces (which stretch a body along the radial direction and crush it along the transverse direction) would mount inexorably until they eventually destroyed his body and then crushed its atoms and their constituent fundamental particles out of existence.

It must be stressed that the picture of a black hole presented above is based on our current understanding of the nature of space, time and matter, and, in particular, on the predictions of Einstein's general theory of relativity—a theory that is by no means universally accepted as valid. Nevertheless, the most serious current rival to Einstein's theory of gravitation (the *scalar-tensor theory of Brans and Dicke*) predicts the existence of black holes with properties practically identical to those described above, so, barring the existence of hitherto unsuspected properties of highly-condensed matter, we can be fairly confident that the above picture is correct.

During the late 1960s, active interest in black holes (which was previously largely confined to theoretical physicists) spread to observational astronomers, who began a systematic search for such objects. This interest was originally stimulated by the discovery of *quasars* (intense radio sources with extremely large red shifts which appear to be of stellar rather than galactic dimensions but whose nature is not yet known) and was given a further boost by the discovery of *pulsars* (compact radio sources that emit regular pulses of radiation) and the subsequent discovery that the latter were in fact the *neutron stars* whose possible existence had been predicted by Oppenheimer and Volkoff in 1939. The search for black holes has taken several forms, the most important to date being the study of binary star systems in which only one object is observed. Since a black hole behaves as if it were an invisible mass surrounded by a gravitational field which at large distances is indistinguishable from that of a material body, there is no reason why such an object should not be a member of a binary star system in which one of the components was originally a massive star which has now completed its evolution (the time taken for a star to complete its evolution decreases rapidly as its mass increases). No black holes have been found among the so-called visual binaries (systems near enough for their individual members to be resolved), but a number of eclipsing and spectroscopic binaries are at present being investigated and some of these may eventually prove to contain black holes. At the time of writing, the most promising candidates appear to be the eclipsing binary ε Aurigae (which Cameron and Stothers have shown may contain a black hole), the spectroscopic binary β Lyrae (an extremely unusual system which Devinney and Wilson believe to contain a black hole surrounded by an opaque or nearly-opaque disc of matter), and a number of one-line spectroscopic binaries which Gibbons and Hawking have shown may possibly contain black holes. It should, however, be stressed that alternative theories of the constitution of all these systems have been put forward by other astronomers and that no definite identification of a black hole in a binary system has yet been made. The chances of discovering a black hole of stellar mass which is not a member of a multiple system appear to be highly remote, since there is no known way in which such an object could be detected unless it happened to pass extremely close to the Solar System—a most unlikely occurrence.

The possibility of the existence of black holes of much greater than stellar mass (up to 10^8 or 10^9 solar masses) is also being investigated by astronomers, and a number of possible methods of detecting such objects have been suggested. Most of these methods involve the detection of *gravitational waves*, which, according to theory, should be emitted in large amounts whenever a material body falls into a black hole or two black holes coalesce. The existence of such waves (which are the gravitational equivalent of electromagnetic waves and travel at the same velocity as the latter) is predicted by Einstein's general theory of relativity, and, although their properties have been studied by theoretical physicists for some time, it is only comparatively recently that any serious attempt has been made to detect them. Since 1966, however, Weber has carried out a series of experiments in which he claims to have detected intense pulses of gravitational waves emanating from the centre of our Galaxy, and, although his results have (at the time of writing) not yet been confirmed by any other workers, they have aroused intense interest among astrophysicists. Field, Rees and Sciama have shown that the enormous intensity of the waves associated with Weber's pulses (assuming that they do in fact exist) indicates that the radiation is almost entirely confined to the plane of the Galaxy, in which our Sun lies. (If radiation of the observed intensity were to be emitted isotropically from a source at the Galactic centre, it would require the source to convert mass into energy at a completely unrealistic rate of 10^3 to 10^6 solar masses per year, which, if sustained, would cause the entire mass of the Galaxy to disappear in a very small fraction of its age.) Misner has suggested that a possible source of such waves is a rapidly-rotating black hole of between 10^4 and 10^8 solar masses located at the centre of the Galaxy, and has shown that highly-beamed gravitational waves of the type and intensity indicated by Weber's results could conceivably be produced by a form of *gravitational synchrotron radiation* as stars and other material fell into such a hole along spiral paths lying in the Galactic plane. The feasibility of this theory has still to be fully examined, however, and, at the time of writing, there is no direct evidence for the existence of a black hole at the Galactic centre, or, indeed, for the existence of any black hole in the mass range under present consideration.

Let us conclude this examination of the properties of black holes with a brief discussion of some of their cosmological implications. In particular, let us consider the problem of what happens to the matter that is squeezed out of existence in a black hole (assuming that this does in fact happen). Some cosmologists have recently suggested that our universe may be *multiply connected* (in the topological sense) either with itself or with one or more other universes, the points at which such connections occur being singularities in space-time. According to their argument, it is therefore conceivable that matter which disappears into the singularity at the centre of a black hole may subsequently re-appear from another singularity (a "*white hole*") in another part of our own universe or in some other universe, possibly at the centre of a quasar or radiogalaxy. In view of the large number of unexplained phenomena recently discovered by astronomers, such speculations should not be dismissed out of hand.

See also: Neutron stars.

Bibliography

ALPIN, P. S. (1972) *Gravitational radiation experiments. Contemp. Phys.* **13**, 3, 283.
BERGMANN, P. G. (1968) *The Riddle of Gravitation.* London: John Murray.
HAWKING, S. W. (1972) Gravitational radiation: the theoretical aspects. *Contemp. Phys.* **13**, 3, 273.
HJELLMING, R. M. (1971) Black and white holes. *Nature Physical Science*, **231**, 20.
LUYTEN, W. J. (1962) White dwarfs, in *Encyclopaedic Dictionary of Physics* (J. Thewlis, Ed.), **7**, 759, Oxford: Pergamon Press.
MESTEL, L. (1961) Degeneracy in astrophysics, in *Encyclopaedic Dictionary of Physics* (J. Thewlis, Ed.), **2**, 274, Oxford: Pergamon Press.
OPPENHEIMER, J. R. and SNYDER, H. (1939) On continued gravitational contraction. *Phys. Rev.* **56**, 455.
PENROSE, R. (1972) Black holes and gravitational theory. *Nature*, **236**, 377.
PENROSE, R. (1972) Black holes. *Scientific American*, **226**, 5, 38.
RINDLER, W. (1962) Relativity theory, general, in *Encyclopaedic Dictionary of Physics* (J. Thewlis, Ed.), **6**, 268, Oxford: Pergamon Press.
TRIMBLE, V. (1971) Supermassive objects in astrophysics. *Nature*, **232**, 607.
WRUBEL, M. H. (1962) Stellar evolution, in *Encyclopaedic Dictionary of Physics* (J. Thewlis, Ed.), **6**, 880, Oxford: Pergamon Press.

H. I. ELLINGTON

BLOCKING PATTERNS FROM CRYSTALS.

If a ball model of a crystal lattice is rotated in one's hand, successive rows and planes of "atoms" can be seen to pass through the line of vision. It is therefore not surprising that the regular arrangement of atoms within crystals can influence the trajectories of protons and heavy ions in their passage through solids.

Suppose a well-collimated beam of, say, protons is incident at some convenient angle onto a target, as illustrated in Fig. 1. Some protons undergo violent collisions with the atoms of the solid and suffer scattering through large angles so that they emerge through the target surface. If the target has no

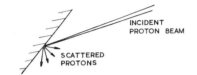

Fig. 1. *Illustration of proton scattering from a target.*

regular atomic arrangement, the scattered protons will emerge in an essentially random manner completely devoid of structure. However, if on the other hand the target does have a regular atomic arrangement, as in the case of a single crystal, the scattered protons will have their trajectories steered by the periodic distribution of interatomic potential within the lattice. Consider, for instance, as shown in Fig. 2, a proton which is scattered in a direction rather close to a close-packed atomic row. Such a proton has its path blocked by the row and is consequently deflected. This situation will be true for all protons which are scattered close to the direction of close-packed rows, with the result that it is virtually impossible for a proton to emerge from the crystal in one of these directions. This phenomenon is known as "blocking" and occurs not only in the case of rows, but also for densely populated planes—which are, after all, only a succession of rows with different interatomic distances. The net effect is to produce a strong "crystallographic" anisotropy in the spatial distribution of protons back-scattered from single crystals.

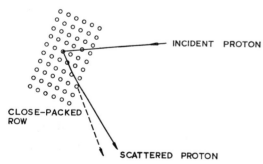

Fig. 2. *Illustration of how a scattered proton's trajectory is blocked by a close-packed row.*

A simple method of studying the phenomenon of blocking is to record the spatial distribution of protons or helium ions by the use of either a fluorescent screen, a photographic emulsion or plastic film,† all of which respond to proton or helium ion irradiation in one way or another. The regular patterns which result from the crystallographic anisotropy of the scattering, are called "blocking patterns". Figures 3 and 4 show two typical patterns, recorded at different energies and by a fluorescent screen and a plastic film respectively.

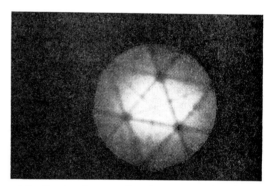

Fig. 3. *A photograph of a proton blocking pattern from the {111} face of a gold single crystal recorded directly on a fluorescent screen held parallel to its surface at a distance of 1 cm. The proton beam was incident at an angle of 70° to the surface normal at an energy of 25 keV.*

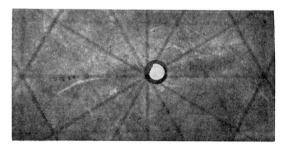

Fig. 4. *A photograph of a helium ion blocking pattern from the basal plane of a titanium single crystal recorded in a cellulose nitrate film placed parallel to its surface at a distance of 5 cm. The helium ion beam was directed at normal incidence (hence, the hole in the film), onto the target at an energy of 1·5 MeV. (Courtesy of J. Poate, Nuclear Physics Division, Harwell.)*

† Providing a helium ion's energy is sufficiently high (above about 1 MeV) then the dissipation of energy in a plastic film results in track formation, such tracks become visible on etching with a suitable acid solution.

Some Applications

Perhaps the simplest way to observe proton-blocking effects is by the use of a fluorescent screen at energies as low as about 20 keV. Crude energy discrimination of the scattered protons is inherent in the method due to the dependence of the phosphor response on energy, this enhances the high energy component of the continuous spectrum of scattered protons, and allows reasonable estimates to be made of blocking angles in this energy range. At 20 keV the scattering is very much stronger than at high energies, with the result that an almost uniform glow is seen on the screen, information can consequently be gained for a very large number of axes and planes with just one exposure.

(i) The orientation of crystals

The orientation dependence of proton scattering from single crystals immediately suggests a quick and reliable method of orienting single crystals. Using the fluorescent screen technique the crystal structures can be observed dynamically and recognized immediately. The crystal to be examined is fixed to a goniometer which can be operated remotely from outside the vacuum system. A proton beam of the desired intensity is collimated to about 1 mm spot size and fired at the crystal in such a way that a blocking pattern is easily visible on the fluorescent screen. Then, in a matter of seconds, the crystal orientation can be adjusted so as to produce the correct pattern on the screen relative to a known mark. The degree of orientation required determines the sophistication necessary; for instance, the fluorescent screen technique as described is perhaps useful only for an orientation accuracy of something less than about 0·5°.

(ii) Epitaxial films and surface layers

One of the main potential uses of blocking patterns is in the study of surface layers of only a few tens to hundreds of atom layers thick. By more conventional techniques films of this thickness are not readily studied, X-rays usually examine depths to the order of millimetres and transmission electron microscopy introduces the disadvantages that the sample must be an isolated film removed from its substrate. However, at 20 keV the majority of protons are scattered from less than 100 Å below the surface and consequently the structure of thin epitaxially grown films can be readily studied whilst still in contact with the substrate.

Returning to the mechanism responsible for blocking let us consider Fig. 5, which shows a single isolated row of atoms, each atom being separated by a regular distance d from the next. Suppose an energetic proton suffers a large angle collision with

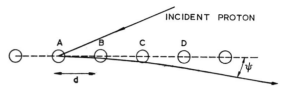

Fig. 5. The trajectory of a proton blocked by a close-packed row.

atom A and is elastically scattered towards its nearest neighbour B close to the line of centres. For practical purposes in such large angle collisions, as with atom A, the scattered proton appears to come right from the centre of the atom, then, providing the proton passes sufficiently close to the nucleus of atom B, it will be deflected further away from the row. Further small angle scattering will occur at atoms C and D and so on until the proton's trajectory eventually emerges from the influence of the row at some small angle, ψ.

The blocking phenomenon can be understood theoretically if we smear out the interatomic potential from the individual atoms into a continuum so that the proton sees regular rows and sheets of high potential. Using this approximation and basing the interaction on the familiar screened Coulomb potential $V(r)$ suggested by Thomas–Fermi,

$$V(r) = \frac{Z_1 Z_2 e^2}{r} \varnothing(r/a)$$

where $\varnothing(r/a)$ is the Fermi function with screening length $a = a_0 \, 0{\cdot}8853 \, (Z_1^{2/3} + Z_2^{2/3})^{-1/2}$, Z_1 and Z_2 are the atomic numbers of the projectile and target atoms respectively, and e is the electronic charge; Lindhard (1965) has shown that the critical blocking angle subtended by a row of equi-spaced atoms is given by

$$\psi_{\min} = \left(\frac{2Z_1 Z_2 e^2}{Ed}\right)^{1/2} \quad \text{if} \quad \psi_{\min} < a/d$$

where E is the particle energy and d is the interatomic spacing. On the other hand, at large Z_2 or when the incident particle's energy is low, the characteristic blocking angle smoothly changes to

$$\psi_{\min} = \left\{\left(\frac{a}{d}\right)\left(\frac{2Z_1 Z_2 e^2}{Ed}\right)^{1/2}\right\}^{1/2} \quad \text{for} \quad \psi_{\min} > a/d.$$

Bibliography

LINDHARD, J. (1965) *Mat. Fys. Medd. Dan. Vid. Selsk.* **34**, 14.

NELSON, R. S. (1968) *The Observation of Atomic Collisions in Crystalline Solids*, Amsterdam: North Holland Publishing Co.

R. S. NELSON

C

CALCULATION OF THERMODYNAMIC PROPERTIES FROM SPECTROSCOPIC DATA. The methods of statistical mechanics permit calculation of thermodynamic properties from a knowledge of partition functions (Mayer, 1962). Molecular parameters required for such calculations are the mass, moments of inertia, symmetry number†, and the fundamental vibrational frequencies of the molecule. The rigour of the calculations depends on the quality of the spectroscopic data available and the depth to which it may be interpreted in terms of rotational and vibrational behaviour of the molecule. Rigorous calculations of the partition function of a monatomic gas may be made since the energy is translational only (except for the inclusion of higher electronic states at higher temperatures) and the entropies of monatomic gases have been calculated using the Sackur–Tetrode equation (Walker, 1962) in good agreement with entropies obtained from specific heat data and the Third Law. Corrections must be made for the non-ideal behaviour of real gases. Calculations of considerable accuracy can also be made for polyatomic molecules which can be described by the rigid rotator–harmonic oscillator approximation. The calculations for non-rigid molecules are subject to uncertainties concerning the barrier to free rotation and its contribution to thermodynamic properties. Clear expositions of the basis for these calculations are given by Herzberg and by Wilson. Standard tables are available for calculating vibrational contributions to the partition function. Computer methods have now greatly reduced the labour of these calculations for complex molecules. Considerable thermodynamic data of an approximate nature have been determined for large hydrocarbons and functionally substituted hydrocarbons by the so-called group additivity methods (Benson et al., 1969). These methods are discussed further.

---

† Symmetry number, here denoted by σ, is the number of ways in which the molecule can be regenerated unchanged by rigid rotations around its symmetry axes, and is characteristic for a poins group. A factor is also introduced in moleculet for which free rotation can occur for the number of indistinguishable ways in which the molecule is regenerated by free rotations.

The formulae for calculating the thermodynamic properties for one mole of an ideal gas at atmospheric pressure are given in terms of the total partition function Q as follows:

$$H° = H_0° + RT + RT^2 d(\ln Q)/dT$$

$$G° = H_0° + RT - RT \ln [Q/N]$$

$$C_p = R + R \frac{d}{dT}\left[\frac{T^2 d(\ln Q)}{dT}\right]$$

$$S° = R(1 - \ln N) + RT\frac{d(\ln Q)}{dT} + R \ln Q$$

where $H°$ is the heat content, $G°$ the free energy, C_p the heat capacity at constant pressure, $S°$ the entropy, all of one mole of a perfect gas, the superscript ° refers to substances in their standard states, subscript $_0$ to the absolute zero of temperature, N is the Avogadro number, and $H_0° = E_0°$, where $E_0°$ is the zero point energy. Different authors use $E_0°$ or $H_0°$ interchangeably in the formulae given above.

Since the total partition function appears in these equations as its natural logarithm, the derived thermodynamic functions can be expressed as a *sum* of contributions from any *factors* in which Q can be expressed. For rigid molecules in which vibrational–rotational, coupling may be neglected, it is customary to factor Q into translational, rotational, vibrational, and electronic components thus

$$Q = Q_{\text{trans}} Q_{\text{rot}} Q_{\text{vib}} Q_{\text{elec}}$$

where Q_{trans} is known as the translational partition function, and the subscripts rot, vib, and elec, denote the rotational, vibrational, and electronic counterparts, and all may be separately evaluated to a good approximation. The formulae for the thermodynamic functions may then be expressed in terms of the separate partition functions and then summed. The formulae for such calculations are usually given in numerical form for the rigid rotator–harmonic oscillator approximation.

Translation

$(H° − H_0°)/T = 20{\cdot}796$

$(G° − H_0°)/T = −9{\cdot}5722\,[3\log_{10} M + 5\log_{10} T] + 30{\cdot}468$

$C_p° = 20{\cdot}796$

Rotation

$(H° − H_0°)/T = 12{\cdot}471 = \dfrac{3}{2} R$

$(G° − H_0°)/T = −9{\cdot}5722\,[\log_{10}(I_a I_b I_c) + 3\log_{10} T − 2\log_{10}\sigma] + 22{\cdot}526$

$C_p° = 12{\cdot}471 = \dfrac{3}{2} R$

Vibration

$(H° − H_0°)/T = 8{\cdot}3143 \sum_j [(x_j/(e^{x_j} − 1))]$

$(G° − H_0°)/T = −8{\cdot}3143 \sum_j [\ln(1 − e^{−x_j})]$

$C_p° = 8{\cdot}3143 \sum_j [x_j^2 e^{x_j}/(e^{x_j} − 1)^2]$

where M is the molecular mass, $I_a I_b I_c$ is the product of the principal moments of inertia of the molecule (molecular parameters in a.m.u. and Å), σ is the symmetry number and $x_j = v_j hc/kT$ where v_j is the wave number of the jth fundamental frequency of the molecule. These equations apply to non-linear molecules, and give data in joules. Values of entropies (so-called spectroscopic entropies) obtained from these equations for several molecules have been compared with calorimetric entropies, showing excellent agreement. Non-agreement of such calculations for molecules with potentially free internal rotation about a particular bond (e.g. ethane) led to a recognition of the energy barrier to free rotation and to one method of estimating such barriers (Wilson, 1961). These barriers are by no means understood and a number of methods have been developed for analysing, measuring and allowing for them in calculation of thermodynamic properties (Wilson, 1959).

Where the bond in question is a multiple bond the barrier to rotation is high and the contribution to the thermodynamic functions is torsional in nature and included with the vibrational part of the partition function. It is possible to observe this torsional frequency in the infra-red or Raman spectrum if the selection rules allow, and to use this value directly in the calculation. Where the rotation is entirely free, the contribution is no longer vibrational but rotational, and a separate partition function for the internal rotation is calculated. Usually, however, the rotation is hindered. No reliable method exists of calculating barriers to free rotation, but a cosine shape of barrier is often assumed and Pitzer and Gwinn have tabulated contributions to thermodynamic functions based on hindered rotation for particular groups (Pitzer and Gwinn, 1942). Other shapes of barrier have also been discussed.

More Approximate Methods

For larger non-rigid molecules with free or partly restricted rotation, exact calculations are not available because the spectroscopic data cannot be obtained but approximate methods have been developed for homologous series of compounds. Janz (1955) and Benson have reviewed all the empirical methods available. The idea of Parks and Huffman that adding a particular group to a molecule added a constant increment to its thermodynamic properties is the basis of group additivity methods. The original idea took no account of the statistical method, but the idea was extended, expressed in terms of partition functions, and can now be applied to a vast number of compounds. The main requisites are (a) fundamental thermodynamic data for a few key molecules of the set (these are frequently measured data) (b) knowledge of vibrational and molecular data for the added groups (which are usually of the type of CH_2 groups in a long carbon chain).

Bibliography

BENSON, S. W., CRUICKSHANK, F. R., GOLDEN, D. M., HAUGEN, G. R., O'NEAL, H. E., RODGERS, A. S., SHAW, R. and WALSH, R. (1969) *Chem. Revs.* **69**, 279.

HERZBERG, G. (1945) *Infrared and Raman Spectra*, Princeton N. J.: D. van Nostrand.

JANZ, G. J. (1955) *Quart. Revs. Chem. Soc.* IX, 229.

MAYER, J. E. (1962) Statistical mechanics, in *Encyclopaedic Dictionary of Physics* (J. Thewlis, Ed.), **6**, 847, Oxford: Pergamon Press.

PITZER, K. S. and GWINN, W. D. (1942) *J. Chem. Phys.* **10**, 428.

WALKER, P. A. (1962) Thermodynamics, third law of, in *Encyclopaedic Dictionary of Physics* (J. Thewlis, Ed.) **7**, 293, Oxford: Pergamon Press.

WILSON, E. B. Jr. (1940) *Chem. Revs.* **27**, 17.

WILSON, E. B. Jr. (1959) In *Advances in Chemical Physics*, Vol. II, p. 367. New York: Interscience.

WILSON, E. B. Jr. (1961) Hindered rotation, in *Encyclopaedic Dictionary of Physics* (J. Thewlis, Ed.) **3**, 701, Oxford: Pergamon Press.

J. C. LOCKHART

CARBON FIBRE. Carbon fibre has been known since 1879 when Edison produced carbon filaments for electric lamps, an application that was short-lived owing to the invention of the metal filament lamp. It was not until about 70 years later that carbon fibres were manufactured in any quantity, and this time used in the form of wool, felt, braid and cloth for insulating high-temperature furnaces operating with non-oxidizing atmospheres. Some fibre was also used for carburizing refractory metals. The advent of space exploration with its requirements for high-temperature resistant materials, and structural materials combining high strength, high stiffness, and low density led to further world-wide interest, further important developments, and a widening of the field of application.

Carbon fibre is a generic term used to describe organic fibre that has been heated to temperatures well in excess of its decomposition temperature. Its structure varies from that of amorphous carbon to the oriented crystalline state (graphite) depending on the chemical constitution of the precursor fibre and the processing conditions. For present purposes commercially available products are designated "carbon fibre" when they have been heated to about 1000°C and "graphite fibre" when heated to higher temperatures—up to 2500°C or more.

Formation of Carbon and Graphite

Knowledge of the conversion of an organic precursor into graphite is far from complete, and there has not been any striking advance since the publications of Rosalind Franklin (1951). Franklin pointed out that polymers rich in hydrogen, poor in oxygen, and possessing an oriented molecular structure usually produce carbon that will graphitize at higher temperatures, while those with opposite characteristics produce non-graphitizing carbons. She also deduced a probable mechanism for the conversion of carbon to graphite, which is summarized below.

When carbons prepared by pyrolysis to 1000°C are heated to higher temperatures the first change in structure is the lateral growth of small graphite-like layers (fused carbon hexagons) at the expense of the "non-organized carbon". The non-organized carbon, which may well include both chain and ring structures and traces of other elements, particularly hydrogen, is attached to the edge atoms of the graphite-like layers, and links neighbouring two-dimensional structures together. With increasing temperature and duration of heating the non-organized carbon is absorbed, atom by atom or at least by small fragments, into the layer planes to which it is attached. Ultimately only a few carbon atoms are left to link neighbouring units.

The next stage in graphite formation is the further growth of these two-dimensional structures and their organization into parallel layers to constitute the three-dimensional graphite crystal structure. If the crystalline growth were determined by the migration of isolated or small groups of atoms from one crystallite to another, it would require the breaking of carbon–carbon bonds in the two-dimensional hexagonal structures and a temperature of at least 1800°C. Since three-dimensional growth occurs at measurable rate from 1000°C upwards it is postulated that layers or groups of layers are displaced intact.

The need to increase the temperature to 2500°C or more may then be interpreted as the need to provide sufficient thermal energy to displace and unite crystallites of continuously increasing size. However, such a mechanism is only feasible if there is a gradual shifting of position and if the layer planes of neighbouring crystallites lie nearly parallel to one another. The shape of the crystallites alone makes a tendency to parallelism of nearest neighbours highly probable and, in fact, it is found that in the graphitizing carbons a preferential orientation of the crystallites exists not only between nearest neighbours but in bulk as well. In non-graphitizing carbons X-ray analysis reveals no trace of preferential orientation.

The so-called non-graphitizing carbons have a relatively high porosity which suggests that the early stages of carbonization are not accompanied by flow which is necessary for the coalescence of structure required for graphitization. Owing to the presence of appreciable quantities of hydrogen which leads to the formation of tarry matter, the graphitizing carbons pass through a viscous state in the temperature range 300–500°C, which prevents the carbon from setting into a rigid structure at low temperature.

Carbon Fibre

For the production of carbon fibre to be a feasible process the precursor fibre must have a melting or softening temperature substantially higher than its decomposition temperature, so that it remains solid throughout the carbonization processes. Even a minor degree of softening can result in the fibres sticking together in which case the product is useless. An additional requirement is that both the theoretical and actual carbon yields of the precursor fibre should be reasonably high.

At the time of the first development work there were only two fibres in commercial production which fulfilled the first of these requirements, namely regenerated cellulose rayon and polyacrylonitrile fibre (e.g. Orlon). The latter was known to become infusible on heating in air for a few hours at 200°C (so-called black Orlon) and it is curious why Union Carbide Corporation originally chose to work with regenerated cellulose, which has a much lower carbon yield. Later work has shown that other fibres may be prevented from softening and flowing if they are first subjected to oxidation in air up to temperatures of 200–300°C, a process which results in the formation of stable crosslinked and/or cyclic structures. Fibres of polyvinyl alcohol and fibres derived from petroleum pitches and coaltar are in this category. A further interesting development is the discovery that fibres of cellulose and polyvinyl alcohol give higher yields of carbon, and carbon fibres with enhanced tensile properties, if they are first heated in hydrogen chloride vapour up to temperatures of several hundred degrees, a process

Table 1. Properties of Carbon and Graphite Fibres compared with Steel and Aluminium

Material	Specific gravity	U.T.S. GN/M^2	Specific U.T.S. GN/M^2	Young's modulus GN/M^2	Specific Young's modulus GN/M^2
(a) Grafil A	1·75	2·3	1·31	200	114
(b) Grafil HT (Type II)	2·80	2·5	1·39	253	141
(c) Grafil HM (Type I)	1·95	2·1	1·08	380	195
Kureha carbon fibre	1·61	1·1	0·683	69	42·8
Steel	7·8	1·0	0·127	210	26·9
Aluminium L65	2·8	0·47	0·168	75	26·9

(a), (b) and (c) = Courtaulds graphite fibres; high strain, high strength, and high modulus types respectively. Average values from manufacturer's data.

which assists in the removal of hydrogen and oxygen without loss of carbon. Perhaps the most interesting process is that developed and now commercialized in Japan by the Kureha Chemical Industry Company Limited, which utilizes a precursor fibre spun from petroleum pitch with a carbon content of 95%. The properties of this fibre and graphitic type fibres are compared in Table 1.

Graphite Fibre

The manufacture of graphite fibre is a more complex matter, not simply because much higher temperatures and more costly furnaces are required, but also because not all organic precursors yield a well-defined graphitic structure even at the highest temperatures, and those that do pass through a viscous state involving loss of fibre form and carbon-containing volatiles. Fortunately, solutions have been found to these problems. Firstly, carbon fibres have been found to graphitize when stretched at temperatures of not less than 2500°C. Secondly, preoxidation processes, such as already mentioned, may be employed to prevent loss of fibrous form and carbon volatiles from graphitizing precursors.

These ideas have been exploited commercially. In 1965 graphite fibres prepared by the hot-stretching of carbon fibres derived by the pyrolysis of regenerated cellulose rayon were marketed by Union Carbide Corp. under the trade name "Thornel"; products of higher strength and modulus, Thornel 50 and Thornel 75, have since appeared. In 1961 staff at the Industrial Research Institute of Osaka, led by Dr. Akio Shindo, oxidized polyacrylonitrile fibres and then further heated them in an inert atmosphere up to 2500°C, but unfortunately the product had a poorly defined graphitic structure and consequently disappointing mechanical properties. At about the same time W. Watt, L. N. Phillips and W. Johnson, working at the Royal Aircraft Establishment, Farnborough, England, were independently exploring the possibilities of polyacrylonitrile as precursor fibre. They applied similar oxidation and heat treatments but held the fibre at constant length during the oxidation stage to avoid shrinkage. This additional feature prevented the chain molecules from relaxing and becoming disoriented both during oxidation and subsequent heating, thereby ensuring that the graphite-like layers would be formed highly oriented along the fibre axis and would readily crystallize into a three-dimensional graphite lattice. This process gave birth to two products, with outstanding mechanical properties. Type I heat treated to 2600°C and possessing high modulus, and Type II, heat treated to 1500°C of higher tensile strength but lower modulus (see Table 1). These two types and a third variation of higher breaking strain are now made under licence from the National Research Development Corporation. Among licencees in the U.K. are Hyfil Limited, Courtaulds Limited and Morganite Modmor Limited; foreign licences have also been granted.

Processes based on cellulose fibres have the disadvantage of a relatively low carbon yield compared with those utilizing acrylic precursors, and they also require stretching at about 2700°C. On the other hand, any process using a precursor fibre which requires an oxidation step is at present limited by the maximum diameter of fibre that can be processed. If the diameter is too great, oxygen fails to penetrate to the centre in a reasonable time and hollow carbon fibres result from volatization of the innermost regions. It is interesting to note that the two processes differ in that one has a pre- and the other a post-carbonization orientation step. The polyacrylonitrile precursor fibre used is a more highly oriented version of the textile fibre Courtelle manufactured by Courtaulds Limited (a copolymer containing 93% acrylonitrile) and the orientation is preserved as indicated above. Regenerated cellulose fibre may also be obtained in a highly oriented version but, in contrast, the orientation cannot be preserved throughout carbonization since the molecular chains contain carbon–oxygen bonds which are severed when oxygen is removed; hence orientation must follow carbonization.

Structure of Carbon and Graphite Fibres

Up to 1000°C it is difficult to distinguish between the types of carbons produced from different precursors. Even after heating to 3000°C, non-graphitizing carbons have no trace of three-dimensional structure, but many do form, as a separate phase, a proportion of well-developed graphitic carbon. Where the graphitic phase is apparently absent some believe that the crystallites are too small to be detected by diffraction techniques, while others believe that crystallites do not exist and the structure is composed of a random spatial arrangement of trigonal and tetrahedral carbon bonds. This type of structure has been proposed for carbons which have vitreous characteristics, such as are obtained by pyrolyzing highly cross-linked phenolic and other resins. This type of carbon has come to be known as "glassy carbon" and may be obtained in fibrous form by extruding and pyrolizing phenol-hexamine polymers; it is said to be particularly resistant to chemical attack and to abrasion.

The properties of carbon fibres are dependent on their temperature of preparation. The tensile strength of graphitizable carbon fibre passes through a maximum at about 1500°C, while its modulus continues to increase with temperature. At this temperature three-dimensional order is poor and diffraction patterns are consistent with layer planes that are only approximately parallel and equidistant; this type of structure has been termed turbostratic. After heating to temperatures approaching 3000°C there is distinct evidence of three-dimensional order and a modulus about half that of the theoretical value is obtained, but strength is at least an order of magnitude less than expected for a graphite crystal. It is not yet known whether the weakness lies between the graphite crystals within fibrils or between the fibrils themselves, but strength decreases with increasing specimen length indicating the presence of flaws on the microscale, flaws which may well be present in the precursor fibre. When the constituent chain molecules of organic fibres are oriented in the direction of the fibre axis by stretching, the modulus and strength of the fibres are markedly increased, and it is interesting to note that this behaviour is followed when carbon fibres are stretched at temperatures in the range 2600 to 3000°C. Such treatment leads to strain-induced crystallization and very probably removes lattice defects as well. However, fibre properties are likely to remain below those of a graphite single crystal owing to imperfections at all levels of structure.

Most of the graphite fibre that has been produced has a filament diameter of about 10 microns (10 μm). Each fibre consists of fibrils of average diameter about 500 Å (50 nm) made up of graphite crystallites that are several hundred Ångstrom units in length, breadth and thickness. The exact arrangement of the crystallites is not yet certain; however, it is known that they are preferentially aligned to the fibre axis and that the better the alignment the higher the modulus.

Applications

Carbon and graphite fibres are supplied in a variety of forms; lengths from less than 1 mm to 10 mm, lengths of 1 m or more, continuous filament tows containing 5000 or 10,000 filaments, loose fibre (wool), felt, paper braid, non-woven and woven cloth. Warp sheets and tape impregnated with resin are also available. High-temperature stability, at least in non-oxidizing atmospheres, has led to their use for thermal insulation in furnaces, and in combination with phenolic resins as ablative materials which will withstand the high temperatures and thermal shock developed on re-entry of vehicles into the Earth's atmosphere. The conductive properties of the fibre in various forms are being exploited as flexible electric heating elements for such diverse purposes as furnaces, blankets, clothing and car seats. Except for strong oxidizing agents, chemical resistance is good up to at least 350°C and carbon fibre in one form or another has found widespread use for packings, glands, seals and catalyst supports. For these purposes it is sometimes combined with nylon or PTFE to improve the sealing and lubrication properties. The frictional properties of graphite have also been exploited particularly in reinforced nylon gear wheels and bearings; lubrication is not required. For many of these applications carbon rather than graphite fibre is adequate and cheaper, though the best choice is not always obvious. Graphite fibre is at least 99·9% carbon, whereas carbon fibre may contain as little as 90% and the impurities may impair certain desirable properties. For example, they may catalyse the degradation of the resin component in ablative materials, and reduce chemical resistance and electrical conductivity.

Space and underwater exploration, and terrestial flight have created demands for materials that are stronger, stiffer and of lower density than conventional structural materials, and this need is being met by making use of the inherently high strength and modulus of graphite fibre. For this purpose graphite fibres are "cemented" together in uniaxial array with epoxy or other thermo-setting resins and cured in a mould under heat and pressure to form a coherent mass in sheet, rod, bar, or other shape. Such composite materials are highly anisotropic, but by laminating layers of fibres in different directions stress-bearing capacity may be optimized in any desired region or direction. Filament winding methods may also be used to produce vessels, pipes, etc. In general the properties of the composite are dependent on the volume fraction of the fibre, the degree of fibre alignment, and the strength of the fibre/matrix bond. Examples of this technology in the aircraft industry are ailerons, flaps, bulkheads, floor sections, wing tips, helicopter blades and rotor shafts, torque tubes, pressure vessels, and experimental fan blades for the RB.211 by-pass engine. In rocketry, vehicle structures and motor cases, nozzle reinforcement, tanks and pressure vessels have been fabricated and used. There are many suggestions for using carbon fibres in marine engineering, the automobile industry, and in sports equipment; some are a reality, but the amount of fibre they consume at the present moment is negligibly small.

More extensive use of carbon fibre is thwarted by high cost which, as with many new developments, is closely linked to volume of production. Kureha carbon fibre is available in this country at around £10 per kg, while graphite fibre of U.K. origin ranges from about £20–£100 per kg, depending on filament length and maximum processing temperature. If a large volume cost-effective application emerges the price of fibre could fall to at least one-third of its present level.

Bibliography

FRANKLIN, R. (1951) *Proc. R. Soc.* **209**, 196.
GILL, R. M. (1972) *Carbon Fibres in Composite Materials*, London, Iliffe.
High-temperature Resistant Fibres from Organic Polymers, Ed. J. Preston. *App. Polymer Symposia*, **9**, 1969.
International Conference on Carbon Fibres, London, The Plastics Institute, 1971.
NATIONAL RESEARCH DEVELOPMENT CORP., Brit Pat. 1,110,791 (24. 8. 68).
Second Conference on Industrial Carbons and Graphite, London, The Society of Chemical Industry, 1970.
WATT, W., PHILLIPS, L. N. and JOHNSON, W. (1966) *Engineer* (London), **221**, No. 5757, 815.

W. G. HARLAND

CHARGE-TRANSFER COMPLEXES. Charge-transfer complexes may be defined as complexes between molecules, radicals, ions or atoms,† having an excited state characteristic of the complex as a whole, in which, relative to the ground state, essentially one electron has been transferred from one moiety (classified as the electron donor) to the other moiety of the complex (classified as the electron acceptor). The association of the electron donor, hexamethylbenzene, with the electron acceptor, 1,3,5-trinitrobenzene, is an example of such a complex. The absorption of electromagnetic radiation involved in the intermolecular charge-transfer transition often corresponds to frequencies within the visible region. This accounts for the colours which characterize many of these complexes. The transition may, of course, occur outside the visible region. In such cases it is usually to higher frequencies where it may be obscured spectroscopically by absorptions of "localized" electronic transitions of the components within the complex.

Although complexes, now termed charge-transfer complexes, were described in the nineteenth century, it was Mulliken in the early 1950s who proposed the intermolecular charge-transfer nature of the electronic absorption referred to above. It was part of an overall valence-bond description of this class of complex, the ground state of which for the simple case of a complex between one molecule of the electron acceptor (A) with one molecule of the electron donor (D) was described by the wave function:

$$\psi_N = a\psi(A, D) + b\psi(A^- - D^+) \quad (1)$$

where $\psi(A, D)$, the so called no-bond function, is the contribution from dispersion, dipolar, quadrupolar and suchlike van der Waals forces; $\psi(A^- - D^+)$ is the dative function and corresponds to a structure in which one electron has been transferred from D to A.

In the majority of charge-transfer complexes the overall energy of interaction is relatively weak compared with most covalent bond energies. In these cases the contribution from the dative structure is considered to be small, i.e. $a \gg b$ in eq. (1). By contrast the excited state characteristic of the complex is described by the wave function:

$$\psi_E = a^*\psi(A^- - D^+) - b^*\psi(A, D) \quad (2)$$

where $a^* \approx a$ and $b^* \approx b$, cf. eq. (1). The charge-transfer absorption corresponds to the transition $\psi_N \rightarrow \psi_E$.

Since 1950 there has been a rapid increase in interest in these complexes. Recent work has led some workers, whilst upholding the charge-transfer nature of the electronic transition characteristic of these complexes, to suggest that the contribution of charge-transfer forces to the stabilization of the ground state of the molecule is very much smaller than had originally been proposed. As a consequence the use of the term "charge-transfer" as an expression to categorize these complexes has been criticized. Furthermore, this same term reflects the valence-bond description chosen by Mulliken in discussing these complexes. Alternative molecular-orbital descriptions are possible. Various other names have been suggested. The most widely used

---

† In the following discussion the complexing species will be referred to as molecules, although it should be understood that they could be radicals, ions or atoms unless otherwise specifically stated.

of these is the expression "electron-donor acceptor" or "EDA" complex. This term, however, covers a more extensive group of interactions including hydrogen-bonded complexes, and has also been applied to the addition products formed by attack of electrophiles and nucleophiles on delocalized systems. The formation of the latter products involves the formation of normal covalent bonds and the products should not be described as "complexes" of any type. The term "charge-transfer complex" is often used but, for the reasons indicated, it must be recognized that the semantics of the term are based on the existence of an intermolecular charge-transfer transition. It should also be emphasized that on occasion there is no sharp division between charge-transfer complexes and other types of complex.

Components

Because the components in the complex behave as an electron donor and an electron acceptor, the archetype charge-transfer complexes involve donor molecules with low ionization potentials, combined with acceptor molecules of high electron affinity. Typical donors include aromatic hydrocarbons and amines, particularly aromatic amines. Typical acceptors include: benzene substituted with strong electron-withdrawing groups, for example: 1,3,5-trinitrobenzene (TNB) and 1,2,4,5-tetracyanobenzene (TCNB); quinones such as tetrachloro-*p*-benzoquinone (chloranil) and 2,3-dichloro-5,6-dicyano-*p*-benzoquinone (DDQ). Other strong acceptors include 7,7,8,8-tetracyanoquinodimethane (TCNQ), tetrachlorophthalic anhydride, tetracyanoethylene (TCNE). Amongst the inorganic electron acceptors are included molecular and atomic iodine and the Ag^+ ion.

However, complexing is not restricted to strong acceptors. Relatively weak donors and acceptors often show identifiable properties. Indeed, the terms donor and acceptor are themselves relative. Thus 4,4'-dinitro-*p*-terphenyl acts as a donor in complexing with tetracyanoethylene but as an acceptor in complexing with *N,N*-dimethyl-*p*-toluidine. Some dimers have been described as charge-transfer complexes—for example, that formed from the radical *N*-ethylphenazyl. In this case one molecular species must act as electron donor and as electron acceptor.

In some molecules containing both electron donor and electron acceptor centres which are not conjugated one to the other, there appears to be an across-space intramolecular interaction between these two centres; for example between the pyridinium and phenyl centres in I.

An interesting situation exists in the case of complexes formed between a molecule in an electronically excited state acting as acceptor (or donor), with another molecule of the same species in the ground state acting as the second component (excimers).

$X = -NH_2, -OH, -OCH_3, -CH_3, -Cl, -H$

I

Electronic Spectra

The transition corresponding to $\psi_N \rightarrow \psi_E$ has been measured for a very large number of systems, and the relationships of the energy of the transition ($h\nu$) with various properties of the donor and acceptor components have been noted, particularly for systems in solution. For many series of complexes with a common acceptor species an empirical linear relationship between ionization potential (ionization energy) of the donor and $h\nu$ is observed. Other linear relationships include those with the half-wave oxidation potential, and the energy of the highest-filled molecular orbital of the donor. Similarly a reasonably linear correlation is observed in many cases between $h\nu$ and the halfwave reduction potential of the acceptor, and between $h\nu$ and the energy of the lowest unfilled orbital of the acceptor for complexes with a common donor species. Such relationships have been used to estimate ionization potentials and electron affinities of molecules. The results must be considered with circumspection. The observed linearity of the plots is to some extent fortuitous, and is partly the result of the limited range of electron donation or acceptance considered. It is most safely used in comparisons of molecules with closely related structures.

In solution the charge-transfer transitions are observed as broad bands. In general there is little if any evidence of fine structure. In the crystalline state the charge-transfer absorptions are generally polarized in such a way that the transition dipole is along the line of centres of the two components. In some cases there is evidence of vibrational structure in the solid phase.

Intermolecular charge-transfer spectra have also been observed in the vapour phase.

On occasion in the crystalline and liquid and vapour phases more than one intermolecular charge-transfer band is observed. These have been assigned either to transitions from the highest and second highest occupied levels in the donor, or to transitions to the lowest and second lowest empty levels in the acceptor (or a combination of both).

Equilibrium in the Liquid and Vapour Phases

In the liquid phase (normally in a diluting solvent) and in the gas phase the complex is in equilibrium with its components. Evidence suggests that at least in solvents of low solvating power the rates of formation and decomposition of the complex are very fast.

Although it is generally assumed that the stoichiometry of the complex in solution is 1 : 1 in respect of A and D, there is good evidence that, at least for some systems, there are significant concentrations of complexes of other stoichiometry, in particular termolecular complexes AD_2 and A_2D.

Structure in the Solid State

In complexes involving planar donor and planar acceptor molecules the structures generally consist of stacks of parallel donor and acceptor molecules placed alternately. In some cases the molecules in one stack are parallel with those in all other stacks; in other cases molecules in adjacent stacks, whilst parallel to others in their own stack, are angled to those in adjacent stacks (herring-bone). In many cases the donor and acceptor pairs, though parallel, are not arranged to maximize the eclipsing of one molecule by the other.

Structures of many crystalline charge-transfer complexes, including linear and non-linear arrangements with planar and non-planar molecules, have been determined particularly by Hassel, Herbstein, Prout, Wallwork, and their co-workers.

Other Properties in the Solid State

To a first approximation solid electron donor–acceptor systems can be divided into (i) those which are "weak" charge-transfer systems, i.e. those in which in terms of eq. (1) $a \gg b$, and (ii) those in which there has been effectively a transfer of a complete electron in the ground state to form, in the case of initially neutral, closed-shell molecules, a pair of ion radicals. This is essentially a redox reaction, and it is a question of terminology as to whether or not the essentially ionic product is described as a charge-transfer complex. It represents the second limit of ψ_N in eq. (1) in which now $a \ll b$ to the point of being totally the second term. These systems have primarily the properties of the two ion radicals as shown by their infra-red, ultra-violet-visible and electron-spin resonance spectra. There is also a sharp distinction between the semiconducting properties of these two types of electron-donor–acceptor interaction. Thus a complex such as aniline–1,3,5-trinitrobenzene, in which $a \gg b$, has a high resistivity (10^{17} ohm cm) and a high activation energy (2·54 eV) compared with 1,6-diaminopyrene TCNQ, in which $a \ll b$, and has a low resistivity (0·5 ohm cm) and a low activation energy (0·28 eV).

Some "weak" complexes such as the complexes of pyrene with tetracyanoethylene, with bromanil and with 1,3,5-trinitrobenzene show photoconductivity. In some cases, at least, the conduction state is not identical with the charge-transfer excited state.

Related to the 1 : 1 systems where there has been an almost complete electron-transfer from the donor to the acceptor moiety are the salts formed from the anion radical of TCNQ with various cations, having the general formula $M^+(TCNQ)^-_n$ where $n = 1$, 1·5 or 2. The quinolinium $(TCNQ)^-_2$ salt, for example, is paramagnetic and has a metallic-like, temperature independent, conductivity. Such salts are not normally classified as charge-transfer complexes.

Possible Involvement of Charge-transfer Complexes

Charge-transfer complexes as derivatives of organic compounds especially of aromatic hydrocarbons, have been used to identify, to separate and to purify them. This use, involving particularly picric acid and 1,3,5-trinitrobenzene as electron acceptors, antedates the nomenclature "charge-transfer complex" by many decades.

More recently electron acceptors have been used in column chromatography as a means of separating electron-donor molecules. 2,4,7-Trinitrofluorenone has often been used for this purpose.

Currently other physical properties of charge-transfer complexes are being exploited, particularly the semiconductive and photosemiconductive behaviour.

There is also considerable discussion as to the possible role of charge-transfer complexes in reaction mechanisms both in ground state, thermal reactions, and in photochemical reactions, including photosynthesis; also in other biochemical processes such as drug-enzyme interactions and chemical carcinogenesis.

Bibliography

ANDREWS, L. J. and KEEFER, R. M. (1964) *Molecular Complexes in Organic Chemistry*, San Francisco: Holden-Day.

BRIEGLEB, G. (1961) *Elektronen-Donator-Acceptor-Komplexe*, Berlin: Springer-Verlag.

FOSTER, R. (1969) *Organic Charge-Transfer Complexes*, London and New York: Academic Press.

FOSTER, R. (ed.) 1973 *Molecular Complexes*, Vol. I, London: Paul Elek.

PERSON, W. B. and MULLIKEN, R. S. (1969) *Molecular Complexes: A Lecture and Reprint Volume*, New York: Wiley.

R. FOSTER

CIRCULAR DICHROISM. Circular dichroism is the absorption counterpart of optical rotation and for an assembly of randomly orientated molecules, the effect is confined to compounds with a dissymmetric or chiral structure.

Biot (1812) found that the plane of linearly polarized light propagated along the principal axis of a quartz crystal undergoes a rotation, and that the angle of rotation is dispersed with respect to wavelength. Subsequently Biot (1815) discovered that natural products in the liquid or vapour phase exhibit a similar optical rotation, indicating that the effect is basically molecular. Fresnel (1825) ascribed optical rotation to the circular birefringence of optically active media. The plane of linearly polarized light, which is composed of a left- and a right-circularly polarized component, undergoes a rotation if one component is retarded relative to the other on propagation through a transparent medium. The angle of rotation, α, in radians per unit pathlength at the wavelength, λ, is given by Fresnel's equation.

$$\alpha = (n_L - n_R)\pi/\lambda \qquad (1)$$

where n_L and n_R are the refractive indices of the medium for left- and right-circularly polarized light, respectively.

As a physical model, Fresnel suggested that the molecules of an optically active medium have a helical form. Generalizing the model, Pasteur (1860) proposed dissymmetry as the structural criterion for optical activity. Dissymmetric or chiral structures are devoid of secondary symmetry elements, a mirror plane, an inversion centre, or a rotation–reflection axis generally, and they are not superposable upon the corresponding mirror-image structure by rotations and translations alone. Dissymmetry is consistent with any primary symmetry element, and chiral molecules may belong to any of the pure rotation point groups. Pasteur's criterion was based upon the observation that the crystal forms of the sodium ammonium salt of leavorotatory tartaric acid, $Na(NH_4)$—$(-)$—tartrate, and of dextrorotatory tartaric acid, $Na(NH_4)$-$(+)$-tartrate, are enantiomorphous, being non-superposable but exhibiting a left-right handed mirror-image relationship. Since the respective optical rotations persisted in aqueous solutions, Pasteur inferred that the structures of the individual molecules of $(-)$- and $(+)$-tartaric acid are similarly enantiomorphous.

In the classical as in the quantum theory of radiation, the circular birefringence of a medium in the transparent wavelength regions implies a corresponding circular dichroism at absorption frequencies. The differential absorption of left- and right-circularly polarized by solutions of chiral molecules was first observed by Cotton (1895), using copper(II) and chromium(III) $(+)$-tartrate. With plane-polarized radiation, Cotton demonstrated that the dispersion of the optical rotation has an anomalous form at absorption frequencies, and that the transmitted light has an elliptical polarization, due to the circular dichroism.

Drude (1892) related a positive circular dichroism at an absorption frequency, and a corresponding dextrorotation at longer wavelengths, to a right-handed helical charge displacement in a dissymmetric molecule under the radiation field, and the negative optical activity of the enantiomorphous molecule to a mirror-image left-handed helical charge displacement. The collinear electric and magnetic dipole moments of a classical helical charge displacement become, in the quantum theory, the corresponding moments of an electronic transition. Rosenfeld (1928) defined the rotational strength, R_{ab}, of an electronic transition between the stationary states, a and b, of a molecule as the scalar product of the electric dipole and magnetic dipole transition moment,

$$R_{ab} = \text{Im}\ \{\langle a\,|\hat{\mu}|\,b\rangle \cdot \langle b\,|\hat{m}|\,a\rangle\} \qquad (2)$$

where $\hat{\mu}$ and $\hat{m}$ are the electric and magnetic dipole operator, respectively, and Im signifies that the imaginary part is to be taken, since the momentum operator involved is imaginary.

The rotational strength, representing the scalar product of a polar with an axial vector, is a pseudoscalar quantity and is antisymmetric with respect to all secondary symmetry operations. A collinear electric and magnetic dipole moment are not concurrently allowed by the selection rules for any transition in a molecule with a secondary symmetry element. A non-zero rotational strength is allowed only in molecules belonging to the pure rotation point groups, in which pseudoscalars are totally symmetric.

Experimentally the rotational strength of a given transition is measured by the area of the corresponding absorption band,

$$R_{ab} = (3hc\ 10^3 \ln 10/32\pi^3\ N)\int [(\varepsilon_L - \varepsilon_R)/\nu]d\nu \qquad (3)$$

where ν is the absorption frequency and ε the decadic molar extinction coefficient (1. M^{-1} cm^{-1}), h, c, and N, being Planck's constant, the velocity of light, and Avogadro's number, respectively. Circular dichroism spectrophotometers employing mechanical, electro-optic, or photo-elastic modulators to give sinusoidal or square-wave $\pm\lambda/4$ retardation currently measure the differential absorption of left- and right-circularly polarized light down to values of the dissymmetry factor, g,

$$g = (\varepsilon_L - \varepsilon_R)/\varepsilon_{av} \qquad (4)$$

of one part in 10^6.

Observed g-factors are generally small, being of the order of the ratio of molecular dimensions to the wavelength of light ($\sim 10^{-3}$). Optical activity is dependent upon a phase difference in the radiation field between spatially distinct regions of a molecule, and it is an effect in which the molecular extension cannot be ignored relative to the wavelength of light. The g-factor attains a value of the order of unity only for macroscopic chiral structures, e.g. crystals belonging to the enantiomorphous space groups, or cholesteric liquid crystals.

Circular dichroism studies in the chemical field are applied primarily to stereochemical problems, although there are significant spectroscopic applications. A large g-factor ($\geq 10^{-2}$) is indicative of a magnetic dipole transition allowed in the zero-order, with an electric moment acquired in the first or higher order by a vibronic mechanism or a static perturbation due to substituent groups. The relative polarization-directions of allowed electric dipole transitions in a chiral molecule belonging to a dihedral point group may be identified from the relative signs and magnitudes of the corresponding bands in the circular dichroism spectrum of the molecule, particularly in cases where the absolute configuration is known.

Conversely, the absolute stereochemical configuration of a molecule, i.e. the particular structure of a mirror-image pair, may be determined from its circular dichroism spectrum and the quantum theory of rotational strengths [eqn. (2)], a knowledge of the particular electronic transition responsible for a given circular dichroism band assisting the analysis. The analysis is facilitated in dimeric and polymeric systems, consisting of chirally disposed light-absorbing groups, or chromophores, by a separate spectroscopic study of the monomeric chromophore, particularly if the polarization directions of the monomer electronic transitions are established. The right- or left-handed helical sense of the charge redistribution in the corresponding electronic transitions of the dimer are then obtained by inspection of a model of the two mirror-image structures of the chiral dimer.

More empirically the absolute configurations of a series of chemically similar molecules are interrelated by the positive or negative sign of the circular dichroism band due to an electronic transition common to the series. In cases where the series consists of chiral molecules containing a common symmetric chromophore, with one or more secondary symmetry elements, sector rules connect the position of a substituent, or substituents, in the coordinate frame of the chromophore to the sign of the circular dichroism associated with a given electronic transition.

The sector rules are based upon either a static or a dynamic coupling mechanism. In the static model it is assumed that the substituents give rise to a static Coulombic field which mixes the electric with the magnetic dipole transitions of the chromophore. The component of the perturbing field which gives the chromophoric transitions a non-zero rotational strength has pseudoscalar symmetry in the point group of the symmetric chromophore, and the pseudoscalar coordinate functions of that group provide the possible sector rules for the series of substituted chiral molecules. In the dynamic model it is assumed that a transient electric dipole is induced in the substituent by the Coulombic field of the transitional charge distribution in the chromophore. The transient induced dipole of the substituent and the transition dipole moment of the chromophore, electric or magnetic, couple to give a non-zero rotational strength if the substituent-chromophore system is chiral. The sector rule correlating the sign of the rotational strength to the substituent position is given in this model by the coordinate function expressing the angular dependence of the Coulombic potential between the leading electric multipole of the chromophore transition and the electric dipole induced in the substituent.

Similar sector rules are given by both mechanisms for a given chromophore, although the dynamic coupling rule is one order higher in the coordinates of the substituent than the simplest static coupling rule. In the case of the magnetic dipole lone-pair to antibonding π-orbital transition of the carbonyl chromophore, which has C_{2v} symmetry, the dynamic model gives an octant rule, XYZ, whereas the simplest pseudoscalar function in C_{2v} is XY, giving a quadrant rule, although the octant rule also emerges from the static model. The present experimental evidence supports an octant more than a quadrant rule for the circular dichroism of chiral ketones due to the carbonyl $n \to \pi^*$ transition near 300 nm.

In a magnetic field all transparent substances rotate the plane of polarized light propagated along the field direction. The effect, discovered by Faraday (1846), has its absorption counterpart in magnetic circular dichroism. Superconducting magnets are commonly employed to generate the fields of several Tesla (10^4 gauss) required to induce in a symmetric molecule a circular dichroism comparable to that exhibited by a field-free chiral molecule.

The rotational strengths governing natural optical activity have magnetic analogues in the Faraday strengths, of which three types have been distinguished. These are the A-, B-, and C-terms. Unlike the rotational strengths, all three Faraday strengths are scalar quantities and conserve parity.

The A- and the C-terms derive from a degenerate excited and ground electronic state, respectively, and exhibit the Zeeman splitting of the degeneracy in a magnetic field. An A-term appears in the spectrum as a temperature-independent pair of oppositely signed circular dichroism bands with equal areas in the frequency region of a given isotropic absorption. A C-term appears as a temperature-dependent circular dichroism band, or as a pair of oppositely signed bands with unequal areas, depending upon the Boltzmann population of the individual Zeeman components of the electronic ground state.

The B-terms arise from the mixing of the zero-order electronic states of the molecule under the perturbation of the external magnetic field. A B-term appears in the spectrum as a temperature-independent circular dichroism band at the frequency of the corresponding isotropic absorption. In the case of an accidental degeneracy, or near degeneracy, between two electronic transitions with a mutually perpendicular polarization the corresponding B-terms appear as a quasi-A-term, consisting of a pair of oppositely signed circular dichroism bands with comparable areas.

The principal applications of magnetic circular dichroism are spectroscopic. The sign and the magnitude of an A- or a C-term identifies the particular symmetries of the electronic states connected by the transition from which the term derives. Where a forbidden electronic transition becomes vibronically allowed, the particular active normal mode may be similarly identified. The relative polarization directions of two electronic transitions of a molecule close together in energy may be inferred from the resulting B-term circular dichroism, but the magnitudes of B-terms, which depend upon a sum over all states, have yielded little information as yet.

The applications of magnetic circular dichroism to problems of molecular structure are limited. Whereas natural optical activity is dependent upon the detailed three-dimensional stereochemistry and electronic structure of a chiral molecule, the Faraday terms are sensitive only to the nuclear and electronic structure of a molecule in the plane perpendicular to the direction of the external magnetic field. Magnetic circular dichroism distinguishes molecules with a threefold or higher rotational symmetry axis, where A- and C-terms are allowed, from those devoid of such a symmetry element and exhibiting only B-terms.

Bibliography

BUCKINGHAM, A. D. and STEPHENS, P. J. (1966) Magnetic optical activity *Ann. Rev. Phys. Chem.* **17**, 399.
CALDWELL, D. J. and EYRING, H. (1971) *The Theory of Optical Activity*, New York: Wiley–Interscience.
CRABBE, P. (1965) *Optical Rotatory Dispersion and Circular Dichroism in Organic Chemistry*, London: Holden-Day.
DJERASSI, C. (1960) *Optical Rotatory Dispersion*, New York: McGraw Hill.
HAWKINS, C. J. (1971) *Absolute Configuration of Metal Complexes*, New York: Wiley–Interscience.
MASON, S. F. (1963) *Quart. Rev. Chem. Soc.* **17**, 20.
MOSCOWITZ, A. (1962) *Adv. Chem. Phys.* **4**, 67.
SCHATZ, P. N. and MCCAFFERY, A. J. (1969) The Faraday effect, *Quart. Rev. Chem. Soc.* **23**, 552.
SNATZKE, G., ed. (1967) *Optical Rotatory Dispersion and Circular Dichroism*, London: Heyden.
TINOCO, I., Jr. (1962) *Adv. Chem. Phys.* **4**, 113.
VELLUZ, L., LEGRAND, M. and GROSJEAN, M. (1965) *Optical Circular Dichroism*, New York: Academic Press.

S. F. MASON

COLOUR TRANSPARENCIES, COPYING OF.
Modern colour films are capable of producing very acceptable reproductions of most subject matter; often the results are extremely pleasing. But when the subject matter, instead of being an original scene, is itself a colour transparency, the resulting picture, if recorded directly on the type of film normally used in cameras, often exhibits disappointing quality, because of excessive contrast ("burnt-out" highlights and "blocked-up" shadows) and loss of colour saturation.

The reason why films, which give good tone reproduction when used to record original scenes, give poor tone reproduction when used for duplicating transparencies arises from the fact that the human eye sees differently in different environments. When transparencies are viewed, they are usually either projected in a darkened room or, if they are of "cut-sheet" sizes, placed over a diffusing surface which is brightly lit from behind. In either case the surround to the picture is usually much less bright than the picture itself. In the case of the projected transparency the rest of the room is usually much darker than the picture on the screen; in the case of the transparency on the diffuser, the difference is not usually so great, but the surround is normally definitely dim compared to the picture (assuming that the transparency is viewed as for normal display, that is without any uncovered part of the illuminated diffuser being visible, so that the surround consists of the rest of the room).

For reasons which are not fully understood, the brightness of the surround affects quite profoundly the response which the eye makes to a picture. It is well known that a dark or dim surround tends to make colours look lighter; but this effect tends to be greater for dark colours than for light colours, and hence a dark or dim surround tends to reduce apparent contrast (Bartleson and Breneman, 1967).

For original scenes, the surround usually has an average brightness similar to that of the scene itself, and hence in this case the surround does not tend to reduce the apparent contrast. We, therefore, have the situation that camera films (or film systems) intended for making pictures which will be viewed with dark or dim surrounds must be made to have high contrasts in order to overcome the contrast-lowering effects which those surrounds produce (Hunt, 1969). But if such films are then used for duplicating colour transparencies, the final contrast is too high, because the dark or dim surround does not have any contrast-lowering effect on the exposure of the film, as it does on the eye. Thus, if a transparency-film intended for camera use is employed for duplicating transparencies, an increase in contrast will have been included twice, once correctly when the original scene was photographed, and a second time incorrectly, when the original transparency was duplicated.

If, on the other hand, a reflection print was photographed on such a transparency-film, then the increase in contrast would only occur once: this is because the original reflection print would have been made without any increase in contrast since it would generally be viewed with a surround of average brightness similar to that of the picture; the increase in contrast in the transparency-film is then indeed required to off-set the apparent reduction in contrast caused by the transparency surround. It is for this reason that when transparencies are made from reflection prints, or other reflecting copy, satisfactory results are often obtained using camera films. Conversely, to make good reflection prints from transparencies requires considerable reduction of contrast, as is well known by graphic reproduction workers.

In Fig. 1 the results of measurements of displayed density plotted against log exposure are shown for commercial systems which give widely accepted results in reflection prints, in cut-sheet transparencies, and in transparencies intended for projection (Hunt et al., 1969). The density presented to the observer is plotted as ordinate: this was measured with a tele-photometer from a typical observer-viewing position and thus the effects of

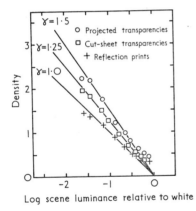

Fig. 1. *The points show the displayed density of a nine-step grey scale plotted against the log luminances of its steps relative to white. Reflection print systems have a gamma of 1.0, cut-sheet transparency systems a gamma of 1·25, and projected transparency systems a gamma of 1·5.*

flare light in the viewing situation are included; for instance, in the case of reflection prints, light reflected from the top-most surface of the print was included and not excluded as is normally the case with good-quality reflection densitometers; and in the case of projected transparencies the image on the screen was measured and thus the effects of flare in the projection lens and ambient light in the projection room were included and not excluded

as in the case with transmission densitometers. The abscissa shows the luminance, on a relative log scale, of the steps of a grey scale in the original scene measured with the same tele-photometer from a point adjacent to the camera position. It is seen that, for the majority of the steps of the scale, the gammas of the systems are 1·0 for the reflection prints, 1·25 for the cut-sheet transparencies and 1·5 for the projected transparencies. The dim surround typical of conditions obtaining when viewing cut-sheet transparencies on an illuminated diffuser necessitates a 25% increase in gamma; while the dark surround typical of conditions obtaining when projecting in a darkened room, necessitates a 50% increase in gamma.

If a system having a gamma of 1·5 were used for duplicating itself, the overall gamma of the duplicate would be equal to 1·5 × 1·5 which is equal to 2·25. However, a film system resulting in a gamma of 1·5 in terms of the picture projected on the screen has to have inherently in itself a much higher gamma at high densities, as shown by the sensitometric curve in Fig. 2, in order to overcome the flare in the viewing situation. If the duplication is then carried out under relatively flare-free conditions, such as occurs in contact-printing or in enlarging with a low-flare lens in complete darkness, then the resulting gamma is even higher (Hunt, 1968). It is for this reason that some advantage can be obtained by giving the film on which the duplicate is going to be made a weak uniform "flash" exposure: this corresponds to the flare light present when the original transparency is viewed

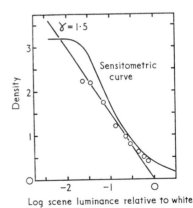

Fig. 2. *The sensitometric curve of a film intended for the production of transparencies for projection (obtained by exposing the film in a sensitometer and measuring the densities in a densitometer) compared with the results (circles) achieved in practice (obtained by measuring both original scene and reproduction with a telephotometer). Camera flare and viewing flare straighten the sensitometric curve and reduce its slope to a value of about 1·5 in practice.*

by projection, and has the effect of lowering the gamma of the original transparency to about 1·5. It is not possible to reduce the gamma much below this figure because the maximum density in the duplicate then becomes too low.

In Fig. 3 is shown the curve (broken line) obtained when film having the characteristic curve in Fig. 2 (shown again in this figure as the full line) is used for duplicating without any flashing or flare light being present.

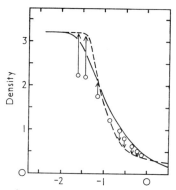

Fig. 3. Sensitometric curve of projection-transparency film (full line) compared to the curve resulting from the use of this type of film for duplicating with no flare (broken line). The circles show the results obtained in practice for an original transparency with camera and viewing flare: the arrow heads show how their densities are distorted if this type of film is used for duplicating.

It is seen that a very high gamma is obtained and that this is accompanied by a reduction in latitude so that more of the steps now come on the low contrast toe and shoulder regions, hence producing the "burnt-out" highlight and "blocked-up" shadow effects already referred to. When projected, duplicates would have their shadow gammas reduced appreciably by viewing flare, but when used on graphic-arts scanners, which are normally fairly free from flare, the effective characteristic would be rather similar to that shown by the broken line in Fig. 3.

In Fig. 4 similar results are shown for the cut-sheet film system; in this case the gamma would have increased from 1·25 to 1·25 × 1·25, which is equal to 1·56, if typical amounts of flare had been present both when exposing the duplicate and when viewing it. The curve in Fig. 4, however, shows the result which would be obtained in the absence of any exposing flash or viewing flare, as would be approximately the case for a contact exposure, or one made in an enlarger with a low-flare lens in complete darkness, and used on a low-flare graphic

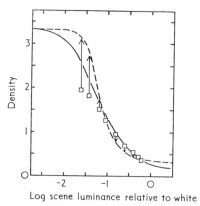

Fig. 4. Same as Fig. 3. but using throughout a film intended for the production of cut-sheet transparencies.

arts scanner. It can be seen that a high gamma and burning out of the highlights and blocking up of the shadows again occur, but not to such a serious extent as in Fig. 3.

What is really required for duplicating is a film which has a gamma of 1·0 and a long enough exposure scale to accommodate the whole density scale of the original transparency. Because transparencies often have density scales of 3·0 or more, this is a difficult task for the photographic manufacturer. Moreover, the colour saturation of reproduced colours decreases with gamma, and hence, other things being equal, when duplicating materials having a gamma of 1·0 are used instead of camera films having a gamma of 1·25 or 1·5, a loss of colour saturation occurs; this loss is in addition to the general loss of colour saturation, referred to at the beginning, which is caused by the inherent limitations of subtractive colour reproduction systems.

The duplication of transparencies is therefore not an easy matter. However, special duplicating films are available having gammas equal to about 1·0; the use of photographic masks for colour and tone correction, if skillfully executed, can overcome the more important losses of quality (Hunt, 1967); and if the duplicate is to be used for reproduction via a graphic arts scanner, similar corrections can be applied in making the separations.

Bibliography

BARTLESON, C. J. and BRENEMAN, E. J. (1967) *Phot. Sci. Eng.* **11**, 254.

CLARKE, F. J. J. (1961) Colorimetry, in *Encyclopaedic Dictionary of Physics* (J. Thewlis, Ed.), **1**, 743, Oxford: Pergamon Press.

HUNT, R. W. G. (1967) *The Reproduction of Colour*, 2nd ed., pp. 234–36, London: Fountain Press.

HUNT, R. W. G. (1968) *Duplication and Conversion of Colour Transparencies*, pp. 5–15, Institute of Printing, London.

Hunt, R. W. G. (1969) *Brit. Kinematog. Sound and Tel.* **51**, 268.

Hunt, R. W. G., Pitt, I. T. and Ward, P. C. (1969) *J. Phot. Sci.* **17**, 198.

Stapleton, R. E. (1962) Photography, Colour, in *Encyclopaedic Dictionary of Physics* (J. Thewlis, Ed.) **5**, 451, Oxford: Pergamon Press.

R. G. W. Hunt

COUNTING TECHNOLOGY, RECENT ADVANCES IN. The last decade has seen considerable advances in nuclear counting techniques. While the most dramatic changes have taken place in semiconductor counting, important advances have been made in the scintillation method and in the applications of gas counting. Reference will also be made in this article to the new and promising techniques of "transition radiation" measurement.

Gas Counting

Proportional and Geiger counters have not themselves changed greatly, although the use of purer materials such as ultra-pure oxygen-free high conductivity copper and gold-plated quartz has prolonged the life of gas-filled, as distinct from gas-flow, proportional counters. Also with the use of modern insulating materials such as Teflon, nylon and certain ceramics the need for guard rings or tubes is no longer so important.

Proportional counters have long been used for low-level counting applications as, for example, in radiocarbon dating. An important technique to reduce the background arising from gamma interactions in the wall was described by Drever, Moljk and Curran who substituted a thin wire screen for the cathode wall. However, the counting efficiency was reduced since the anticoincidence counters were incorporated in the same counting volume. Oeschger and others have overcome this by interposing a very thin electrically conducting but gas-tight wall between the main and anticoincidence counting regions. However, the counter is very delicate and somewhat difficult to use.

Multiwire proportional (and also multiwire spark) counters have become important in nuclear and cosmic ray physics and show promise in the medical physics field since they permit the location of the tracks of ionizing particles and X-rays over large areas to an accuracy of a fraction of a millimetre. The mode of construction varies, but a typical counter has three grids of fine parallel wires, the direction of the wires in the two outer grids being orthogonal, thus providing a rectilinear system of coordinates. There are various ways of identifying the signals from the two mutually perpendicular wires nearest to the event. The signals may be lead to separate amplifiers, but this involves complicated electronic circuitry, and various methods have been used to simplify the system. For example, Gilland and Emming showed how the use of external electromagnetic delay lines can be used to identify the wires so that simple timing circuits can be used to provide spatial information. One method is to measure the difference in time of arrival of the signal at each end of a delay line to which all the wires in one grid are attached.

Charpak *et al.* have also described the use of multiwire counters in the spark mode. The signal is large and independent of the size and nature of the ionizing event. No amplification is needed and the signal is visible to the eye. Attenberger *et al.* have described a spark chamber of the more common plate type which is capable of supporting a number of tracks simultaneously. On one of the plates of the spark chamber are an insulating layer (0·025 cm mylar) and a resistance layer (0·025 cm loaded vinyl). The resistance layer provides isolation between sparks and the dielectric provides electrical energy storage to enhance the brightness.

Scintillation Counting

A considerable number of crystalline scintillators are now available and the most used, sodium iodide activated with thallium, has increased tremendously in size while the resolution has also improved considerably. It is now possible, for example, to obtain crystals up to 75 cm in diameter while resolutions (in smaller crystals) around 7% for ^{137}Cs gammas (660 keV) and peak-to-valley ratios of 20 to 1 for the two γ-rays of ^{60}Co (1·17 and 1·33 MeV) are quoted.

A polycrystalline scintillator, trade-named Polyscin, in which NaI crystals are imbedded in a matrix, is much more resistant to thermal and mechanical shock and may be substituted in some cases for large single NaI crystals.

Anthracene and stilbene crystals still find a place, the latter being of special value in the technique of pulse shape discrimination (P.S.D.) to which reference will be made later. CsI(Na) and CsI(Tl) are used for particle identification by rate of energy loss and are more resistant to mechanical and thermal shocks than is NaI(Tl). LiI(Eu), enriched in ^{6}Li, is widely used as a neutron detector using the ^{6}Li (n, α) ^{3}H reaction. CaWO$_4$ has been used a little for γ-ray measurements because of the high atomic number of tungsten, but is very slow and has a low light output.

Turning to liquid scintillators, these have the advantage of being available in large volumes (up to hundreds of litres) and in shapes limited only by the container geometry. Liquid scintillators have in general very short decay times, down to 2·6 nsec, and this is of great value in many nuclear physics applications, as are also deuterated scintillators. Boron and gadolinium loaded phosphors are useful as neutron detectors while lead and tin loaded liquid scintillators are also available for γ- and X-ray measurements. Other liquid-scintillators can be used for P.S.D., although here deoxygenation of the scintillator is specially important.

Other scintillators can be used to count samples dissolved in water. For example, using highly purified dioxan, 2 ml of aqueous solution can be counted in 20 ml total volume. This is especially useful for biological and medical applications.

Solvents commonly used at the present time are m-xylene, p-xylene, toluene, isopropyl biphenyl, phenylcyclohexane and dioxan. When used in large volumes, the rather low flash points of toluene and xylene present a hazard and it has been reported that pseudo-cumene (1,2,4-trimethylbenzene) which has a considerably higher flashpoint has also a higher light output than the best toluene or xylene scintillators.

Typical primary solutes, that is the scintillation agents, are referred to by the abbreviations TP, PPO, PBO, PBD, butyl-PBD, BBOT, BIBUQ, etc. The full chemical names of these compounds will be found in manufacturers' catalogues (e.g. Nuclear Enterprises Ltd., Koch-Light Laboratories Ltd., amongst others) or in the references in the bibliography at the end of this article. Secondary solutes, that is spectrum shifters used to change the wavelengt of the emitted light to a value more suitable for photomultipliers, include POPOP, BBO, PBO and α-NPO.

Automatic liquid scintillation counting systems are now available from a number of firms. Typical units automatically introduce up to several hundred samples, one at a time, by a light-tight lock into the counting region for either a preset time or preset number of counts, the counting data being printed out, together with sample identification, on a typewriter, punched tape or other suitable system. Simple data processing is sometimes also included. The units are usually equipped with two or three channels which can be preset to suit the spectrum characteristics of different radioactive species, for example ^{3}T, ^{14}C and ^{32}P, all much used in biological and other scientific work.

Plastic scintillators play a prominent role because of the ease with which they can be handled, the size—both large and thin—in which they can be obtained, the ease with which they can be machined to different shapes and their high light transmission. Some have decay constants shorter even than those of liquid scintillators (down to 1·7 nsec). They are much used in high-energy physics and can be used for fast neutron detection as well as for alpha, beta and gamma counting. ZnS loaded plastic is sometimes used for fast neutron counting or, with B or ^{6}Li added, for slow neutron detection. Plastic phosphors capable of retaining their shape up to 150°C have been developed.

Glass scintillators have also come into wider use with the considerable increase in light output which has been achieved. These can be used over a wide temperature range—for example Boreli et al. have examined the effect of temperature from -180°C to $+250$°C while typical glasses melt at ~ 1200°C—and can be used for vacuum seals as, for example, in the Daly-type mass spectrometer scintillation ion beam detector. They are resistant to most chemicals and can be used under extreme environmental conditions. Li loaded scintillation glasses are available for neutron detection and some scintillation glasses have been used for P.S.D.

Semiconductor Detectors

The first semiconductor detectors in common use were diffused junction and surface barrier silicon counters and these continue to be used, especially for alpha and other charged particle detection and measurement. However, the development of Li drifted germanium and silicon detectors has provided the nuclear physicist, the radiochemist and others with one of their most useful tools. Ge(Li) detectors are especially useful for the measurement of the energy and intensity of γ-rays because of their very high resolution and they are now available in volumes in excess of 120 ml. They have the disadvantage of requiring to be used and stored at liquid nitrogen temperatures. Lithium drifted silicon detectors, on the other hand, can be kept and used at room temperature (although their characteristics are improved at low temperature), but, because of the much lower atomic number, are not normally employed for gamma spectrometry. They are, however, used for charged particle and X-ray measurements.

The resolution of these detectors is, as mentioned previously, very high and provides, for γ-rays, an improvement of more than two orders of magnitude over sodium iodide detectors. This is extremely useful for many purposes. In nuclear physics, for example, the very much higher resolution makes possible the investigation of very complicated decay schemes, while in neutron activation analysis the need for chemical separation of specific elements in an irradiated sample is very much reduced. A resolution of better than 2 keV can be achieved with Ge(Li) detectors in the ^{60}Co γ-ray region (1·17, 1·33 MeV). Also thin germanium and silicon detectors for X-ray work have shown a peak width of less than 300 eV for the 14·4 keV γ-ray of ^{57}Co and 105 eV at 1·74 keV. Semiconductor detectors for α-particles can give a resolution as low as 13 keV for 5·5 MeV α-particles while a similar peak width can be obtained with 1 MeV β-particles. On the other hand they are, in general, less sensitive and more expensive than scintillation detectors and, in the case of Ge(Li) detectors, strict housekeeping is essential to ensure that the liquid nitrogen supply never runs dry.

Intrinsic germanium crystals, using extremely pure germanium with impurity concentrations of less than 1×10^{10} per cc (l in 10^{12} to 10^{13}), have been prepared in recent years in volumes of several ml. These have the advantage of rapid fabrication and can be stored at room temperature although they must be used at low temperatures. However, an accidental warming up to ambient temperature is no longer a disaster.

A. H. F. Muggleton has written a useful review article on semiconductor detectors to which the reader is directed for further information (see Bibliography). Also E. M. Gunnerson reviewed the semiconductor properties of a large number of materials—diamond, cadmium sulphide, zinc sulphide, zinc oxide, silicon carbide, silicon, germanium, cadmium telluride, gallium arsenide, gallium phosphide, gallium antimonide, indium antimonide, indium arsenide and lead sulphide.

The efficiency of Ge(Li) detectors can be increased by use of well or hollow cylindrical type crystals, with the source placed inside. Also the ratio of photoelectric peak height to Compton distribution can be considerably enhanced by use of a NaI(Tl) or plastic scintillator anticoincidence shield around it.

Germanium and silicon detectors are rather sensitive to radiation damage and typical irradiation exposures after which diffused junction counter performance begins to deteriorate are of the order of 5×10^8 α-particles, 5×10^{10} protons, 10^{13} electrons, 10^{12} fast neutrons, 5×10^8 R of γ-radiation and 5×10^8 fission fragments. Lithium drifted devices are about 100 times more sensitive to radiation damage than are junction counters. Silicon carbide crystals are about one hundred times less sensitive to radiation and can be used up to 700°C, but have not been made in large sizes as yet.

Neutrons can be detected by coating semiconductor detectors with B or ^{235}U while, for threshold detectors, ^{232}Th and ^{238}U have been used. Neutron spectrometry can be carried out using a hydrogenous (e.g. polythene) layer in contact with the detector or ^{6}Li or ^{3}He in a sandwich construction.

One- and two-dimension position sensitive detectors have been constructed by Hofken et al., Lauterjung et al., Eccleshall and Yates, Owen and Awcock, Laegsgard, Gunnersen et al. and others, using semiconductor detectors. Techniques vary, but they consist of parallel rows of semiconductor detectors (two orthogonal rows for two dimensions) and are in fact in principle rather like the multiwire counter to which reference was made earlier.

The use of Ge(Li) detectors in low level counting has recently been discussed by Walford, Aliaga-Kelly and Gilboy.

Pulse Height Analysis

Whereas one or two hundred channel analysers were the most commonly used before semiconductor counters became available, the latter have made necessary the use of 1000, 2000 and 4000 channel analysers. Modern pulse height analysers are sophisticated systems with many extra features built in—internal amplifiers, power supplies for counters, coincidence facilities, background subtraction, channel marker facilities, memory splitting, read-out to X–Y plotters, electric typewriters, line printer, page printer, paper tape read-in and read-out, and so on.

However, it has become increasingly common to use a small computer for this purpose and a number of suitable models are available. Typically the detector signal is passed to an analogue-to-digital convertor interfaced to the computer. The latter not only carries out the pulse height analysis but can also process the data according to the experimental requirements.

Photomultipliers, Pulse Shape Discrimination

There have been advances in the design of photomultipliers and various types now available include tubes with fast anode pulse rise times (~1·5 nsec), large diameter (30 cm), able to withstand severe environmental conditions including use at high temperature—one on the market can be operated at 150°C—and tubes with low potassium content and thus low background.

As mentioned earlier, the fact that certain scintillators have slightly different pulse shapes for different radiations e.g. neutrons, gammas, protons, etc., provides a mechanism for differentiating between these. There are a number of different methods for analysing the pulse shape (for example, the zero cross-over, charge comparison and last dynode-anode space charge saturation techniques) and reference may be made to the bibliography. The scintillators employed for this purpose are mainly stilbene and certain liquid scintillators, although others show some effect.

Other Counting Methods

Cerenkov counters continue to be used in special applications and various materials are used—gases, PbF_2, glass, perspex, liquids (including liquid hydrogen and, of course, water). Spectrum shifters can be added as in the case of scintillation counters. Cerenkov detectors can be used to count high-energy β-radiation in aqueous solutions. Apart from the discrimination the technique affords against lower energy background, it permits the counting of 100% aqueous solutions and not ~10% as in the case of liquid scintillation counting.

A possible new technique which is commanding a good deal of interest at present depends on the use of "transition radiation". When a charged particle passes from one dielectric medium to another photons are given off and the number of photons, mainly in the X-ray region, is directly proportional to the particle energy. Thus the combination of such a detector with a measurement of momentum would distinguish particles of different mass. The problem is to detect the photons which travel in the forward direction. Yuan et al. have used a series of multiwire counters as described by Domcovsky et al. in the path of the particles and X-rays and this system helps to distinguish the photons from the signal in the counters due to the original ionizing particle. This technique is still at an early stage, however.

Bibliography

ATTENBERGER, S. et al. (1972) Spark chamber with multi-track capability, *Nucl. Instr. and Meth.* **107**, 605.

BIRKS, J. B. (1964) *The Theory and Practice of Scintillation Counters*. Pergamon.

CHARPAK, G., RAHM, D. and STEINER, H. (1970) Some developments in the operation of multiwire proportional counters, *Nucl. Instr. and Meth.* **80**, 13.

DEARNALEY, G. and NORTHRUP, D. C. (1964) *Semiconductor Counters for Nuclear Radiations*, Spon.

DYER, A. (1971) and CROOK, M. A. et al. (1972) *Liquid Scintillation Counting*, Vols. 1 and 2, Heyden.

DOMCOVSKI, Z. (1971) High energy charged particles separation by means of a cascade of multiwire proportional chambers, *Nucl. Instr. and Meth.* **94**, 151.

GROVE, R. et al. (1970) Electromagnetic delay line readout for proportional wire chambers, *Nucl. Instr. and Meth.* **89**, 257.

GUNNERSON, E. M. (1967) *Rep. Prog. Phys.* **30**, 27.

MCBETH, G. W. et al. (1970) *Pulse Shape Discrimination with Organic Scintillators*, published by Koch Light Laboratories Ltd.

MUGGLETON, A. H. F. (1972) Semiconductor devices for gamma ray, X-ray and neutron radiation detection *Phys. E* **5**, 390.

SIEGBAHN, K. (1965) *Alpha, Beta and Gamma-Ray Spectroscopy* Chapters 5 and 6. North-Holland.

WALFORD, G. V. et al. (1972) *I.E.E.E. Trans. on Nucl. Sci.* **NS-19** (1), 127.

WINYARD, R. A. et al. (1971, 1972) Pulse shape discriminations in inorganic and organic scintillators, *Nucl. Instr. and Meth.* **95**, 141; **98**, 525.

YUAN, L. C. L. et al. (1972) Relativistic particle separation by means of transition radiation utilising multiwire proportional chambers, *Phys. Lett.* **40B**, 689.

H. W. WILSON

CRITICAL POINT. Figure 1 typifies the behaviour of a single-component fluid in the neighbourhood of its gas–liquid critical point—the variation of pressure p with volume V is plotted as a series of isotherms, $a_1a_2a_3a_4$, b_1b_2, etc. The curve d_1d_2, which corresponds to the highest temperature, approximates the hyperbola characteristic of a perfect gas, but at a lower temperature, such as corresponds to $a_1a_2a_3a_4$, discontinuities are found because liquefaction occurs. If the initial state of a fluid is represented by a_1 and it is then compressed isothermally the volume decreases continuously with gradual increase in the pressure until its state corresponds to the point a_2 on the *saturation dome* $a_2c_2a_3$. At a_2 a discontinuity of slope of the isotherm occurs and further compression causes condensation, which takes place at constant pressure, until point a_3 is reached where the fluid is entirely in the liquid state. At a_3 there is again a discontinuity and further reduction in volume is accompanied by a very steep

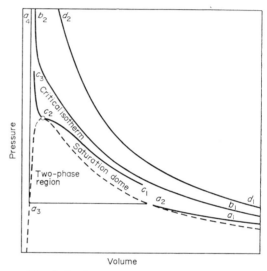

Fig. 1. Isotherms of a real fluid.

increase in pressure along the line a_3a_4. As the temperature of compression is increased the volumes of the saturated vapour V_v and of the saturated liquid V_L (the *orthobaric volumes*, which may be expressed as either molar or specific quantities) are found to approach one another until a temperature is reached at which they are equal, and the points of discontinuity of slope merge into a single horizontal point of inflexion on the critical isotherm $c_1c_2c_3$. This is the *critical point*, defined for each substance by characteristic values of *critical temperature* T_c, *critical pressure* p_c, and *critical volume* V_c (or *critical density* $\varrho_c = 1/V_c$). (As a rough guide, for low-boiling liquids $T_b/T_c \approx 2/3$ where T_b is the normal boiling temperature—values of T throughout this article being expressed on the thermodynamic scale—and $\varrho_c \approx \varrho_b/3$ where ϱ_b is the density of the liquid at its normal boiling point; for organic liquids the value of p_c seldom exceeds 5 MPa but for some inorganic liquids it is considerably greater, e.g. for water $p_c \approx 22$ MPa.) At temperatures equal to or greater than the critical temperature isotherms exhibit no discontinuity of slope, though there may still be, at temperatures only a little in excess of the critical temperatures, a non-horizontal point of inflexion. At the critical point the first and second partial derivatives of both pressure (at constant temperature) and temperature (at constant pressure) with respect to volume are zero, but the partial derivative $(\partial p/\partial T)_{V=V_c}$ remains finite. An alternative way to define the critical temperature implicit in the abore is that it is the temperature above which it is impossible to liquefy a gas by compression.

The properties of the two phases become identical at the critical point and, experimentally, the simplest method of determining T_c is to observe the temperature at which the meniscus between vapour and liquid disappears and reappears in a sealed glass tube containing the amount of substance corresponding to its critical density. In his classical work on carbon dioxide, in which he produced the first diagram of the type of Fig. 1, Andrews used a tube open at its lower end and confined the substance over mercury so that the pressure could be transmitted to a pressure-measuring instrument and so that the volume could be varied by adjustment of the amount of mercury in the tube. Isotherms, vapour pressures and orthobaric volumes in addition to the critical temperature and critical pressure could be determined with this apparatus which, modified only in detail, remains in use today.

With the accumulation of data in the period from 1860 to 1950 various anomalies associated with the critical point appeared—most discussion revolved around the shape of the *saturation dome* or *coexistence line*, the boundary of the two-phase region. Since the critical isotherm is horizontal at the critical point, i.e. the compressibility is infinite, the experimental difficulties in the way of determining the critical volume (or density) are severe, and the coexistence line, which is well defined to within a few kelvins of the critical temperature, must usually be completed by interpolation.

The coexistence line is generally plotted in terms of density against temperature (Fig. 2). According to two empirical relationships

$$(\varrho_L + \varrho_V)/\varrho_c = 2 + x(1 - T_r) \qquad (1)$$

and

$$(\varrho_L - \varrho_V)/\varrho_c = y(1 - T_r)^n \qquad (2)$$

where ϱ_L, ϱ_V are the orthobaric densities respectively of the liquid and vapour, ϱ_c is the critical density, T_r is the reduced temperature T/T_c, x and y are constants, and the exponent $n \approx 1/3$. It is experimentally convenient for ϱ_L and ϱ_V to be determined at the same temperature, and if a series of values for different temperatures is measured eq. (1), the *law of rectilinear diameters*, allows ϱ_c to be determined. Addition and subtraction of eqs. (1) and (2) allows ϱ_L and ϱ_V to be calculated for the region where measurements cannot be made, and so the coexistence line may be completed as a continuous curve. Experiments made as near to the critical point as possible suggested however that in fact the coexistence line had a flat top, but this anomaly was resolved by

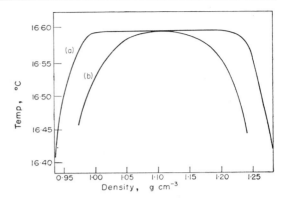

Fig. 2. Vapour–liquid coexistence curves of xenon obtained with tube 19 cm long of 14 mm internal diameter; (a), with tube held vertically; (b), with tube held horizontally (after Weinberger and Schneider, 1952, Can. J. Chem. 30, 422).

Weinberger and Schneider who showed that such results are due to the effect of gravity: the two lines they obtained for xenon in a tube 19 cm long and 14 mm internal diameter are shown in Fig. 2, curve (a) (exhibiting a flat top) for the tube held vertically and curve (b) (with no discontinuity) for the tube held horizontally. It has since been clearly demonstrated, e.g. by studies of refractive index, that in the usual relatively tall glass tube there is a density gradient in the fluid when it is in the critical region even though only one phase is present. Other experiments had tended to show that hysteresis occurred as the temperature was cycled up and down through the critical value and that there was a promontory at the top of the coexistence curve, but the claims for the existence of these phenomena were probably made because there was a lack of awareness of the time required for equilibrium to be reached in these conditions—which may well be measured in days rather than minutes. Currently there seems to be no doubt that eqs. (1) and (2) give a qualitatively correct indication of the shape of the coexistence curve in the critical region although in practice plots of $(\varrho_L + \varrho_V)$ against $(1 - T_r)$ and $(\varrho_L - \varrho_V)$ against $(1 - T_r)^{1/3}$ are frequently slightly curved so that there is some uncertainty about the correct extrapolation of the lines to $T_r = 1$, and the critical density (or volume) cannot be determined with as high a precision as can the critical temperature and pressure.

The equation of state proposed by van der Waals

$$\left(p + \frac{a}{V^2}\right)(V - b) = RT, \qquad (3)$$

where a and b are constants characteristic of each fluid, reproduces the general shape of curves such as are shown in Fig. 1. Although it is inadequate for the representation of exact measurements it provides the starting point for all discussions of the critical point and subsequent development of more exact methods of representing the behaviour of real fluids. It can be shown that for a fluid whose behaviour is described by eq. (3) the exponent in eq. (2) is $1/2$, yet as closely to the critical point as this exponent can be determined it proves to be approximately $1/3$. A value of $1/2$ for the exponent is consequent not only upon eq. (3) but also on the theoretical assumption generally made that the free energy is an analytic function at the critical point. According to current thinking this may not be true, but there is no conclusive evidence for the suggestion available yet: anomalies associated with the critical point remain but they are of a more subtle kind than those which have been discussed above.

According to the *principle of corresponding states*, first enunciated by van der Waals as a consequence of eq. (3), if pressure, volume and temperature are expressed in reduced form, i.e. as p/p_c, V/V_c, and T/T_c, the pVT diagrams for all substances (typified by Fig. 1) are superposable. While qualitatively correct the principle failed as a quantitative guide to the properties of real fluids, especially in relation to vapour pressure, which often differs several fold at the normal boiling point from that predicted by application of the simple corresponding states principle. The introduction by Pitzer of an additional parameter, the *acentric factor* ω, transformed the principle into a valuable unifying and predictive method. The acentric factor is defined by

$$-\log(p/p_c) = 1 + \omega \quad \text{for} \quad T_r = 0.7 \quad (4)$$

and the value for any substance of ω is then a measure of the amount by which the reduced vapour pressure at $T_r = 0.7$ differs from that of the noble gases (because substances with simple spherical molecules such as the noble gases do obey the principle of corresponding states with fair accuracy and their vapour pressures at a reduced temperature of 0.7 are about one tenth of their critical pressures). Pitzer's tables, which relate pVT properties to the value of ω, allow estimation of those properties, including vapour pressure, for a wide variety of non-polar substances with useful accuracy. Another parameter which allows practical application of the principle of corresponding states is the critical compression factor $z_c = p_c V_c / RT_c$ and this has been widely used by engineers for the purpose of estimating pVT properties.

Critical points are also associated with other phase transitions: that most closely analogous to the gas–liquid transition is found when two partially miscible liquids are mixed. If the initial composition of the mixture of components A and B is chosen correctly the mixture separates into two phases, one containing a preponderance of A with some of B and the other containing a preponderance of B with some of A. As the temperature is raised the difference between the compositions of the two coexisting phases decreases and at the *critical solution temperature* the two phases become identical in composition and the solution forms a single phase. The shape of the coexistence curve when concentrations are plotted against temperature is similar to the density-temperature curve just discussed and the variation with temperature of the difference between the two concentrations at the same temperature may be described by an equation analogous to eqn. (2) in which temperature appears in a term $(T_c - T)^{1/3}$. Some solutions also exhibit a lower critical solution temperature and there is therefore for these solutions a temperature range in which partial miscibility occurs while above and below this range there is complete miscibility.

Observation of the critical point in both gas–liquid and liquid–liquid systems may be made difficult by the appearance of the *critical opalescence*, a strong scattering of light which makes the system opaque: study of this light scattering is one experimental technique by which these critical phenomena may be elucidated.

Other critical points occur in the magnetic behaviour of materials, the occurrence of superconductivity in metals and superfluidity of helium. Modern studies suggest that the theoretical explanations of all these phenomena will have many similarities.

Bibliography

BOWDEN, S. T. (1961) Miscibility of liquids, in *Encyclopaedic Dictionary of Physics* (J. Thewlis, Ed.), **4**, 691, Oxford: Pergamon Press.

DIN, F. (1962) Pressure–Volum–Temperature relations, in *Encyclopaedic Dictionary of Physics* (J. Thewlis, Ed.), **5**, 648, Oxford: Pergamon Press.

EGELSTAFF and RING (1968) Chapter 7 in *Physics of Simple Liquids* (ed. by Temperley, Rowlinson and Rushbrooke). Amsterdam: North Holland.

ENDERBY, J. (1962) Superconductivity, in *Encyclopaedic Dictionary of Physics* (J. Thewlis, Ed.), **7**, 104, Oxford: Pergamon Press.

FRIEDMAN, A. S. (1961) Equations of state, in *Encyclopaedic Dictionary of Physics* (J. Thewlis, Ed.) **3**, 2, Oxford: Pergamon Press.

LEWIS and RANDALL (1961) *Thermodynamics*, 2nd edition revised McGraw-Hill. New York:

MACKINNON, L. (1962) Superfluid, superfluidity, in *Encyclopaedic Dictionary of Physics* (J. Thewlis, Ed.), **7**, 112, Oxford: Pergamon Press.

REID and SHERWOOD (1966) *Properties of Gases and Liquids: Their Estimation and Correlation*, 2nd edition. New York: McGraw-Hill.

ROWLINSON (1969) *Liquids and Liquid Mixtures*, 2nd edition, London: Butterworth.

SENGERS and SENGERS (1968) *Chem. Engng. News* **46**, June 10, 104.

SNYDER, W. (1962) Van der Waals equation of state, in *Encyclopaedic Dictionary of Physics* (J. Thewlis, Ed.) **7**, 583, Oxford: Pergamon Press.

D. AMBROSE

CRYSTAL STRUCTURE DETERMINATION, DIRECT METHODS OF. The primary objective of the direct method of X-ray structure analysis is to determine the location of the atoms in the unit cell of a crystal by means of a direct mathematical analysis of the diffraction pattern. This is in contrast to other procedures which rely heavily on the introduction of auxiliary chemical, physical and structural information and may involve a considerable amount of trial and error. One of the virtues of the direct method is that it permits the determination of structures of unknown formula or chemical composition, thereby establishing these desiderata as one of the consequences of the analysis.

X-ray structure analysis provides molecular information concerning stereoconfiguration, conformation, bond lengths and angles, and charge distributions. It also provides intermolecular information such as hydrogen bonding, molecular complexes, charge transfer complexes, coordinations and clathrate formation.

An X-ray diffraction pattern from a single crystal is comprised of numerous diffraction maxima to each of which a set of four numbers may be assigned. Three of the numbers are integers which are the Miller indices identifying the crystal plane associated with a maximum. The fourth number represents the intensity of scattering from the plane. The data from a diffraction experiment therefore consist of sets of four numbers associated with the diffraction peaks and auxiliary information such as cell parameters and space group.

The maxima of the electron density function $\varrho(\mathbf{r})$ locate the positions of the atoms in a unit cell. The electron density at a point with position vector $\mathbf{r}$ in a unit cell of volume v may be represented by a three-dimensional Fourier series,

$$\varrho(\mathbf{r}) = v^{-1} \sum_{\mathbf{h}=-\infty}^{\infty} F_{\mathbf{h}} \exp(-2\pi i \mathbf{h} \cdot \mathbf{r}), \qquad (1)$$

where the coefficients

$$F_{\mathbf{h}} = |F_{\mathbf{h}}| \exp(i\varphi_{\mathbf{h}}) \qquad (2)$$

are the crystal structure factors whose magnitudes are proportional to the square roots of the corresponding observed X-ray intensities. The components of $\mathbf{h} = (h, k, l)$, the Miller indices, and $|F_{\mathbf{h}}|$ may be considered to represent the set of four numbers associated with each diffraction maximum. The angle $\varphi_{\mathbf{h}}$ is the phase associated with $F_{\mathbf{h}}$.

If the phases $\varphi_{\mathbf{h}}$ were obtainable from a diffraction experiment, eq. (1) could be computed immediately to give the crystal structure. Only the magnitudes $|F_{\mathbf{h}}|$ are ordinarily obtainable from experiment, whereas the $\varphi_{\mathbf{h}}$ are not, which gives rise to the so-called phase problem of crystal structure analysis. However, owing to the fact that the electron distribution functions for the individual atoms in crystals are known to a good approximation, it is possible to compute, in principle, the phases from the measured X-ray intensities. This provides the basis for the direct mathematical analysis of a diffraction pattern which is usually successful in practice.

A Fourier inversion of eq. (1) may be performed which expresses the $F_{\mathbf{h}}$ in terms of the electron density distribution, $\varrho(\mathbf{r})$. The usual integral expression may be reduced to a sum of contributions from each of the discrete atoms comprising the electron density giving

$$|F_{\mathbf{h}}| \exp(i\varphi_{\mathbf{h}}) = \sum_{j=1}^{N} f_{j\mathbf{h}} \exp(2\pi i \mathbf{h} \cdot \mathbf{r}_j), \qquad (3)$$

where $f_{j\mathbf{h}}$ is the atomic scattering factor for the the jth atom in a unit cell containing N atoms and $\mathbf{r}_j$ is its position. The form of eqs. (3) omits an exponential damping term which would occur under the sum sign as a consequence of the thermal motion. This damping may be largely corrected for in the treatment of the intensity data and it is assumed that this has been done. In the simultaneous eqs. (3), the unknown quantities are the phases $\varphi_{\mathbf{h}}$ and the atomic positions $\mathbf{r}_j$. The known quantities are the magnitudes of the structure factors $|F_{\mathbf{h}}|$ and the atomic scattering factors $f_{j\mathbf{h}}$ which have been tabulated. For each $\mathbf{h}$ there are two equations, one for the real part and one for the imaginary part. Analysis of Eqs. (3) shows that the system of simultaneous equations is greatly overdetermined by the number of data, $|F_{\mathbf{h}}|$, available from the sphere of scattering for X-ray radiation from a copper target. For crystals possessing a centre of symmetry, the overdeterminacy amounts to a factor of about 50 for each atomic coordinate and for crystals which do not possess a centre of symmetry the overdeterminacy factor is about 25 for each atomic coordinate. Because of the high degree of overdeterminacy it may be expected that simple phase determining relations exist, having probable validity, which yield phase information with a high degree of reliability.

Since the objective is to locate the atomic positions $\mathbf{r}_j$ in the unit cell, the question arises concerning why these quantities are not determined directly, thus averting the need for any concern with the phases $\varphi_{\mathbf{h}}$. At the present time it is often more readily possible to determine phases and then obtain the structure from the computation of eq. (1) than to find the atomic coordinates directly.

A general approach to the problem of obtaining atomic positions without the use of phases is afforded by the Patterson function,

$$P(\mathbf{r}) = \sum_{\mathbf{h}=-\infty}^{\infty} |F_{\mathbf{h}}|^2 \exp(-2\pi i \mathbf{h} \cdot \mathbf{r}). \qquad (4)$$

Equation (4) can be readily calculated since the coefficients of the Fourier series are proportional to the measured X-ray intensities. The maxima represent the interatomic vectors in a crystal. Analysis of $P(\mathbf{r})$ is limited by a lack of resolution of the interatomic vectors. Nevertheless, many complex structures have been determined with the aid of the

Patterson function, particularly those possessing heavy or moderately heavy atoms, since the interatomic vectors associated with such atoms can be relatively easily identified. The positions of the heavy atoms may be deduced from the interatomic vectors. Once these are found, methods are available for developing the complete structure.

Although lack of resolution may limit the facility with which Patterson functions may be analysed, the great overdeterminacy of the system of eqs. (3) implies that an alternative procedure of general applicability may facilitate the analysis and permit a large range of structure problems to be readily solved from the X-ray scattering data. The mathematical background and nature of such a procedure involving the determination of phase values will now be described. It may be mentioned that under special circumstances, there are other powerful techniques available for evaluating phases. Examples are isomorphous replacement and anomalous dispersion. Because of the special nature of these techniques and the limited opportunity to employ them, they will not be discussed here. Isomorphous replacement does play a key role however in the study of protein structures.

Phase-determining Relations

The intensity data from an X-ray diffraction experiment are subjected to a data reduction procedure before a phase determination is carried out. The objective of the data reduction is to produce a set of numbers which resembles as closely as possible an ideal set of intensities which would be obtained from motionless point scatterers as opposed to that actually arising from moving atoms with spatially distributed electron densities. Other corrections are also made, for example, for Lorentz and polarization factors, extinction and absorption. The basis for the transformation of the measured X-ray intensities to those corresponding to motionless point scatterers is an analysis of the statistical properties of the $|F_h|^2$ by means of probability theory. It is found from such studies that it is possible to define a normalized structure factor E_h, having the average property that $\langle |E_h|^2 \rangle_h = 1$, in terms of the corresponding F_h,

$$E_h = F_h / [(m_2^0 + m_0^2) s_{2h}/m]^{1/2}, \quad (5)$$

where

$$s_{2h} = \sum_{j=1}^{N} f_{jh}^2, \quad (6)$$

n is the number of symmetry related atomic positions in the unit cell, the symmetry number, and

$$m_i^k = \int_0^1 \int_0^1 \int_0^1 \xi^l \eta^k \, dx \, dy \, dz \quad (7)$$

where $\xi = \xi(x, y, z; \mathbf{h})$ and $\eta = \eta(x, y, z; \mathbf{h})$ are the real and imaginary parts, respectively, of the trigonometric portion of the structure factor as defined, for example, in the *International Tables for X-ray Crystallography* for the particular space group of interest. For pure real reflections (centrosymmetric) $m_0^0 = 0$, for pure imaginary reflections $m_2^0 = 0$ and for general complex non-centrosymmetric reflections $m_2^0 = m_0^2 \neq 0$. It is assumed in writing eq. (5) that the F_h have been corrected for vibrational motion. The quantity s_{2h}, defined in terms of the atomic scattering factors, eq. (6), effectively corrects the F_h for the effect of the spatial distribution of electron density. As a consequence it is the normalized structure factor magnitudes $|E_h|$, the end products of the data reduction procedure, which represent the scattering from a structure consisting essentially of motionless point scatterers. The phase determining formulas and their associated probability measures are conveniently defined in terms of the E_h. Their phases are the same as those for the corresponding F_h.

A phase-determining formula for centrosymmetric crystals which has found considerable application is the sigma-2 relation,

$$s E_h \simeq s \sum_k E_k E_{h-k}, \quad (8)$$

where s means "sign of". A plus sign corresponds to a phase of zero and a minus sign corresponds to a phase of π, the only two possible phase values for crystals possessing a centre of symmetry. For non-centrosymmetric crystals the general reflections have phase values ranging from $-\pi$ to π. Appropriate phase determining formulas for non-centrosymmetric crystals are

$$\varphi_h \simeq \langle \varphi_k + \varphi_{h-k} \rangle_{k_r}, \quad (9)$$

the "sum of angles formula", and

$$\tan \varphi_h \simeq \frac{\sum_k |E_k E_{h-k}| \sin(\varphi_k + \varphi_{h-k})}{\sum_k |E_k E_{h-k}| \cos(\varphi_k + \varphi_{h-k})}, \quad (10)$$

the "tangent formula". The symbol k_r means that the k vectors entering into the average are restricted to those associated with the $|E_k|$ and $|E_{h-k}|$ of large magnitude. In applying eq. (9), an ambiguity can arise concerning whether $+2\pi$, -2π or 0 should be added to each contributor $\varphi_k + \varphi_{h-k}$. This causes no essential difficulty in procedures for phase determination.

Examination of eqs. (8), (9) and (10) shows that in applying these relations, the values of certain phases are required to be known in order to determine the values of additional ones. Probability measures and auxiliary formulas have been introduced to facilitate these calculations. In the course of application of eqs. (8) to (10), it has been found that the need to have preliminary phase information does not impose a severe restriction and routine procedures involving these relations and appropriate probability measures have been developed. One reason for this is although eqs. (8)–(10) imply sums or averages over large numbers of contributors, single terms are often quite reliable when the $|E|$ values are large enough, as indicated by the appropriate probability measures. Another important reason is that the number of basic phases, in terms of which the remaining ones are defined by means of the phase determining relations, can be quite limited. A basic set of phases can be composed of those phases whose values are known since they may be specified in order to fix the origin and, when appropriate, enantiomorph and reference frame, and those to which unknown symbols, representing the values of additional phases, may be assigned in the course of the phase determination. The procedure based on a set of symbolic phases has been named the symbolic addition procedure. There is generally a certain amount of ambiguity associated with a phase determination owing to the fact that the values for some of the symbols may not be established in the course of the procedure. In the case of centrosymmetric crystals, unknown symbols can assume either of two values and since there are generally fewer than three or four to consider, it is very easy to examine the possibilities with modern computing facilities. If the symbols represent phase values for general non-centrosymmetric reflections, there are more possibilities to consider since they must be permitted to assume values at smaller intervals, about 45°, rather than the 180° separation which applies to centrosymmetric and pure imaginary reflections. The number of ambiguities can nevertheless still be kept within practical limitations.

The probability measure to be associated with eq. (8) is

$$P_+(\mathbf{h}) \simeq \frac{1}{2} + \frac{1}{2} \tanh \sigma_3 \sigma_2^{-3/2} |E_\mathbf{h}| \sum_\mathbf{k} E_\mathbf{k} E_{\mathbf{h}-\mathbf{k}} \quad (11)$$

where $P_+(\mathbf{h})$ represents the probability that the sign of $E_\mathbf{h}$, as determined by eq. (8), is positive and

$$\sigma_n = \sum_{j=1}^{N} z_j^n \quad (12)$$

where Z_j is the atomic number of the jth atom in a unit cell containing N atoms.

A suitable probability measure which can be associated with eqs. (9) and (10) is the variance of the phase angle $\varphi_\mathbf{h}$ whose value is determined from these equations using known values of other phases. The formula for the variance, V, is

$$V = \frac{\pi^2}{3} + [I_0(\alpha)]^{-1} \sum_{n=1}^{\infty} \frac{I_{2n}(\alpha)}{n^2}$$

$$- 4[I_0(\alpha)]^{-1} \sum_{n=0}^{\infty} \frac{I_{2n+1}(\alpha)}{(2n+1)^2}, \quad (13)$$

where I_n is the Bessel function of imaginary argument of order n,

$$\alpha = \left\{ \left[\sum_\mathbf{k} \varkappa(\mathbf{n}, \mathbf{k}) \cos (\varphi_\mathbf{k} + \varphi_{\mathbf{h}-\mathbf{k}}) \right]^2 \right.$$
$$\left. + \left[\sum_\mathbf{k} \varkappa(\mathbf{n}, \mathbf{k}) \sin (\varphi_\mathbf{k} + \varphi_{\mathbf{h}-\mathbf{k}}) \right]^2 \right\}^{1/2} \quad (14)$$

and

$$\varkappa(\mathbf{n}, \mathbf{k}) = 2\sigma_3 \sigma_2^{-3/2} |E_\mathbf{h} E_\mathbf{k} E_{\mathbf{h}-\mathbf{k}}|. \quad (15)$$

There is a variety of auxiliary formulas which can aid a phase determination. Certain ones are restricted to determining phases for centrosymmetric reflections from a knowledge of structure factor magnitudes, sigma-1, or from combinations of known phases for certain centrosymmetric reflections combined with structure factor magnitudes, sigma-3. An example of a formula of potentially general applicability, but, because of accuracy limitations, best reserved for computing the phases of real or pure imaginary structure factors, is $B_{3,0}$ (modified),

$$|E_{\mathbf{h}_1} E_{\mathbf{h}_2} E_{-\mathbf{h}_1-\mathbf{h}_2}| \cos (\varphi_{\mathbf{h}_1} + \varphi_{\mathbf{h}_2} + \varphi_{-\mathbf{h}_1-\mathbf{h}_2})$$
$$\simeq C(\mathbf{h}_1, \mathbf{h}_2) \langle (|E_\mathbf{k}|^p - \overline{|E|^p}) (|E_{\mathbf{h}_1+\mathbf{k}}|^p - \overline{|E|^p})$$
$$\times (|E_{\mathbf{h}_1+\mathbf{h}_2+\mathbf{k}}|^p - \overline{|E|^p}) \rangle_\mathbf{k}, \quad (16)$$

where $C(\mathbf{h}_1, \mathbf{h}_2)$ and the average term can be computed from the observed normalized structure factor magnitudes and p ordinarily assumes a value in the range 0·5 to 2. For special choices of $\mathbf{h}_1$ and $\mathbf{h}_2$, depending on the space group involved, the sum $\varphi_{\mathbf{h}_1} + \varphi_{\mathbf{h}_2}$ may be known, even though the separate phases may not be. The remaining phase $\varphi_{-\mathbf{h}_1-\mathbf{h}_2}$ is then defined by eq. (16) in terms of the structure factor magnitudes derivable from experiment. Evidently, if the values of $\varphi_{\mathbf{h}_1}$ and $\varphi_{\mathbf{h}_2}$ are known, $\varphi_{-\mathbf{h}_1-\mathbf{h}_2}$ is again defined in terms of these values and quantities known from experiment. The greater are the values of $|E_{\mathbf{h}_1} E_{\mathbf{h}_2} E_{-\mathbf{h}_1-\mathbf{h}_2}|$ and the number of terms contributing to the average, the greater is the accuracy with which eq. (16) can be computed.

Procedure for Phase Determination

In the initial stages of a phase determination, attention is restricted to the normalized structure factors of largest magnitude since they are the structure factors associated with optimal probability measures. The origin in a crystal is fixed by assigning phases to an appropriate set of $|E_\mathbf{h}|$ as determined by the nature of the origins which are suitable for a given space group and the extent to which a particular $\mathbf{h}$ enters into the combinations required by the right sides of eqs. (8)–(10). For noncentrosymmetric space groups it is also necessary to specify one of two enantiomorphs when they are distinct and a reference frame when the structures arising from changes in certain axis directions are distinct. These additional requirements are readily fulfilled by assigning a sign to the magnitude of the imaginary part of a suitable $E_\mathbf{h}$, once the origin in the crystal has been fixed. An enantiomorphous structure can be formed from a given structure by reflection through the origin. Evidently enantiomorphs are not distinct for crystals having a centre of symmetry. Even when they are distinct, both enantiomorphs ordinarily give the same set of X-ray diffraction intensities. There are special techniques for distinguishing enantiomorphs which make use of the anomalous dispersion of X-rays, thus leading to the determination of the absolute spatial configuration of a structure. The absolute spatial arrangement of atoms can be, for example, correlated with a particular direction of rotation for polarized light. The description of the details of origin, enantiomorph and reference frame specification for all the space groups is available in the literature and appears in the province of the theory of semin variants, those linear combinations of phases whose values or magnitudes, depending upon the space group involved, are determined by the observed X-ray intensities, once a functional form has been chosen for the structure factor. The values or magnitudes of the seminvariants, as the case may be, are independent of the choice of permissible origin.

Assuming now that appropriate specifications of phase values have been made (never exceeding four in number for any space group), the phase determination proceeds, employing eq. (8) or (9), and some additional symbols representing phase values are assigned to other large $|E_h|$ as required. The phases of most of the remaining larger $|E_h|$, for example, those larger than 1·5, are then defined by eq. (8) or (9) in terms of the specified ones and the unknown symbols. The corresponding probability measures are employed with the phase-determining formulae. For centrosymmetric crystals eq. (11) is used to determine the probability that each newly obtained phase is correct. The restriction that $P_+(\mathbf{h}) \geqq 0.98$ can generally be readily met throughout a phase determination. For non centrosymmetric crystals a working rule for accepting a phase indication is for the variance, eq. (13), to obey the condition $V \leqq 0.5$. The need to assign additional unknown symbols in the course of a phase determination arises from the impossibility of proceeding without violating the conditions set for the probability measures.

There are several ways to reduce the number of unknown symbols after the phase determination involving the larger $|E_h|$ has been completed. Relationships develop among the symbols in the course of the procedure permitting some symbols to be defined in terms of others. Auxiliary phase determining formulae may be used. Restrictions on the values of the symbols may be based upon known structural features and the extent to which certain values for the symbols may give rise to disagreements among the individual contributors to the phase-determining relations. The assignment of alternative values to the small number of unknown symbols remaining determines the number of Fourier maps which need to be examined.

In the case of non centrosymmetric crystals, eq. (9), is used to obtain the first fifty to one hundred phases, after which alternative values are assigned to the remaining unknown symbols, and the phase determination is continued with each of the alternatives employing the tangent formula, eq. (10). The phase determination is usually carried out until about twenty to thirty phases per atom in the asymmetric unit are obtained. Initial Fourier maps are then computed from these phases using eq. (1) with the exception that the coefficients are composed of the $|E_h|$ and the associated phases instead of the $|F_h|$. These maps are called E-maps and have the advantage of enhancing the resolution of the atomic positions. A correct structure can usually be recognized because it makes chemical sense, having appropriate bond lengths and angles. The structure is subsequently refined employing the complete set of data by means of least-squares procedures. The close agreement of the structure factor magnitudes computed from the refined structure with the measured ones affords a final confirmation of the correctness of the structure.

In carrying out the phase determination for many non centrosymmetric crystals, the first acceptable E-map that is obtained often contains only a portion of the structure. It is readily possible to develop the partial structure into a complete structure by means of the tangent formula. Another source of partial structures for further development by the tangent formula, particularly heavy atom partial structures, is the Patterson map. Phases are computed from the known atomic positions by a formula such as eq. (3) with an additional approximate correction for vibrational motion. A phase so computed is retained if the associated structure factor magnitude indicates that the partial structure makes a significant contribution to the measured magnitude and if the associated $|E|$ is large. The set of phases so obtained can be further expanded by use of the tangent formula, eq. (10), and this expanded set is used to compute a new E-map. More of the structure continues to appear as the process is continued until the determination is complete.

As a brief illustration of the manner in which a phase determination proceeds, the following sequence is taken from an investigation of the structure of arginine:

h	k	l	φ	h	k	l	φ
3	0	1	$\pi/2$	$\bar{3}$	0	10	π
3	3	0	$-\pi/2$	6	3	1	0
6	3	1	0	3	3	11	π
$\bar{3}$	0	$\bar{1}$	$-\pi/2$	3	3	0	$-\pi/2$
3	3	11	π	$\bar{3}$	0	10	π
0	3	10	$\pi/2$	0	3	10	$\pi/2$

In each of the four sums, the Miller indices above the line represent the vectors **k** and **h** − **k**, and their sum is evidently **h**, cf. eq. (9). The phases of reflections (3, 0, 1), (3, 3, 0) and (3, 0, 10) were known either from previous specification or determination. By following the sequence of combinations, it is seen how a phase value is obtained successively for reflections (6, 3, 1) and (3, 3, 11) and how these lead to two corroborative values for the phase of reflection (0, 3, 10). Good internal consistency may be expected when the reflections are associated with the larger $|E|$ values, as was the case here. In a step-by-step fashion, additional phases are determined and as the procedure continues the determinations involve an increasing number of contributors. When phases are defined in terms of symbols, relations among the symbols may develop if different symbols are involved in separate phase evaluations for the same reflection.

Several variations of the symbolic addition procedure for phase determination which has been outlined here are in use. They mainly concern the attempt to replace the initial use of symbols with numerical phases. This facilitates the immediate application of the tangent formula, eq. (10). The procedures involve the use of auxiliary formulas such as sigma-1 and $B_{3,0}$ in order to obtain numerical phases as well as the immediate use of alternative numerical values for certain of the phases. The use of symbols is designed to provide an efficient way to handle the large number of ambiguous alternatives which usually pertain to noncentrosymmetric crystals. If alternative numerical values are used exclusively, their number is limited by the number of alternatives that it is convenient to handle with a computer, given, of course, certain probability measures and rejection criteria for decreasing the number of E-maps which need to be ultimately examined. The use of symbols is also designed to avoid the difficulties that arise when the value obtained for a phase from an auxiliary formula is not sufficiently accurate. If this is used at the start of a determination, it will generally lead to an incorrect result. If it is used at the end of a phase determination to assign a value to a symbol, the error will be apparent and an alternative value can be given to the symbol without the need to repeat the entire phase determination. The manner in which symbols, numerical values, auxiliary formulas and computer programs are employed is at present a matter of personal preference in the various laboratories carrying out the direct determination of crystal structures. These direct methods are now widely employed particularly for structures involving atoms which do not differ greatly in atomic number.

Bibliography

AHMED, F. R. (Ed.), HALL, S. R. and HUBER, C. P. (Coeds.), (1970) *Crystallographic Computing*, Copenhagen: Munksgaard.

International Tables for X-Ray Crystallography, Vol. IV, Birmingham, England: The Kynoch Press (1974, in press).

KARLE, J. (1969) The phase problem in structure analysis, *Advances in Chemical Physics*, Vol. XVI, Prigogine, I. and Rice, S. A. (Eds.) London: Interscience Publishers.

STOUT, G. H. and JENSEN, L. H. (1968) *X-ray Structure Determination*, London: Macmillan.

JEROME KARLE

D

DETECTION AND IDENTIFICATION OF INDIVIDUAL ATOMS AND IONS IN MASS SPECTROMETRY. This article is concerned with atoms and ions of comparatively low energy (a few kilovolts or less) as encountered in conventional mass spectrometry and field-ion microscopy.

Detection Methods

Atoms and ions in this energy range have poor phosphor efficiencies, poor ionizing efficiencies in the gas phase, and very small penetration depths in solids. Consequently it is not possible to use the majority of detectors available to the nuclear physicist or the X-ray worker, such as the scintillator–photomultiplier combination, ionization or spark chambers, or surface barrier or solid-state detectors. It is necessary to find a detection method based on the physical processes occurring when the atom or ion strikes a solid surface. Such a process is the generation of secondary electrons; this has been intensively studied in recent years (Krebs, 1968). The attractions of this process for particle detection are threefold:

  (i) High primary detection efficiencies can be obtained at suitably prepared surfaces.
  (ii) The primary event can readily be amplified into an electron cascade, since the initial secondary electrons will themselves produce further secondaries upon acceleration and reimpact with the active surface.
  (iii) The resultant pulse output from the detector is readily compatible with modern digital or analogue electronic data-handling equipment.

Detectors based on this process are known as *windowless electron multipliers*. A wide variety of potentially useful electron emissive surfaces can be prepared, and a considerable range of detector geometries is also available. It suffices to illustrate this by reference to some examples of the more commonly available types. Linear-focused, "circular cage", "boxcar" and "venetian blind" detectors consisting of a number of discrete electrodes (dynodes) held at increasingly positive potential, have been available for a number of years (Morton, 1962, Pichanick, 1967). The electrons emitted from the surface of one stage are focused on to the next, and so on. The

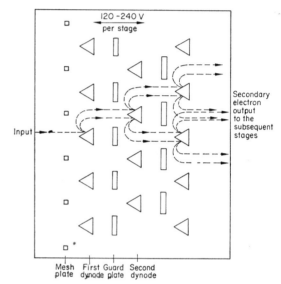

Fig. 1. Schematic diagram of Johnston MM-type particle detector (cross-section).

dynode material is usually an activated copper–beryllium compound. The overall gain of such a multiplier is dependent on the number of stages (usually 12–20) and the overall accelerating potential (usually 1000–3000 volts), and may be as high as 10^8. Another electrostatically focused detector with a discrete dynode structure, which also employs Cu–Be as the secondary emissive surface material, is the Johnston MM type electron multiplier, Fig. 1 (Barnett and Ray, 1970). This usually has twenty stages and is operated at 2000–5000 volts. A very high, stable gain is reported. The opaque dynode structure of this detector is claimed largely to prevent positive ion feedback, which is one of the main causes of instability at very high gains.

Continuous dynode surfaces in the form of strip or tube are employed in several detectors, the secondary electrons being constrained to make repeated collisions with the same surface. Since it is necessary to accelerate the secondary electrons between impacts, a high voltage must be maintained across the dynode surface. A metallic surface coating such as Cu–Be is clearly unsuitable for such an application, and it is customary to employ a highly resistive, semiconducting material such as a chemically reduced lead oxide glass. An example of this kind of detector is the channel electron multiplier (Wiley and Hendee, 1962; Adams and Manley, 1965; Cuthill, 1971), which is based on a straight or curved glass tube, with the addition of a fluted entrance horn to enhance the collection area. This relies on simple electrostatic acceleration to produce multiple collisions with the walls of the tube, and provides gains of 10^6–10^9 with applied potentials of 1·5 to 4·0 kV. A limit is imposed on the operation of these multipliers at high count rates because of space charge effects: the gain current should not exceed a few per cent of the standing (ohmic) current in the tube walls for optimum effectiveness. Another example of a continuous dynode detector is the magnetic strip multiplier (Heroux and Hinteregger, 1960; Cuthill, 1971), which employs crossed electric and magnetic fields to generate secondary electron trajectories in the form of a series of hops across the dynode surface. This type of detector has a good gain, a very fast transit time, and produces extremely sharp pulses, but a problem with its design (and with that of the channel electron multiplier) is that open dynode structures facilitate positive ion feedback, which in turn produces spurious after-pulses and background noise in the detector output (Macar et al., 1970).

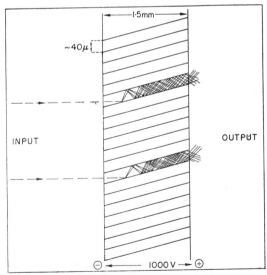

Fig. 2. Schematic cross-section through microchannel plate showing retention of positional information during multiplication process.

One very interesting recent development of the channel electron multiplier is the micro-channel plate, Fig. 2 (Guest et al., 1969). This is a two-dimensional array consisting of a vast number of microscopic channel electron multipliers, and is manufactured by techniques very similar to those used for the production of fibre optics faceplates, except that in this case the plate is deliberately cut *obliquely* to the axis of the channels, so that particles incident normal to the plate do not pass straight through without detection. Each channel is, typically, about 40 microns diameter and 1·5 mm long, and the overall plate diameter may be up to 12·5 cm. A conductive coating (nichrome) is evaporated onto each face of the plate, and the resulting combination has a gain of 10^3–10^4 at an applied potential of 1000 volts (though this gain is reduced by prolonged exposure to ion bombardment). Much higher gains may be obtained if two or more plates are operated in series, with a small vacuum gap between them (Owen et al., 1971). The detection efficiency is limited by geometric factors: a maximum of 68–70% "open area" has been achieved so far, though this may be increased somewhat by the introduction of square or conical tubulation. Ion feedback is a serious problem in individual plates, since each microtube is straight, but in series operation the effect can be reduced by arranging that the channels of the separate plates should be skew with respect to one another, thus eliminating the straight-through path for the feedback ions. The advantage of this kind of detector is its extreme sensitivity to the position of impact of the primary atom or ion, a sensitivity analogous to that of fibre optics. This positional information is lost in the detector systems described previously, unless some collimating arrangement is made which, of necessity, cuts out much of the incident beam. The positional information is of great interest, for instance in studies of atom- and ion-scattering processes, and in mass spectrometry.

Identification Methods

Practical methods are based on refinements of the well-known techniques of mass spectrometry (Palmer, 1962). Essentially two methods of mass separation of the ions may be employed—spatial or time dispersion. The first of these, spatial dispersion, is encountered in conventional magnetic sector or quadrupole spectrometers. Usually these instruments are provided with a simple collector-electrometer amplifier combination for measurement of ion currents. Recently, many instruments have become available in which the collector is replaced by one of the windowless particle detectors described above. These provide increased sensitivity and decreased noise levels, when operated in integrating (d.c.) mode, with electronics of a relatively long time constant. For ultimate sensitivity the detector should be operated in pulse mode, in which fashion the arrival of single ions of a particular mass number may be detected. The problem with this kind of spectrometer is that for accurate determination of mass, the incident beam collimation and detector system slit must be very fine. Hence the system has a very small effective aperture, and there is practically no possibility of identifying a *particular* incident ion unless, by chance, the system is prealigned at the correct mass number. It is in principle possible to use an array of detectors, individually prealigned for different mass numbers or a position-sensitive detector with appropriate output facilities, but at the time of writing systems of this kind having single particle sensitivity are still under development. For optimum accuracy, almost all modern spectrometers use a fixed detector geometry, and tune the electric and magnetic fields to align particular mass numbers at the detector aperture. Thus detector arrays are not really compatible with these instruments.

The alternative principle of mass spectrometry, time dispersion, is more promising for use at the single particle level. If the source ions of charge ne are accelerated through a well-defined potential E, they acquire kinetic energies $\tfrac{1}{2}mv^2$ given by

$$\tfrac{1}{2}mv^2 = neE$$

or

$$m/n = 2eE/v^2.$$

If these ions are now allowed to pass through an aperture in the cathode of the accelerating system, and enter a drift tube of length l, the times of flight along the tube t are related to the mass numbers of the ions by the formula

$$m/n = 2eEt^2/l^2.$$

Thus if a gating arrangement is provided at the entrance to the drift tube, such that ions can enter only at certain well-defined time intervals, and a particle detector is placed at the end of the tube (Fig. 3), a system is available which effectively has an infinitely wide "aperture" (in time) and which permits the identification of the mass number of any *particular* ion selected by the operation of the aperture and gating system. For ions of a few kilo-volts energy, the flight times along a metre-length path will be of the order of 1–10 microseconds. Times of this order may be measured using time-to-amplitude converters, or by means of a fast crystal-controlled digital clock. To minimize timing uncertainties it is desirable to use a detector having a relatively flat primary dynode surface, aligned normal to the incident beam. The main problems in quantitative work arise when more than one ion is admitted to the drift tube during a single opening of the gate. If two ions have the same mass number, they will reach the detector at the same instant, and will be recorded as a single, rather than a multiple, event. If the mass numbers are only slightly different, the second (heavier) ion may not be recorded, since there is usually a short dead-time in the electronics, introduced by the data transfer and storage operations. If the two ions are well separated in mass, these difficulties should not arise, in any particular gating cycle, but it is still necessary to apply statistical techniques in order to obtain a quantitative analysis from a completed spectrum.

The Atom Probe Mass Spectrometer

The most exciting embodiment of the principles discussed above is the atom probe mass spectrometer, invented by Professor E. W. Müller (Müller *et al.*, 1968). This instrument permits the imaging of the atoms on the surface of a solid specimen, using the principle of the field-ion microscope (Müller, 1961; Brandon, 1966) and the chemical analysis of selected individual surface atoms using pulsed field evaporation in conjunction with time-of-flight mass spectrometry. This analytical tool has potentially wide applications in surface studies and in materials science, and is described in detail elsewhere in this volume (*see*: Field-ion microscopy and the atom-probe FIM).

Bibliography

ADAMS, J. and MANLEY B. W. (1965) *Electronic Engineering* **37**, 180.
BARNETT, C. F. and RAY, J. A. (1970) *Review of Scientific Instruments* **41**, 1672.
BRANDON, D. G. (1966) Field-emission and field-ion microscopy; recent developments in, in *Encyclopaedic Dictionary of Physics*, Suppl. **1**, 103, Oxford: Pergamon Press.
CUTHILL, J. R. (1971) Soft X-ray spectroscopy, in *Encyclopaedic Dictionary of Physics* (J. Thewlis, Ed.), Suppl. **4**, 412, Oxford: Pergamon Press.
GUEST, A., HOLMSHAW, R. T. and MANLEY, B. W. (1969) *Mullard Technical Communications*, **97**, 210.
HEROUX, L. and HINTEREGGER, H. E. (1960) *Review of Scientific Instruments*, **31**, 280.
KREBS, K. H. (1968) Electron ejection from solids by atomic particles with kinetic energy, *Fortschritte der Physik*, **16**, 419–490.

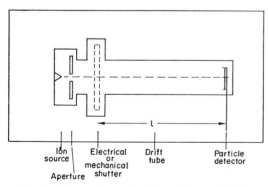

Fig. 3. *Schematic principle of time-of-flight spectrometer.*

MACAR, P. J., RECHAVI, J., HUBER, M. C. E. and REEVES, E. M. (1970) *Applied Optics*, **9**, 581.
MORTON, G. A. (1962) Photomultiplier tubes, in *Encyclopaedic Dictionary of Physics* (J. Thewlis, Ed.), **5**, 483, Oxford: Pergamon Press.
MÜLLER, E. W. (1961) Field emission, in *Encyclopaedic Dictionary of Physics* (J. Thewlis, Ed.), **3**, 120, Oxford: Pergamon Press.
MÜLLER, E. W., PANITZ, J. A. and McLANE, S. B. (1968) *Review of Scientific Instruments*, **39**, 83.
OWEN, L. D., SOLOMON, D. E., FLOREK, R. S. and THOMAS, N. C. (1971) *Bendix Corporation Electro-Optical Division Technical Report.*
PALMER, G. H. (1961) Mass spectrograph, mass spectrometer, in *Encyclopaedic Dictionary of Physics* (J. Thewlis, Ed.), **4**, 508, Oxford: Pergamon Press.
PICHANICK, F. M. J. (1967) in *Methods of Experimental Physics*, Vol. 4A, Eds. V. W. Hughes and H. L. Schultz, New York and London: Academic Press.
WILEY, W. C. and HENDEE, C. F. (1962) *I.R.E. Transactions of Nuclear Science*, N.S. **9**, 103.

G. D. W. SMITH

DIGITAL LEARNING COMPUTERS. A learning computer is a device that processes information without having been programmed to do so. Instead of programming the system, one gives it (during a "training" phase) examples of the processing task to be performed. A machine of this kind is only useful if it is capable of generalizing, that is if it can apply the acquired "knowledge" not only to the examples received during training but also to tasks that are similar in some way.

Digital Learning Elements

The "building brick" of digital learning machines is a device known as a SLAM (Stored-Logic Adaptable Microcircuit, Fig. 1). A 3-bit binary pattern is supplied to the input terminals. This provides the address of one of the eight storage elements into which a 0 or a 1 may be written via the "teach" terminal. There is one bit of storage for each possible input pattern. The setting of 0s or 1s into the storage locations is the teaching process. In use, the incoming patterns address their respective storage locations, and the stored 0 or 1 is "read out" at the output terminal of the element. An aspect of this device is that, even though it is basically a simple storage circuit, in use, it becomes a variable-function logic element, the function of which may be adjusted during training. The eight storage locations of SLAM-8 may contain 256 (2^8) different messages, each corresponding to an input/output function.

In a broad sense, these devices have variable-function properties similar to those of brain neurons with the exception that purely digital techniques are used and the devices do not generalize. A microcircuit SLAM-8 contains 189 metal–oxide–silicon elements. It is now feasible to manufacture a 4-input 16-bit SLAM-16, capable of 65,536 logic functions.

Networks of Learning Elements

These devices would be rarely used alone; it is the behaviour of networks containing them that is of great interest. A particular network that has been studied is the single-layer system shown in Fig. 2. The SLAM devices are connected at random to a "retina" which receives a pattern of black and white (i.e. 0 and 1) inputs. As patterns are applied to the retina during training, each SLAM "sees" a 3-bit random sample (i.e. 3-tuple) of the entire pattern. In a particular mode of training, the net may be started with all 0s in its stores, and then taught to respond with all 1s at its output for one set of input patterns. During subsequent use, each element of the net responds to a test pattern with a 1 if its sample has been "seen" during the training run.

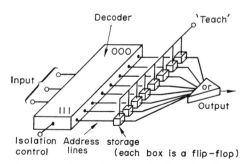

Fig. 1. *SLAM-8 is a purely digital variable-function logic circuit that may be used in learning nets equally well as an electronic neuron.*

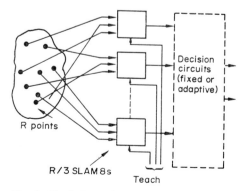

Fig. 2. *Single-layer learning network; this network has the property of generalization, since it responds correctly to patterns similar to those seen during training.*

Through this process the network generalizes. If the training patterns are similar, in the sense that the inputs to only a few SLAMS differ for any two training patterns, the "unseen" patterns that cause the net to respond with all 1s are similar to those in the training set in the same way. Typically, simulation has shown that an 80-SLAM-8 net, after being trained on sixty handwritten 3s, responds with either 79/80 or 80/80 1s as its output for unseen 3s, where no 4 causes such a high response. If, however, the training patterns are very dissimilar, even though they themselves will always be correctly recognized, strange combinations of these patterns will be clasfied in the same way. If such strange combinations are at all meaningful and likely to appear at the input of the machine, it could result in an error. Clearly, if the performance of the machine is satisfactory, the contents of its storage may be recorded and a fixed non-learning machine may be built to perform the same task.

Nets with Feedback

A well-known characteristic of data-processing machines where the output depends on sequences of input events (i.e. sequential machines) is that they can always be modelled as simple logic circuits (viz. combinational networks with no temporal behaviour) with feedback loops. These loops provide the temporal action and the delays in the loops determine the timing of this action. Such circuits can generate long sequences of messages even if the input is held constant.

The electronic learning nets discussed above, in common with learning nets developed by others, are generally combinational. It can be shown that by the addition of a set of feedback channels with built-in delay, sequential properties are added to the net. In other words, combinational nets behave in a "Stimulus-Response" way whereas the Response of feedback nets is a function of a sequence of stimuli. However, feedback may be explained in terms of "Stimulus-Response" behaviour by postulating an "Inner Response" which is fed back to the input and used as a subsequent stimulus. This is quite common in biological systems. For example, during speech perception an "Inner Response" consists of nerve signals that are transmitted to speech motor muscles. These signals are also perceived, forming a closed information loop. It is likely that much of what we call "thought" is due to this process. For example, when human subjects recognize patterns it is likely that after seeing a pattern, say a hand-printed A, recognition consists of an "Inner Response", that is, some inner generation of a model A which is compared with the seen A, and the decision on whether the input is or is not an A is based on this comparison.

Experiments have been performed that show just this type of behaviour in electronic networks of SLAMs where the output is fed back to the input. For example, a system has been trained to respond to perfect Ts and Hs. When tested with a distorted T the system responds with a less distorted T which when fed back brings about a response of an even less distorted T and so on until a perfect T is generated. Similarly, if a distorted H is presented, the system "oscillates" until a perfect H is generated.

The Processing Ability of Digital Learning Computers

Probably the most interesting form of behaviour being investigated in feedback nets is their ability to learn to generate very long sequences of data. This is a result of the fact that the fed-back "Inner Response "generates a new "Inner Response" which generates another and so on. Such sequences may be "learnt" from the environment and the learning of speech, language and music become clearly feasible in electronic equipment. Indeed, such electronic models can also explain the phenomenon that psychologists call *attention*, that is, a *wilful* ability to control one's thoughts.

A simple example of this type of processing is shown in Fig. 3. A net (SOPHIA, a 12-SLAM single-layer net) is trained by the continuous application of a vertical bar pattern shifted from left to right across its "retina". In subsequent testing the net responds to this sequence by outputting one very much like it (Fig. 3a). If the sequence is applied backwards, the system does not respond in a coherent way (Fig. 3b). It is seen that, during use, the net is triggered into an "oscillation" by the incoming information; we can say that recognition is taking place. If the input is not recognized, the net goes into an incoherent "oscillation". This may be analogous to the recognition of speech in humans where, in the first case, a heard message is "understood" and in the second case it leads the listener to say "What?" It also provides a mechanism analogous to the recall (in humans) of an entire song after only the first few bars have been heard.

A larger machine, MINERVA, is being equipped with 256 feedback loops and this leads to an estimate

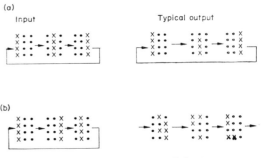

Fig. 3. Sequence sensitire behaviour.

that it will be able to remember about 10^{20} different sequences; about 10^5 "words" long where each "word" consists of 256 bits. Not only does this processing power outstrip the ability of very large computers at a fraction of their cost, but also it provides an interesting comparison with the processing power of the human brain.

It is estimated that in the lifetime of a human brain the total information "throughput" is about 2×10^{22} bits. We note that this is comparable to the capacity of MINERVA which is probably a tiny baby in comparison to the learning machines that could be built in the future.

Applications

Digital learning nets take the electronic machine into the realm of application to non-analytic problems in which Man has traditionally reigned supreme as the sole problem-solver. Typical of such problems are the recognition of writing and speech, the control of systems with complex dynamics (including artificial limbs), and all forms of pattern recognition. The latter means a great deal more than the recognition of, say, postal codes. It deals with processing of visual and photographic data of very many kinds. For example, it may be of medical use in the spotting of early signs of cancer and tumors, or of industrial use in the analysis of data from special instruments such as spectrometers and chromatographs. Many of these tasks are not only economically important in industry, but would also be of some benefit to people for whom it would be quite impossible to bear the cost of conventional computers, paraplegics being typical of such cases.

Applications in automatic control may also be reduced to pattern recognition tasks; the measurements taken on a plant being the "patterns" and the recognition signals being the desired control action on the plant. Typically, the "teacher" would have to be an existing controller, possibly human.

Digital learning computers, particularly those with feedback, have been most useful in providing a language with which to describe, in electronic terms, rather vague concepts such as "thought" and "awareness". It may well be that electronic "thought" may become as commonplace a piece of hardware as is the "memory" of conventional digital computers.

Specialist Terms

Combinational net: a net whose output is an immediate function of its present input. Such a net has no paths between internal outputs and inputs not available to the observer.

Feedback net: a net containing paths between internal outputs and inputs which the net connected to overall outputs and inputs. Such a net stores information and has its output dependent on the past history of the inputs.

Variable-function logic elements: an element which contains a memory. The message stored in the memory may be varied causing a variation of the input/output function of the element.

Bibliography

ALEKSANDER (1971) !*Microcircuit Learning Computers*, London: Mills & Boon.
ALEKSANDER (1973) Random logic nets: stability and adaptation. *Int. J. Man-Machine Studies*, **5**, 115–13.

I. ALEKSANDER

E

ELECTRET TRANSDUCERS. Are electrostatic transducers whose bias is supplied by a semipermanent electrostatic charge embedded in a dielectric (electret), whereas conventional electrostatic transducers require an external source of typically 200 volts as bias. An example of the latter is the condenser microphone which is used as a laboratory standard for acoustic measurements because of its excellent time and temperature stability, accurate frequency response and low harmonic distortion. The electret transducer is a member of the same family and has thus inherently the same capabilities. However, electret technology has not progressed to the stage where laboratory standard quality can be claimed.

The first electret transducer was made from one of Eguchi's (1919) carnauba wax electrets by Nishikawa and Nukiyama (1928; and Sessler and West, 1973). Applications as a microphone and electrostatic loudspeaker were also proposed a few years later by Gemant (1934, 1935) and Rutherford (1935). Dating back to the same early period are a proposed electret phonograph pickup (Parker, 1931, Euler, 1959) an and electret tape recorder (Rutherford, 1932). Electret transducers have also been reported (Gutman, 1948) during the early 1940s in Japanese field telephones.

However, interest in electret transducers has remained low, until recently, because of their poor stability and the cumbersome methods available to make carnauba wax electrets. In later years, electret transducers were made experimentally from thermoplastic materials (Gutman, 1948; Wiseman and Linden, 1953; Wieder and Kaufman, 1953; Euler, 1959), ceramics such as lead zirconate titanate (Gubkin and Skanavi, 1958, and more recently, aluminium oxide (Davies and Collins, 1969).

The needed breakthrough came in 1962 when Sessler and West reported their thin polyester film electret microphone (Sessler and West, 1962). Although the polyester film electrets still proved to be insufficiently stable in humid environments, work in this direction eventually led to the use of fluoroplastic film materials for the electret. High volume resistivity ($>10^{17}$ ohm-cm), high surface resistivity ($>10^{16}$ ohm/square), low water absorption ($<0.01\%$ per 24 hr) and a low rate of water vapour transmission (<0.4 g/100 in.^{2}/24 hr/mil) are common characteristics of the fluoroplastics such as fluorinated ethylene propylene (FEP), polytetra-fluoroethylene (PTFE) and chlorotrifluoroethylene (CTFE), which were found to give the most stable electrets (Sessler and West, 1966; Murphy and Fraim, 1968; Reedyk, 1971). Good results have also been reported with polyvinylidene fluoride (PVF_2) (Takamatsu and Fukada, 1969). Unfortunately, these materials are also subject to a high degree of stress relaxation so that special care must be taken in transducer design to compensate for this effect (Warren, 1971; Djuric, 1972; Madsen, 1973).

The first commercial application of electret transducers finally appeared in 1968 in the form of a built-in microphone in a portable tape recorder (*Electronics*, 1968). Low sensitivity to solid borne vibrations made it practical to mount the microphone and the tape-drive mechanism in the same case (Nomura and Imai, 1971). In the telephone industry, the electret microphone is of interest as a replacement for the carbon microphone in order to reduce power consumption and distortion and to improve stability. Field tests of electret microphones combined with a semiconductor amplifier in standard telephone sets (Reedyk, 1968, 1971) resulted in 1972 in the first application of the electret transducer in the telephone industry: a noise-cancelling electret microphone for telephone operator's headsets (Reedyk, 1973). Experimentally, electret key transducers have been made with a view to using these in electronic push-button dials and keyboards (Sessler and West, 1971; Reedyk, 1972). In these, an electret is placed between the plates of a capacitor, partially filling up the gap; a signal capable of operating an electronic switch results when the gap is closed by pushing on one of the plates of the capacitor. Also, in the late 1960s and early 1970s, electret transducers found their way to the market in the form of microphones for studio applications, hearing aids, sound-level measurements and earphones in a hi-fi-stereo headset.

Electret transducers offer, depending on the application, several advantages over other equally capable transducer systems: they can be made small yet have good sensitivity and signal-to-noise ratio; they have a wide dynamic range (typically from 15 dB to 150 dB ref. 2×10^{-5} Nm^{-2}) and they have a smooth response extending over a frequency range of several decades. Light plastic materials can be used and the construction of electret transducers is basically very simple so that they are easy to manufacture.

Another commercial application is a digitizer for graphic information. Two ultrasonic electret transducers in the form of strips are placed at right angles along the edges of a board. A special pen produces a spark when it touches the surface or a sheet of paper placed on top of the board. The X and Y coordinates of the pen tip are derived from the arrival times of the spark induced ultrasonic wave at the electret transducers along the edge of the board (Graphpen®).

A modern version of the electret phonograph pickup has also been described recently (Kawakami, 1969). The stylus operates via a mechanical lever, a small bellows, which is connected by means of a capillary, to a small cavity in front of the diaphragm of an electret pressure transducer. This arrangement provides the necessary mechanical advantage and acoustic impedance transformation to convert the movement of the stylus into a displacement of the electret diaphragm.

These applications of electret transducers were reviewed recently in a presentation to the Acoustic Society of America (Sessler and West, 1973; Reedyk, 1972). Other applications in research reviewed in some detail in the same papers are pressure transducers used in gas analysis, in shock tube experiments and to measure jet engine noise, infrasonic electret microphones with a lower cut-off of 10^{-3} Hz, ultrasonic electret transducers for the excitation and detection of waves up to 400 MHz in liquids and solids and a square array of 200×200 ultrasonic electret microphone elements used in acoustic holography.

The principle of operation of electret transducers follows from the theory for electrostatic transducers (Hunt, 1954). The equations of motion of electrostatic transducers contain the potential difference across the airgap as parameter. This is a quantity which cannot be measured in electret transducers but must be derived from the electret charge density (Cross, 1950; Sessler, 1963). It should be mentioned here that the electrodes of an electret transducer at rest are at the same potential due to the finite insulation resistance of the materials used to separate backplate and conductive diaphragm. The potential drops across the airgap and the electret will be equal and of opposite polarity in equilibrium. A simple method to determine the electret charge density, and from this with the aid of electrostatic field theory the potential drops across airgap and electret, is to apply an equivalent bias voltage (Reedyk and Perlman, 1968) of such polarity that the field in the airgap is cancelled (see Fig. 1). Equivalence of the electret surface charge density and bias voltage is simply observed with a null indicator (for example, an amplifier and oscilloscope) while the moving electrode or diaphragm is acoustically or mechanically excited. It follows that the same equivalent bias voltage (but of opposite polarity) applied to an uncharged transducer will result in the same sensitivity as when the dielectric layer of the transducer carries the equivalent surface charge density as shown in Fig. 1. A simple expression [eqn. (1)] to relate the electret surface charge density (σ) and the equivalent bias voltage (V_B) results when plan parallel operation is assumed and fringing is neglected.

$$\sigma = \frac{\varepsilon_r \varepsilon_0}{b} V_B. \qquad (1)$$

Here, b = thickness of the electret, ε_r = relative dielectric constant of the electret material and ε_0 = permittivity of free space. It should be noted that σ is the *net* surface charge density and is the result of all charges, polarization charges and real charges. The fluoroplastic materials used at present to make electrets exhibit very little dielectric absorption and therefore have practically speaking no long-term "frozen-in" polarization charges so that the electret charge is almost entirely due to a space charge as a result of trapped real charges. The effect of polarization charges may thus be ignored for all practical purposes in present electret transducers. Materials such as carnauba wax, polyesters and polycarbonates exhibit considerable dielectric absorption and as a result electrets made of these materials frequently show a reversal of polarity several days after fabrication (Gross, 1961).

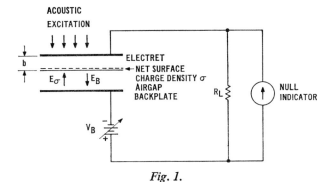

Fig. 1.

From eq. (1) and electrostatic field theory follows the relation between the voltage drop V_a across the airgap, the equivalent bias voltage V_B and the electret surface charge density σ:

$$V_a = -\frac{\varepsilon_r a}{b + \varepsilon_r a} V_B = -\frac{ba}{b + \varepsilon_r a} \frac{\sigma}{\varepsilon_0} \qquad (2)$$

where a = thickness of the airgap. Equation (2) then makes it possible to relate the equations of motion for the electret transducer [(eqs. (3), (4), (5) and (6)] with those for the externally biased electrostatic transducer.

It is useful to review the different forms in which electret transducers can be made, before discussing the equations of motion. The three basic configurations are:

1. The electret is attached to the fixed electrode. This is historically the oldest form in which carnauba wax electret transducers were made. A present example is the earlier mentioned electret earphone, which has 1-mm thick electrets held against the fixed electrodes without an airgap; a thin metallized polyester diaphragm is placed in the centre of the gap between the electret-electrode assemblies in a push–pull arrangement. The electrodes and electrets are perforated to make them transparent for sound. This configuration has the advantage that the mechanical and electrical requirements can be separated. The material with the best electrical properties is selected as the electret material, whereas a suitable material for the diaphragm is selected for its mechanical properties (Reedyk, 1971).

2. The electret is a thin metallized film which also serves as the diaphragm and the moving electrode (Sessler and West, 1962). This is the form in which most electret transducers available today are being made. It has the advantage that the electret film can be charged in a continuous process (Cresswell and Perlman, 1970). On the other hand, special care must be taken in the design (Warren, 1971; Djuric, 1972; Madsen, 1973) of these transducers to compensate for stress relaxation of the fluoroplastic films, which are selected for their stability as an electret. Usually, the backplate has a number of raised points or ridges of from 0·0005 in. to 0·002 in. high to support the diaphragm. In this manner the diaphragm can be stiffened without applying a high tension to prevent collapsing due to the electrostatic forces of attraction.

3. The electret is not metallized and is suspended in the gap between two fixed electrodes (Brit. Patent, 1968), which for electroacoustical applications, are perforated (see Fig. 2). This is not only the most general form from which the other configurations can be derived by letting one of the gaps go to zero and allowing one electrode to move with respect to the other, it also has the distinction of being a perfectly linear transducer; there are no harmonics of the fundamental frequency generated as will follow from the discussion below.

Although theoretically the electric fields and forces have been studied, and a model to demonstrate the principle was easily made by sandwiching a 0·001-in. TFE electret between two sintered bronze disks, there appear to be no existing applications.

The equations of motion can be derived from the theory for the push-pull electrostatic loudspeaker (Hunt, 1954) by letting the time constant for the charging circuit go to infinity; however; it is more straightforward to derive the theory directly from the electrostatic field equations (Gross, 1950; Wiseman and Feaster, 1957; Sessler, 1972) and the balance of mechanical and electrical forces acting on the electret diaphragm. The equations of motion following the notation explained in Fig. 2 and below are then:

$$e = LS \frac{d^2 \sigma_{i1}}{dt^2} + RS \frac{d\sigma_{i1}}{dt} + \frac{\sigma}{\varepsilon_0}\left(a_2 + \alpha \frac{b}{\varepsilon_r}\right)$$
$$+ \frac{\sigma}{\varepsilon_0} x + \frac{d}{\varepsilon_0} \sigma_{i1}, \qquad (3)$$

$$f_a = l_m \frac{d^2 x}{dt^2} + r_m \frac{dx}{dt} + \frac{x}{c_m} + \frac{\sigma S}{\varepsilon_0} \sigma_{i1} + \frac{\sigma^2 S}{2\varepsilon_0}. \qquad (4)$$

Here, $x = x(t)$, the displacement of the diaphragm, and $\sigma_{i1} = \sigma_{i1}(t)$, the charge density induced on electrode 1, are both functions of time; L and R are self-inductance and resistance in the external

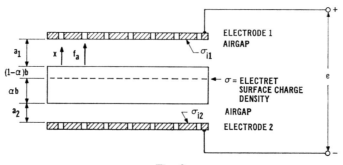

Fig. 2.

circuit; f_a = applied force due to acoustic pressure; l_m = mass of the diaphragm; r_m = mechanical resistance; c_m = mechanical compliance, S = area of diaphragm. σ_{i2}, the charge density induced in electrode 2, has been eliminated by means of the expression $-\sigma_{i2} = \sigma + \sigma_{i1}$ which follows from the field equations.

Equations (3) and (4) take into account that the polarity of the induced charge densities σ_{i1} and σ_{i2} is the opposite of the polarity of the electret charge density σ.

It is assumed here that the volume distribution of the electret charge can be represented by a uniform surface charge density σ at some depth below the surface. Ideally, for this transducer, the volume charge density should be uniform or the surface charge density σ concentrated in the mid-plane of the foil. In that case, α would be equal to $\frac{1}{2}$ so that the transducer would be perfectly symmetrical. Noting also that $d = a_1 + a_2 + b/\varepsilon_r = $ constant, it is seen that the equations are linear so that no harmonic distortion of the applied signal is introduced by the electromechanical coupling.

In a practical situation, the electrical eq. (3) includes the inductance and resistance term and the force eq. (4) the inertia of the diaphragm and its airload as well as mechanical damping. Since these terms do not play a role in the actual transduction process, they may be left out and reintroduced at a later stage [see eqs. (11), (12) and Fig. 3]. Higher-order terms do appear, however, in the following expressions for the single-sided versions of the electret transducer. For example, by letting $a_2 = 0$ and allowing one of the electrodes to move the equations of motion are found to be:

$$e = \frac{\sigma \alpha b}{\varepsilon_0 \varepsilon_r} + \frac{\sigma_{i1}}{\varepsilon_0}\left(a + \frac{b}{\varepsilon_r}\right) - \frac{\sigma_{i1} x}{\varepsilon_0}, \quad (5)$$

$$f_a = \frac{x}{c_m} - \frac{S}{2\varepsilon_0}\sigma_{i1}^2. \quad (6)$$

As a result of the second-order terms with $\sigma_{i1} x$ and σ_{i1}^2 higher harmonics are generated. However, the effect of this non-linearity is small and is known to intro-

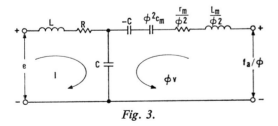

Fig. 3.

duce less than 1% second harmonic distortion for standard condenser microphones at sound pressure levels of 135 dB (ref. 2×10^{-5} Nm^{-2}) (Sessler and West, 1962; Djuric, 1972; Beranek, 1949).

Returning now to the push–pull transducer, it is recognized that eqs. (3) and (4) have well-known solutions, which are found by expressing the variables in a Fourier series (Hunt, 1954). Because the equations are linear, there are only the zero-order and first-order solutions, which give the conditions for static equilibrium:

$$\sigma_{10} = -\frac{\frac{2}{d}\left(a_2 + \alpha \frac{b}{\varepsilon_r}\right) - \frac{S\sigma^2 c_m}{\varepsilon_0 d}}{1 - \frac{S\sigma^2 c_m}{\varepsilon_0 d}} \frac{\sigma}{2}, \quad (7)$$

$$x_0 = \frac{S\sigma^2 c_m}{\varepsilon_0 d} \frac{a_2 + \alpha \frac{b}{\varepsilon_r} - \frac{d}{2}}{1 - \frac{S\sigma^2 c_m}{\varepsilon_0 d}} \quad (8)$$

Here, $S\sigma^2 c_m/\varepsilon_0 d$ is the electromechanical coupling factor (Hunt, 1954), which must be less than 1 or the system becomes unstable and the diaphragm collapses against one of the electrodes. Notice that the static displacement x_0 is identically zero if $\alpha = \frac{1}{2}$, and can be made equal to zero by adjusting the airgaps a_1 and a_2 if $\alpha \neq \frac{1}{2}$. The conditions for small signal operation, which follow from the first order solutions of eqs. (3) and (4), are:

$$e = \frac{\sigma}{j\omega \varepsilon_0} v + \frac{d}{j\omega \varepsilon_0 S} I, \quad (9)$$

$$f_a = \frac{1}{j\omega c_m} v + \frac{\sigma}{j\omega \varepsilon_0} I, \quad (10)$$

where $I = d\sigma_{i1}/dt$, S is the current in the electrical mesh and $v = dx/dt$ is the average velocity of the diaphragm. The mechanical force equation (10) can be expressed in terms of a voltage f_a/φ and a current φv, where $\varphi = \sigma S/d$ is the transduction coefficient (Hunt, 1954). Setting also $\varepsilon_0 S/d = C$, the capacitance of the transducer and adding serially inductance L, resistance R, the mechanical mass l_m of the diaphragm and airload and mechanical damping resistance r_m a general electrical equivalent results in eqs. (11) and (12), from which the equivalent circuit of Fig. 3 can be drawn:

$$e = \left(j\omega L + R + \frac{1}{j\omega C}\right) I + \frac{1}{j\omega C}\varphi v, \quad (11)$$

$$\frac{f_a}{\varphi} = \frac{1}{j\omega C} I + j\omega \frac{l_m}{\varphi^2} + \frac{r_m}{\varphi^2} + \frac{1}{j\omega \varphi^2 c_m}$$
$$- \frac{1}{j\omega C}\varphi v. \quad (12)$$

From the conditions for electromechanical stability, it follows that $\varphi^2 c_m$ must be smaller than C so that the negative capacitance $-C$ can be combined with the electrical equivalent $\varphi^2 c_m$ of the compliance in an effective electrical equivalent compliance $\varphi^2 c'_m$

by setting

$$\frac{1}{\varphi^2 c_m'} = \frac{1}{\varphi^2 c_m} - \frac{1}{C}.$$

For the single-ended transducer, a similar solution is obtained (Sessler, 1963). The capacitance C, however, is then modified by the static displacement x_0, whereas in the push–pull version, C is independent of the displacement of the diaphragm.

Bibliography

BERANEK, L. (1949) *Acoustic Measurements*, Wiley, N.Y., p. 220.
Brit. Patent 610,297 (1968).
CRESSWELL, R. A. and PERLMAN, M. M. (1970) *J. Appl. Phys.* **41**, No. 6, 2365.
DAVIES, L. W. and COLLINS, R. E. (1969) *Electronics Letters*, **5**, No. 19, p. 462.
DJURIC, S. V. (1972) *J. A. S. A.*, **51**, 129.
EGUCHI, M. (1919) *Proc. Phys. Path. Soc. Japan*, ser. 31, 326.
Electronics (1968) **41**, No. 26, 133.
EULER, J. (1959) *Elektrotechn. Zeitschr.-B*, Bd. **11**, H. 9, 359.
GEMANT, A. (1934) Brit. Pat. 438,672.
GEMANT, A. (1935) *Phil. Mag.* S. 7, **20**, No. 136, Suppl., p. 929.
Graphpen®, Science Accessories Corp.
GROSS, B. (1950) *Brit. J. Appl. Phys.* **VI**, p. 259.
GROSS, B. (1961) Electrets, in *Encyclopaedic Dictionary of Physics* (J. Thewlis, Ed.), **2**, 623, Oxford: Pergamon Press.
GUBKIN, A. N. and SKANAVI, G. I. (1958) *Bull. Acad. Sciences*, USSR Phys. Series, **22**, 327.
GUTMAN, F. (1948) *Rev. Mod. Phys.*, **20**, No. 3, 457.
HUNT, F. V. (1954) *Electroacoustics*, Harvard University Press, Cambridge, Mass., John Wiley, N.Y.
KAWAKAMI, H., 1969 Audio Eng. Soc., Preprint No. 693 (B-3), also U.S. Patent 3,649,775.
MADSEN, H. S. (1973) *J. A. S. A.*, **53**, No. 6, 1616.
MURPHY, P. V. and FRAIM, F. W. (Oct. 1968) *J. Audio Eng. Soc.*
NISHIKAWA, S. and NUKIYAMA, D. (1928) *Proc. Imp. Acad. (Tokyo)* **4**, 290–91.
NOMURA, M. and IMAI, N. (1971) *Jap. Electr. Ind.* **18**, No. 11, 50.
PARKER, O. B. (1931) U. S. Patent No. 1,804,364.
REEDYK, C. W. (1968) *Electrochem. Technol.*, **6**, No. 1–2, 6.
REEDYK, C. W. (1971) *IEEE Trans. Audio and Electroac. AU-19*, No. 1, 1
REEDYK, C. W. (1972) 83rd Meeting Ac. Soc. America.
REEDYK, C. W. (1973) *J. A. S. A.*, **53**, No. 6, 1609.
REEDYK, C. W. and PERLMAN, M. M. (1968) *J. Electrochem. Soc.* **115**, No. 1, 49.
RUTHERFORD, R. T. (1932) U.S. Patent No. 1,891,780.
RUTHERFORD, R. T. (1935) U.S Patent No. 2,024,705.
SESSLER, G. M. (1963) *J. A. S. A.*, **35**, 1354.
SESSLER, G. M. (1972) *J. Appl. Physics*, **43**, No. 2, 405.
SESSLER, G. M. and WEST, J. E. (1962) *J. A. S. A.*, **34**, No. 11, 1787.
SESSLER, G. M. and WEST, J. E. (1966) *J. A. S. A.*, **40**, No. 6, 1433.
SESSLER, G. M. and WEST, J. E. (1971) *Proc. 7th Int. Congress on Acoustics*, p. 413.
SESSLER, G. M. and WEST, J. E. (1973) *J. A. S. A.*, **53**, No. 6, 1589.
TAKAMATSU, T. and FUKADA, E. (1969) *Rep. Progr. Pol. Phys. Japan*, XII, 417.
WARREN, J. E. (June 1971) Doctor's Thesis, Purdue Univ.
WIEDER, H. H. and SOL KAUFMAN (1953) *Electrical Engineering*, **72**, 511.
WISEMAN, G. G. and LINDEN, E. G. (1953) *Electrical Engineering*, **72**, 869.
WISEMAN, G. G. and FEASTER, G. R. (1957) *J. Chem. Physics*, **26**, No. 3, 521

C. W. REEDYK

ELECTRIC MOTORS USING TRANSVERSE FLUX. The history of rotating electrical machines is largely a story of shape and during the years 1831 (the date of Faraday's discoveries in electromagnetism) to 1888 (when Tesla invented the induction motor) the pioneers of electrical engineering tried many different physical dispositions of magnets and conductors. Faraday himself used the phrase "magnetic circuit" to describe the path taken by magnetic fluxes and went on to comment that the design of magnetic circuits could be likened to the problem of designing electric circuits using bare copper wire in a bath of salt water. He was fully aware that there was no magnetic equivalent of the electric "insulator" and that magnetic flux was free to leak from all parts of the circuit. As a result of this fact alone the electric circuits of machines have, in the main, consisted of multi-turn coils of relatively thin wire whilst the magnetic circuits have all consisted of a single turn, short and thick.

By the turn of the century the shape of electric motors had become virtually fixed and cylindrical symmetry was an essential part of it. The magnetic circuits lay in planes normal to the shaft of the machine whilst the electric circuits substantially conformed to the surface of the cylinder which formed the magnetic poles. The idea of splitting and "unrolling" a rotating-machine structure to produce straight-line motion directly in a "linear motor" was first proposed by Wheatstone in 1845 and although he even made such a machine, an economically practical application was not found for linear motors until the advent of the liquid metal pump for the moving of radio-actively contaminated sodium/potassium mixtures in nuclear reactors.

The earlier linear motors were almost invariably copies of their rotary counterparts and one important feature of electromagnetism not readily exploitable in rotary machines was thereby neglected and it was this: There are no rules of electromagnetism which state that either the electric or the magnetic circuit of a machine should lie in a plane.

From this point of view it could be argued that all conventional rotary machines and their linear equivalents are essentially making use of their magnetic flux in only one dimension, i.e. the flux either *enters* or *leaves* the secondary surface normally.

The exceptions to this argument occured from time to time, as it were, by accident, rather than as the result of any deliberate attempt to exploit the other two dimensions. An example of this occurred during World War II when aircraft generators were required to run at higher and higher speeds, for greater output power/weight ratio. By leading the flux into the rotor normally across the airgap and extracting it *axially* across a second disc-like airgap at the end of the machine it was possible to locate both exciting coil and output coils on the stationary part of the generator, thus enabling a much more robust and simple construction to be used for the rotating member, which could accordingly withstand higher centrifugal stresses.

During the 1960s, interest in the use of linear motors as propulsion units for high-speed ground transport (speeds of the order of 250–300 m.p.h. being envisaged) was revived (it having been first proposed in 1905 by Zehden).† The only large linear induction motor which had previously been built was that machine known as the "Electropult" built by Westinghouse in the U.S.A. and used to launch aircraft at speeds of up 225 m.p.h. In this case, as in almost all subsequent transport proposals, the primary coil system was mounted on the vehicle to reduce both the amount of input power and the cost of the track to a minimum. The Electropult developed some 10,000 h.p. and is fully described elsewhere.‡

The speed of a travelling magnetic field is determined only by the coil spacing in the primary winding and by the frequency of the a.c. supplied to it (f). Thus the field pattern travels one wavelength λ in one cycle of the supply and the field speed v_s is thus given by $v_s = \lambda f$ or as is more commonly used, $v_s = 2pf$, where p is the pole pitch in the direction of motion. In the driving of high-speed vehicles considerable power (several MW) is required and it is clearly desirable to use existing power supplies directly (50 Hz in the U.K., 60 Hz in the U.S.A.). For 250 m.p.h., however, this involves motors having a pole pitch of 1·5 m or more and the end windings of the primary coils dominate the design, not only presenting an acute problem in how to accommodate such a useless bulk of copper but how to reduce the excessive primary leakage flux from such portions of the electric circuit.

Relief from these problems was initially sought through the use of higher frequencies, the input mains supply being "processed" before being fed to the linear motor by means of solid-state electronic circuits which could convert, for example, d.c. or single-phase a.c. at 50 Hz into variable frequency 3-phase a.c. for the motor feed. Initial designs of such invertors were disappointing however, an invertor for a 30-ton vehicle weighing at least some 13 tons, and electrical engineers turned their attention back to the problem of changing the basic shape of the linear motor.

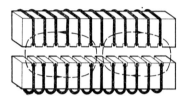

Fig. 1. Magnetic circuits and coil arrangement in a double-sided linear motor with Gramme-ring windings and longitudinal flux.

In France there was a return to the Gramme-ring type of primary coil in which the return conductors were taken around the "outside" or non-active surfaces of the machine as shown in Fig. 1. This topological change had the effect of easing the problem of primary winding design but increasing primary leakage flux still further. In addition it did nothing to help the magnetic circuit (the weaker of the two) and the whole of the flux from one pole still had to be accommodated in the core of the machine (also as shown in Fig. 1) resulting in large masses of track steel (in a single-sided motor) and a very low primary power/weight ratio.

The breakthrough came when it was realized that the electric circuit of a "conventional" linear motor could be made acceptable for pole pitches of 1·5 m by considerable "short-pitching" of the primary coils. (This gives rise to additional primary resistive losses but these can be dissipated by careful design.) In this case the *magnetic* circuit could be turned through 90° to give a transverse flux arrangement as illustrated in Fig. 2.

In (a) the plan view of the poles of a linear motor shows that the conventional pole structure calls for the whole of the flux in pole N to be transferred to the area S via a path parallel to the direction of motion, but in (b) there are effectively two motors, side-by-side, and the flux from each pole is transferred *laterally* from N. to S. pole and the resulting core depth is thus entirely independent of pole pitch (i.e. wavelength) and therefore of speed, even though a frequency of 50 Hz is retained for the supply.

---

† Zehden, A. (1905), U.S. Patent 732,312.

‡ A wound rotor motor 1400 ft long, *Westinghouse Engineer* (1946, **6**, 160).

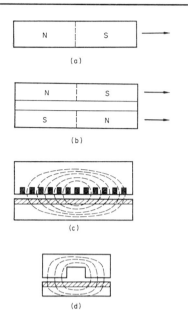

Fig. 2. Development of a transverse flux motor. (a) Plan view of the poles of a conventional longitudinal flux machine. (b) Plan view of the poles of a transverse flux motor. (c) Flux paths in the core of a longitudinal flux motor (single-sided) viewed from a side parallel to the direction of motion. (d) End view of a transverse flux, single-sided motor.

Figure 2(c) shows a side view of the conventional motor illustrating the depth of steel core, behind the slots, which is needed to carry the total flux from a complete primary pole. This depth is required in both primary and track secondary, the latter being by far the more expensive since the full depth would be required all along the route. Contrast this with the end view of a transverse flux machine as shown in Fig. 2(d) in which the core depth above the slots is only large enough to contain the flux from the *width* of the pole which may be only 15 cm. From Fig. 2(c) the core depth required is proportional to the pole pitch (about 1·5 m) and the amount of track steel for the transverse flux motor is but a small fraction of that which would be required in the longitudinal flux, i.e. conventional design.

Other attractive features of transverse flux machines include their ability to produce repulsive forces between primary and secondary over certain ranges of speed and in particular, at low speed, the primary may be capable of lifting more than its own mass above the track. In such circumstances the machine can be so designed that the levitated member is stable in pitch, roll and yaw and in lateral and vertical displacement. A further feature is that it is possible to design transverse flux machines so that the end windings of the primary coils are embedded in primary steel and thus contribute to the induction process without the introduction of large leakage fluxes which give rise to low power factor and poor performance.†

In one design, the primary unit contains no slots whatever, the whole of the primary conductors being located in the airgap as an interwoven "mat" of conductors, none of which protudes beyond the extent of the primary or secondary steel.†

The steel core of a transverse flux motor should be transversely laminated, a feature which can sometimes be relaxed in the case of secondary track in which the core depth may be of the order of 1 cm or less, in which case a solid sheet of steel may be used without too great a penalty in magnetizing current due to skin effect.

The whole philosophy of transverse flux design is based on the single idea that the magnetic circuit is *weaker* than the electric and must therefore be designed to exploit the most favourable axis of the motor. This, of course, applies equally well to all types of electromagnetic device but manufacturing restrictions often outweigh the advantages which could otherwise be enjoyed. Figure 3, for example, shows the elaborate design for a transverse flux rotary machine, in which the "width" (normally referred to as "the length" of the active part of the pole system) is greater than the pole pitch. The transverse flux arrangement is therefore seen to be unprofitable, whilst in contrast the linear motor lends itself so well to the transverse flux technique that restrictions on magnetizing current and primary leakage reactance can be relaxed. This in turn allows the designer to seek other gains, such as weight or track-cost reduction which can be achieved by designing for the magnetic circuit to be saturated to a considerable degree and to contain solid, as well as laminated, parts.

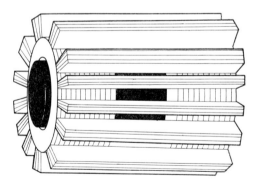

Fig. 3. Arrangement of a transverse flux rotary motor.

---

† Eastham, J. F. and Laithwaite, E. R. (1973) Linear-motor topology, *Proc. IEE*, **120**, 337.

Transverse Flux Tubular Motors

So far, only flat linear motors with transverse flux have been described. But it is also possible to apply the technique to advantage in the design of tubular linear motors, an invention due to Dr. J. F. Eastham in 1971.† In the conventional tubular motor

Fig. 4. A longitudinal flux tubular motor consisting of a row of simple coils.

as shown in Fig. 4, the primary windings consist of a row of coaxial coils, suitably phase-spaced to give the required synchronous speed. Since flux emanates in all directions on the *outside* of the windings, no steel is required to complete the external magnetic circuit in most cases, for the area through which flux may return is very large.‡ However, this beneficial two-dimensional spreading of the flux from the outer cylindrical surface carries with it a penalty also, for the secondary core *inside* the windings must be laminated in two dimensions and the core should therefore consist of a bundle of wires of ferromagnetic material rather than a stack of steel strips, as in transformer constructions.

The worst feature of a conventional tubular motor, however, is illustrated in Fig. 5 from which it can be appreciated that the whole of the flux from a cylindrical pole must be returned *axially* along the core so that the ratio of peak working (radial) flux density B_g to peak core flux density B_c, for sinusoidal excitation, is given by the equation

$$(2/\pi)(2\pi rp) B_g = \pi r^2 B_c$$

or

$$B_g/B_c = \pi r/4p. \tag{1}$$

A transverse flux machine which produces both twist and axial thrust is shown in Fig. 6. Tracing out a path around the circumference of the winding at any section reveals a two-pole rotating magneto-motive force pattern, and although the phase relationship around the section A (shown dotted) is not the same as that at section B, nevertheless the flux inside the winding is at all times radial in all positions implying that this fact can be exploited in two ways:

(i) The core can consist of a series of coaxial laminar discs, as is the case in a conventional rotating machine.
(ii) The core section is independent of pole pitch and the dimensions of the machine are not restricted by equation (1) above.

Outside the machine the flux has the possibility of returning in both radial planes and in planes

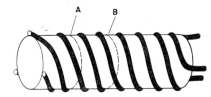

Fig. 6. Windings of a transverse flux tubular motor which produces both torque and axial thrust.

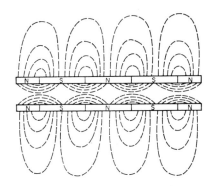

Fig. 5. Cross-section through a longitudinal flux tubular motor showing the constriction of the flux paths inside the windings.

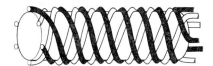

Fig. 7. A transverse flux primary consisting of a double-helical winding which produces only axial thrust.

---

† Eastham, J. F., U.K. Patent Application Nos. 59,138/71 and 59,162/71.

‡ Laithwaite, E. R. (1968) Some aspects of electrical machines with open magnetic circuits, *Proc. IEE*, **115**, 1275.

Fig. 8. One coil of a transverse flux tubular motor.

normal to the axis, making the effective reluctance outside the windings even lower than that of the conventional tubular motor.

Tracing out a path along the primary coil surface parallel to the axis also reveals a travelling m.m.f. pattern which gives rise to the thrust. Where thrust alone is required, the primary coil system can be constructed as a double layer in which the one layer has an oppositely-handed helical form in relation to the other, as shown in Fig. 7. This machine still exhibits a totally transverse flux pattern within the winding but produces no twisting action.

Instead of a double helical winding, the primary conductor system may be built from a number of elliptical coils, such as that shown in Fig. 8, subsequent coils of the system being lapped one over the other, as in a normal a.c. machine two-layer winding.

Transverse tubular motors have been built and shown to give many times the mechanical thrust per input current loading, and per weight of core, of their conventional tubular motor counterparts.

Bibliography

EASTHAM, J. F. and ALWASH, J. H. (1972) Transverse flux tubular motors, *Proc. IEE*, **119**, 1709.

LAITHWAITE, E. R. (1971) *Linear Electric Motors*, Mills & Boon.

LAITHWAITE, E. R., EASTHAM, J. F., BOLTON, H. R. and FELLOWS, T. G. (1971) Linear motors with transverse flux, *Proc. IEE*, **118**, 1761.

E. R. LAITHWAITE

ELECTROGRAVIMETRIC ANALYSIS

Introduction

This is the oldest form of electro-analysis and involves the quantitative deposition of an element onto an electrode from an aqueous solution of its ions. The elements are predominantly metals and include: copper, silver, gold, mercury, bismuth, antimony, arsenic, tin, lead, nickel, cobalt, cadmium, zinc, platinum, palladium, rhodium, iridium and tellurium. Selenium has been co-deposited with copper; molybdenum, vanadium and uranium have been deposited as hydroxides on the cathode, and lead, manganese and thallium as higher oxides on the anode. Halogens also have been deposited as silver halides on a silver anode and sulphur as sulphide on a cadmium anode. The principal uses, however, involve the deposition of the metal cathodically. Such procedures are usually clean with a minimum of chemical manipulation and reasonable accuracy.

General Principles

For analytical purposes the deposition must be complete and must involve only a single deposition process, i.e. no other species should be co-deposited within the range of potentials employed. When electrolysis is carried out in a suitable cell the small current (residual current due to impurities, dissolved oxygen, etc.) suddenly beings to increase rapidly at the Decomposition Voltage V (see Fig. 1).

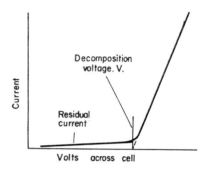

Fig. 1.

If the solution is unstirred the current will quickly fall due to depletion of the metal ions, M^{z+}, in the region of the cathode and an increase in voltage is necessary to maintain the current. This disturbance of potential is known as concentration polarization. This effect is progressive and eventually hydrogen will be evolved leading to spongy deposits which may be non-cohesive. Such deposits will also introduce adsorption errors. To overcome concentration polarization the solution is vigorously stirred either by some form of paddle or by rotating the cathode or anode. The latter procedure is generally adopted. This stirring increases the mass transfer to the electrode surface, keeps the potential steadier, allows higher currents to flow and a greatly reduced time of electrolysis. The potential, V, has no theoretical significance though it may have practical value. It is related to the individual electrode potentials (say E_1 and E_2) and the potential drop across the cell, as follows:

$$V = E_2 - E_1 + iR, \qquad (1)$$

i = current in amps, R = resistance of solution in ohms.

Metal deposition on the cathode commences at a potential E given by the Nernst equation

$$E = E^0 + RT/ZF \ln aM^{z+} \qquad (2)$$

where E^0 is the standard electrode potential on the Standard Hydrogen (S.H.E.) Scale, R is the ideal gas constant, T is the temperature (Kelvin), Z is number of electrons involved per atom, F is the Faraday and a M^{z+} the activity of the metal M^{z+} which is deposited.

The cathode potential E must be made more negative (by increasing V) for a reasonable rate of deposition of M (see Fig. 2). As the metal M deposits so the activity of M^{z+} decreases and the potential becomes more negative, in accordance with equation (2). The current–cathode potential curve will then move towards more negative potential values as shown in Figs. 2 and 3 and a situation can arise in which a second metal N can be co-deposited with M.

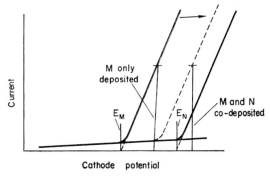

Fig. 2.

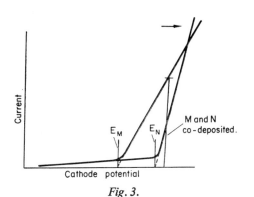

Fig. 3.

It is apparent, therefore, that some control must be exercised on the cathode potential to avoid this situation and the work of H. J. S. Sand (1940), using a third electrode (often a saturated calomel electrode [SCE]), as a reference sensor in a separate circuit, enabled the method to be considerably widened in scope and made possible a number of separations which are not otherwise feasible (see separation of metals, later).

It also laid the basis for the modern techniques of controlled potential electrolysis (see later).

Electrolysis at Constant Current

This was the original method of deposition wherein the current was maintained constant and the voltage V was continually adjusted until deposition was complete (see Fig. 2 for drift of potential as depletion occurs). A rapid change of voltage at the point of depletion occurred and it was necessary to know the potential at which hydrogen would commence evolution so that this value would not be exceeded. Thus a copper solution in dilute HNO_3 requires at least $V = 1.3$ volts before deposition occurs. Hydrogen is not liberated below 2·3 volts so that a reasonable rate of deposition can be maintained up to 2·2 volts. The apparatus can be simply represented in Fig. 4.

A is an ammeter, R_1 and R_2 are high-wattage low resistances. The 12 volt d.c. can be supplied from a 12-V car battery or a d.c. mains supply (240 or 110 V) with suitable series resistance. The method has been satisfactory for copper determination but is limited in scope to a single metal and is generally unsuitable for separations which involve careful control of potential. Since the currents involved are not usually very high the method tends to be time consuming. In consequence it has largely been superseded by electrolysis at constant potential.

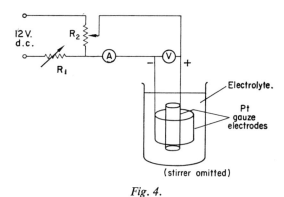

Fig. 4.

Electrolysis at Constant Potential

In this method the cathode potential is monitored by an additional electrode and the voltage V adjusted so as to maintain the cathode potential at a predetermined value. Under these conditions the current falls with time, as the ionic concentration of the species decreases, in accordance with the equation

$$i_t = i_o \times 10^{-kt} \qquad (3)$$

where i_o is the original current, i_t the current after time t and k is a constant given by

$$k = 0.43 \frac{DA}{\delta V} \qquad (4)$$

where D is the diffusion coefficient of the depositing ion, A is the area of the electrode, δ is the thickness of the diffusion layer and V the volume of the electrolyte.

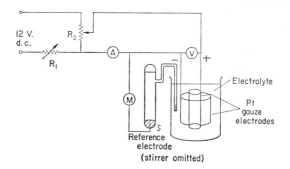

Fig. 5.

When the current falls to about 0·001 of its original value the deposition can be considered complete.

The apparatus, which is represented in Fig. 5, is similar to that in Fig. 4 except that the additional reference electrode S is involved with a separate meter M which monitors the cathode potential. With vigorous stirring high currents (10 amps) can be involved, a close control of cathode potential is possible and, in some separations, very necessary (to ±5 mV). In the earlier work of Sand and co-workers (1940) the control was manual. More recently the control has been done electronically by means of a potentiostat. This works on a pre-set potential between the reference and the cathode and from the error signal readjusts the potential V, across the cell, so as to restore the pre-set cathode potential. The process is, therefore, automatic once the apparatus is switched on. Several potentiostats are now commercially available. One that has been used for metal depositions is the "Wadsworth Controlled Potential Electro-Depositor", manufactured by Southern Analytical Ltd. (Heringshaw and Halfhide, 1960; Wadsworth, 1960). Industrial instruments are also available for carrying out electrolyses at controlled potential such as the B.T.L. "Electrolytic Analysis Apparatus" and the Griffin and George "Electrochemical Analysis Apparatus". These are complete with constant speed stirring motors, electrodes and appropriate meters, and designed to run from 220–240 V a. c. mains.

Electrodes

For the deposition of metals, platinum gauze (52, 48 or 45 mesh) is used as cathode which gives a large surface area consistent with mechanical strength. It is generally in the form of a hollow cylinder. Usually pure platinum is not used but an alloy with 10% iridium or rhodium gives increased rigidity. Since many metals, in particular lead, bismuth, tin, zinc, arsenic and antimony, tend to alloy with the platinum and cause progressive disintegration of the electrode when deposits are removed, it is best to deposit a coating of copper first. On stripping in nitric acid the copper is removed with the other metals and the platinum is unharmed.

The anode is also frequently made of platinum alloy but some care is necessary in selecting the anions present in the solution. No problem arises with sulphates, nitrates or perchlorates, since oxygen is evolved at the anode during electrolysis. However, with chlorides, chlorine can be generated and will rapidly attack the platinum causing dissolution and some deposition on the cathode, and general disintegration of the anode. Cathodic depositions from chloride, however, are often more rapid, efficient and give better deposits than from other media. Chlorine evolution can be avoided by loading the solution with an anodic depolarizer such as hydroxylamine or hydrazine salt which absorbs the chlorine and the attack on the platinum is thus prevented.

The electrodes should always be kept clean and free from grease and not handled with the fingers, and heated to a dull red in a flame before use. Mercury has occasionally been used as a cathode and is particularly useful for scouring the solution free from copper, silver, gold, arsenic, antimony, bismuth, tin, lead, zinc, cadmium, iron, cobalt, nickel, platinum, palladium, rhodium, iridium and thallium so that calcium, magnesium, tantalum, zirconium, tungsten, vanadium, uranium and aluminium can be determined by other methods. This method utilizes the Melaven cell (Melaven, 1930) (see Fig. 6).

The cell allows for a rapid separation of the solution from the cathode. After electrolysis, with platinium or mercury electrodes, the electrodes are left connected in circuit and the electrolyte removed, the electrodes washed with water and removed. The current should not be switched off with the electrodes in solution since rapid galvanic action can occur and this causes serious errors. Washing and drying (acetone or alcohol followed by ether and a current of air) should be rapid to minimize the risk of oxidation.

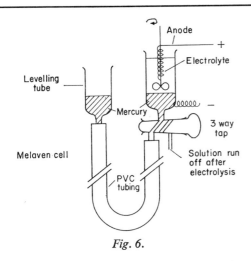

Fig. 6.

Physical State of the Deposit

For analytical purposes the deposit should be firmly adherent, dense and smooth so that it can survive the electrolyte process and subsequent washing and drying without loss. Deposits which are powdery, spongy or granular tend to become detached and have poor adhesion. Efficient stirring, appreciably high current densities and increased temperatures tend to favour this condition. In general, also deposition from solutions containing complexing ions give appreciably better deposits than from simple salts. Thus silver gives a more adherent deposit from a cyanide solution where it exists as the argentocyanide ion,

$$Ag(CN)_2^- + e \rightleftharpoons Ag(s) + 2CN^- \qquad (5)$$

than from a silver nitrate solution. Similarly zinc and cadmium give better deposits and use is made of this in the electroplating industry where the finish of the product is very important. Nickel can also give a good deposit from strongly ammoniacal solutions where the $[Ni(NH_3)_6]^{2+}$ ion exists. Lead gives a satisfactory deposit from tartrate solution whereas that from a nitrate solution forms loose crystals.

The important factor is the grain size which depends on several factors. Grain size is governed by the rate of formation of nuclei and the growth of the nuclei to crystals. If few nuclei are formed then large grains are formed or dendritic growths occur at the cathode. If too many nuclei arise then the deposit tends to be powdery. A nice balance is, therefore, required between the various factors to produce the best results.

Anodic Deposits

A few metals may be deposited as higher oxides on the anode. In particular lead can be determined as $PbO_2 \cdot xH_2O$ from a solution containing it but requires an appreciable concentration of nitric acid. The latter also acts as a cathodic depolarizer and is reduced at a lower potential than lead so that no deposit of lead occurs on the cathode. The anode is washed and dried at 120°C (393·2 K) but complete removal of the hydrated water is not possible at this temperature and a conversion factor of 0·864 instead of 0·866 (theoretical) is used. A number of metals interfere such as mercury, silver, arsenic, antimony, tin, chromium and tellurium.

Other oxides such as manganese and thallium have also been deposited anodically but difficulties have arisen with the adhesion of the deposit and the need to maintain the pH above a certain minimum value.

Applications

(a) *Single metal determinations.* Selected details of metals, standard electrode potentials, types of media employed and potential against a reference (S.C.E.) are summarized in Table 1. For fuller details of individual metal determinations see Sand (1940) and Vogel (1962).

(b) *Separation and determination of metals.* An examination of potentials in Table 1 will show that some metals have values which are close together and cannot, therefore, be separated even by careful control of the cathode potential. To effect such a separation there must be a difference of at least 0·3 volt in their potentials or co-deposition can occur. Thus silver is easily separable from copper and copper is easily separable from lead in nitric acid solutions but bismuth, antimony, and copper cannot be separated from each other in solutions of simple ions. However, by use of complexing ions such as cyanide or tartrate in alkaline solution copper forms a stable complex which requires an appreciably greater negative value to reduce it.

$$Cu(CN)^{2-} + e \rightleftharpoons Cu(s) + CN^- \qquad (6)$$

In consequence bismuth can be separated at a lower negative potential, copper at a more negative potential and antimony is not deposited under alkaline conditions.

Lead may be separated from bismuth, antimony and tin as the oxide $PbO_2 \cdot xH_2O$ on the anode after conversion of the latter metals to stable fluorides.

Arsenic can only be quantitatively deposited from chloride solution and with copper. The deposit is, in fact, copper arsenide. However, the potential must be carefully controlled to avoid the formation of arsine AsH_3.

Lead and tin have similar potentials but can be separated by the use of fluoride ions which forms complexes with the tin which are especially stable. The lead is deposited as metal and the solution is converted to the borofluoride and then to the ammonium oxalate compound from which the tin can then be deposited.

Table 1

Metal (symbol)	Standard electrode potential (volts) (S.H.E. Scale)	Medium	Cathode potential (S.C.E.) volts
Au	+1·30	Complex HCl + KCl + $NH_2OH \cdot HCl$	+1·0
			+1·0
		KCN	+0·1
Ag	+0·80	HNO_3	+0·5
		KCN	+0·2
Hg	+0·80	HNO_3	+0·2
		KI + NaOH	−0·2
Cu	+0·34	$H_2SO_4 + HNO_3$	−0·5
		KCN	−0·6
Bi	+0·20	HNO_3	−0·5
		Na tartrate	−0·4
Sb	+0·1	HCl + $NH_2OH \cdot HCl$	−0·4
		Tartaric acid + $N_2H_4 \cdot HCl$	−0·5
Sn	−0·1	HCl + $NH_2OH \cdot HCl$	−0·6
		oxalic acid, NH_4 oxalate + $NH_2OH \cdot HCl$	−0·9
Pb	−0·12	Tartaric acid + NH_4OH	−1·0
		HCl + $NH_2OH \cdot HCl$	−0·8
Ni	−0·22	HCl + excess NH_4OH	−1·0
		NH_4 oxalate	−1·0
Cd	−0·40	H_2SO_4	−0·8
		Na acetate + acetic acid	−1·0
Fe	−0·43	NH_4 oxalate	−2·0
Zn	−0·76	Acetic acid + excess NaOH	−1·4
		KCN + NaOH	−1·5

Although nickel and cobalt have less negative standard potentials than cadmium they require greater potentials to deposit them, probably due to the formation of complex ions. Thus cadmium is deposited alone from a sulphuric acid solution of the ions.

Similarly cadmium is deposited from a hydrochloric acid containing hydroxylamine hydrochloride but not zinc and a separation is thus possible. The zinc can subsequently be removed after addition of KCN and NH_4OH.

A very large number of analyses have been published referring to the separation and determination of metals in brasses and bronzes, white metal alloys, nickel bronzes, aluminium and magnesium alloys, solders and steels. Details of such methods are given in Ewing (1960) and Sand (1940), Chapters 3, 4, and 5. See also Vogel (1962).

The method has also been developed on the micro scale with a modified Pregl apparatus using a controlled potential procedure (Lindsey and Sand, 1935).

Internal Electrolysis

This term was applied by Sand (1940) to deposition analysis in which a base metal (e.g. zinc) was used as anode but no external e.m.f. was applied. The cathode was platinum and the arrangement functioned as a short-circuited battery so that noble metals such as silver, mercury or even copper were deposited. Usually the currents are much smaller and thus the process is more time consuming than the controlled potential method. Problems also arise if more than one metal can be deposited. The method has, therefore, had a limited success but has been applied to silver in galena and pyrites, mercury in brass, copper in steels, bismuth and copper in lead, galena and lead tin alloys and traces of cadmium in zinc.

Present Position of the Method

Although the method employs relatively inexpensive equipment and often gives relatively accurate results it tends to be somewhat time consuming and has been largely replaced during the last decade by more rapid and elegant methods, especially by coulometry or polarography. The former involves measurement of the number of coulombs consumed either by a suitable voltammetric cell or by an electronic current integrator whereas the latter involves a current–voltage curve—the polarogram—which gives a step or wave for the deposition process, the height of which is a measure of the concentration in the test solution. These methods, therefore, avoid the manipulations involved and are also much wider in scope.

Bibliography

DELAHAY, P. (1957) *Instrumental Analysis*, Chap. 6. New York: Macmillan.
EWING, G. A. (1960) *Instrumental Methods of Chemical Analysis* 2nd ed. Chap. 12. New York: McGraw-Hill.
HERINGSHAW, J. F. and HALFHIDE, P. F. (1960) *Analyst*, **85**, 69.
LINDSEY, A. J. and SAND, H. J. F. (1935) *Analyst*, **60**, 739, 744; (1938) **63**, 159.
MELAVEN, A. D. (1930) *Ind. Eng. Chem. Anal. Ed.* **2**, 180.
SAND, H. J. S. (1940) *Electrochemistry and Electrochemical Analysis*, Vol. II, Chap. 3. London and Glasgow: Blackie & Son Ltd.
VOGEL, A. I. (1962) *Quantitative Inorganic Analysis*, Chap. VI.
WADSWORTH, N. J. (1960) *Analyst*, **85**, 673.
WILLARD, H. H., MERRITT, L. L. and DEAN, J. A. (1965) *Instrumental Methods of Analysis* p. 628. New York: Van Nostrand.

J. V. WESTWOOD

ELECTROMAGNETIC GUN. The vast majority of electric motors are designed to produce rotary motion but there is an increasing interest in electrical machines which produce thrust and motion in a straight line. The term "linear motors" is applied to all such machines and they are often described by using the idea that a rotating machine, as represented in Fig. 1(a), can be considered to have been split along a radial plane and "unrolled", to produce the basic flat linear motor shown in (b). If, however, this planar motor is considered to be capable of being re-rolled into a cylinder whose axis is perpendicular to that of the original machine, the resulting pole array, as shown in Fig. 1(c), indicates that the new topology is that of a tubular motor in which the motion takes place along the axis of the machine. "Electromagnetic gun" is an alternative name for the tubular type of linear motor.

The two changes in topology shown in Fig. 1 create no restrictions as to the *type* of electromechanical device being considered. It is possible to construct, in tubular form, an induction motor, a synchronous motor, a reluctance motor or indeed any type of magnetic or electromagnetic device, working from either a.c. or d.c. The earliest tubular motors were reluctance devices in which a row of coaxial coils was energized in sequence by means of a rotary switch, and a steel slug was attracted by each coil in turn towards its centre, the inertia of the moving mass serving to ensure that it was "handed on" to the next coil at the right time. The usual application which the early pioneers had in mind was that of firing shuttles across weaving looms, although no commercial exploitation has ever been successful in this field.

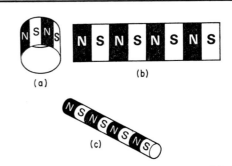

Fig. 1. Conventional motor pole array: (a) is unrolled to give flat linear motor (b) and re-rolled to make tubular motor or electromagnetic gun (c).

In 1912 a Frenchman, Emile Bachelet, patented a levitation device using a.c.-carrying, vertical-axis coils to float an aluminium sheet, and other horizontal-axis coils to propel the sheet and, although the propelling mechanism consisted only of a single coil at each end of a track, he proposed the system as a high-speed transport mechanism with many propelling coils, and as such, his invention is properly classified as a tubular motor. This work was followed by a proposal by Birkeland in 1918 to use a tubular reluctance motor as a silent "cannon", but the venture never proceeded beyond the model stage.

With the advent of atomic energy, tubular motors were first exploited commercially as pumps for liquid metal. Mixtures of sodium and potassium in the liquid state were used to remove heat from the cores of nuclear reactors and, since the liquid itself was radioactively contaminated, it was necessary for a pump to be sufficiently reliable to be left unattended for years. The conductivity of sodium–potassium mixtures is of the order of that of solid aluminium at normal temperatures and the liquid can itself therefore be used as the "rotor" of a machine, either as a d.c. motor or induction motor. This type of pump, having no mechanical moving parts, was extremely reliable and was known as an "annular linear induction pump" (ALIP). Even so, it was occasionally necessary to replace the primary coil system and this involved breaking the pipeline. In the case of a flat linear induction pump (FLIP) such a replacement could be effected without a break, and tubular pumps fell into disfavour, although it is interesting to note the existence of another early type of tubular pump in which a primary coil system produced a purely *rotating* field around the pipe but baffles were included in the latter to constrain the liquid to move in a spiral (and thus with a linear component of motion). This motor was known as SIP (spiral induction pump).†

---

† Blake, L. R. (1957) Conduction and induction pumps for liquid metals, *Proc. IEE*, **104**A, 49.

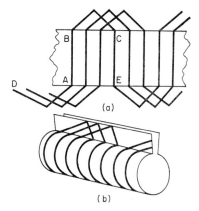

Fig. 2. Elimination of stator end windings by the use of tubular construction.

The principal advantage of the tubular arrangement concerns the end windings† of both primary and secondary members. Figure 2(a) shows a flat linear a.c. motor primary winding, which clearly has its rotary counterpart, it being only necessary to perform the topological change shown in Fig. 1 in reverse, i.e. from (b) to (a), to obtain the primary winding of the conventional a.c. machine. Although the action of such machines is perhaps only properly expressed in terms of "flux linkages", i.e. neither electric current nor magnetic flux having any meaning in the absence of a closed circuit, it is also customary to explain the action of a machine as though the force reaction on the primary winding could be located in specific parts of that winding. Thus in Fig. 2(a), portions of the coils such as AB are regarded as the "useful" part of the primary, since the current, flux and resulting force are all at right angles to each other, whilst the portions such as AD and BC only serve to connect together pairs of useful portions of the winding and are hence regarded as non-useful portions and described as "end-windings". It is true that they provide opportunity for additional forced cooling in a machine for they are not embedded in an insulated slot and their surface is readily available for the release of heat generated in the useful portions of the coils as well as in the end windings themselves but they also generate components of magnetic flux which do not link both windings and this gives rise to "end-turn leakage reactance", a feature which can raise problems for the designer, especially in windings of large pole pitch. In some cases, the problem of accommodating

---

† In motors in which the secondary consists of a homogeneous mass of conductor, either liquid or solid, the expression "end windings" must be interpreted, in the case of the secondary, as meaning those portions of the conductor in which the current flow is parallel to the direction of motion.

the sheer bulk of the end windings increases the mass and cost of other parts of the machine, and there are many designs of machine in which the end windings are an embarrassment rather than a bonus.

Figure 2(b) shows that when the flat winding of Fig. 2(a) is rolled into a cylinder to form an electromagnetic gun, the end windings become entirely unnecessary, since each conductor such as AB in (a) becomes a self-contained circular loop. Thus the only end windings necessary in the primary are single conductors joining together windings of the same phase, whereas in arrangements as in (a) *every conductor* in the coil ABCE is required to have a useful and a non-useful part.

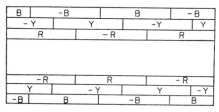

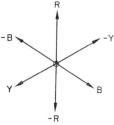

Fig. 3. Winding arrangements for tubular motors. (a) A row of simple coils. (b) Cross-section through a distributed three-layer winding.

This simplification of the primary winding of tubular motors is generally exploited in one of two ways:

(a) simple bobbin coils are placed coaxially as shown in Fig. 3(a)

or

(b) a "layered" winding of two or more layers is wound on to a tubular former as in (b). In a three-layer arrangement, each layer may consist of windings related to one phase only which are distributed across whole pole pitches. This makes the construction extremely simple, the winding being continuous and merely reversed at intervals of a pole pitch (half wavelength) as shown in Fig. 4.

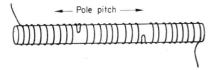

Fig. 4. One layer of the winding shown in Fig. 3.

In the case of secondary conductors alternate rings of iron and copper threaded on to an iron core constitute the tubular equivalent of the "squirrel-cage" rotor of a conventional induction motor, whilst an even simpler secondary construction, which can be used with either induction or d.c. motors, consists of a steel rod completely surrounded by a conducting cylinder.

Several commercial designs of tubular motor have exploited the simple copper clad rotor and at the same time adapted the stator winding to mass production techniques. Figure 5 shows a cross-section through an early type of tubular actuator in which the stator consisted of three standard parts only: coils, steel plates and C-cores as shown in Fig. 6. The C-cores serve the dual purpose of closing the magnetic circuit on the outside of the motor and of protecting the winding mechanically. The plates, of which parts form the stator "teeth" (for they extend radially inwards virtually to the surface of the rotor itself), are extended outwards for a considerable radial distance beyond the outside surface of the C-cores and again for a dual purpose. Firstly, these parts of the plates act as cooling fins for the removal of heat from the primary and secondary conductors. Secondly, they provide a path for magnetic flux to return across the air spaces between neighbouring plates. Because the area of such space is large, the reluctance of this path is comparable with that of the path through the C-cores. The plates also provide location holes for the bolts which hold together the entire structure. From such designs forces of the order of one newton per 8 watts of input may be obtained.

Figure 7 shows an alternative commercial arrangement in which a stack of slotted punchings is held together by an epoxy resin and then drilled from end to end to make the hole which is to contain the secondary member. Stator coils are then pushed into place in the slots and located with their central hole in line with the hole in the steel. The fact that the ferromagnetic stator material surrounds the secondary entirely does not constitute a magnetic short circuit for the flux as any section may enter the secondary from any direction in a cross-sectional plane and it matters little whether the hole is close to the sides of the stator or not so long as the total cross-section of steel is sufficient to contain the flux per pole.

A most interesting feature unique to tubular motors is that the effective magnetic circuit reluctance *outside* the stator can be very low, even though no ferromagnetic material is incorporated in the stator.

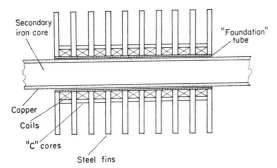

Fig. 5. Cross section through one form of commercial machine.

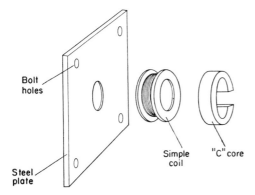

Fig. 6. Component parts of the machine shown in Fig. 5.

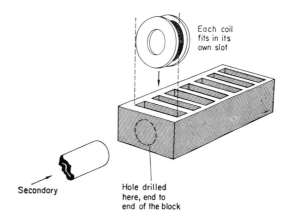

Fig. 7. Construction of an alternative commercial design.

To appreciate the reasons for this it is recommended to consider first the equivalent effective airgap of a single-sided flat linear motor with no ferromagnetic material above its pole surface, as shown in Fig. 8. At first sight, since most of the flux above the surface is contained in paths whose length is greater than a pole pitch, p, it might appear that the effective airgap exceeds p. Yet, unlike the flux in a rotary machine the linear motor flux can spread to infinity and the integrated effect of this spreading results in an equivalent pole pitch of p/π. This value, whilst still too high for most single sided motor applications, indicates the benefits to be derived from allowing the flux to curve in two-dimensional patterns, as opposed to the one-dimensional constraint on the gap flux of "conventional" machines (either rotary *or* linear).

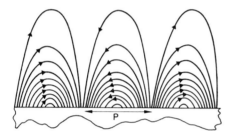

Fig. 8. Magnetic field pattern above the surface of a flat linear motor.

In tubular motors, the flux outside the stator is allowed to spread into the third dimension and the effective airgap outside the stator is reduced below the value p/π by a factor depending only on the bore radius to pole pitch ratio a/p. Figure 9 shows this factor plotted against a/p. Useful tubular motors can therefore be manufactured without stator steel, provided the ratio a/p is small.

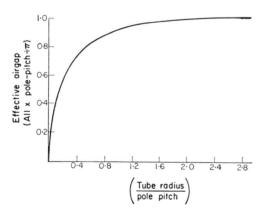

Fig. 9. Relationship between external reluctance (effective airgap) and the ratio of tube radius to pole pitch.

The principal factor which limits the force available from a given size of electromagnetic gun is that of secondary core saturation. The magnetic circuit benefits accruing from an expanding three-dimensional space outside the machine turn to penalties inside the core where space diminishes rapidly and it is seen that the whole of the flux emanating from one cylindrical pole (of area $2\pi ap$) must be capable of being contained in the secondary core (of area πa^2). The ratio of the peak pole flux density (B_g) to the peak secondary core density (B_c) is, for sinusoidal distribution, thus given by

$$B_g/B_c = \frac{\pi a}{4p}. \qquad (1)$$

This limitation on thrust/core area remained a problem until the development of the transverse flux tubular motor by Eastham and Alwash.† This invention is fully described under LINEAR MOTORS USING TRANSVERSE FLUX where it is shown that all flux inside the core can be made entirely radial at any section and therefore unrestricted by core area. At the same time it becomes unnecessary to laminate the core in two dimensions, i.e. by using a bundle of wires, but a conventionally stacked pile of disc stampings can be used, as in any rotary machine. Earlier commercially manufactured tubular motors had usually dodged the problem by using solid-steel secondary cores and accepting loss of effective area through skin effect and increased effective airgap, both from skin effect and the need to use a thicker coating of conductor to ensure a sufficiently high Goodness factor.‡

Bibliography

LAITHWAITE, E. R. (1966) *Induction Machines for Special Purposes*, Newnes.
LAITHWAITE, E. R. (1971) *Linear Electric Motors*, Mills & Boon.

E. R. LAITHWAITE

ENTHALPIMETRY. Although this term should strictly be applied to any constant pressure calorimetric measurement it has, by custom, been confined to a group of analytical techniques in which the heat change in the analytical reaction is used to measure the extent or degree of advancement of the reaction. If the molar enthalpy change is ΔH J for each mole of product formed, then the formation of n moles of product will be associated with a heat evolution of q joules where

$$q = -n \Delta H. \qquad (1)$$

---

† Eastham, J. F. and Alwash, J. H. (1972) Transverse-flux tubular motors, *Proc. IEE*, **119**, 1709.
‡ Laithwaite, E. R. (1965) The goodness of a machine, *Proc. IEE*, **112**, 538.

If excess of a reagent B is added to a solution of a second reagent A the amount of product formed will be proportional to the amount of A present, as will the heat q evolved. The measured heat change will include the heat of dilution of B as it is added to the solution of A, and also thermal effects arising from any difference in the temperatures of the solutions being mixed. Analytical applications of eqn. (1) will clearly be simpler if these effects can be neglected; Sajó (1955, 1956, 1964) has done much to develop enthalpimetric methods of analysis by devising "thermoneutral" reagents. In these, the solution of B is mixed with another component (or components) whose heat of dilution balances as closely as possible that of B. If, in addition, the reagents are in thermal equilibrium before mixing, the measured heat output is effectively q, and the value of this for addition of excess B can be obtained. To obtain n it is necessary to know ΔH, or to calibrate the system with solutions containing known amounts of A. Experimentally, the simplest procedure is to add excess of the reagent B *rapidly*. The reaction heat is evolved rapidly, and the maximum temperature rise is closely related to q, heat losses being negligible in such a rapid, pseudo-adiabatic process; the technique is often called *direct injection enthalpimetry* (Jordan, 1964, 1966).

An alternative procedure is to add successive small amounts (Δn_B) of B to a solution containing the reagent A which is initially in excess. The heat output (Δq_B) for this addition is, unless the reaction is incomplete, proportional to Δn_B and to the sum of the heat of reaction and the heat of dilution. The ratio $\Delta q_B/\Delta n_B$ remains constant until all the A has reacted. It then changes sharply to a different value since Δq_B then depends only on the heat of dilution of B. A plot of $\Delta q_B/\Delta n_B$ against the amount of B added takes the form shown in Fig. 1a. If the reaction is incomplete at the end-point, then $\Delta q_B/\Delta n_B$ changes less sharply at this point, as shown by the broken curves. Alternatively, the total heat liberated $q_B (= \Sigma \Delta q_B)$ can be plotted against the amount of B added. For a reaction which goes to completion, the resulting enthalpogram is a straight line passing through the origin, whose slope changes sharply at the end-point (Fig. 1b). If the reaction is incomplete at the end-point, the change in slope occurs gradually and the shape of the enthalpogram is a function of the initial concentration of A and the dissociation constant of the reaction product (Tyrrell, 1967). Careful analysis of such curves can give this dissociation constant, and hence the standard free energy change ΔG^0 in the reaction, as well as the standard enthalpy change, ΔH^0 (Izatt, 1966). In practice the heat evolved or absorbed in a reaction is not measured directly. Usually heat losses from the reaction vessel are minimized, and the change in temperature of the reaction vessel is taken as a measure of the heat liberated or absorbed in the reaction. If the temperature change of the vessel contents is plotted against the amount of B added, the "thermograms" produced have the form shown in Fig. 1b, the ordinate then being the temperature of the system. The experiment is essentially a titration whose course is followed by thermometric means, and it can be described as a thermometric titration. The term "thermometric titrimetry" has been used as an alternative to "enthalpimetry", and has provided the title for two monographs dealing with this general field. (Tyrrell and Beezer, 1968; Bark and Bark, 1968).

Many reactions involve quite substantial heat changes, and enthalpimetry or thermometric titrimetry provides in principle an almost universally applicable analytical technique which has been used to a small extent for many years. The prospect of using it for routine analytical purposes depended, however, on the development of a robust, sensitive, and inexpensive instrument for detecting changes in temperature, the thermistor. A typical thermometric titrator consists of a thermistor, a motor-driven stirrer, and a burette tip immersed in a solution contained in an insulated vessel (Linde, 1953). The burette is usually motor-driven and the contents

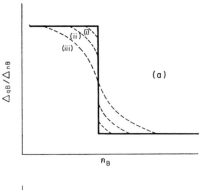

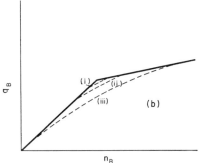

Fig. 1. Idealized enthalpograms: (a) derivative, (b) integral (full lines—complete reaction: broken lines—incomplete reaction). Formation constant of product decreases from (i) to (iii). A poor end-point is obtained in case (ii) and no detectable end-point in case (iii).

are maintained at a constant temperature. The thermistor is placed in one arm of a Wheatstone bridge which is adjusted to be in balance at the beginning of the experiment. The burette is started, the temperature of the reaction vessel contents changes as the reaction proceeds, the resistance of the thermistor alters, and the bridge goes out of balance. For *small* temperature changes, the out-of balance signal is proportional to the temperature change in the vessel, and can be recorded as a function of time on a potentiometric recorder. Since the reagent is delivered from the burette at a constant rate, the recorder trace is, in effect, a thermogram. The volume of reagent added at the end-point can be obtained from the chart record showing the beginning and end of the titration, provided that the chart speed and the burette delivery rate are known. Changes in the heat capacity of the system during the titration are minimized by adding the reagent in the form of a concentrated solution using an efficient stirrer to ensure rapid mixing. Heat losses are minimized by using high burette delivery rates (up to 10 ml min^{-1}), and, for these reasons, recorder chart speed must also be high. The system can be adapted to direct injection enthalpimetry by replacing the motor-driven burette by a one-stroke injection burette designed to add excess reagent as rapidly as possible. The important parameter is then the chart deflection, the chart speed being unimportant, and a fast-response recorder is desirable.

The magnitude of the temperature change at the end-point is, within wide limits, independent of the volume of the reagent solution being analysed. If the heat capacity of the vessel, stirrer, etc., when empty is C_0, and the heat capacity per unit volume of solution is C_s, the total volume being v, then the temperature rise at the end point is related to the number of moles of reagent present (n_A) by the equation:

$$\Delta T = -n_A \Delta H/(C_0 + vC_s). \qquad (2)$$

This assumes heats of dilution to be negligible. Generally, $C_0 < vC_s$ and, to a first approximation, equation (2) reduces to

$$\Delta T \sim -n_A \Delta H/vC_s = -c_A \Delta H/C_s \qquad (3)$$

where c_A is the total molar concentration of A in the reaction vessel. On this approximation, ΔT is independent of the volume of the solution in the reaction vessel and small amounts of material can therefore by analysed by using small volumes. The lower limit to this miniaturization is set by the physical problem of arranging the temperature probe, stirrer, and burette tip in a small volume, and by the requirement that $C_0 < vC_s$. In practice, if direct injection enthalpimetry is used, analyses can be done (Jordan, 1966) on as little as 1 μm of reagent with $v \sim 2$ ml. The lower limit of c_A can be estimated from eq. (3). If $\Delta H \sim 40$ kJ mol^{-1}, and $C_s \sim 4$ kJ deg^{-1} l^{-1}, then for $c_A \sim 10^{-2}$ M, $\Delta T \sim 0 \cdot 1$°C.

A simple Wheatstone bridge with a single thermistor coupled to a 1-mV recorder can give a full-scale deflection for such a temperature change.

The titration procedures can be made more convenient by providing a means of automatic end-point detection. This may be done by electronically differentiating the output from the bridge. The differentiated output changes with the amount of titrant added in a manner similar to that shown in Fig. 1a. The differentiated output falls suddenly from one value to another at the endpoint, and this sudden change can be made to operate a relay which stops the burette. The volume added up to the end-point can then be read off with as much convenience as the burette design allows.

The procedures so far described have been based on classical adiabatic calorimetry, and are not very suitable for "on-line" analyses or as a basis for automated techniques intended to deal with large numbers of routine analyses. For these flow techniques hold much more promise. The two reagent solutions are pumped at known rates into a mixing chamber, react, and then go to waste. In one form (Priestle, 1965) of flow calorimeter the inlet and effluent temperatures were monitored continuously and the difference was a measure of the heat output rate in the reaction vessel. Since the system was in a steady state, the signal was time-independent. If the reaction went to completion in the mixing chamber, and one reagent (B) was in excess, then this signal would be proportional to the concentration of A in the appropriate inlet stream. It is not difficult to arrange the circuitry to give a direct digital readout of the concentration of A. Such designs can be developed to give great sensitivity coupled with very short response times (Strafelda, 1968; Desnoyers, 1969).

Another class of flow calorimeter utilizes the principle of a heat-flow calorimeter. Instead of minimizing heat losses from the reaction vessel, and measuring temperature changes, the heat developed is allowed to flow to a heat sink through thermoelements (usually p–n semiconductor junctions). An electrical potential proportional to the rate of heat flow across the junction is generated. This signal is proportional to the rate of reaction in the mixing chamber averaged over the residence time therein. If a reagent A is allowed to react with an excess of reagent B in such a calorimeter, and if the reaction is first order with respect to A, then the time-independent signal from such a calorimeter is proportional to the concentration of A in the inlet solution. A successful instrument is currently available, and can detect heat output rates of the order of 5×10^{-7} J sec^{-1} (Wadsö, 1968). This makes it possible to extend enthalpimetric methods of analysis to biological systems.

The advantages of enthalpimetric methods have not always been properly appreciated partly because of the belief that such a universally applicable technique lacked the power to discriminate between consecutive or concurrent reactions. This is not always so. One characteristic of the method is that it can be used with reactions that are not highly favoured in a thermodynamic sense. In aqueous solution it is possible, for example, to get a reasonably sharp thermometric end-point when titrating occurs with $pK_a \sim 10$, e.g. boric acid, the ammonium ion (Tyrrell, 1967); this is not possible by other means. Acid strengths of still weaker acids can be measured, since, on titration with base, curves of type (iii) in Fig. 1b are obtained. From these both the pK_a of the weak acid and the enthalpy of dissociation can be found. In the titration of a mixture of a strong and weak acid with a base, the stronger acid is neutralized first. Since its heat of neutralization is usually greater than that of the weak acid, a change in the slope of the thermogram occurs when all the strong acid has been neutralized. A second change of slope marks the end of the neutralization stage of the weak acid unless its heat of neutralization is so small that it cannot be distinguished from the heat of dilution of the base. For example, two distinct end-points are found when a mixture of hydrochloric acid and ammonium chloride is neutralized with a strong base (Jordan, 1963). The first corresponds to the complete neutralization of the hydrochloric acid, and the second to that of the ammonium ion behaving as a cationic acid. Similar results have been obtained for the neutralization of a mixture of a strong acid and a hydrolysable metal cation (Miller, 1959).

An important area of application lies in the field of non-aqueous titrimetry since a thermistor is not affected by the solvent in the way that electrode systems or chemical indicators may be. An example of the power of the method in this respect was provided by its application to the analysis of aluminium and zinc alkyls, important as Ziegler catalysts. A mixture of R_3Al, R_2AlCl, $RAlCl_2$ and R_2AlH could be analysed by titration with di-n-butyl ether to give the total of all components except the last. Addition of excess ether followed by a second titration with benzophenone gave the R_2AlH content, all the end-points being detectable by thermometric means. The method is simpler and faster than available alternatives (Everson, 1967).

Even in cases where direct thermometric titration is not satisfactory an indirect method may be practicable. Very weak organic acids cannot be estimated by direct thermometric titration with a strong base dissolved in methanol or isopropanol since the heat of reaction is too small. If, however, the acid is dissolved in dry acetone and titrated with base, little change in temperature occurs until the acid has been neutralized and the solution ceases to be acidic (Vaughan, 1965). The acetone then begins to react exothermically to give diacetone alcohol and the temperature rises rapidly. The onset of this rapid rise marks the end-point of the neutralization reaction. An improved procedure giving a higher sensitivity and without the need for such careful exclusion of water involves the replacement of acetone by acrylonitrile, or acrylonitrile mixed with up to 40% aromatic hydrocarbon (Greenhow, 1972). At the end-point, base-catalysed cyanoethylation of the alcohol in the titrant accompanied by anionic polymerization of the acrylonitrile occurs with a very rapid rise in temperature. This technique of using a "thermometric indicator" to mark the end-point of a reaction seems capable of considerable development. Cobalt can be analysed at trace levels in the presence of nickel by allowing the cobalt to catalyse the decomposition of hydrogen peroxide solution (Sajó, 1967). The rate of this reaction, and hence the rate of heat evolution, is proportional to the cobalt concentration. Any species, such as the cyanide ion, which "poisons" the catalyst, can be analysed by measuring the change in heat output rate of the cobalt-catalysed hydrogen peroxide decomposition at a fixed cobalt composition in the presence of varying amounts of cyanide ion.

The great sensitivity of modern flow calorimeters makes it practicable to carry out a number of biochemical assays. For example, enzyme activities can be estimated from the rate of heat output in the presence of excess substrate, while substrate concentrations can be found by working at a fixed enzyme activity and low substrate levels (Brown, 1969; Beezer, 1972). Because of the high specificity of many enzymes a specific substrate can be determined in the presence of many other substances. Since, in addition, enzyme systems are sensitive to low concentrations of specific inhibitors, the reduction of the heat output rate associated with the enzyme reaction when inhibitor is added can be used to determine the inhibitor concentration (Brown, 1969).

Enthalpimetry is an analytical tool with wide potential applications. For many purposes quite simple instrumentation suffices, and ingenuity in applying known analytical procedures can make it applicable to many problems for some of which it clearly becomes the method of choice. Modern developments in microcalorimetric instrumentation make it possible to use enthalpimetric methods in biological fields where their potentials have only begun to be realized.

Bibliography

BARK, L. S. and BARK, S. M. (1968) *Thermometric Titrimetry*, Oxford: Pergamon Press.
BEEZER, A. E. and TYRRELL, H. J. V (1972) *Science Tools*, **19**, 13.
BROWN, H. D. (1969) *Biochemical Microcalorimetry*, New York, London: Academic Press (Chapter VII).
DESNOYERS, J. E., PICKER, P. and JOLICOEUR, C. (1969) *J. Chem. Thermodynamics*, **1**, 469.
EVERSON, W. L. and RAMIREZ, E. M. (1965) *Analyt. Chem.* **37**, 806, 812.

GREENHOW, E. J. (1972) *Chemistry and Industry*, 422.

IZATT, R. M., HANSEN, L. D., CHRISTENSEN, J. J. and others (1966) *J. Am. Chem. Soc.* **88**, 2641; *J. Phys. Chem.* **70**, 2003.

JORDAN, J. (1963) *J. Chem. Educ.* **40**, A5; (1964) *Analyt. Chem.* **36**, 2162 (with Wasiliewski, W. C. and Pei, P. T. S.); (1966) *Microchem. J.* **10**, 260 (with R. A. Henry).

LINDE, H. W., ROGERS, L. B. and HUME, D. N. (1953) *Analyt. Chem.* **25**, 404.

MILLER, F. J. and THOMASON, P. F. (1959) *Analyt. Chem.* **31**, 1498.

PRIESTLEY, P. T., SEBBORN, W. S. and SELMAN, R. F. W. (1965) *Analyst*, **90**, 589.

SAJÓ, I. (1955) *Acta Chim. Acad. Sci. Hungary*, **6**, 233; (1956) *Zeit. analyt. Chem.* **153**, 128; (1964) *Zeit. analyt. Chem.* **202**, 177 (with Ujvary, J.); (1967) *Mikrochim. Acta*, 248 (with Sipos, B.).

STRAFELDA, F. and KROFTOVA, J. (1968) *Coll. Czech. Chem. Comm.* **33**, 3694.

TYRRELL, H. J. V. (1967) *Talanta*, **14**, 843.

TYRRELL, H. J. V. and BEEZER, A. E. (1968) *Thermometric Titrimetry*, London: Chapman & Hall.

VAUGHAN, G. A. and SWITHENBANK, J. J. (1965) *Analyst*, **90**, 594.

WADSÖ, I. and MONK, P. (1968) *Acta Chem. Scand.* **22**, 1842.

H. J. V. TYRRELL

F

FERROELECTRICS — RECENT APPLICATIONS OF. Ferroelectricity was first discovered in Rochelle salt by J. Valasek in 1921. Since then a great variety of materials have been discovered which show characteristic ferroelectric behaviour, that is, they show a non-linear hysteresis between polarization and applied electric field. This behaviour is a consequence of the occurrence in the material of electrostatic dipoles exhibiting long-range order to give a net dipole moment. The dipoles are usually the result of small displacements of atoms from points of high symmetry in the structure. In ferroelectric structures these displacements are parallel, in the so-called anti-ferroelectrics they are antiparallel. The existence of ferroelectricity in a substance very often means that other important effects are also to be found in the same material. Thus, ferroelectrics usually display a high dielectric constant, and piezoelectric, pyroelectric, electro-optic effects etc. For this reason, a discussion of the uses of ferroelectrics must involve a discussion of the use of these other effects.

Considerable confusion is found in the literature regarding the nature of phase transitions in ferroelectrics. Ideally, the material is said to be ferroelectric over a certain temperature range delimited by transition (or Curie) points; above (and below) these points no ferroelectric behaviour is to be observed and the material is then referred to as *paraelectric*. However, this scheme is often complicated by the existence of transitions to anti-ferroelectric phases, which may share some of the properties of the ferroelectric phase (such as the possession of a high dielectric constant). To confuse matters further, some ferroelectrics have transitions to phases which cannot be truly classified as above. For example, in the large class of compounds known as perovskites, with general formula ABX_3 (A, B cations; X anion), transitions are often found which lead to phases whose crystal structures consist of a crumpling of the anion framework. This crumpling is best dealt with in terms of the tilting of the anion octahedra and does not involve the formation of real dipoles. It is difficult to know how to classify these phases; in the literature they have been variously referred to as paraelectric or anti-ferroelectric. In addition, since materials can be piezoelectric without being ferroelectric, it is possible to have ferroelectric to non-ferroelectric transitions where both phases possess non-zero piezoelectric coefficients, as in Rochelle salt.

For these reasons, in any discussion of ferroelectrics, it is useful to include also those non-ferroelectric materials which show some similarities in their properties.

The occurrence of a "square" P-E hysteresis loop implies that there are two stable polarized states and that an applied electric field can be used to switch between them. This, therefore, raises the possibility of using ferroelectrics as binary switches in computer applications. Such devices for information storage have been constructed, the storage being accomplished by a process of matrix addressing, as shown in Fig. 1. Here, a ferroelectric crystal, such as barium titanate, has orthogonal sets of X–Y electrodes on opposite surfaces. These electrodes overlap to define a series of cells, which are subjected to voltages applied simultaneously, but with opposite sign, to the X–Y electrodes. In this way, the polarization state of the addressed cell can be read; the cell may or may not produce a switch signal depending on its state at the time. This type of store, unfortunately, has a number of disadvantages, such as low switching speeds and a limited life due to crystal fatigue.

The pyroelectric effect has been used in thermal-imaging systems. The usual method for this is

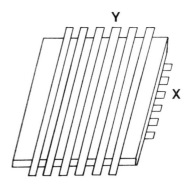

Fig. 1. Arrangement of electrodes for a ferroelectric matrix store.

shown in the Vidicon tube arrangement illustrated in Fig. 2. An infra-red image of the object is formed on a thin triglycine sulphate (TGS) target after passing through a silicon window. This produces a voltage pattern at the rear of the TGS target because of its large pyroelectric coefficient. The voltage pattern is read by a scanning electron beam. The output signal is then converted to a conventional television display, from which, after calibration, it is possible to make accurate temperature measurements of the surface of the object. This technique has proved useful in many fields where temperature measurement is important; for instance it has been used in cancer research where it is known that there is a connection between malignant growth and temperature distribution in the skin.

A search is at present being conducted for other materials to replace TGS as a pyroelectric detector. If a ferroelectric is used near its Curie temperature, a large temperature-dependent pyroelectric coefficient results, together with rapid changes in the dielectric constant. For infra-red applications, a high pyroelectric coefficient with a low dielectric constant are necessary and therefore a compromise must be made. The most promising materials in this respect appear to be the lead zirconate/titanate(PZT) solid solutions, particularly when doped with lanthanum (PLZT), and considerable research is at present being conducted into the possible use of this type of material.

The piezoelectric effect has obvious applications in electromechanical transducers. Originally barium titanate was the sole material used, but, more recently, higher piezoelectric coefficients and coupling factors have been obtained with PZT ceramics, and these have now almost completely replaced barium titanate.

A recent development at Philips Eindhoven Laboratories (M. de Jong, 1972) has been a 10-MHz ceramic filter element made from a disc polarized radially and vibrating in a radial thickness shearmode. With a suitable disc and electrode geometry, it is possible to amplify modes below a specific frequency and suppress those above it.

Another type of electrical filter makes use of surface waves which can be excited in piezoelectric materials by means of an interdigital array of electrodes. The characteristics of such a filter are determined by the material itself as well as the layout of the electrodes. Filters of this type find most application in broadband filtering where fairly large bandpass losses can be tolerated.

Other uses of the piezoelectric effect have been in high-voltage generators for spark-ignition, gramophone pick-ups, microphones, ultrasonic transducers and in underwater sonar devices.

Perhaps the most rapidly growing area of application is in the field of *electro-optics*. Research at Bell Telephone Laboratories showed that it is possible to store information, in the form of holograms, in single crystals of poled, undoped lithium niobate. A typical arrangement for recording the holograms is shown in Fig. 3. The laser light is split into two; one beam shines directly on to the crystal and the other passes through a photographic transparency of the object before it reaches the crystal. The holographic image is stored in the crystal by the modulation of the refractive index caused by the intense laser-light interference fringes. If the crystal is rotated through fractions of a degree it is possible to store as many as 1000 different holograms at once. An early problem was that the reading operation carried out by blocking out the object beam, tended to erase the stored image. One recently reported way of solving this is to fix the hologram thermally after storage, by heating the crystal to about 100°C. Heating to 300°C is used to erase the hologram. Optical memories based on holography have many advantages

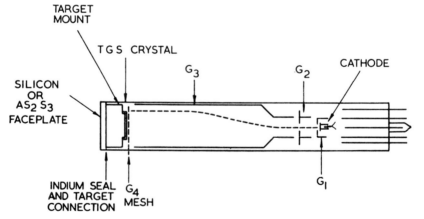

Fig. 2. Schematic diagram of English Electric Valve Co. Pyroelectric Vidicon. (Courtesy of E. H. Putley.)

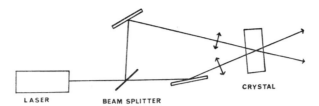

Fig. 3. Arrangement for recording and reading holograms. The c-axis of the $LiNbO_3$ crystal is aligned normal to the bisector of the two beams.

over direct optical storage. For example, in the holographic method, blocks of information (10^4 bits, say) can be selected from within a larger storage unit. Also, there is a high tolerance to scratches and dust on the optical surfaces.

In a variation of this technique, information-storage is obtained by using FE–PC (ferroelectric–photoconductive) sandwiches as page composers. The page composer is an array of light valves in which a complete optical pattern, representing a page of information, can be formed, usually by electrical addressing. The pattern is then transferred to an optical storage medium, such as the holographic system already described. The FE–PC sandwich consists of a ferroelectric layer against a photoconducting layer, both sandwiched between transparent electrodes. By careful choice of the dark impedance of the photoconductor and the ferroelectric, the latter can be successfully isolated from the effect of the applied voltage, except where the sandwich is illuminated. This feature is of utmost importance in the process of matrix-addressing a portion of a large memory.

In the optical-display field, the electro-optic effect has again been used in large-screen displays. Figure 4 illustrates a projection system developed in France called the TITUS tube. Here, light from a xenon lamp passes through a polarizer and is then modulated by a potassium dideuterium phosphate (KD*P) crystal and reflected through a crossed polar. The KD*P target is coated on the front with a transparent conducting film and on the rear with a dielectric mirror. The video signal voltage is applied between the conducting coating and a grid at ground potential; the voltage pattern is transferred to the crystal by a scanning electron beam which produces secondary electron emission. By this means it has been possible to project an image onto a large screen 5 metres square with a brightness equivalent to that found in normal cinema practice. Operation at −60°C further allows the device to be used in storage mode, making it suitable for slow-scan radar, computer displays and optical pattern recognition systems.

In a more recent adaptation, the KD*P crystal is addressed by photoconduction. This device, called Phototitus, produces a picture with a resolution of

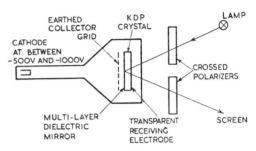

Fig. 4. Schematic diagram of the TITUS tube (Courtesy of G. Marie).

20 line pairs/mm with a writing time of 10 μsec, and allows image addition and subtraction.

Many investigations have been carried out into the use of ferroelectrics in alphanumeric displays for calculating machines, instrument panels, etc. These usually exploit the way the refractive index changes with applied fields. When viewed between crossed polars, large colour changes are observed, depending on the applied field; the most useful materials seem to be PLZT ceramics which have the advantage of being transparent.

The use of ferroelectrics in optical display systems has not become widespread, however, for a number of reasons, including their very limited lifetime (during the switching from one state to another the material experiences a large converse piezoelectric effect which tends to cause fatigue), their inability to transmit and switch a wide range of wavelengths when used with a broadband incoherent source, and the difficulty of providing a large viewing angle. Because of this, liquid-crystal displays have become highly competitive. However, in a recent device produced by Siemens (München), PZT ceramic has been incorporated into a liquid crystal system in order to overcome the problem of the crosstalk, which occurs when orthogonal X and Y electrodes are placed either side of a liquid crystal film for addressing purposes. The PZT layer is interposed between the liquid–crystal film and one set of electrodes. The non-linear capacitance of the PZT allows a selected element only to be activated into dynamic scattering, when a particular threshold voltage is applied. The device can be used in another mode: by initially storing the information to be displayed in the PZT layer and then by applying a suitable pulse the information can be displayed by the liquid–crystal film.

Very recently, a new type of electro-optic lens has been developed at the Sandia Laboratories, Albuquerque, using a layer of PLZT between crossed polars. The PLZT rotates the plane of polarized light so that, under normal conditions, the arrangement appears reasonably transparent, allowing 21% of the incident light intensity through. If the incident intensity exceeds a certain threshold value, as measured with an array of five photodiodes, a voltage is applied to a set of electrodes attached to the PLZT. This is arranged so that the resulting change in the angle of polarization causes almost all (about 99·99%) of the light to be absorbed by the polars. When the light intensity drops, the electrical stimulus is removed and the lens clears. The apparatus can be battery-operated and is capable of stopping a brilliant flash of light within 50 μsec. This makes it useful as a protection to the eyes during welding operations and as a protective shutter for sensitive light detectors, such as Vidicons.

In high-power lasers, ferroelectrics—in particular KDP—are successfully employed as Q-switches. This is extremely effective where laser damage would normally be expected.

Finally, it is worth mentioning a field where use is made of materials that, while not strictly ferroelectric, show some of the properties of ferroelectrics. These are important as dielectric resonators, especially in the microwave region. The materials are perovskites, and therefore closely related to true ferroelectrics, and are used in ceramic form. In this application, ceramics with permittivities between 25–100, low-temperature coefficients of capacitance, and low losses, are required. Ferroelectricity is accompanied by high dielectric losses; it is for this reason that non-ferroelectrics are of more use here. R. C. Kell at the G.E.C. Hirst Research Centre, Wembley, has been investigating the possible use of various zirconate and titanate mixtures for this purpose. Calcium zirconate and strontium zirconate have positive temperature coefficients of capacitance, whereas barium zirconate and strontium and calcium titanate have negative coefficients. Because they are perovskites they are capable of forming homogeneous solid solutions and it is possible, by ad-

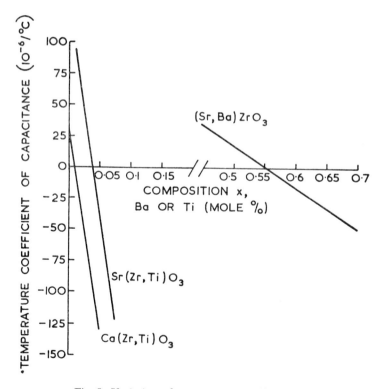

Fig. 5. Variation of temperature coefficient of capacitance with composition for various zirconate and titanate ceramics. (Courtesy of R. C. Kell.)

justing the compositions, to obtain very small temperature coefficients (Fig. 5). Typical permittivities of 35 with low losses ($Q = 2000$) can be obtained with a temperature coefficient of capacitance selected from a range between $+50 \times 10^{-6}/°C$ and $-100 \times 10^{6}/°C$.

Bibliography

BURFOOT, J. C. (1967) *Ferroelectrics*, London: Van Nostrand.
GRENOT, M., PERGALE, J., DONJON, J. and MARIE, G. (1972) *Appl. Phys. Lett.* 21, 83.
KELL, R. C., RICHES, E. E., BRIGGINSHAW, P., OLDS, G. C. E., THOMAS, A. J., MAYO, R. F. and RENDLE, D. F. (1970) *Elect. Lett.* 6 (19), 614.
MARIE, G. (1969) *Philips Tech. Rev.* 30, 292.
Proceedings of the 1971 IEEE Symposium on Applications of Ferroelectrics, published in (1972) *Ferroelectrics*, 3, Nos. 2/3/4.
Proceedings of the 2nd European Congress of Ferroelectricity (Dijon 1971), published in (1972) *J. de Physique*, 33, C-2.

A. M. GLAZER

FIBRE REINFORCEMENT OF CEMENT AND CONCRETE

Cement and concrete products are notable for their weakness in tension unless reinforced by steel bars and for their lack of toughness which gives rise to frequent cracking under impact loads, thermal shock or dimensional changes due to humidity variation. The use of fibres to overcome such deficiencies is traditional in building and asbestos cement is one fibrous composite which has already found wide acceptance. Recently interest has been expressed in glass, steel or plastics fibres as reinforcement for cement paste or for concrete.

In a composite with continuous fibres aligned parallel to the direction of the load, the Young's modulus of the composite E_c is given by a simple mixture theory:

$$E_c = E_f V_f + E_m V_m$$

where E and V are Young's modulus and volume fraction respectively, and the subscripts c, f and m refer to composite, fibre and matrix. A similar relation applies to the ultimate strength, σ.

$$\sigma_c = \sigma_f V_f + \sigma_m V_m.$$

For composites containing chopped fibres not aligned with the load it is necessary to introduce efficiency factors n_l for length and n_o for orientation.

$$\sigma_l = n_o n_l \sigma_f V_f + \sigma_m V_m.$$

Krenchel (1964) has shown by a statistical model that

$$n_l = 1 - \frac{\sigma_f d}{2\tau l}$$

where l is the fibre length, τ is the interfacial bond strength and d the fibre diameter. Krenchel also derives n_o for various typical situations; $n_o = 1/5$ for fibres distributed randomly in a volume and 3/8 for fibres distributed randomly in a plane.

More detailed analyses will be found in Holister and Thomas (1966) or Kelly (1966). The above treatment does however bring out the importance of the aspect ratio, l/d, of the fibres and the degree of alignment in the direction of the load.

In the case of fibre reinforced cement or concrete, the strength and stiffness of the matrix are governed principally by the ratio of mixing water to cement and the effectiveness of the compacting procedure in removing air voids. Mixing of fibre with cement paste or concrete is very difficult without increasing the water/cement ratio, thus weakening the matrix. Also as the amount of fibre is increased it forms a springy mat, preventing the elimination of voids. Consequently the amount of fibre in a cement based composite is relatively small, less than 10 per cent and more probably in the region 2·5–5% by volume.

Glass Fibre Reinforced Cement

A special difficulty in the case of reinforcement with glass fibre is that the A glass or E glass commonly available as fibre is rapidly attacked by the alkali released during the hardening of cement. A more resistant glass containing zirconia has been developed; an alternative approach is to coat the glass with an alkali-resistant size. It is also possible to use a cement of lower alkalinity for the matrix (Biryukovich, 1964). High alumina and pozzolanic cements have been used, or pulverized fuel ash (pfa) may be mixed with Portland cement to reduce the degree of corrosion.

Glass wool, which consists of single filaments, has been tried as a reinforcement but it is very difficult to mix into cement paste. The usual reinforcing element is a roving consisting of a number of ends or bundles lightly stuck together, each end being composed of many (commonly 204) individual fibres. The adhesive coating, applied during the spinning process, can be varied by the manufacturer so that the roving is either filamentizing or integral, that is it does or does not break down readily into individual filaments. The continuous roving may be used for filament winding or may be chopped into convenient lengths. It may also be converted into an open network known as crenette.

When glass fibre is used in conjunction with concrete containing coarse aggregate greater than 5 mm, the most important advantage is the 4–5 times increase in impact resistance, compared with plain concrete, for a 2–3% addition of fibre. There is also a small increase in bending strength, but compressive strength and splitting tensile strength are reduced. A more interesting material is obtained by glass fibre reinforcement of neat cement or a paste of cement plus fine filler or sand. The fine filler is conveniently pfa, as this reduces the alkalinity of the matrix.

An effective method of making thin sheets of this material is the spray-suction process (Ali and Grimer, 1969). A slurry of water/binder ratio 0·5–0·6 is pumped through a spray-head on which is mounted a glass-chopper. The stream of glass from the chopper meets the spray about half-way between the spray head and the paper covered perforated base of a suction mould. Adjustable screed boards round the edges enable sheets of various thickness to be made. After the required thickness has been built up excess water is extracted by suction. The mould is then reversed and used as a suction pad to carry the sheet to a demoulding station. At this stage the sheet is flexible and can be remoulded, avoiding sharp bends, and in this way corrugated or folded sheets, pipes, ducts and tubes are made. Decorative surfaces are obtained by trowelling coloured fine aggregate into the surface. The products have the fibres randomly orientated in two directions.

For other forming processes the materials must be pre-mixed. The fibre is first dispersed in the mixing water preferably with a suitable proportion of a thickening agent. The mixed fibrous paste may be filled into moulds in layers about 25 mm thick and tamped and vibrated. Alternatively, mixes may be pressed, injection moulded or extruded. Fluid mixes are used for pressing, excess water draining away during the pressing process. Both the pressing and extrusion method produce a certain amount of alignment of fibres, but the alignment factor is not so high as is the case with spray-suction fabrication.

Table 1. Properties of composites made by spray-suction.
Glass content 5% of 34 mm fibres. 28 days air storage

Modulus of elasticity, GN/m^2	20–22
Tensile strength, MN/m^2	15–18
Impact strength (Izod) Nmm/mm^2	22–25
Density, g/cm^3	2·0–2·1
Drying shrinkage, %	0·32–0·35
Moisture expansion, %	0·10–0·15
Thermal expansion, mm/mm/°C	$10·6 \times 10^{-6}$

By extrapolation from long-term tests the life of composites made of cement and alkali-resistant glass has been judged to be satisfactory under normal conditions of exposure (Steele, 1972). For use in water-saturated conditions, an addition of pfa to the cement is recommended. Alternatively high alumina cement may be employed or a polymer may be incorporated with the cement. Provided a suitable matrix is specified, fibre reinforced cement behaves very well in fire. Sufficient cohesion is provided by the reinforcement to keep the composite intact and prevent the penetration of flame. The stress-strain curve of glass fibre reinforced cement consists of two portions; a linear portion which terminates at about the cracking strain of the unreinforced matrix, and a quasi-plastic region which extends to ultimate failure (Allen, 1971). Creep behaviour is similar to that of concrete and fatigue behaviour is good. These properties of glass fibre reinforced cement suggest its use in thin sections where good impact behaviour and fire resistance are required. Prototypes of permanent shuttering to concrete, cladding panels, floor units, window units, pipes and ventilating ducts have been made. Other uses include boats and pontoons.

Steel Fibre Reinforced Concrete

In contrast to glass fibre steel fibre is used in conjunction with concrete or mortar and not neat cement paste. The range of wire sizes is from 0·15 mm to 0·5 mm diameter and 5 to 50 mm in length; the steel is usually high tensile grade (more than 2000 MN/m^2) since it is produced by a drawing process. The bond to concrete is poor and as a consequence some makers produce indented wires.

Although steel fibre mortar has been sprayed by the Gunite process, the concrete is produced by the normal mixing and casting procedure. It is preferable to add the wires last, shaking them in slowly by means of a sieve or vibrating feeder. Depending on the aspect ratio of the wires, only a limited amount can be mixed in before the wires start to form segregated balls of fibre (Hannant, 1972). Compaction of steel fibre reinforced concrete is usually possible with normal techniques such as table vibration, shutter vibration, poker vibration or by slip form paving. The workability decreases rapidly as the fibre content and the length/diameter ratio increase and the speed of compaction is greatly reduced at fibre volume concentrations greater than 1·0%.

It has been suggested (Romualdi and Batson, 1963) that steel fibre reinforcement prevents small cracks propagating through the composite and enables the concrete itself to carry tensile loads. However, cracking of the matrix sets in at a similar extension to that for the unreinforced concrete (Shah and Rangan, 1971). Typical properties of steel fibre reinforced concrete as quoted by the patentees (Battelle Development Corporation, 1963) are given in Table 2, but these values will vary in practice depending on the mix proportions and percentage of wire.

The frost resistance of the reinforced concrete is better than that of the control. On exposure the wires in the surface of the concrete rust causing some staining but at any appreciable depth in the

Table 2. *Properties of steel fibre reinforced concretes*

Property	Range	Advantage over unreinforced control concrete
Flexural strength (first crack)	6–12·5 MN/m²	2×
Flexural strength (ultimate)	6–17·5 MN/m²	3×
Compressive strength	35–84 MN/m²	Significant increase
Shear strength	5·1 MN/n²	2×
Modulus of elasticity	26·8 GN/m²	none
Impact strength	–	3×
Abrasion resistance	–	2×
Thermal expansion	10–11 × 10⁻⁶ mm/mm/°C	none

concrete the wires are protected by the alkaline nature of the matrix.

Potential applications for steel fibre reinforced concrete include roads, runways, floor slabs and floor surfacings, thin section structural components, pipes, railway sleepers and various marine and high temterature applications.

Concrete Reinforced with Plastics Fibres

Nylon and polypropylene are both unaffected by the alkali released by cement during hardening, and have been used for fibre reinforcement. They can be drawn into fibres but these are relatively expensive. The cheapest form of plastics fibre is fibrillated polypropylene. Thin film is cut into tapes about 18 mm wide which are then stretched in a stream of hot air. This orientates the molecules giving a high longitudinal strength while the lateral weakness causes it to split into fibrils. The fibrillated material also has a higher modulus of elasticity— 7·7 as compared with 4·9 GN/m² for the drawn fibre, although these modulii depend on loading rate.

Because the elastic modulus is very much lower than that of cement or concrete, the use of plastics fibre in a cement composite usually leads to lower tensile strengths. For instance, 2% of nylon fibre lowers the flexural strength of cement mortar by 12%. However, a very considerable increase in impact strength is obtained for quite small percentages of plastics fibres; a 26 times increase has been reported for nylon and 32 times for polypropylene (Goldfein, 1963; Williamson, 1965).

Fibrillated polypropylene has been used successfully to increase the impact resistance of pile shells.

From 0·15 to 0·25% of 12,000 denier material with 40–45 twists per metre, cut to lengths of 40 mm, is mixed with the concrete in a normal mixer. The concrete has maximum size of aggregate 10 mm and a water/cement ratio of 0·35. Air-entrained polypropylene reinforced mortar is also used for modelling the surface of concrete units. Among the uses suggested for the material is the fabrication of concrete to resist explosive shocks.

Acknowledgement

The work described has been carried out as part of the research programme of the Building Research Establishment of the Department of the Environment and this paper is published by permission of the Director.

Bibliography

ALI, M. A. and GRIMER, F. J. (1969) *J. Mats. Sci.* **4**, 389.
ALLEN, H. G. (1971) *J. Composites Mat.* **5**, 194.
BATTELLE DEVELOPMENT CORPORATION (1963) British Patent 1,068,163, Dec.
BIRYUKOVICH, K. L. BIRYUKOVICH, Y. L. and BIRYUKOVICH, D. L. (1964) *Glass-fibre Reinforced Cement*, Kiev, Budevelnik. Trans. No. 12. London: Civil Engineering Research Association.
GOLDFEIN, S. (1963) *Plastic Fibrous Reinforcement for Portland Cement*. Washington U.S. Clearing House for Federal Scientific and Technical Information AD 427,342.
HANNANT, D. J. (1972) Steel fibre reinforced concrete. *Proc. Intl. Conf. Prospects for Fibre-reinforced Construction Materials*, Olympia, London, 1971. Building Research Establishment: Garston, Watford.
HOLISTER, G. S. and THOMAS, C. (1966) *Fibre-reinforced Materials*. Amsterdam: Elsevier.
KELLY, A. (1966) *Strong Solids*. Oxford: Clarendon Press.
KRENCHEL, H. (1964) *Fibre Reinforcement*. Copenhagen: Akadamisk Förlag.
ROMUALDI, J. P. and BATSON, G. P. (1963) *J. Engineering Mechanics Division of ASCE* **89**, 147.
SHAH, S. P. and RANGAN, B. V. (1971) *ACI Journal Procs.* **68**, 126.
STEELE, B. R. (1972) Glass fibre reinforced cement. *Proc. Intl. Conf. Prospects for Fibre-reinforced Construction Materials*. Olympia, London, 1971. Building Research Establishment: Garston, Watford.
WILLIAMSON, G. R. (1965) *Fibrous Reinforcement for Portland Cement Concrete*. Washington U.S. Clearing House for Federal Scientific and Technical Information AD 465,999.

R. W. NURSE

FIELD-ION MICROSCOPY AND THE ATOM-PROBE FIM.

The field-ion microscope (FIM), introduced by Müller in 1951, still is the most powerful microscopic imaging device known, as it remains the only instrument capable of routinely showing individual atoms as the building blocks of the specimen. In the early development (Müller, 1960) the principle of using ions for radially projecting the surface of an evenly curved metal tip onto a phosphor screen, taken over from the preceeding invention of the field emission microscope, has been shown to produce a million times magnified image with a resolution of 2–3 Å. The information is carried from the specimen to the screen by ions which originate after the application of a high field at a critical distance of about 4 Å above each protruding metal atom of the specimen surface. Neutral molecules or atoms of an externally supplied "image gas", usually hydrogen, helium, neon or argon, are field-ionized at atomic sites of locally enhanced field strength of the order of 2 to 5 V/Å, and the ions are accelerated in radial trajectories towards the screen where they image as a pattern of bright spots. In an optical analogue, the emitting tip surface is the light source as well as the magnifying lens. Thus the tip (of typically 200–2000 Å radius of curvature) must be nearly hemispherical and smooth down to atomic dimensions. This is automatically achieved by applying a slightly higher voltage than that required for imaging whereby excessively protruding tip sections "field-evaporate" preferentially. After viewing and photographing the image, field evaporation may again be applied to remove in a controlled manner one atomic layer after another, so that by this sectioning, structural details in the depth of the specimen become accessible at the imaged surface.

The atomic resolution capability makes the FIM a unique tool for the direct observation of point defects in the crystal lattice of the metal tip. Individual vacancies, interstitials and impurity atoms can be located and counted, they may be introduced by *in situ* experimentation such as bombardment with medium energy ions or with α-particles, and their annealing behaviour can be followed quantitatively. Dislocations with their core structure, grain boundaries and slip bands introduced *in situ* by the field stress have been observed in atomic details. After this pioneering work in the fifties (Müller, 1960) the next decade of field-ion microscopy saw a proliferation of research in the direction of more quantitative investigations of the basic subjects of earlier exploration. While applications in the metallurgical field were broadened, the elementary processes of both field ionization and field evaporation were thought to be reasonably well understood. Details of image intensity or contrast on specimens of pure metals as well as alloys were explained in terms of local ionization probability and image gas sypply as well as charge transfer to, and preferential evaporation of one atomic species in alloys. Refined methods of image interpretation proved successful in the investigation of the ordered state of binary alloys such as PtCo and Pt_3Co, in the study of the defect structure of hexagonal close-packed metals and the observation of precipitate nuclei in alloy steels. The intricate atomic configurations of dislocations or stacking faults intersecting the tip hemisphere was elucidated by using computer simulation of spherical cuts through a lattice containing the defects. FIM work has also contributed significantly to basic problems of surface physics with the measurement of binding energies and surface migration energies of single atoms evaporated onto a perfect net plane, or the investigation of the interaction between two like or different atoms on such a substrate. Details of these techniques are described in text books by Müller and Tsong (1969), by Bowkett and Smith (1970), and by Müller and Tsong (1973).

The best FIM images are obtained from refractory metals using helium ions and a tip temperature below 30 K. Other metals field evaporate below the best imaging field of helium, 4·5 V/Å, and must be imaged with gases of lower ionization potential at some loss in resolution (Müller, 1960). Hydrogen ionizes at much lower fields, but the image quality is poor due to complications by the simultaneous appearance of H^+ and H_2^+ ions, the latter partly dissociating before leaving the high-field region near the tip, and due to the promotion of field evaporation and surface rearrangement by adsorbed hydrogen. Neon as the imaging gas most suitable for the common transition metals has been used since 1957. The exceedingly low phosphor screen efficiency when excited by neon ions made the experimentation difficult until photoelectronic image intensification became available to permit the imaging of gold and copper (McLane *et al.*, 1964). The recent incorporation of a microchannel plate, a device originally developed for photoelectronic night viewing, is a most helpful technical innovation in field-ion microscopy (Turner *et al.*, 1969). After converting the ion image into a secondary electron image amplified about 1000 times by cascading in the 50-μ diameter, two-dimensional channel arrays, the screen image is bright enough to reduce exposure times to fractional seconds. High-quality neon images of iron, gold and copper specimens are now almost routinely obtained. However, the hope of imaging even less stable specimens with argon, albeit with some sacrifice in resolution, has not been fulfilled. Only marginal images of aluminium or silver have been reported. Surface rearrangements due to the adsorption of residual gases or possibly argon itself, as well as the yield of the soft metals to the field stress during field evaporation seem to be limiting factors.

Without making use of its imaging capability, the field-ion emitter has been successfully employed as an ion source of a mass spectrometer. Field-ioniza-

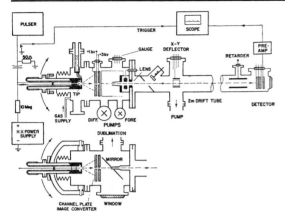

Fig. 1. The atom-probe field-ion microscope.

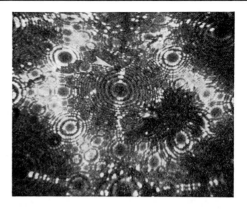

Fig. 2. Field-ion micrograph of a tungsten crystal. The probe hole of the atom-probe FIM is aimed at an impurity atom at the grain boundary.

tion mass spectra of organic vapors are surprisingly simple due to the absence of the dissociating electron impact of conventional ion sources, and a variety of surface reactions have been studied (Beckey, 1971). Experimentally more difficult is a mass spectrometric analysis of field evaporation products because the tip is quickly consumed by drawing an ion current of sufficient magnitude to exceed the noise level. This problem is radically solved in the atom-probe FIM introduced by Müller in 1967 (Fig. 1). A single atomic site is chosen by the observer of the FIM image by tilting the tip so that the desired image spot falls onto a probe hole in the screen. The selected atom is field evaporated by superimposing a nanosecond high-voltage pulse V_p over the dc image voltage V_0. It travels as an ion of charge ne with a constant speed v, calculated from the kinetic energy equation

$$\tfrac{1}{2}mv^2 = ne(V_0 + V_p),$$

through the 1- to 2-meter-long drift tube beyond the probe hole. The ion velocity is measured by determining the time of flight with a fast oscilloscope which is triggered by the evaporation pulse and which records the time signal of the arrival of the ion at the detector at the far end of the drift tube. The noise problem commonly encountered with the detection of a single particle is circumvented by the close time correlation between the evaporation and the detection event, which are typically some 1 to 10 μsec apart. The field evaporated particle is identified by its atomic mass m and its charge ne, where the multiplicity of the elementary charge may be $n = 1, 2, 3$ or 4.

The unique capability of identifying a single particle selected from the atomically resolved image of the specimen makes the atom-probe FIM by far the most sensitive microanalytical tool known today. At the present time only three or four instruments are known to be in operation. They are being used to determine the nature of bright image spots under various conditions on field-ion micrographs, such as those due to adsorption, or due to impurity atoms segregated at dislocation cores or grain boundaries (Fig. 2). The composition of intermetallic compounds forming the nuclei of precipitates has been analysed and the state of gaseous absorbates has been observed. Ordinarily, carbon monoxide is not dissociated upon adsorption. However, in the high field it is found to come off in the form of doubly and singly charged carbon ions and partially as a metal-carbide ion.

There are three fundamental observations made with the atom-probe which were totally unexpected and necessitate a reconsideration of the basic processes of both field ionization and field evaporation (Müller, 1971). (1) The noble image gases helium and neon are found to be adsorbed at the emitter up to a temperature of 150 K, indicating a field induced binding energy 10 to 20 times larger than the van der Waals forces. The adsorbed noble gas atom is located at the apex of the imaged metal surface atom, where it facilitates field ionization of a contacting second gas atom. (2) Field evaporation of the refractory metals occurs in the form of fourfold and threefold charged ions, rather than doubly charged ones as expected theoretically. The relative abundance of the various charge states depends upon evaporation rate and temperature. (3) In the presence of helium as imaging gas the substrate metal field evaporates in the form of a multiply charged metal-helium molecular ion, and in the presence of hydrogen, as a metal-hydride ion.

In view of these new experimental results it is clear that the current theories of field ionization and field evaporation are incomplete. Earlier quantitative interpretations of experiments involving the determination of binding and surface migration energies must be revised to take account of the quite complex situation at the metal surface.

Bibliography

BECKEY, H. D. (1971) *Field Ionization Mass Spectrometry*, Pergamon Press.

BOWKETT, K. M. and SMITH, D. A. (1970) *Field Ion Microscopy*, Amsterdam–London: North Holland Publishing Co.

BRANDON, D. G. (1966) Field-emission and field-ion microscopy; recent developments in, in *Encyclopaedic Dictionary of Physics* Suppl. **1**, 103, Oxford: Pergamon Press.

McLANE, S. B. et al. (1964) *Rev. Sci. Instrum.* **35**, 1217.

MÜLLER, E. W. (1960) *Adv. Electronics Electron Phys.* **13**, 83.

MÜLLER, E. W. (1961) Field emission, in *Encyclopaedic Dictionary of Physics* **3**, 120, Oxford: Pergamon Press.

MÜLLER, E. W. and TSONG, T. T. (1969) *Field-Ion Microscopy, Principles and Applications*. New York, London, Amsterdam: Elsevier.

MÜLLER, E. W. and TSONG, T. T. (1973) Field-ion microscopy, field immigration and field evaporation, in *Progress in Surface Science*, Vol. 4, part 1, pp. 1–139 (Ed. S. Dawson), Oxford, Pergamon Press.

MÜLLER, E. W. (1971) *Ber. d. Bunsenges.* **75**, 979.

TURNER, P. J. et al. (1969) *J. Sci. Instrum.* **2**, 731.

ERWIN W. MÜLLER

FOUR-COLOUR PROBLEM. The four-colour problem is one of the most famous unsolved problems in mathematics. The problem is to find how many colours are needed to colour any map, drawn on a plane surface or on the surface of a sphere, such that no two regions with a common border will have the same colour. It is obvious that at least four colours are necessary and it is conjectured that four colours will prove to be sufficient. The problem was first formulated by a student at Edinburgh University, Francis Guthrie, in 1852 but did not appear in print until 1878. Since then it has been worked on by many famous mathematicians but without success. The lines on which a proof may be constructed to show that five colours are sufficient is given below but to prove more relies heavily on graph theory. The best recent result using such an approach has been to show that four colours are sufficient for a map of less than thirty-nine regions.

Concepts

Any map on the surface of a sphere can be transformed to the equivalent plane map by the use of a suitable projection. The map to be considered is then represented by a planar graph, consisting of regions, edges and vertices, where the edges separate adjacent regions and meet only at vertices.

The unique *dual* of this planar graph may be considered as that graph obtained by taking an interior point of each region as a vertex in the dual, and joining each pair of these vertices which lies in adjacent regions, by an arc which crosses the common boundary once. These arcs, which form the edges of the dual, must not intersect.

Sufficiency of Five Colours

The proof uses the fact that the average number of regions adjacent to any region is less than six. This may be seen on constructing the graph in the following way. Starting with two adjacent regions (see Fig. 1) A and B, region C is constructed by connecting a point on the boundary of A to a point on the boundary of either A or B by an arc. This arc adds one region and not more than three edges to the total (the maximum is obtained if two new vertices, one on the boundary of A and one on the boundary of B, are created). Further arcs are drawn in a similar fashion and if $n - 3$ arcs are drawn there will be a total of n regions (including the outside region) and a total of not more than $3n - 6$ edges (including the initial 3 edges bounding A and B). Since each edge separates two regions there is a total of not more than $6n-12$ adjacencies of regions, i.e. in a general map there are not more than an average of $6-12/n$ adjacent regions to any given region. Hence there will be at least one region which has five or less adjacent regions.

The sufficiency of five colours for a planar map is proved by induction on the dual, i.e. the vertices are coloured. From the above argument there is at least one vertex V that is connected to five or less other vertices. Remove V from the graph and assume the theorem is true for the remaining $n - 1$ vertices. Taking the worst possible case and assuming that there are five adjacent vertices to V each of which is coloured in a separate colour, respectively (for clockwise rotation of the vertices about V, say), c_1, c_2, c_3, c_4 and c_5, it may be shown that at least two of the vertices V_1, V_2, V_3, V_4 and V_5 may be coloured with the same colour, hence leaving one colour free to colour V when it is reintroduced into the graph.

Consider the graph of vertices coloured with c_1 and c_3 only, the colours associated with V_1 and V_3.

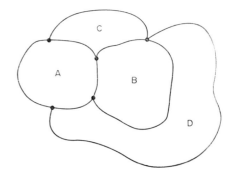

Fig. 1

Either there is a path through this graph from V_1 to V_3 or there is not. In the latter case V_3 can be assigned colour c_1, leaving c_3 free for V. Otherwise, consider the graph of vertices coloured with c_2 and c_4. There can be no path through this graph from V_2 to V_4 since such a path would have to intersect the path joining V_1 to V_3. Hence c_2 may be assigned to V_4 leaving c_4 free to colour V.

Hence the theorem is true for n regions and the sufficiency of five colours follows.

Many equivalent formulations of the four-colour problem using graph theory concepts have been put forward in the last few years. Detailed accounts of the more important of these are to be found in Ore (1967).

Surprisingly, perhaps, the equivalent problems have been solved for more complicated surfaces. Six colours have been proved to be sufficient for the Moebius strip, the Klein bottle and the projective plane, and seven colours for the torus.

Bibliography

BUSACKER, R. G. and SAATY, T. L. (1965) *Finite Graphs and Networks*, McGraw-Hill.
HEAWOOD, P. J. (1890) Map-colour theorems, *Quart. J. Pure Appl. Maths.* **24**, 322–38.
MOON, J. W., MOSER, L. and ZYKOR, A. A. (1963) *Theory of Graphs and its Applications*, Proceedings of Symposium in Smolenice.
ORE, O. (1967) *The Four-Colour Problem*, Academic Press.
TUTTE, W. T. (1967) Graph theory, in *Encyclopaedic Dictionary of Physics* (J. Thewlis, Ed.), Suppl. **2**, 110, Oxford: Pergamon Press.

JOHN H. THEWLIS

FRACTIONAL HORSEPOWER MOTORS. At first sight there might appear to be nothing to say about small electric motors which had not already been said about large motors of various types individually, e.g. about induction motors, synchronous motors, d.c. machines, etc., but this would be analogous to saying that there was no feature of an ostrich which had not been disclosed on sparrows, hawks, humming birds and the like. For an ostrich cannot fly and there is a *fundamental* reason why this is so, which may be described purely in terms of engineering principles. The lifting force available from a wing of given shape is proportional to its area, that is to the *square* of a linear dimension. The mass of a bird of given shape is proportional to its volume, i.e. the *cube* of a linear dimension. Even despite the apparent ingenuity of the evolutionary processes, there emerges a size beyond which unassisted flight is physically unrealizable. The physical properties which decide what size this shall be involve the viscosity and density of air and the nature of muscles and of the inherent power/weight ratio of living things, which arguments lead eventually to the structure of the carbon atom and the dependence of what we term "life" on the provision of water.

It is interesting and instructive to discover what physical properties determine the minimum size of insect which can propel itself in still air, the maximum size to which an insect (or a mammal) may grow and the maximum size of creature which can survive a fall under gravity from any height.

What is at issue in the case of electric motors is whether there is a maximum or a minimum size beyond which, for a given type of motor, manufacture becomes unprofitable because their properties have deteriorated as the result of size changes alone. For this purpose, all electric motors must be divided into two distinct classes:
1. Those which radiate energy electromagnetically across the inevitable space between moving member and fixed member.
2. Those which merely induce "magnetism" in the secondary member.

It is convenient to call the first of these classes "electromagnetic machines" and the second "magnetic machines".

In a paper published in 1965† it was shown mathematically that machines in the first class are inherently better the *bigger* they are, whilst those in the second are better the *smaller* they are. This "watershed" in what might appear to be a collection of "like" objects cannot be in any way compromised.

What must appear, at first, strange to the student of electrical engineering is that there are definable size limits beyond which certain types of machine not only become unprofitable but fail to operate at all. Thus there is a minimum size of d.c. shunt generator which can be self-exciting at a given speed, and because the strength of steel has a certain value, there is an overall minimum size *at any speed* which does not burst the rotor by centrifugal action.

Electromagnetic actions can be measured in terms of the forces between charges which have either a relative velocity, a relative acceleration, or both, the first giving rise to the so-called "flux-cutting rule" for e.m.f. generation and the second to the "flux-linking" rule. The ratio of the electrostatic force between two charges q_1 and q_2 and the electromagnetic force due to a relative velocity v between q_1 and q_2 is given by the expression $[1 - (v/c)^2]^{-\frac{1}{2}} - 1$, where c is the velocity of electromagnetic waves. The effective relative velocity between the electrons which constitute electric currents can be shown to be dependent only on current density and at the order of current densities approved for household wiring, electrical machines and transmission apparatus the value of v is of the order of 1 cm/sec. Therefore of the apparently available electrostatic force the percentage available to make an electro-

---
† Laithwaite, E. R. (1965) The goodness of a machine, *Proc. IEE*, **112**, 538.

magnetic machine is only

$$\{[1 - 1/(3 \times 10^{10})^2]^{-\frac{1}{2}} - 1\} \times 100. \quad (1)$$

However, the number of conduction electrons in 1 cm³ of copper carry a total charge of about 8000 coulombs, so that the apparently "available" charge in a cube of copper of 1 cm side, if placed 1 metre away from an equal charge would develop a force of

$$64 \times 10^6 \times 36\pi \times 10^9 \text{ newtons} \quad (2)$$

which is equivalent to the force of gravity on a mass of $7 \cdot 38 \times 10^{17}$ kg. One may now pose the philosophical question: Having multiplied the force (2) by the expression (1) above, the resulting electromagnetic forces are seen to be such as to allow machines to be built which compete on almost level terms of power/weight with engines which are driven by steam, oil or petrol. Is this pure chance, or do human beings have appropriate dimensions by what Sir William Bragg described as "the Nature of Things"?

It is not the purpose of this article to debate this question but the purpose in raising it was to emphasize that induction, synchronous and a.c. and d.c. commutator motors show a deterioration in performance (particularly in power/weight ratio—as of course do human creatures) as attempts are made to manufacture smaller and smaller models.

In fractional h.p. sizes the "magnetic" machines (consisting of hysteresis motors and reluctance motors) become competitive. Provided the power output is not reduced to a fraction of a watt, however, the electromagnetic machines just survive although factors other than pure size militate against them also. For example, fractional h.p. motors are required to drive record players, tape recorders and other household appliances where only a single-phase supply is available. Thus the induction motor requires a starting mechanism (usually a shaded pole arrangement) which does not, at best, give a high starting torque.

The action of the various types of machine will now be described and the particular exploitations on the one hand, and limitations on the other will be discussed separately as the design of each is explained.

Induction Motors

The difficulties which the designer of small induction motors faces may be listed simply under the headings

(i) Absence of starting torque unless a specific mechanism for starting is included.

(ii) Deterioration of the Goodness factor† if all dimensions are in the same proportions as those of a large machine.

---

† Laithwaite, E. R. (1965) The Goodness of a machine, *Proc. IEE*, **112**, 538.

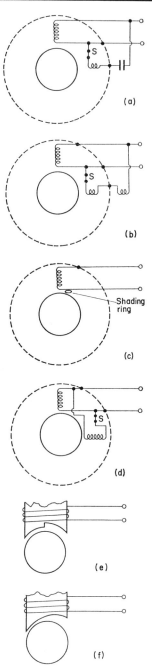

Fig. 1. Starting mechanisms for single-phase induction motors. (a) Auxiliary winding with "capacitor start". (b) Auxiliary winding using inductor start. (c) "Shaded pole" motor. (d) Auxiliary winding with inherently different R/X ratio from that of main winding. (e) "Reluctance start" motor. (f) Continuously changing airgap type of reluctance start motor.

Starting mechanisms are all aimed at developing a component of magnetomotive force which is displaced both in time and in space relative to the main exciting m.m.f. Some common methods of achieving this feature are illustrated in Fig. 1. The schematic diagrams in (a) to (d) show how separate windings are included in the quadrature axis and in each case the current which they draw from the supply can be seen to be out of phase with that drawn by the main winding: (a) is the "capacitor-start" motor in which the phase of the current in the auxiliary winding is of leading phase. Type (b) follows naturally in that an inductor can be seen to produce a lagging current, giving reversed torque compared with that of (a). Where the machines are "large", i.e. approaching 1·0 h.p. or more, the auxiliary windings are often "short-term rated", which design technique exploits the fact that the auxiliary winding is only needed at slow speeds. In such cases the phase-splitting component is automatically open circuited by a centrifugal switch, s, which opens when a certain predetermined speed is reached.

Among motors in which the phase-splitting device remains connected at all times the commonest type is the "shaded-pole motor" shown in (c). Simple theoretical explanations of the action of this device usually go to some length to show that, in a magnetic circuit arrangement as shown in Fig. 2, the effect of secondary current in the shading ring is to cause the flux which threads it to lag on the flux in the unshaded part of the pole. Such explanations are, however, less than a "half truth" for they take no account of the rotor conductors and yet the rotor current loading at standstill is likely to be larger than that in the shading ring.

What really occurs is evidenced by the type of starting mechanism illustrated in Fig. 1(e) in which a "step" in the airgap is sufficient to cause the flux crossing the short airgap to lag that which crosses the large. Such machines are known as "reluctance start motors". One explanation of the action of the step may be offered in terms of the effective Goodness factor of the separate parts of the motor, the flux linking the secondary conductors in a path of lower reluctance having a higher Goodness factor and therefore a larger angle of lag. Thus it is obviously necessary to examine the action at the edges of the poles of a simple shaded-pole motor, as shown at A and B in Fig. 3. The fact that the rotor conductors themselves shade the pole results in substantial torques being developed in the directions shown. The effect of the shading ring can be shown to be mainly that of reducing the flux density at edge B and therefore allowing the travelling field at A to become effective. This improved physical explanation was first put forward by Eastham.†

Sometimes a combination of stepped gap and shading ring is employed, giving two stages of phase shift. Such a motor is called a "triple-start" machine.‡ The ultimate in stepped-gap technique is achieved when the airgap length varies continuously as shown in Fig. 1(f).†

It should be emphasized that in all shaded-pole motors the airgap flux never becomes a uniformly rotating field, i.e. of constant amplitude with constant space-phase distribution. Rather the field pattern may be described as "elliptic" consisting of a purely *rotating* component super-imposed on a purely *pulsating* component. When running, the action of the rotor currents is to enhance the rotating field at the expense of the pulsating and in an "ideal" machine the latter would tend to zero as the rotor approached synchronous speed. This fact troubled Nicola Tesla, the inventor of the induction motor, for he feared (before he ever made a successful machine) that a truly rotating field was required and

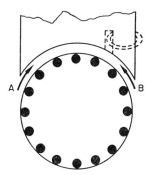

Fig. 3. The self-shading action of the edges of a pole.

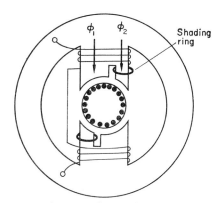

Fig. 2. Schematic layout of a shaded pole motor illustrating the phase shift obtained by the short-circuited rings.

† Eastham, J. F. and Williamson, S. (1973) Generalized theory of induction motors with asymmetrical airgaps and primary windings, *Proc. IEE*, **120**, 767.

‡ Veinott, C. G. (1959) *Theory and Design of Small Induction Motors*, McGraw-Hill Book Co., New York.

yet this only appeared possible when the machine was delivering zero output, i.e. at synchronous speed.

In electromagnetic devices, the main action is subject to the Goodness equation and therefore improves with increase in size. But some of the imperfections in these machines are purely magnetic, i.e. those associated with the paths of leakage flux, and these tend to allow imperfections to dominate the performance in small machines. One criterion of design has been stated as follows:

$$B_g/J_s = k(p/g)$$

where B_g is the airgap flux density, J_s the stator current loading, p the pole pitch or half wavelength of the travelling field and g the airgap length.† This means that in fractional h.p. motors, a limit is reached where it is no longer possible to reduce the mechanical clearance so as to scale the length of the magnetic circuit in proportion to the rotor periphery. Thereafter good designs of fractional h.p. motors are seen to have wide slots and narrow teeth to give a high value of J_s and to accept the physical limitations imposed by size alone as a relatively low value of airgap flux density compared with that of a large induction motor. Indeed the teeth of small induction motor stators are often as narrow as can be punched out of steel sheet without their breaking off in the process. Some liberties, however, can be taken with the rotor slots and it is not uncommon for the cage of conductor to be totally embedded in steel, as shown in Fig. 4(a), for reasons of mechanical strength, the magnetic short-circuitry "bridges" over the slots, as at X, being very highly saturated, and therefore ineffective as magnetic short circuits, when the machine is in operation. Similarly the stators of shaded-pole motors are often punched out in the shape shown in Fig. 4(b), again appearing, at first sight, to constitute magnetic short circuits.

Single-phase Commutator or "Universal" Motors

If *either* the field current or the armature current of a d.c. motor be reversed, the direction of the torque is reversed, so that if *both* are reversed the direction of thrust and of motion remains unchanged. Hence a machine designed as a d.c. motor should be capable of running on a.c. supply. However, there are several features of a.c. operation which limit severely the applications of the a.c. commutator motor.

First, the inductance of the field winding of a shunt motor is far too high to allow even a "trickle" of field current to flow at a reasonable armature voltage (the latter is limited by the voltage between adjacent commutator segments which will not be capable of producing flash-over). Single-phase machines are therefore limited to series operation. The speed/torque curve of a series machine is almost of rectangular hyperbolic shape. The machine delivers its maximum torque at standstill and on no-load its theoretically attainable speed is infinite. In practice the no-load speed of medium-sized motors is high enough to fling the rotor conductors from their slots, and speed cut-outs are an essential feature in these motors.

Commutation is a very serious problem in a.c. machines. As with d.c. machines, the inductance of the rotor windings delays the current reversal point until the current continues beyond the position at which a brush leaves a commutator segment and, if no corrective measures are taken, considerable arcing and burning occurs. Armature reaction aggravates the problem, but unlike the d.c. machine, the a.c. motor cannot easily be fitted with interpoles, for the m.m.f. which they produced would be out of phase with that required to achieve good commutation. Furthermore, the degree of phase shift is a function of load, and although several ingenious phase-splitting circuits have been designed, none has ever been popular with manufacturers. Furthermore, the commutator is called upon to reverse $\sqrt{2}$ times the current in the equivalent output d.c. machine since the peak value of a.c. is $\sqrt{2}$ times its r.m.s.

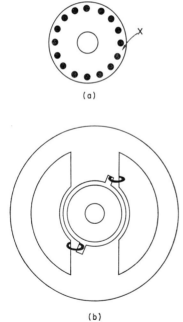

Fig. 4. Magnetic "short-circuits" which are saturated in normal use. (a) A cage rotor. (b) A shaded pole stator.

---

† Laithwaite, E. R. and Barwell, F. T. (1969) Application of linear induction motors to high-speed transport systems, *Proc. IEE*, **116**, 713.

value. Special compensating windings have been used successfully to combat armature reaction and although high powered single-phase commutator motors have been applied to rail traction the following additional design features have been found necessary:

(i) Laminated iron for both stator and rotor magnetic parts.
(ii) Multi-polar (usually twelve or more) construction, giving a low value of flux/pole.
(iii) Single-turn armature coils necessitating a large number of commutator segments (perhaps two to three times the number in an equivalent power, d.c. motor).
(iv) Long commutator segments to compensate for the narrowness introduced by (iii), in order to maintain the brush contact area necessary to pass the armature current.

When the features have been incorporated the machines are seen to be limited to low-voltage supplies as a consequence of (iii). The difficulty in commutation is responsible for most of the above design features, for in addition to the difficulties of current reversal in a d.c. machine, the armature coil undergoing commutation receives a directly induced e.m.f. by *transformer* action from the primary winding. This e.m.f. is proportional to (flux/pole) $\times$ (turns/coil) $\times$ (frequency of supply). Large a.c. traction motors have only been run successfully when the supply frequency has been reduced to 50/3, i.e. $16\tfrac{2}{3}$ Hz by mercury arc divider.

But with fractional h.p. commutators most of the above limitations can be relaxed. In the first instance the windage torque alone is sufficient to prevent the motor from bursting due to overspeed on no-load. In this it resembles the insect for which flying becomes more difficult the smaller are its dimensions. Efficiency is not too serious a consideration in a small motor and the provision of high resistance brushes is alone sufficient to reduce the short-circuit current introduced into armature coils as they undergo commutation to a safe value. This same high resistance reduces the armature coil leakage time constant to a low value and machines driven from a supply at 50 Hz become possible.

Of course a machine designed for a.c. will run superbly on d.c. and this was probably the meaning behind a famous remark attributed to Steinmetz that, "For an a.c. commutator motor, the ideal frequency is zero", rather than suggesting that no attempt should be made to design an a.c. machine.

Reluctance Motors

The principle of a reluctance machine is illustrated in Fig. 5. If used as a generator the flux threading the stator output coil is seen to vary as the result of changes in reluctance of the magnetic circuit as the result of movement of the salient ferromagnetic rotor. In Fig. 5 the rotor is shown at the positions of maximum and minimum reluctance. The flux threading the output coil is limited by the saturation level of the iron. The minimum flux is limited by the topology of the magnetic circuit at the position shown in Fig. 5(b). As the size of the machine is

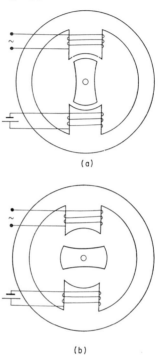

Fig. 5. Schematic magnetic circuit layouts illustrating the principle of a reluctance motor.

increased the reluctance of the circuit in this position is seen to reduce linearly with increase in dimension, being proportional to

$$\frac{\text{path length}}{\text{area}}.$$

Since the maximum flux density cannot increase indefinitely, the ratio of maximum to minimum flux decreases with increase in size and therefore the output does not increase in proportion to the volume of the machine as is the case with electromagnetic machines. The above argument is merely a particular case of the overall rule that magnetic machines improve with size reduction.

When used as a motor the d.c. excitation is usually omitted, it being sufficient to allow the machine to seek positions of minimum reluctance at the times at which the flux is a maximum, and this may be seen to occur at speeds 0, 1500, 3000, 4500, ... r.p.m. for the structure shown in Fig. 5, fed from a supply of 50 Hz. The torque at the lowest speed is usually the greatest and is therefore the designed speed.

Relatively recently a new rotor topology due to Lawrenson and Agu† has improved the ratio of maximum and minimum reluctance for a given size of machine. The basic arrangement of these improved machines is shown in Fig. 6.

Multi-polar reluctance motors are virtually universally adopted for mains-fed electric clocks, the basic arrangement being shown in Fig. 7. The corners of the teeth on rotor and stator are rounded to prevent sudden reversal of the machine on a single

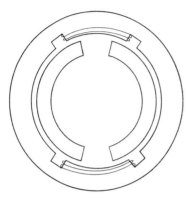

Fig. 6. Magnetic circuit diagram of a relatively new type of reluctance motor.

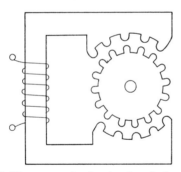

Fig. 7. The magnetic circuit of a clock motor.

tooth lock, for, like a single-phase induction machine without starting mechanism, the single-phase reluctance machine may run in either direction but may be self-starting if switched on when the rotor occupies a torque-favourable position and is of low inertia. Electric clock motors are extremely cheap to produce, both stator and rotor having magnetic parts which consist often of no more than three or four rather thick stampings.

Hysteresis Motors

The action of a hysteresis motor is quite different from that of any other type of electric motor in

---

† Lawrenson, P. J. and Agu, L. A. (1964) Low inertia reluctances machine, *Proc. IEE*, **111**, 2017.

that there is no induced e.m.f. in the secondary and the latter has no salient parts so that reluctance action is not responsible for the torque between rotor and stator. The output torque is rather the result of the inherent lag of the magnetic induction B on the applied m.m.f., which phenomenon is known as "hysteresis". A physical explanation of the mechanism is attempted with reference to Figs. 8 and 9, the former showing the B–H relationship for the rotor steel and the latter a disc of this steel which is being subjected to a relatively moving m.m.f. which, in the simple case shown, consists of a horseshoe magnet rotating about an axis normal to the plane of the diagram so that the poles (N. and S.) rotate parallel and in close proximity to the surface of the rotor sheet.

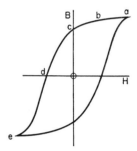

Fig. 8. Magnetic characteristics of the steel used in hysteresis motors.

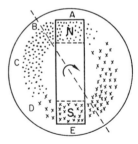

Fig. 9. Flux induced in a steel disc as the result of rotation of a horseshoe magnet.

The position A on the rotor corresponds, at the instant shown, to the condition a on Fig. 8, since at this point the rotor experiences maximum m.m.f. At B, the flux is scarcely reduced, having just been under the influence of maximum m.m.f. and therefore at a condition specified by b on Fig. 8. In the zero m.m.f. position C, a large proportion of the maximum flux remains, corresponding to the remanence point c on Fig. 8. Only when the oppositely sensed m.m.f. has built up to a considerable value (the coercivity, shown at d) does the rotor flux fall to zero (D in Fig. 9). Thereafter, it builds up rapidly to its maximum reversed value at E.

The rotor flux has been represented by dots and crosses on Fig. 9 so as to give a visual indication that the axis of the flux (shown dotted) lags behind that of the applied m.m.f. and therefore the rotor is subject to an accelerating torque in the same manner as is the synchronous motor, the induction motor, etc., all of which exhibit a lag between the axes of primary m.m.f. and secondary flux. What is most important to note, however, is that the magnitude of the lag, and therefore of the torque, in the case of the hysteresis motor is in no way speed-dependent, but only depends on the shape and size of the B–H loop for the steel being used. The conditions shown in Fig. 9 therefore obtain when there is *no* relative velocity between applied m.m.f. and rotor. It is only necessary for there to *have* been, at some earlier time, a relative velocity in the direction indicated by the arrow. The speed-torque curve of the motor is therefore as shown in Fig. 10, being the d.c. compound-wound-motor designer's idea of perfection. It is sad indeed that the use of hysteresis motors is economically restricted to fractional h.p. sizes.

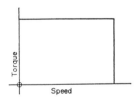

Fig. 10. Speed-torque characteristic of a hysteresis motor.

For ease of production the rotors of hysteresis motors consisting (as they must) of highly hysteretic steel are usually solid and will therefore exhibit eddy currents in the presence of a relative m.m.f. velocity. Such currents prevent secondary flux from penetrating deep into the steel (skin effect) and the conventional rotor therefore consists of a thin-walled steel cylinder.

Recent research, however, has shown that greater values of torque/inertia ratio can be obtained using rotors consisting of a set of thin discs mounted on a common shaft and spaced apart by distances depending on the disc diameter† (the larger the diameter, the larger the gap between successive discs). Such a rotor can be driven by a short linear primary system, as shown in Fig. 11, the action then being analogous to that of a mechanical rack and pinion. The gain in acceleration is possible because the flux emanating from the stator is able to spread in all directions and in particular to distribute itself axially so as to enter the *faces* of the discs and such flux

Fig. 11. A "rack-and-pinion" type of hysteresis motor.

is much more important than that which enters the *edges* of the discs, indeed, in a theoretically "ideal" design the disc thickness should reduce gradually to zero with increasing radius.

Bibliography

LAITHWAITE, E. R. (1966) *Induction Machines for Special Purposes*, London: Newnes Ltd.
LAITHWAITE, E. R. (1971) *Linear Electric Motors*, London: Mills & Boon.
VEINOTT, C. G. (1959) *Theory and Design of Small Induction Motors*, New York: McGraw-Hill Book Co.
VEINOTT, C. G. (1970) *Fractional and Subfractional-horsepower Electric Motors*, New York: McGraw-Hill Book Co.

E. R. LAITHWAITE

---

† Laithwaite, E. R. and Hardy, M. T. (1970) Rack and pinion motors: hybrid of linear and rotary machines, *Proc. IEE*, 117, 1105.

G

GEOMETRODYNAMICS. Geometrodynamics is a new point of view of a particular branch of Einstein's theory of general relativity: the study of empty (matter-free) space-times. Standard general relativity can be characterized as a description of the gravitational field, and its interaction with matter fields, by means of a four-dimensional curved space-time of simple topology. By contrast, geometrodynamics deals exclusively with the sourceless gravitational field, interacting only with itself, viewed as a time sequence of three-dimensional curved spaces of possibly complicated topology. These two views of general relativity have Einstein's field equations as a common basis, and disagree only about the realm of applicability of these equations. The dynamics of empty but highly curved space is so rich and varied that it can imitate many of the large-scale properties of matter. But if Einstein's equations allow a geometrical explanation of a phenomenon, it would be redundant to have in addition another non-geometrical explanation. For this reason geometrodynamics replaces *all* "external and physical" matter by intricately curved empty space.

The close relationship between geometrodynamics and Einstein's general relativity implies that there can be no experimental distinction between the two views in the realm of large-scale gravitation physics. Justification of geometrodynamics might come from its quantized applications to small-scale physics. From this "quantum-geometrodynamics" it should be possible to derive the properties of matter fields (such as electromagnetism or elementary particles) as large-scale descriptions of small-scale *geometrical* properties of space-time. Classical geometrodynamics is particularly useful in the mainstream of general relativity research as a simple, complete part of the theory in which new mathematical methods and physical ideas can be discovered and tried out.

Geometrodynamics was founded in the 1950s by John Wheeler as a logical continuation of Einstein's work in general relativity. He accepted Einstein's basic conviction that physics is geometry, but he also took seriously the failure to find a satisfactory unified field theory. In the realm of pure gravitation, Einstein was eminently successful in reducing physics to geometry: He explained "gravitation without gravitation", as a consequence of space-time curvature, and for particles he obtained "equations of motions without equations of motions", as a consequence of his field equations. What was lacking was a geometrical interpretation of matter: it enters standard general relativity only as the source of the gravitational field, through its mass-energy, momentum, and stress. Wheeler proposed to attack this problem in a spirit of "daring conservatism": conservative in retaining Einstein's equations in their original 1916 form as description of gravity, he daringly extended this description of the sourceless field to spaces of all types of geometrical and topological complexity. It may well be asked how this procedure could be expected to reproduce, even approximately, the standard theory *with* sources, since for any given solution of Einstein's equations the source is uniquely determined: it can be computed from the curvature of the corresponding geometry. However, this is true only if one knows the details of the solution everywhere. Given some gravitational field, or curvature, in only a portion of space, no one can deduce the exact nature of the source of the curvature. Due to the nonlinearity of Einstein's equations, curvature in one region may produce more curvature in other regions of space. Thus, not only matter but also curvature can be an effective source of gravity. It is therefore reasonable to suppose that the source of *all* curvature is curvature itself, and that what appears as matter on a large scale is really curvature of empty space-time on a smaller scale.

An illustration of these ideas is the geometrodynamical model of a "mass without mass", or "geon". A geon is a massive object constructed out of small-scale ripples in the gravitational field, i.e. gravitational waves.† Gravitational waves are disturbances in the gravitational field which propagate in a fashion similar to electromagnetic waves. But whereas the waves of Maxwell electromagnetism can be superimposed linearly without interaction, the nonlinearity of Einstein's equations demands

---

† The geon was first introduced as a massive object made from zero restmass electromagnetic waves, held together by their own gravitational attraction. But in the spirit of geometrodynamics, electromagnetic waves themselves should be explained as a property of geometry. Hence we here confine attention to purely geometrical, "gravitational" geons.

that gravitational waves attract each other. The attraction can be sufficiently strong that the waves are bent around into the persistent configuration of a ring. This ring of gravitational waves held together by their own gravitational attraction is an example of a geon. Although such a geon is not stable towards decay or collapse, it can last a long time compared to its size. Thus the gravitational wave curvature can last long enough to produce further curvature at large distances from the geon. This large-scale curvature is the same as what would be produced if the geon were replaced by a ring of real matter. Test particles respond to the curvature by performing Keplerian motions, like in the field of a real mass. Likewise in other large-scale effects the geon behaves like a real mass, a "mass without mass".

It is important to remember that the geon is merely one classical model of a massive body, and has nothing directly to do with elementary particles. Similarly, other classical models have no such connection to the quantum world, since in the geometrodynamic view, the quantum principle is to be added to general relativity rather than derived from it, as Einstein would have hoped. Thus, only quantum geometrodynamics could be expected to yield "elementary particles without elementary particles". However, one would expect that classical geometrodynamics should prepare the way for such features of elementary particles which already show up on the classical level. Such a feature is the electrical charge.

Charge can be viewed as a source or sink of lines of force. The standard approach views it as a special physical "substance" whose ability to spew forth or to swallow lines of force is not further explained. Wheeler's geometrodynamic model of "charge without charge" explains this ability by the multiply connected nature of space, illustrated in Fig. 1 by

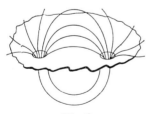

Fig. 1.

a two-dimensional analog: lines of force do not end at the charge but flow into the handle or "wormhole". They emerge at the other wormhole mouth, which represents a charge of the opposite sign. Thus, to understand charge and other properties of elementary particles it is essential to apply Einstein's equations—eventually in quantized form—to multiply connected spaces. The multiple connectedness is the major new feature introduced by geometrodynamics into general relativity; the intention is to use the great variety of possible topologies to explain a great variety of physical effects, and possibly all of physics. The details of this program today still await implementation.

If it is valid to consider all of physics as geometry, then it is important to recognize in geometry the dynamical nature familiar from other field theories: one needs to know what variables can be initially freely specified, and what equations govern their time development. (These typically geometrodynamical questions of course have implications for all of general relativity, and workers both within and without the geometrodynamic persuasion have contributed to the solution.) The original formulation of Einstein's theory does not immediately answer these questions, since a solution is a single four-dimensional manifold rather than a sequence of geometries in time. In order to discover dynamics in a solution, one must slice the four-manifold by a sequence of three-dimensional spacelike surfaces. Each spacelike surface is invariantly characterized by the intrinsic geometry within the surface and the extrinsic curvature which describes how the surface is imbedded in the four-manifold. The intrinsic and extrinsic geometries change as one proceeds from surface to surface, and this change constitutes the time development.

It may appear that the slicing of the four-manifold into spacelike three-manifolds and a time direction is in conflict with the principle of general covariance (equivalence of all coordinates). But the slicing is arbitrary: the dynamical description of one four-manifold is the collection of time developments corresponding to all ways of spacelike slicing. This is a covariant notion in which no particular slicing is preferred, similar to the covariant description of a tensor by the collection of its components in all possible coordinate systems.

Intrinsic and extrinsic geometry of one spacelike slice may be likened to initial position and momentum in particle mechanics: if these initial data are given, the dynamical equations determine the entire motion (i.e. the entire four-manifold). But the geometric initial data cannot be specified arbitrarily, since four of Einstein's ten equations ("constraints") involve the initial data only, and the remaining equations specify the time development ("dynamical equations"). Thus four of the ten metric components of the four-manifold are not determined by the dynamical equations; this arbitrariness describes precisely the freedom of choice of the successive spacelike slices. On the other hand, the four constraints reduce the two times six components of the intrinsic and extrinsic geometry to eight arbitrary components. Four of these are arbitrary due to the freedom of choice of initial surface and coordinates on it, and the remaining four numbers per space point describe the physical degrees of freedom of the gravitational field. By carrying out the mathematical reduction suggested by the above numerics one can show that the gravitational field has its own degrees of freedom in two polarizations, like electromagnetic and other transverse zero restmass fields. These gravitational degrees of freedom are the freely specifiable variable at an initial time. The rest of the gravitational field and its time development is then determined by the constraints and the dynamical equations. Thus one has a complete picture of the geometry as a dynamical system.

The real test of geometrodynamics as a description of nature beyond ordinary gravitational phenomena comes in the quantized version. Here the dynamical description is essential, because the quantum wave functional assigns a probability amplitude not to conjugate pairs of dynamical variables, but to only one of the conjugate sets. For example, in a "position"-type representation the description of geometrodynamics in terms of spacelike surfaces is unavoidable: Each three-dimensional (spacelike) geometry has some probability amplitude in the quantum state, and only in a classical limit do all the three-geometries mesh to form a single four-manifold.

An important non-classical feature expected of quantum geometrodynamics is the possibility of topology change between different three-geometries in a quantum state. Dimensional arguments suggest that such topology change is most likely to take place on the scale of the Planck length $(hG/c^3)^{1/2} = 10^{-33}$ cm. Wheeler has likened quantum geometry on this scale to a "foam" of great topological complexity. In a pure geometrodynamical world everything is made of this foam. At present important mathematical obstacles need to be overcome in order to discover the relevant properties of the quantum foam and to acertain whether the full vision of geometrodynamics can be realized.

Bibliography

FLETCHER, J. (1962) Geometrodynamics, in *Gravitation: An Introduction to Current Research* (L. Witten, Ed.). New York: John Wiley.

WHEELER, J. A. (1962) *Geometrodynamics*. New York and London: Academic Press.

WHEELER, J. A. (1968) *Einstein's Vision*. Berlin, Heidelberg, New York: Springer-Verlag.

D. BRILL

H

HELIUM, NEGATIVE ION OF. The existence of a metastable atomic negative ion of helium was suggested by pre-war mass spectrograph data (Hiby, 1939). It has now been established that the $(1s2s2p)$ configuration of He$^-$ is bound by 0·08 eV (Oparin et al., 1970), and that the ^{4}P state of this configuration is metastable against both radiative decay and autoionization through electrostatic inter electron coupling. The $^4P_{1/2}$ and $^4P_{3/2}$ substates autoionize by spin-orbit mixing with short-lived doublet states of the same configuration; the $^4P_{5/2}$ substate decays more slowly through the tensor spin-spin interaction between the electrons.

A number of attempts have been made to measure the lifetimes τ_J of the three substates (Simpson et al., 1971). Their approximate values are

$$\tau_{1/2} \approx \tau_{3\,2} \approx 10 \text{ μs}, \quad \tau_{5/2} \approx 0.5 \text{ ms},$$

which are long enough to permit precise measurements on the radio-frequency fine-structure transitions between the substates (Mader and Novick, 1972). Such measurements could critically test three-electron wave functions in the manner of the classic work on the hydrogen atom and more recent measurements in He.

The negative ions ^{4}He$^-$ and ^{3}He$^-$ have important applications as the primary sources of α-particle and τ-particle beams from tandem electrostatic accelerators. The negative ion beams are most commonly produced with the aid of charge changing collisions between fast He$^+$ ions and alkali metal atoms (usually lithium or potassium) in a vapour cell. The dominant production mechanism is a two-step process involving the nearly resonant formation of metastable He* atoms in the $(1s2s)^3$S state, and an electron transfer collision between He* and an alkali atom (Donnally and Thoeming, 1967).

Bibliography

DONNALLY, B. L. and THOEMING, G. (1967) *Phys. Rev.* **159**, 87.
HIBY, J. W. (1939) *Ann. der Physik* **34**, 473.
MADER, D. L. and NOVICK, R. (1972) *Phys. Rev. Lett.* **29**, 199.
OPARIN, V. A., IL'IN, R. N., SERENKOV, I. T., SOLOV'EV, E. S. and FEDORENKO, N. W. (1970) *Zh. Eksper. Teor. Fiz. Pis. Red.* **12**, 237 (translated in *JETP Lett.* **12**, 162).
SIMPSON, F. R., BROWNING, R. and GILBODY, H. B. (1971) *J. Phys. B: Atom. Molec. Phys.* **4**, 106.

I. G. MAIN

HELIUM DILUTION REFRIGERATION

1. Introduction

The helium dilution refrigerator is a device which can maintain temperatures typically in the range 10 mK to 500 mK, having cooling powers at 100 mK typically in the range 10 μW to 100 μW.

The first suggestion that liquid mixtures of helium-3 and helium-4 could be used for low temperature refrigeration was made by H. London (1951). A decade later, London, Clarke and Mendoza (1962) published data on the key property of osmotic pressure together with detailed proposals for a refrigerator to work below 1K. The first reported attempt by Das, de Bruyn Ouboter and Taconis (1964) was dis-

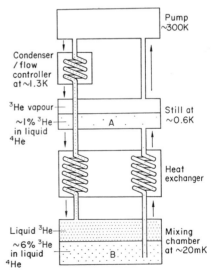

Fig. 1. Diagram of the type of dilution refrigerator in general use, with labelled components and fluid phases.

appointing for reasons which are now understood. However, two groups, Hall, Ford and Thompson (1966) and Neganov, Borisov and Liburg (1966), soon reported success. Since that time helium dilution refrigeration and the associated literature have proliferated until at present (1973) most low-temperature laboratories have at least one such refrigerator in operation and indeed a number of other research areas (e.g. nuclear orientation) use them as off-the-shelf tools. The type illustrated in Fig. 1 and described in this article is that in general use, although a number of variations have been suggested and in some cases successfully tested (London, Clarke and Mendoza, 1962; Pennings, Taconis and de Bruyn Ouboter, 1971; Taconis et al., 1971). The literature is large, but the most useful general references are probably those by Betts (1968), Wheatley, Vilches and Abel (1968) and Wheatley, Rapp and Johnson (1971).

The principles of operation depend on the properties of helium isotope quantum fluid mixtures which are outlined below. For greater detail, suitable reviews are given by Betts (1971), Radebaugh (1967), Wheatley (1968, 1970) and Ebner and Edwards (1971).

2. Properties of Liquid ^{3}He–^{4}He Mixtures

Liquid mixtures of ^{3}He and ^{4}He form a single phase with any degree of concentration at temperatures down to 0·87 K. Below that temperature some concentrations become unstable and separate into two distinct phases, as shown in Fig. 2, the concentrated (^{3}He-rich) phase floating on the denser dilute (^{4}He-rich) phase, with a visible surface between them. The concentrations in the two phases are functions of temperature, and at absolute zero it appears that the concentrated phase is pure ^{3}He, whereas the dilute phase contains about 6% ^{3}He. To be more precise, for $T \lesssim 0·2$ K, it is found that the mole fractions of minority atoms are given to a close approximation by

$$X_4 \text{ (upper phase)} = 0·85\, T^{3/2} \exp(-0·56/T) \quad (2.1)$$

and

$$X_3 \text{ (lower phase)} = 0·064(1 + 10·8\, T^2) \quad (2.2)$$

The complete phase diagram is shown in Fig. 3.

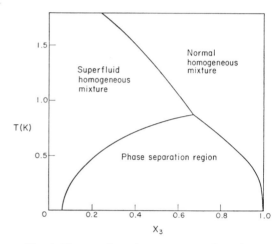

Fig. 3. The complete phase diagram of ^{3}He–^{4}He mixtures showing that below 0·87 K, there exists a region representing two liquid phases in equilibrium.

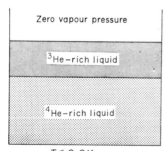

Fig. 2. A phase-separated mixture at $T \lesssim 0·2$ K. According to eq. (2.1) the upper phase is almost pure ^{3}He: according to eq. (2.2) the lower phase contains about 6·4% of ^{3}He.

At temperatures much below 1 K dilute solutions of ^{3}He in liquid ^{4}He behave in many ways like gaseous ^{3}He, but with an effective mass m^* for the atoms which is greater than the normal mass m_3: $m^* \simeq 2·5\, m_3$. The reason for this is that the ^{4}He has negligible entropy itself, so that it behaves simply as a background solvent (except that its presence contributes to the magnitude of m^*). It is crucial to the argument which follows to point out that a dilute solution of ^{3}He at sufficiently low temperatures behaves like an ideal Fermi–Dirac gas. The degeneracy temperature for a 6·4% mixture is readily calculated and comes out to be 0·38 K, so that in making quantitative analyses of behaviour of such a mixture below, say, 0·2 K, it is appropriate to use Fermi–Dirac expressions in their low-temperature degenerate form.

Some cooling could be achieved even without the property of phase separation because when a degenerate Fermi–Dirac gas is expanded adiabatically, cooling results according to the equation

$$T_f = T_i (V_i/V_f)^{2/3} \quad (2.3)$$

where the subscripts i and f refer to initial and final temperatures and volumes. A dilute mixture of ^{3}He in liquid ^{4}He could therefore be cooled by diluting it with superfluid ^{4}He introduced through a super-leak, the process being equivalent to an adiabatic gas expansion. Such methods have been tried but there is also another different and ultimately more efficient cooling mechanism which exists by virtue of the phase-separation boundary.

3. The ^{3}He Cryostat Analogy

The easiest way to understand this mechanism is by analogy with an ordinary evaporation refrigerator, such as a ^{3}He cryostat. Consider a volume containing liquid ^{3}He in equilibrium with its vapour at, say, 0·3 K where the vapour pressure is 0·25 Pa (1·88 × 10^{-3} torr). The equilibrium is of course determined by the condition that the chemical potential be the same in both phases. However the entropy is quite different, being much higher in the vapour phase. Now suppose atoms are slowly removed by pumping from the vapour phase, causing a reduction in vapour pressure and in chemical potential. The liquid–gas system will respond by allowing a net transfer of atoms from the liquid phase to the gas phase in an attempt to re-establish equilibrium. If while the process continues a rate of heat input $\dot{Q}$ is required to keep the temperature constant as atoms are removed at a molar rate $\dot{n}_3$, then we should like to relate $\dot{Q}$ to $\dot{n}_3$ and the relation, assuming reversibility is clearly

$$\dot{Q} = \dot{n}_3 T \Delta S \quad (3.1)$$

where ΔS is entropy difference per mole in the vapour and liquid phases at temperature T. Thus

$$\dot{Q} = \dot{n}_3 T [S_v(T) - S_l(T)] \quad (3.2)$$

where the subscripts v and l denote vapour and liquid respectively.

For a recirculating ^{3}He cryostat, $\dot{Q}$ will have to include whatever heat the returning ^{3}He brings (which depends ultimately on the efficiency of heat exchange) plus stray heat leaks. The heat brought to the ^{3}He cryostat by the returning ^{3}He is given up in an isobaric process so that $\dot{Q}_3 = \dot{n}_3 \Delta H$ where H signifies the molar enthalpy. Thus

$$\dot{Q}_3 = \dot{n}_3 [H_l(T_r) - H_l(T)] \quad (3.3)$$

where T_r is the temperature of the returning ^{3}He.

The heat balance equation, including an external power input $\dot{Q}_{ext}$ and a heat leak $\dot{Q}_{hl}$ is then

$$\dot{Q}_{ext} = \dot{n}_3 T [S_v(T) - S_l(T)]$$
$$- \dot{n}_3 [H_l(T_r) - H_l(T)] - \dot{Q}_{hl}. \quad (3.4)$$

In practice $\dot{n}_3$ is determined by ordinary pumping considerations and the lowest achievable temperature for a given apparatus corresponds to the condition $\dot{Q}_{ext} = 0$.

4. The Mixing Chamber of a Dilution Refrigerator

Figure 1 is a diagram of the whole dilution refrigeration with labelled components. In this and later sections, individual components will be discussed and analysed, beginning with the mixing chamber where the cooling actually occurs. The relevance of the ^{3}He-cryostat analogy of section 3 will be seen by reference to Fig. 2 which shows that the mixing chamber of a dilution refrigerator is very much like a ^{3}He-cryostat but *upside-down*, with liquid ^{3}He floating on "gas-like" ^{3}He as explained above. Leaving aside for the moment the question of how ^{3}He may be "pumped" downwards across the interface, let us suppose that it moves at a rate $\dot{n}_3$ moles/sec, then the heat balance equation, of which (3.4) is the analogue is

$$\dot{Q}_{ext} = \dot{n}_3 T_{mc} [S_d(T_{mc}) - S_c(T_{mc})]$$
$$- \dot{n}_3 [H_c(T_r) - H_c(T_{mc})] - \dot{Q}_{hl}. \quad (4.1)$$

Here the subscripts d and c refer to dilute (lower phase in Fig. 4) and concentrated (upper phase) ^{3}He solutions, T_{mc} denotes the mixing chamber temperature and it is assumed that only ^{3}He atoms are circulated. Usually about 5% of ^{4}He is also circulated, but this does not materially affect the present argument. Now since both upper and lower phases are degenerate Fermi fluids, we can express the entropies and enthalpies in convenient explicit forms which, on numerical substitution yield

$$\dot{Q}_{ext} = \dot{n}_3 (94 T_{mc}^2 - 12 T_r^2) - \dot{Q}_{hl} \text{ watts.} \quad (4.2)$$

This numerical form, due to Radebaugh (1967), is quoted by Ebner and Edwards (1971) who also give, in appendix C of their paper, a more general and thorough thermodynamic treatment than that given here of heat calculations for various processes taking place in the dilution refrigerator. Thus the cooling power, according to eq. (4.2), depends on $\dot{n}_3$, T_r and $\dot{Q}_{hl}$. Its lowest temperature is determined by the condition $\dot{Q}_{ext} = 0$.

For non-recirculating versions, or where heat-exchange with the returning ^{3}He stream is assumed perfect, we may set $T_r = T_{mc}$ so that eq. (4.2) reduces to

$$\dot{Q}_{ext} = 82 \dot{n}_3 T_{mc}^2 - \dot{Q}_{hl} \text{ watts.} \quad (4.3)$$

Experience tends to confirm this result. For example, Vilches and Wheatley (1967) found that a non-recirculating dilution refrigerator with a circulation rate of $\dot{n}_3 = 10^{-5}$ moles/sec and a lowest recorded temperature of 4·5 mK had a cooling power at 11 mK of $\dot{Q}_{ext} \approx 0·1$ μW, a performance which is consistent with eqn. (4.3) with $\dot{Q}_{hl} \approx 20$ nW. Recirculating dilution refrigerators commonly have base temperatures of about 20 mK which from eqn. (4.2) implies, for comparable values of $\dot{Q}_{hl}$ and $\dot{n}_3$, that $T_r = 54$ mK $= 2·7 T_{mc}$. From this it is clear that in practice, heat exchange is a most important consideration which we shall return to after discussing the factors which govern $\dot{n}_3$.

5. *The Stimulation of ^{3}He Circulation; the Still*

The basic task is to devise a way of removing ^{3}He from the dilute lower phase in the mixing chamber, since a reduction in number density there reduces the chemical potential, creating a non-equilibrium situation which the system will tend to correct by allowing a net transfer of ^{3}He atoms from the upper phase. This is solved by connecting the lower phase through a capillary to a still as shown in Fig. 1. The choice of still temperature is made in the following way. Wheatley, Vilches and Abel (1968) give a table of partial vapour pressure p_3 and p_4/p_3 for a range of still temperatures from 0·5 K ($p_3 = 1\cdot9$ Pa [0·014 torr]; $p_4/p_3 = 1\cdot1 \times 10^{-3}$) to 0·85 K ($p_3 = 17$ Pa [0·13 torr]; $p_4/p_3 = 1\cdot8 \times 10^{-1}$). From the point of view of overall rate of helium circulation, the higher temperature would seem most appropriate because 17 Pa is a reasonably large vapour pressure which should allow a good rate of circulation. However that circulation would contain 18% of ^{4}He which is considered to be undesirably high on two grounds. Firstly, it places on the pumping system an extra burden without producing any refrigeration and, secondly, after recondensation it will at some stage reach a temperature in the heat exchanger where phase separation occurs according to eqn. (2.1). This is accompanied by a release of heat of mixing which reduces the efficiency of the exchanger. Fortunately this latter effect is less important at lower temperatures and 5% or less of ^{4}He is generally acceptable. On the other hand, if too low a still temperature is chosen the vapour pressure will be low and will hold back the pumping rate. Moreover the table referred to does not include the effects of ^{4}He film flow for which the lowest flow rates are in the region of 3×10^{-4} moles/s/m. Consequently, a compromise temperature is chosen, generally in the region of 0·6 K, and precautions are taken to inhibit the flow of ^{4}He II film by pumping through a small hole in a thin diaphragm in the pumping line or by more ingenious means.

Incidentally the concentrations in the liquid phases A and B of Fig. 1 are not the same because the temperatures are different. The concentration at the mixing chamber is given by eqn. (2.2) and approximates to 6·4% ^{3}He but that in the still is lower, typically about 1% ^{3}He, and is largely determined by a balance of osmotic pressures in the two chambers, heat flush effects being relatively slight because the number of ^{4}He excitations (phonons and rotons) at 0·6 K has fallen well below the number of ^{3}He atoms in a 1% mixture. Thus the still at 0·6 K has a vapour pressure of 4·5 Pa (0·034 torr) largely composed of ^{3}He, and the designer's task is now merely to build a pumping system with as high a speed as possible, consistent with tolerable size and expense. Given the restricted choice of still temperature, it is the pumping system which in practice determines $\dot{n}_3$ and typical values are in the range 10^{-5} to 10^{-4} moles/s. The still temperature has to be maintained by applying heat at a rate which is readily estimated by taking into account the various heat contributions as for the mixing chamber. The cooling power at the still is $\dot{n}_3 T_s (S_v - S_l)$ and this is generally more than large enough to remove stray heat leaks and the heat contained in the returning ^{3}He (assuming that it returns in liquid form). A typical necessary heating rate is $\dot{Q}_s = 40 \dot{n}_3$ watts, and a precise choice is usually made by the operator after initial testing.

6. *Condenser, Flow Controller and Phonon Stopper*

These are fairly minor items with simple functions but there are a variety of ways of constructing and siting them and home-made designs may require some alterations following experience.

(a) The condenser merely ensures that the returning ^{3}He is liquefied, preferably completely so, before arrival at the still. Within the still the return line should also be designed to encourage heat exchange with the liquid in the still, so that unnecessary load on the main heat exchangers is avoided.

(b) The flow controller, which may be a coil of capillary or a piece of sintered copper sponge, is intended simply to absorb the pressure drop Δp between the outlet side of the pumping assembly and the pressure in the mixing chamber, close to zero. Of course the main heat exchanger has some flow resistance but this is usually minimized to avoid viscous generation of heat. Thus the flow controller is generally sited somewhere above the heat exchange unit and has an impedance Z chosen so that

$$Z = \Delta p / \eta \dot{n}_3 V_3 \qquad (6.1)$$

where η is the viscosity of liquid ^{3}He and V_3 its molar volume.

(c) The phonon stopper is often not a specially designed item, but the point is this: the still and mixing chamber are connected by a continuous column of the superfluid dilute phase. Now superfluid pure ^{4}He is an extremely effective conductor of heat, the mechanism being internal convection. However the presence of ^{3}He atoms strongly inhibits this particularly since, as mentioned above, the ^{3}He atoms have a large numerical supremacy over the ^{4}He excitations. It is possible (Betts, 1968) readily to estimate a characteristic length over which heat-carrying phonons from the still are stopped and is typically well below 0·1 mm for commonly used capillary sizes.

7. Heat Exchange Units

The good design and operation of these items is of crucial importance since they largely control the lowest temperature achievable in the mixing chamber. This was illustrated by the numerical example given in section 4 where it was argued that a base temperature of 20 mK implies that the returning ^{3}He is a factor of nearly 3 higher in temperature than the mixing chamber. With ideal heat exchange this factor would be unity, but a number of difficulties prevent such perfectly efficient performance. The analysis of heat exchangers for dilution refrigerators has been examined in detail in two papers by Siegwarth and Radebaugh (1971, 1972) in which machine calculated efficiency curves are presented as guides to design. The main design considerations are that the liquid volume be as small as possible so that equilibrium can be established rapidly when the temperature is changed, the impedance to flowing liquids be small, the thermal conductivity of the exchanger between the two streams be adequate, and the heat transfer areas be as large as possible to overcome the effects of the Kapitza resistivity. This resistivity is a surface effect and is defined by

$$\varrho = A \Delta T / \dot{Q} \tag{7.1}$$

where $\dot{Q}/A$ is the heat current density and ΔT the discrete temperature jump at the interface between two materials. It is generally of great significance at the lowest temperatures where, for pure liquid ^{3}He/copper, it is given by

$$\varrho_c = \frac{20 \times 10^{-9}}{T_c^3} \, m^2 \, K/\mu W \tag{7.2}$$

and for saturated (6·4%) mixture/copper by

$$\varrho_d = \frac{7 \times 10^{-9}}{T_d^3} \, m^2 \, K/\mu W. \tag{7.3}$$

A serious designer would naturally turn to Siegwarth and Radebaugh (1971, 1972) who include all the factors mentioned above in a systematic way, but a very simple model presented by Betts (1968) may help to convey some idea of the importance of having an adequate contact area for the returning ^{3}He below, say, 250 mK. The heat capacity per atom of ^{3}He in the dilute phase is known to be an order of magnitude greater than in the returning concentrated phase, so the model assumes for simplicity that the dilute counterflow does not warm appreciably as it absorbs heat in the exchanger, but remains at T_{mc}. The heat brought into the mixing chamber by the ^{3}He is given as in equation (4.1) by

$$\dot{Q}_3 = \dot{n}_3 [H_c(T_r) - H_c(T_{mc})]. \tag{7.4}$$

Heat is supposed to flow perpendicular to the fluid flow, from concentrated to dilute ^{3}He through an exchanger at uniform temperature T_e. Figure 4 illustrates the model and we wish to use it to estimate

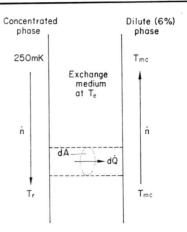

Fig. 4. Diagram of the heat-exchange model analysed in the text.

the contact area A required to cool the returning ^{3}He from the arbitrary temperature of 250 mK to T_r and hence from eqn. (4.2) to find the lowest achievable T_{mc} in terms of A.

For an element of area dA the rate of heat flow $d\dot{Q}$ is given by

$$\frac{d\dot{Q}}{dA} = \frac{T_c - T_e}{\varrho_c} = \frac{T_e - T_{mc}}{\varrho_d}. \tag{7.5}$$

Elimination of T_e gives

$$T_c - T_{mc} = \frac{d\dot{Q}}{dA}(\varrho_c + \varrho_d). \tag{7.6}$$

But in the concentrated flow,

$$d\dot{Q} = \dot{n}_3 \, dH_c = 25 \dot{n}_3 T_c dT_c \text{ watts} \tag{7.7}$$

where we have for simplicity assumed that the low-temperature limiting form for enthalpy given by Radebaugh (1967) is a reasonable approximation up to 250 mK. Hence, eliminating $d\dot{Q}$ from eqs. (7.6) and (7.7), integrating over temperature limits $T_c = T_r$ to 250 mK, and using the forms of eqs. (7.2) and (7.3) for ϱ_c and ϱ_d, we obtain an explicit (though algebraically long) relation between $A/\dot{n}_3$, T_r and T_{mc}. T_r can be eliminated by use of eqn. (4.2) with $\dot{Q}_{hl} = 0$ for simplicity. We thus relate T_{mc} with $A/\dot{n}_3$ and Table 1 is an illustration of the numerical results.

Table 1

T_{mc} (mK)	$A/\dot{n}_3$ (m.s/mole)
10	$6·6 \times 10^4$
20	$8·9 \times 10^3$
30	$2·6 \times 10^3$
40	$1·0 \times 10^3$
50	$4·5 \times 10^2$

The model predicts that for $\dot{n}_3 = 10^{-5}$ moles/s, an exchange area of nearly a tenth of a square metre is necessary to achieve 20 mK, whereas for 50 mK relatively small areas suffice. These calculations are broadly in line with experimental evidence. This model does not include any terms representing viscous dissipation, but in practice it is of course essential to keep the flow impedance of the heat exchange unit as low as possible, consistent with other requirements (such as physical size), and a suitable criterion is that viscous dissipation must be much less than both the mixing chamber cooling power and the total rate of heat exchange in the unit.

Several types of successful heat exchange units have been described in the literature, including continuous tubular constructions with various geometries by Neganov, Borisov and Liburg (1966) and Ehnholm et al. (1968), and discrete "stepwise" constructions (usually with four to six units) using small units of sintered copper powder by Wheatley, Vilches and Abel (1968) or using small copper foil units by Roubeau and Varoquaux (1970). Probably the most systematic experimental study is contained in the article by Wheatley, Rapp and Johnson (1971).

8. The Ultimate Limit

Wheatley, Rapp and Johnson (1971), having mentioned that many workers find the performance of carefully designed refrigerators fall short of expectations, list and discuss three possible causes which are generally neglected in first-order descriptions. Firstly convection, due to unfavourable temperature distributions in heat exchangers, can seriously degrade their performance. Fortunately convective instabilities can be inhibited by restricting channel diameters to well below 1 mm. These instabilities are particularly likely to occur at temperatures near 0·1 K in the dilute phase. Secondly, improper pressure distribution on the concentrated side may allow the formation of vapour bubbles which carry latent heat to colder regions, thus counteracting the normal process of heat exchange. Thirdly, the presence of superfluid on the concentrated side may arise because of being pumped from the still as explained in section 5 or by film creep from the mixing chamber. It is undesirable for the reason given in section 5 and also because its efficient heat transfer can interfere with the temperature distribution in and between heat exchange units.

Even if heat exchange were perfect (or unnecessary as in the single-shot refrigerator of Vilches and Wheatley (1967)) there will always remain the unavoidable problem of stray heat leaks. However, Wheatley, Vilches and Abel (1968) discuss another effect in which, when ^{3}He in dilute solution flows out of the mixing chamber through a circular tube of diameter d, heating is produced by viscosity. A temperature gradient is established in the tube and consequently heat flows back into the mixing chamber. Analysis shows that, in the absence of other limiting effects, this would prevent T falling below 4 mK $(1 \text{ mm}/d)^{1/3}$. This result is certainly in line with the result of Vilches and Wheatley (1967) and since d can hardly be made significantly larger then a few millimetres the practical limit is probably in the region of a few millikelvin.

Bibliography

BETTS, D. S. (1968) *Contemp. Phys.* **9**, 97.
BETTS, D. S. (1971) *Contemp. Phys.* **12**, 153.
DAS, P., DE BRUYN OUBOTER, R. and TACONIS, K.W. (1964) *Proceedings of the Ninth International Conference on Low Temperature Physics* Part B, 1253. New York: Plenum Press.
EBNER, C. and EDWARDS, D. O. (1971) *Physics Reports*, **2C**, 77.
EHNHOLM, G. J., KATILA, T. E., LOUNASMAA, O. V. and REIVARI, P. (1968) *Cryogenics*, **8**, 136.
HALL, H. E., FORD, P. J. and THOMPSON, K. (1966) *Cryogenics*, **6**, 80.
LONDON, H. (1951) *Proceedings of the International Conference on Low Temperature Physics*, p. 157. Oxford.
LONDON, H., CLARKE, G. R. and MENDOZA, E. (1962) *Phys. Rev.* **128**, 1992.
NEGANOV, B. S., BORISOV, N. S. and LIBURG, M. YU. (1966) *Zh. Eksp. Teor. Fiz.* **50**, 1445 (English translation in *Soviet Physics JETP* **23**, 959).
PENNINGS, N. H., TACONIS, K. W. and DE BRUYN OUBOTER, R. (1971) *Physica*, **56**, 171.
RADEBAUGH, R. (1967) *U.S. National Bureau of Standards Technical Note* No. 362.
ROUBEAU, P. and VAROQUAUX, E. J. (1970) *Cryogenics*, **10**, 225.
SIEGWARTH, J. D. and RADEBAUGH, R. (1971) *Rev. Sci. Instrum.* **42**, 1111.
SIEGWARTH, J. D. and RADEBAUGH, R. (1972) *Rev. Sci. Instrum.* **43**, 197.
TACONIS, K. W., PENNINGS, N. H., DAS, P. and DE BRUYN OUBOTER, R. (1971) *Physica*, **56**, 168.
VILCHES, O. E. and WHEATLEY, J. C. (1967) *Phys. Letters*, **24A**, 440 and *Phys. Letters*, **25A**, 344.
WHEATLEY, J. C. (1968) *Am. J. Phys.* **36**, 181.
WHEATLEY, J. C. (1970) *Progress in Low Temperature Physics*, Vol. 6, ch. 3. North-Holland Publishing Company.
WHEATLEY, J. C., RAPP, R. E. and JOHNSON, R. T. (1971) *J. Low Temp. Phys.* **4**, 1.
WHEATLEY, J. C., VILCHES, O. E. and ABEL, W. R. (1968) *Physics*, **4**, 1.

DAVID S. BETTS

HIGH-ENERGY HEAVY IONS. The history of the study of the properties of heavy ions and their interactions with matter is as old as nuclear physics itself. In 1899 Rutherford identified a component of the radiations emitted by uranium-bearing minerals that only weakly penetrated matter. This "soft" component, named "α-rays" by Rutherford, was shown to consist of ionized helium atoms by Rutherford and Royds in 1909. Until the discovery of spontaneous fission in 1940, α-decay was the only known type of radioactive disintegration in which heavy particles are emitted by the parent nucleus. Some understanding of the energy loss mechanisms when heavy charged particles penetrate matter was obtained from studies of the range of fission fragments in and the stopping power of various materials (Katcoff et al., 1948; Segrè and Weigand, 1946), but little information of the interaction of these highly charged particles could be obtained.

Since the identification of energetic heavy ions as a component of the galactic cosmic radiation in the late 1940s, and the subsequent investigation of their properties, there has been increasing interest in their application to both fundamental and applied research in many disciplines. This application was made possible by the availability of heavy ions accelerated in the laboratory. Several types of accelerators may be operated with heavy ions. Thus electrostatic accelerators, such as the tandem Van de Graaf, linear accelerators and cyclotrons have all been successfully employed in many institutions around the world. For example, at the University of California, Berkeley, three cyclotrons are available for the production of ions up to ^{4}He. A 27-in cyclotron can accelerate ^{3}He to a maximum energy of 18 MeV; an 88-in cyclotron can accelerate α-particles up to 130 MeV (32·5 MeV/nucleon); while a 184-in synchrocyclotron can accelerate α-particles to 910 MeV. Heavier ions, as heavy as argon, are accelerated to energies of 10 MeV/nucleon in the Hilac and recent improvements have given it the capability of accelerating ions as heavy as uranium to energies of 8·5 MeV/nucleon at an intensity of 10^{11} ion/sec.

Purser (1971) and Main (1971) have reviewed the heavy ion facilities available up to about 1971. Up to that time energies up to about only 10 MeV/nucleon were available. Even with this severe energy limitation fields as diverse as cosmic ray physics, cosmology, cellular radiobiology, nuclear chemistry and nuclear physics were able to apply energetic heavy ions to the solution of many fundamental and practical problems (White, 1971). These relatively low-energy heavy ion beams have also found industrial application in a field generally referred to as ion-implantation. This technique is now widely used in the manufacture of semiconductor devices, but there is a growing interest in such techniques to modify the chemical and physical properties of solids in general (see ION IMPLANTATION).

The modest energies available until 1971 limited progress in many areas of research and it became clear that higher energy heavy ions were needed. In 1966 the Lawrence Berkeley Laboratory proposed the construction of an accelerator, specifically designed for the acceleration of heavy ions. This accelerator, named the Omnitron, was to have a capability of accelerating ions as light as helium and as heavy as uranium to energies as high as 500 MeV/nucleon. Although this accelerator was generally recognized to be well designed it was, because of economic reasons, unfortunately never constructed. Continued demand for higher energy particles finally led to the modification of existing proton synchrotrons for heavy ion acceleration. To date (Summer 1973) two proton synchrotrons have successfully accelerated heavy ions. In 1970 the Princeton Particle Accelerator successfully accelerated deuterons to 2·4 GeV and α-particles to 4·8 GeV. On July 10th, 1971, a beam of N^{5+} ions was obtained at an energy of 279 MeV/amu (total kinetic energy of 3·9 GeV), followed somewhat later by the acceleration of N^{6+} ions (White et al., 1971). Within a month deuterons, α-particles, and N^{7+} had been successfully accelerated at the Bevatron, the proton synchrotron of the Lawrence Berkeley Laboratory of the University of California, to a kinetic energy of 36 GeV (Grunder et al., 1971).

The acceleration of heavy ions to such high energies is an extremely important scientific achievement and has opened many new and exciting possibilities.

Discovery of High-energy Heavy Ions in the Cosmic Radiation

In 1948 the first report was published describing the discovery of heavy particles in the primary cosmic radiation (Freier et al., 1948). This discovery was made possible by particle detectors (nuclear emulsions) flown by balloons at altitudes of ~30,000 m. Figure 1 shows three photomicrographs of the track of one of these heavy particles as it passes through nuclear emulsions. By studying the characteristics of the track it is possible to show that it was produced by a particle with atomic number greater than 40 and with kinetic energy greater than 100 GeV.

Confirmation of the existence of these heavy particles was obtained using other detectors operated at high altitudes. Their discovery immediately led to an interest in studying their reaction with matter.

Investigation of the properties of energetic heavy ions by means of the galactic cosmic radiation is difficult and tedious. Even with the great improvements in experimental techniques made possible with the advent of earth-orbiting satellites there are severe limitations to the resolution of charge and energy of the incident particles and their interaction products, and the rate at which data are obtained is extremely slow.

Nevertheless, despite the difficulties, these investigations are sufficiently important that extensive studies followed the first discovery of the heavy ions. The composition of the galactic cosmic radiation might be expected to give information on the

Fig. 1. Three photomicrographs of the track of a heavy nucleus of very high energy, detected in nuclear emulsions exposed at approximately 30,000 m. Pictures (a), (b) and (c) show the initial track and the track at depths of 5 and 10 g/cm² in the emulsion respectively. The incident particle was estimated to have charge number greater than 40 and energy greater than 100 GeV (courtesy E. J. Lofgren).

Table 1. Comparison of the estimated charge distributions in galactic radiation and galactic matter

Element group	Intensity* (particles/cm² sec)	Atomic abundance (% by number)	Universal abundance (% by number)
Hydrogen (protons)	3·6	88	90
Helium (α-particles)	4×10^{-1}	9·8	9
Light nuclei (Li, Be, B)	8×10^{-3}	0·2	10^{-4}
Medium nuclei (C, N, O, F)	3×10^{-2}	0·75	0·3
Heavy nuclei ($10 \leq Z \leq 30$)	6×10^{-3}	0·15	0·01
Very heavy nuclei ($Z \geq 31$)	5×10^{-4}	0·01	10^{-5}
Electrons and photons ($E > 4$ BeV)	4×10^{-2}	1	

* At solar minimum.

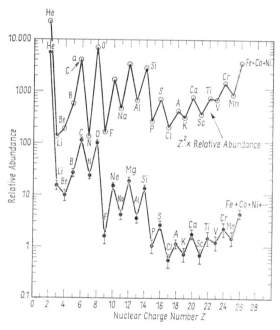

Fig. 2. Relative abundance of the elements in the galactic cosmic radiation at 150 MeV/nucleon (Normalized to oxygen = 100). [Data due J. A. Simpson and M. Garcia-Munoz, measured May 1967 to May 1968 by University of Chicago telescope aboard the IMP 4 satellite.]

elemental composition of the galaxy. Extensive investigations of the naturally occurring heavy ions may be summarized by saying that outer space is irradiated by cosmic radiation of galactic origin. This radiation consists of stripped nuclei in the energy range from $\approx 10^8$ to $\approx 10^{19}$ eV, having a mass distribution similar to that of the universe (Table 1). In free space these particles are distributed isotropically and are present at an intensity of about 4 particles/cm² sec. Table 1 summarizes the intensity of and mass distribution of these nuclei, from which we see that the principal component of galactic cosmic radiation consists of protons. Alpha-particles are present at only 10% of the proton intensity, and all other heavier nuclei contribute only 10% of the α-particle intensity. Photons and electrons are present at an intensity comparable to that of the heavy nuclei ($\approx 1\%$ of the proton intensity).

Figure 2 summarizes experimental data, obtained from satellite flights, of the relative abundance of heavy ions in the cosmic radiation as a function of charge, Z. The relatively high abundance of ^{3}He, Li, Be and B is believed to be due to their production by spallation reactions of C, N, and O ions with intergalactic matter. Elements in the iron group also appear to be present in high quantities. Measurements with emulsions and other detectors have shown the existence of small numbers of highly charged ions, up to Pb, Th, and U. The origin of these massive particles is not yet fully understood.

Scientific Interest in Studies with High-energy Heavy Ions

Cosmic-ray physics and cosmology

Close to the earth the composition of the cosmic radiation is determined first by the production mechanisms and acceleration processes, but ultimately by its fragmentation due to interaction with the galactic material (principally hydrogen) during its journey toward the earth. The interstellar material consists principally of hydrogen and studies of the fragmentation of ions in hydrogen and air would prove invaluable in the interpretation of the chemical composition of the galactic cosmic radiation near the earth and also testing theoretical models of the origins of the galactic radiation.

Data are particularly required for three important groups of reactions:
 (a) The ^{4}He + P reaction.
 (b) Light (C, N, O) nucleus production and loss.
 (c) Fragmentation of medium through heavy (iron-group) nuclei.

Studies in the latter group would indicate whether the observed cosmic-ray abundances may be accounted for by the fragmentation of very heavy (e.g. U) nuclei alone, or by some spectrum of particles over a wide range of mass.

Studies of the fragmentation of ^{4}He and the light nuclei (C, N, O) on hydrogen will indicate whether the existence of D and T in the cosmic radiation is due to α-stripping and also whether the spallation of C, N, and O can account for the observed intensities of Li, Be, and B (Dauber, 1971).

An intriguing possibility is the use of radionuclides that may decay only by K-capture (e.g. ^{7}Be, ^{53}Mn) as cosmic age indicators. Thus, for example, by determining the ratio of ^{53}Mn to ^{53}Cr as a function of particle energy in the range 50 to 200 MeV/nucleon the age of the parent cosmic radiation may be determined (Reames, 1970).

Nuclear physics

The availability of energetic heavy ions in the laboratory can lead to many exciting and interesting areas of research of which only a few can be mentioned here.

Energy loss phenomena when heavy particles penetrate matter are not yet completely understood. At particle energies above ~ 1 MeV/amu Bethe–Bloch theory should describe these mechanisms accurately. The Bethe–Bloch description is known to be good for mesons, protons, deuterons, and α-particles over a wide range of energies, but has not been adequately tested experimentally with energetic particles of very high charge, Z. At these energies (≥ 1 MeV amu) the energy loss per cm should scale as Z^2 for particles of the same velocity. At lower particle energies (velocities), however, this simple scaling does not apply because the particle can capture electrons from the medium it is traversing. This phenomena of electron capture lowers the rate of energy loss and thus causes an increase in the particle's penetration in matter. No adequate theoretical description of the electron capture (and loss) phenomena yet exists. Energy is also lost to the absorbing medium as a result of nuclear interactions. Some theoretical studies of nuclear stopping have been reported (Linhard *et al.*, 1963), but experimental studies are needed. The production of heavy ion beams of high charge will facilitate the experimental investigation of these energy loss mechanisms, and facilitate a better understanding of radiation damage phenomena in solid materials.

The total interaction cross section of heavy ions with nuclei is of interest. From a study of the interaction of heavy ions in the cosmic radiation in nuclear emulsion it was proposed that the total interaction cross-section, σ_{tot}, could be represented well, for all target nuclei by:

$$\sigma_{tot} = \pi R_0^2 (A_{inc}^{1/3} + A_{target}^{1/3} - b)^2$$

where A_{inc} is the mass number of the incident ion, A_{target} is the mass number of the target nucleus, and R_0 and b are constants.

Experiments with heavy ion beams in the laboratory would enable the determination of total cross-sections with some accuracy.

Other diverse uses for which heavy ions have been proposed include Coulomb excitation studies, investigation of "fireball" phenomena in particle production, and the production of neutron beams of good momentum resolution. White (1971) has suggested that a considerable contribution to our understanding of quantum electrodynamics could arise from the study of the Lamb shift in the hydrogen-like spectra from fully-stripped heavy ions which have captured an electron. From these examples the uses to which heavy ions may be put in nuclear physics are seen to be diverse and exciting.

Nuclear chemistry

Energetic heavy ions are of great utility in nuclear chemistry studies. They are capable of producing a wide variety of nuclear reactions and are particularly valuable when projectiles with high nuclear charge and/or large orbital angular momentum are required.

Elastic scattering studies give information on the nuclear radius and to some extent the outer region of the nucleus. Inelastic scattering enables the investigation of the level structure within the nucleus and the transition probabilities between various states. Heavy ions, with their high angular momentum, are able to populate high spin states not accessible with lighter particles. The high charge of heavy ions is utilized in Coulomb excitation studies where the target nucleus is excited by electromagnetic radiation emitted by the highly charged particle passing close by. Details of nuclear structure, the distribution of neutrons and protons within the nucleus and the possible existence of nuclear clusters (e.g. α-particles) within the nucleus may be obtained from the so-called transfer reactions, in which a part of one of the interesting nuclei is transferred to the other.

High energy heavy ions are particularly valuable in the production of transuranium elements. In many cases the use of heavy ions represents the only practical way of producing such elements. This seems to be increasingly true for heavier transuranic elements. Thus nobelium and lawrencium have been produced only in this manner. Table 2 summarizes the transuranic nuclides first synthesized by heavy ion bombardment.

Theoretical studies have predicted the possible existence of an island of stability in the periodic table analogous to the rare earths, for elements above the atomic number 112. The possible production of element 112 in tungsten targets subjected to a long exposure by 24 GeV protons was reported in 1971 (Marinov *et al.*, 1971). It was suggested that the production of element 112 was probably effected by the interaction of an energetic tungsten recoil nucleus with another tungsten nucleus. Although this report is not generally thought to have definitely established the existence of element 112 it has given impetus to the use of high-energy particles—particularly heavy ions—as projectiles to produce these "superheavy" elements. It seems probable that the bombardment of uranium with heavy ions energetic enough to surmount the coulomb barrier will be very effective in producing new elements.

Radiobiology and radiotherapy

There is increasing interest in studying the biological effects of heavily ionizing particles. It is not yet completely known to what extent biological repair mechanisms militate against the damage produced by the irradiation of living organisms. There is increasing evidence, however, that repair mechanisms may be extremely efficient at low doses and dose rates when the radiation is lightly ionizing. The evidence is less convincing, however, for densely ionizing radiation. Such effects must be understood if these radiations are to be safely and efficiently exploited.

The use of heavy ions in radiobiological investigations, then, is of great interest both in a fundamental and practical sense. Radiation injury in mammals at the cellular level is strongly dependent upon the density of ionization of the incident particle and the presence (or absence) of oxygen during irradiation. Anoxic cells are about three times as resistant to radiations with low ionization density (e.g. x-rays, γ-rays, electrons). [This damage ratio has been designated the Oxygen Enhancement Ratio (OER).]

The central regions of large tumors are, in general, deprived of nutrients including oxygen. The presence of anoxic cells in a tumor can therefore lead to a substantial increase in the radiation exposures necessary for therapy. Unfortunately, the most convenient and frequently used types of radiation presently

Table 2. Transuranic nuclides first synthesized by heavy ion bombardments

Isotope	Target	Projectile	Half-life	Decay mode
^{246}Cf	^{238}U	^{12}C	35.7 hr	6.75- and 6.71 MeV α-decay
^{245}Es	^{240}Pu ^{237}Np	^{10}B ^{12}C	75 sec	7.65-MeV α-decay
^{246}Es	^{238}U	^{14}N	7.3 min	7.35-MeV α-decay; also electron capture
^{248}Fm	^{240}Pu	^{12}C	0.6 min	α-decay
^{249}Fm	^{238}U	^{16}O	150 sec	7.9-MeV α-decay
^{250}Fm	Uranium	^{16}O	30 min	7.43-MeV α-decay
^{257}Md	^{252}Cf	^{13}C	3 hr	7.1 MeV α-decay
^{253}No	^{246}Cm ^{244}Cm	^{12}C ^{13}C	10–15 sec	8.8 MeV α-decay
^{254}No	^{246}Cm	^{12}C	3 sec	8.3-MeV α-decay
^{255}No	Mixture of californium isotopes	^{12}C	15 sec	8.2-MeV α-decay
^{256}No	^{238}U	^{22}Ne	8 sec	α-decay
^{257}Lw	Mixture of californium isotopes	Boron ions	8 sec	8.6-MeV α-decay

used in radiotherapy are lightly ionizing. (e.g. γ-rays from ^{60}Co sources). It is also unfortunate that the most frequently used types of radiation (x-rays, γ-rays, and electrons) are strongly absorbed in tissue so that the energy absorbed in the tumor is far lower than the energy absorbed in exterior healthy tissues. Several techniques have been proposed for at least redressing this unfortunate imbalance and improving the relative damage to cancer cells. The use of any heavy charged particle can improve the energy absorption distribution as may be seen from Fig. 3. The oxygen effect may be circumvented in

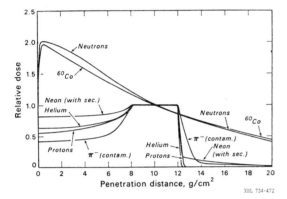

Fig. 3. Depth–dose curves of various types of radiation which might be used for radiotherapy in a hypothetical case, in which the aim is to deliver energy to a volume of tissue in the body extending from 8 to 12 cm in depth. All the curves have been normalized to unity at a depth of 10 cm (g/cm²). A circular incident beam of 10 cm diameter is assumed. The most suitable radiation is that for which the least energy is absorbed outside the specified volume. Charged particles are clearly to be preferred to ^{60}Co γ-rays or neutrons (from Tobias et al.).

two ways. Some work in the United Kingdom has been directed toward the use of hyperbaric chambers, in which anoxic cells are artificially oxygenated. An alternative procedure is to use heavily ionizing particles. As the energy loss (corresponding to the density of ionization) increases, the OER falls, from a value of about 3 at 0.5 keV/μ to unity at about 165 keV/μ. The artificial acceleration of high energy nitrogen ions now makes it possible, for the first time, to perform the necessary fundamental radiobiological studies which necessarily precede clinical trials. Radiobiological experiments with beams of heavy ions of energy 10 MeV/nucleon are encouraging, but it is vital that such studies be repeated with particles whose ranges are comparable to the dimensions of the human trunk. (The range of 200 MeV/nucleon nitrogen ions is 10 g cm^{-2} of water.) Physical studies of ion fragmentation in tissue-like materials, energy loss mechanisms as well as radiobiological studies will be necessary before any attempt at human treatment is undertaken. Theoretical studies are extremely encouraging in suggesting, however, that some specific tumors may respond well to irradiation by heavy ions, perhaps as heavy as neon.

Radiobiological studies with densely ionizing radiations have hitherto been limited by the energy of the particle beams available. The availability of penetrating, densely ionizing radiation would permit studies of radiation effects following whole body irradiation of fairly large animals.

Of particular and pressing interest is the possible biological consequences to those engaged in space missions. Following the reports of light-flashes in the eyes of the astronauts of the Apollo 12 and 13 missions, several attempts have been made to reconstruct those effects in the laboratory (Wick, 1972). Alternative postulates suggest the light flashes to be due to Cerenkov radiation or due to interaction of heavily ionizing particles in the retina of the eye. Hitherto subjects had exposed themselves to x-rays, neutrons, pions, and α-particles and these experiments tend to favour the latter theory. The development of heavy ion beams has made possible the further investigation of these studies. Studies of these effects are pressing because heavy ions may also be capable of permanently damaging the non-proliferating nerve cells of the brain. It has been estimated that in the course of a 1000-day mission into deep space between 1% and 10% of these cells may be lost by an astronaut and serious impairment of function could result.

The Production of High-energy Heavy Ions in the Laboratory (1971)

As we have seen, until 1971 heavy ion beams available in the laboratory were limited to energies of about 10 MeV/nucleon. With the increasing scientific demand for higher energies already described and with no construction of any accelerator specifically designed for the purpose in the next several years it was inevitable that attempts would be made to modify existing experimental facilities to produce high-energy heavy ions. The synchrotron has several advantages over other types of particle accelerators: these in principle are that

(i) Particles having a wide range of charge-to-mass (e/m) ratio may be accelerated.
(ii) Particles may be produced over a wide range of energies and useful particle beams of high quality are produced at essentially all energies in this energy range of interest.
(iii) Particle beams with a good duty cycle—as high as 50% if required. (Under many experimental conditions uniform beam intensity with time is of great value, e.g. when particle

counters with finite resolving time are used). However, should it be required (e.g. for time of flight studies) beam may also be produced in discrete bunches, determined by the radio-frequency supply of the accelerating cavities.

(iv) Relative economy, the cost of a synchrotron capable of accelerating heavy ions to energies of several GeV/nucleon comparing extremely favourably with other accelerators.

With these specific advantages it was natural therefore that attempts would be made to accelerate heavy ions in existing proton synchrotrons. This was first done over the two years 1970–1971.

Synchrotrons designed for the acceleration of protons ($e/m = 1$), in general, require some modification before particles of other charge-to-mass ratios may be accelerated. There are three major technical problems to overcome:

(a) The injector system must be modified to produce a beam of heavy ions of adequate intensity and with suitable optical properties for injection into the synchrotron.

(b) The vacuum system of the synchrotron chamber must maintain a gas pressure sufficiently low that recombination is not severe.

(c) The range of the frequency swing required of the radio-frequency system is considerably larger when heavy ions are accelerated, than is required for protons.

The first and last of these difficulties are technical in nature, the second, however, is fundamental. Beam loss during acceleration in the synchrotron vacuum chamber is almost entirely due to collisions of the ions with residual gas molecules. In such a collision there is a possibility that the accelerating ion will change charge (and therefore e/m) and become asynchronous. Due to the lack of total charge exchange cross-section data as a function of velocity, already referred to, it was not possible to calculate these effects precisely but rather to rely heavily on theoretical predictions. These difficulties are fortunately not insuparable and several detailed descriptions of the way in which they have been overcome at two synchrotrons have been published in the literature (Grunder et al., 1971; White et al., 1971). It is worth remarking that for the forseeable future high-energy heavy ions will be produced at modified proton synchrotrons and consequently these papers are of good general interest to accelerator physicists.

Although the honour of the first acceleration of heavy ions to energies comparable with those found in the galactic cosmic radiation goes to the Princeton Particle Accelerator (White et al., 1971), this accelerator is no longer in operation.

During August 1971 an attempt was made to demonstrate the feasibility of accelerating heavy ions in the Bevatron. On August 2nd deuterons were readily accelerated and within only 8 hr of the initial attempt they were extracted with an energy of 2·1 GeV/nucleon at intensities of 3×10^{10} (subsequently increased to approximately 10^{11}) deuteron/pulse.

On August 6th α-particles at the same energy were extracted at an intensity of 5×10^9 particles/pulse followed by a reduction in energy to 1 GeV/nucleon on August 12th. August 17th saw the successful acceleration of nitrogen ions to 2·1 GeV/nucleon at an intensity of 2×10^5 particles/pulse. This was followed by extraction of beams at 250 MeV/nucleon; α-particle beams were produced on August 22nd and a nitrogen ion beam produced on August 24th.

Improvements to the accelerator have now made it possible to consistently produce O^{8+} beams at intensities of about 2×10^6 particles/pulse and ions as highly charged as neon have now been accelerated at low intensity (Summer 1972).

Table 3. Heavy ion beams produced at the Bevatron during the summer of 1971. Extracted beam intensities given in particles/pulse

Particle	Energy (GeV/nucleon)			Ion source used
	$E = 2\cdot1$	$E = 1$	$E = 0\cdot28$	
Deuterons	10^{11}	not attempted	2×10^{10}	D*
Alphas	5×10^9	5×10^9	10^9	D, P
Nitrogen ions	7×10^5	not attempted	$1\cdot4 \times 10^5$	P

*D, Duoplasmatron; P, Penning.

Experimentations with Heavy Ions in the Laboratory

In the brief time that energetic heavy ions have been available, substantial work in many disciplines has been accomplished. Only a brief review is attempted here.

One of the first experiments to have been performed was an investigation of heavy-ion interactions in emulsions. Figure 4 shows an interaction of N^{7+} ion in nuclear emulsion. These interactions are generally quite spectacular.

Cross-section measurements with heavy ion beams in the laboratory have confirmed that the total cross-section has the form deduced by Bradt and Peters from a study of the interaction of cosmic ray interaction. If the total cross-section, σ_{tot}, has the form:

$$\sigma_{tot} = \pi R_0^2 (A_{inc}^{1/3} + A_{target}^{1/3} - b)$$

measurements by Heckman et al. have determined the value $R_0 = 1\cdot26 \times 10^{-13}$ cm. The parameter b changes somewhat with projectile and target.

In recent years several theoretical physicists—Chew, Feynman, and Yang, for example—have suggested

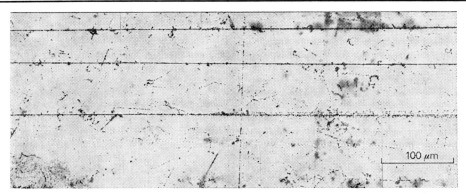

Fig. 4. A 29 GeV N^{7+} ion incident from the left fragments into two He and three H isotopes.

that the study of multiparticle reactions may yield fundamental information in the nature of the hadron.

The theory of multiparticle reactions at high energy has suggested that heavy-ion interactions may be interpreted in terms of the concepts of *limiting fragmentation* which suggests that, at high energies, the production cross-section for the ith fragment is independent of energy. *Factorization* suggests that the total cross-section, σ_{ab}, for interaction between particles a and b may be factored into quantities C_a, C_b that are respectively functions of a and b only. Thus:

$$\sigma_{ab} \propto C_a C_b.$$

A direct consequence of this factorization in the region of limiting fragmentation is that the modes of fragmentation of a heavy ion are independent of the target nucleus. Preliminary investigations with carbon and oxygen ions give strong support for the concepts of factorization. Fragmentation in the forward direction proceeds by a mechanism that strongly prefers that the fragment has the same velocity as the incident particle. (Heckman *et al.*, 1972) (see Fig. 5). Figure 6 gives examples of the charge spectra at two different energies. Chew has suggested that the use of energetic heavy ions may reveal important effects related to the concepts of limiting fragmentation and factorization not readily apparent in nucleon-nucleon interaction studies. These fragmentation studies also yielded much incidental information. Thus, for example, an important experimental check was obtained on theoretical calculations of the range and energy loss of heavy ions in silicon. It was found that Bethe–Bloch theory describes particle energy loss to an accuracy of 1% or better.

The properties of a neutron beam produced by stripping a 5·8 GeV/c deuteron beam have been reported (Leeman *et al.*, 1972), using a beryllium target the momentum spread was ±3·5% and for uranium ±3%—better than predicted using an internal momentum distribution derived from the Hulthen function. By counting the stripped proton in co-

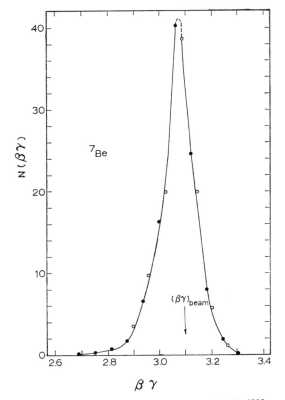

Fig. 5. The observed velocity spectrum of ^7Be produced by fragmentation of N ions. The value of $\beta\gamma$ for the incident beam is indicated by the arrow. (β represents the velocity of the particle in units of the velocity of light and γ is defined by $\gamma = 1 - \beta^2$) [from Heckman et al. 1971].

incidence with the neutron, beams with momentum resolution better than ±1% may be obtained.

Coherent pion production in heavy-ion nucleus interactions is of great interest—the production of

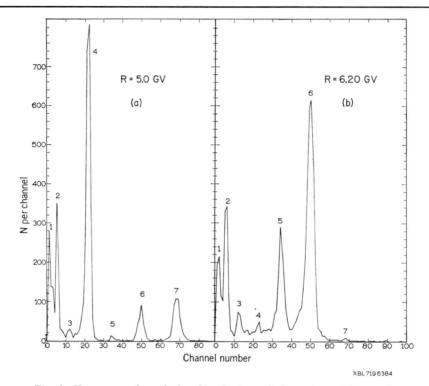

Fig. 6. Two examples of the distribution of charged particles produced by the fragmentation of 29 GeV nitrogen ions on carbon. Rigidities (momentum per unit charge) of 5·0 and 6·2 GeV for the fragments were selected. The elements hydrogen ($Z = 1$) through nitrogen ($Z = 7$) are indicated by their charge numbers. The ordinate represents the relative number of counts detected in each channel. Each channel corresponds to 6·8 MeV (from Heckman et al., 1971).

poins with momenta greater than the momentum possessed by a single nucleon was first reported by Baldin et al. (1971).

In the field of nuclear chemistry, track registration techniques were used to study the binary (and higher order) fission cross-section for 29 GeV nitrogen ions in heavy targets. For Ag, Au, Bi, and U cross-sections some three times larger than those obtained with incident protons were measured. Ternary fission was detected at a rate a hundred times lower than binary fission (Katcoff, and Hudis, 1972). In addition, a search is being made for the production fragments with $Z > 10$ and energy greater than 5 MeV/nucleon produced by ^{14}N bombardment of Au and U targets.

Some preliminary activation studies of 250 MeV/nucleon nitrogen ions in Be targets showed the production of ^{11}C and ^{13}N. These radionuclides are observed at well-defined depths in the target suggesting that they were produced in a completely ionized state as the result of fragmentation of the primary ions and are then stopped in the target at a depth corresponding to the charge state of the fragments. Other target materials are being investigated (Tobias et al., 1971).

Studies of the radiological physics characteristics of heavy ion beams suggest they will be suitable for radiotherapy (Todd, 1971). Many interesting radiobiological results have been reported (Donner, 1971). Typical of the radiobiology experiments is one carried out by a team at the Ames Laboratory of NASA in which some 56 pocket mice (*Perognathus longimembris*) were irradiated by the 250 MeV/nucleon nitrogen ion beam. In most cases the Bragg peak was located in the brains of the animals or in adjacent tissue. Both animals that were examined were found to be radioactive after irradiation. Other experiments included:

(a) Study of chromosomal damage to irradiated leucocytes.
(b) Effects of irradiation by heavy ions on the skin and retina of small animals.
(c) Determination of the OER for cultured mammalian cells.

The acceleration of heavy ions to high energies has

made studies of the effects of whole-body irradiation of mammals feasible and studies of leukemia incidence in mice are already underway.

The production of high-energy heavy ions in 1971 has opened up several avenues of research with countless possibilities of exploitation. Intensive efforts are now underway to improve the potency of the Bevatron for this new area of research. Exciting results seem almost guaranteed.

See also: Ion implantation.

Bibliography

BALDIN, A. M. *et al.* (1971) *Observation of High Energy Pions Produced by Relativistic Deuterons Interacting with Nuclei*, Joint Institute for Nuclear Research, Dubna, Report R 1–5819.
DAUBER, P. M. (1971) Isotopic composition of the primary cosmic radiation (Ed.) *Proc. Symposium held in Lyngby, Denmark*, Danish Space Research Institute.
DONNER LABORATORY STAFF (1971) *Radiobiological Experiments with Accelerated Nitrogen Ions at the Bevatron*, Lawrence Berkeley Laboratory Report, LBL-529.
FREIER, P. *et al.* (1948) *Phys. Rev.* **74**, 213.
GRUNDER, H. A. *et al.* (1971) Acceleration of Heavy ions at the Bevatron, *Science*, **174**, 1128.
HECKMAN, H. H. *et al.* (1971) Fragmentation of ^{14}N Nuclei at 2·1 GeV per nucleon, *Science*, **174**, 1130.
HECKMAN, H. H. *et al.* (1972) Fragmentation of ^{14}N nuclei at 29 GeV, *Phys. Rev. Letters*, **28**, 926.
KATCOFF, S., MISKEL, J. A. and STANLEY, C. W. (1948) *Phys. Rev.* **74**, 631.
KATCOFF, S. and HUDIS, J. (1972) Fission of U, Bi, Au, and Ag induced by 29 GeV ^{14}N ions, *Phys. Rev. Letters*, **28**, 1066.
LAWRENCE RADIATION LABORATORY STAFF (1966) *The Omnitron—A Multipurpose Accelerator*.
LEEMAN, C. W. *et al.* (1972) Deuteron stripping at 6 GeV/c and production of a tagged neutron beam, Paper presented at the XVIth International Conference on High Energy Physics, Chicago, Illinois.
LINHARD, J., SCHARFF, M. and SCHIØTT, H. E. [1963] *Kg. Danske Videnskab Selskab. Mat-Fys. Medd* **33**, No. 14, 1.
MAIN, R. M. (1971) *Trans. Nucl. Sci.* NS-**18**, 1131.
MARINOV, A., BATTY, C. J. and KILVINGTON, A. (1971) *Nature*, **229**, 464.
PURSER, K. H. (1971) *IEEE Trans. Nucl. Sci.* NS-**18**, 1121.
REAMES, D. V. (1970) ^{53}Mn *and the Age of Galactic Cosmic Rays*, NASA/Goddard Space Flight Center Report, X-611-70-5.
SEGRÈ, E. and WEIGAND, C. (1946) *Phys. Rev.* **70**, 808.
TOBIAS, C. A. *et al.* (1971) *Phys. Rev. Letters*, **37A**, 119.
TOBIAS, C. A., LYMAN, J. T. and LAWRENCE, J. H. (1972) Some considerations of physical and biological factors in radiotherapy with high LET radiations including heavy particles, pi Mesons, and fast neutrons, in *Prog. in Atomic Medicine: Recent Advances in Nuclear Medicine*, Vol. 3, p. 167, edited by J. H. Lawrence, Grune and Stratton, Inc.
TODD, P. (1971) Spatial distribution of biological effect in a 3·9 GeV nitrogen ion beam, *Science*, **174**, 1125–1127.
WHITE, M. G. (1971) The acceleration of heavy ions to very high energies and their scientific significance. *Proc. Particle Accelerator Conference*, Chicago, Illinois, p. 1115.
WHITE, M. G. *et al.* (1971) Acceleration of heavy ions to 7·4 GeV in the Princeton Particle Accelerator, *Science*, **174**, 1121.
WICK, G. L. (1972) Cosmic ray detection with the eye, *Science*, **175**, 615.
ZUCKER, A. and TOTH, S. (1968) Heavy-ion induced nuclear reactions in *Nuclear Chemistry*, L. Yaffe (ed.) p. 409. New York: Academic Press.

RALPH H. THOMAS

HOLOGRAPHY, ENGINEERING APPLICATION OF. The uses of holography in engineering lie in the field of inspection and measurement. Holography's value arises from its ability to record and reconstruct exactly a three-dimensional image of an object. Subsequently this image may be compared with either the original object or with a second holographic image of that object, to detect whether any change of shape has occurred.

Recording and Reconstructing a Hologram

To record a hologram (Fig. 1a) the object is illuminated with light from a laser and a photographic plate is positioned to receive light scattered from the object. At the same time, another beam, the "reference beam", is split off from the laser, spread out and directed to uniformly illuminate the photographic plate. If these two beams are coherent where they intersect at the photographic plate, an interference pattern will be formed and can be recorded on the plate. This is a hologram.

In the reconstruction stage (Fig. 1 b) the hologram is illuminated with a replica of the reference beam, making sure the angle of incidence is the same as before. The interference pattern recorded on the hologram will diffract the light into a wavefront which is exactly similar to that which was reflected from the object. So on looking back through the hologram one sees a three-dimensional image, formed by the hologram, of the original object.

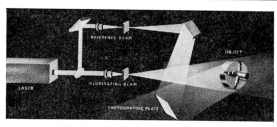

Fig. 1a. Recording a hologram.

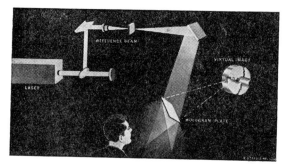

Fig. 1b. Reconstructing an image from a hologram.

Conditions Necessary for Holography

In nearly all holographic work, a laser, usually a continuous wave gas laser, is used as the source of coherent light. Most lasers have a fairly short coherence length—about 20 cm for a He–Ne gas laser—and the paths of the two beams to the photographic plate must not differ by more than this amount if a good interference pattern is to be formed. For this pattern to be faithfully recorded it must not move across the plate during the exposure; movement will result in either a weak reconstruction or none at all. To prevent this happening the path-length difference between the two beams must not vary by more than a quarter of a wavelength of light. This means the components in the holographic system, including the object, must be mounted so as to be free from drift and vibration. A quiet location or a table on good anti-vibration mounts is essential.

The stability of the system is much less of a problem if short exposures can be used. This can be achieved with those types of pulse lasers which have been developed for holographic purposes. The very short duration of the pulse from some of these lasers even permit holograms of moving objects to be recorded.

The photographic plate must, of course, be sensitive to the wavelength of the light used and have sufficient resolution to record the interference pattern. With an angle of 40° between the object and reference beams, this resolution needs to be at least 1000 lines per millimetre; hence special high-resolution emulsions are necessary which are consequently much slower than normal photographic films.

Interference Holography

The hologram in effect stores the wavefront reflected from the object. It contains information about the shape of that object at the time that the hologram was made and can be used subsequently to detect changes in shape that may occur. If the reconstruction is superimposed on the object any large change will be obvious. Small changes, of the order of a few wavelengths of light, will cause dark bands to appear on the object when it is viewed through the hologram. These are due to interference between the light reflected from the object and the reconstructed beam and are similar to the fringes seen in optical interferometers. This effect occurs irrespective of the shape of the object or the nature of its surface.

The reconstructed image can most easily be superimposed on the object by accurately replacing the hologram in the recording system. This can be achieved if a kinematic mount is used to locate the photographic plate. Any strain on the object which results in a change of shape can be seen as a fringe pattern on the object. As the strain is changed, the pattern will be observed to move, hence the name "live" is given to these fringes. The method is very useful for following the development of the fringe pattern and determining the direction of the distortion producing the pattern. Permanent records of the fringe pattern can be made by photographing the object through the hologram.

A permanent record of the fringe pattern can be recorded by the hologram itself if a double-exposure technique is used (Fig. 2). An exposure of half the normal duration is made on the holographic plate with the object in the unstrained condition and then, after the strain has been applied, a second exposure is made. The reconstructed image will have "frozen" into it the interference pattern due to the strain. This method is easier than the live fringe method since no repositioning of the hologram is necessary. Indeed the reconstruction may be examined at a quite different location with a reference beam from another laser.

Analysis of both live and frozen fringe patterns is similar. The fringes represent contours on the component of the displacement in the direction bisecting the angle between the incident illumination on the object and the direction of viewing. To put an absolute value on the displacement it is necessary to know the order numbers of the fringes. It is usually fairly easy to determine this in the case of live fringes, but in a double-exposure hologram there must be reference points; these may conveniently be the points of support.

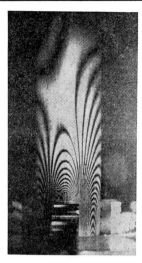

Fig. 2. Interference fringes showing deformation of a tube held in a four-jaw chuck. The reconstruction is from a double-exposure hologram.

If complete analysis of the fringe pattern is to be attempted, considerable care in the way the object is mounted and loaded will be necessary to avoid the complication of unwanted movements. However, even when these are present it may be possible to plot contours of strain across the object and so determine the stress.

Comparison of Components

While it is easy to measure deformations of a particular object by holographic interferometry, the comparison of two objects, say a production component with a master, is not possible except under certain conditions. This is because the difference in the detailed surface structure of the two components gives rise to very fine interference fringes in which are lost the broader fringes due to overall differences in shape. However, this problem can be overcome if the two surfaces to be compared have a smooth finish and are illuminated and viewed at such an oblique angle that they appear polished. Figure 3 shows the fringe pattern obtained looking down the bore of a small cylinder and comparing it with the holographic image of a similar master cylinder. Because of the oblique angles used the sensitivity was reduced, the interval between contours was in this case 1·8 μm.

Vibration

In 1965 Powell and Stetson showed how holography can also be used to study small amplitude vibrations. The method is particularly useful in studying the resonant modes of vibration of an object, for example a turbine blade. To do this a hologram of the object is recorded while it is vibrating, the exposure being long compared to the period of the oscillation. On the reconstructed image there will be a fringe pattern of "time-averaged" fringes (Fig. 4). The brightest fringes will occur where the object was stationary, i.e. at the nodes. Successive orders of the fringes will fall in brightness but indicate where the amplitude of vibration is approximately $n\lambda/4$, where n is the order number of the fringe. The fringe intensity distribution is in fact proportional to

$$\left| J_0\left(\frac{4\pi m}{\lambda}\right) \right|^2,$$

where m is the amplitude of vibration, and this must be taken into account for accurate interpretation of the fringe pattern.

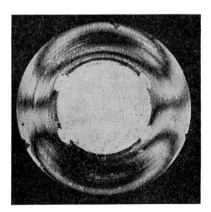

Fig. 3. Interference fringes resulting when a cylinder was compared with hologram of master. The pattern indicates curvature of the axis of the cylinder under test.

Fig. 4. Reconstruction showing time-averaged fringes on turbine blade.

Stroboscopically viewing the live fringes in synchronism with the vibration may also be used to investigate the nature of vibration. Although experimentally this is more difficult than the time-averaged method and indeed may not be possible at all with objects that creep, where it is possible, a fuller investigation of the form of vibration results. Changing frequency makes it possible to observe where resonances occur; varying the phase between the illumination and drive enable a complete cycle of vibration to be followed. Despite these advantages of the stroboscopic method, the simplicity of the time-averaged method has led to its much wider application.

Non-destructive Testing

Being a completely non-contacting technique, holography is finding increasing uses in non-destructive testing. It is being used, for example, in the examination of tyres to determine whether there are any faults in the laminated structure of the body of the tyre. A double-exposed hologram is recorded, the pressure inside the tyre being changed between exposures. The location of the defect reveals itself as a local anomaly in the frozen-fringe pattern, looking like a contour map of a small hill. The use of a powerful laser and rapid processing of the hologram make this a viable process for very large expensive tyres.

Figure 5 shows the presence of a fault in the bonding between the skin and the honeycomb structure in a lightweight panel. In this case the panel was vibrated by an electromagnetic driver at a frequency at which the unbonded area resonates; the fringes shown are time-averaged. An alternative method which worked very well with this object was to observe live fringes while heating the panel with a stream of warm air from a hair dryer. Kinks and loops in the fringe pattern revealed the defect.

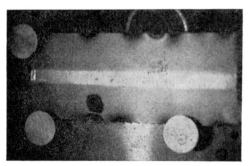

Fig. 5. The fringes show where a fault occurs in the bonding between the skin and the honeycomb structure in a light-weight panel.

Contouring

The shape of an object may be investigated by generating contours over its surface. Holography may be used to accomplish this by recording two holographic images successively on the same plate but varying slightly the wavelength of the light between the exposures. If the laser itself gives out light at two different wavelengths, these can be used, but an alternative way is to change the refractive index of the medium surrounding the surface. Then contours will be generated where the contour interval is $\lambda/\{2(r_1 - r_2)\}$, r_1 and r_2 being the refractive indices of the two media. Figure 6 shows the contour fringes on a metal former, part of an electrostatic lens. The former was filled with liquid and a flat cover glass placed over the top. A holographic exposure was made, the liquid replaced by another with a different refractive index and a second exposure made. The contours on the reconstructed image are at 221-μm spacing.

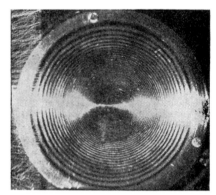

Fig. 6. The fringes represent contours with a contour height of 221 μm.

Conclusion

The examples given indicate the range of application holography is finding in engineering. It is still in the main confined to the laboratory and the inspection department, but with the increased availability of pulsed lasers it will become used in more adverse environments. Very short exposures make

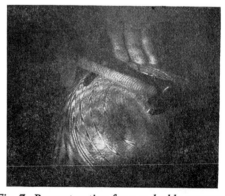

Fig. 7. Reconstruction from a double-exposure hologram of a motor-cyclist's crash helmet showing the effect of a hammer blow. The hologram was recorded with a pulse laser having a pulse duration of 25 nsec and a pulse separation of 25 μsec.

it possible to take good holograms of moving objects. The use of two very short pulses separated by a short time interval permits the study of transient effects. Figure 7 shows an example of this. The crash helmet was being worn at the time the hologram was taken, but the fringe patterns due to transient deformations are clearly visible.

Acknowledgements

The examples given in this contribution are all based on work carried out at the National Physical Laboratory. All the illustrations are Crown Copyright.

Bibliography

ARCHBOLD, E., ENNOS, A. E. and BURCH, J. M. (1967) Application of holography to the comparison of cylinder bores, *J. Sci. Instrum.*, **44**, 489.

ENNOS, A. E. and ARCHBOLD, E. (1968) Vibrating surface viewed in real time by interference holography, *Laser Focus*, **4** (19), 58.

GATES, J. W. C., HALL, R. G. N. and ROSS, I. N. (1972) Holographic interferometry of impact-loaded objects using a double-pulse laser, *Optics and Laser Technology*, **72**.

POWELL, R. L. and STETSON, K. A. (1965) Interferometric vibration analysis by wavefront reconstruction, *J. Opt. Soc. Am.* **55**, 1593.

ROBERTSON, E. R. and HARVEY, J. M. (1970) *The Engineering Uses of Holography*, Cambridge University Press.

SHARPE, R. S. (1970) *Research Techniques in Nondestructive Testing*, Academic Press.

E. ARCHBOLD

I

INFRA-RED GALAXIES. Galaxies which are strong emitters of infra-red radiation at wavelengths ranging from 1 to 1000 μm. The strongest infra-red galaxies may be emitting in this waveband as much as 100–1000 times the energy output of the brightest optical spiral galaxies and could probably be among the most luminous galaxies in the Universe.

The first infra-red data on galaxies at wavelengths $\lambda \leq 3.5$ μm were obtained by H. L. Johnson in 1966. No surprising results emerged from observations over this limited infra-red waveband. It appeared possible that the observed energy distributions could be understood in terms of integrated fluxes from cool stars. Far infra-red spectra of galaxies extending up to $\lambda \cong 22$ μm in several cases and up to 100 μm in one case have been obtained by F. J. Low and his collaborators from 1967 to the present day.

Seyfert galaxies. A class of galaxies known as the Seyfert galaxies stands out by including exceptionally intense sources of infra-red radiation. These galaxies, which number 1–2% of all galaxies, possess exceptional properties even at optical wavelengths. They have compact, relatively bright starlike condensations at their centres with spectra quite dissimilar to those of ordinary stars. A typical spectrum shows a large number of strongly forbidden emission lines, e.g. [O I], [O II], [O III], [N I], [N II], [S II], [S III], [Fe VII], [Ne III], [Ne V], [Ar III] in addition to the permitted lines of He I, He II and the Balmer series of hydrogen. The lines of H and He are very broad and have extended wings. The emission line spectra generally indicate that gaseous atoms are present in a high state of excitation. The presence of dust in Seyfert nuclei has been inferred from measurements of differential extinction of sulphur lines. There is also evidence from Doppler shifted spectral lines of the existence of gas clouds and shells of material ejected at speeds of several hundred km/s from the galactic nuclei.

Figure 1 shows the infra-red spectra (flux density as function of frequency) of several Seyfert galaxies, and Fig. 2 shows infra-red data of galaxies and quasi-stellar objects (QSO's) combined with radio observations at longer wavelengths. Several features may be noted:

(a) the infra-red spectra bear a striking similarity to one another;

(b) there is a high frequency turn over in the spectra of Seyfert galaxies occurring at a wave length close to 2 μm;

(c) the spectra (flux densities as functions of frequency) of all objects in the 2–22-μm waveband are well represented by the power law

$$F_\nu \alpha \nu^{-2};$$

(d) the spectra appear to indicate a peak flux at $\nu \cong 3 \times 10^{12}$ Hz.

From the fairly complete data available for the case of the Seyfert galaxy NGC 1068 the minimum energy output of this source would appear to be $\sim 2 \times 10^{12} L.$—about 100 times more intense than the optical emission from our galaxy (L. stands for the energy output of the sun at all wavelengths, $\sim 4 \times 10^{33}$ erg/s). A similar energy output would

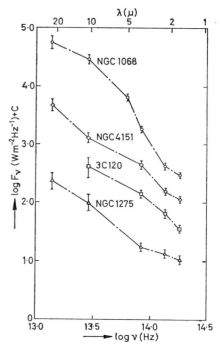

Fig. 1. Spectra (flux densities) of Seyfert galaxies in the wavelength range 2–20 μ. Units of ordinates: $(Wm^{-1}Hz^{-2} + const.)$.

apply in the case of NGC 4151, if its spectrum is similar to that of NGC 1068 near $\lambda \sim 10^2$ μm. The range of luminosity spanned by Seyferts could be in the range $\sim 10^{10}$–$10^{13} L_\odot$.

Galactic centre. At optical and ultra-violet wavelengths the centre of our galaxy is obscured by dust and therefore unobservable. The dust is, however, transparent to infra-red radiation and radio waves. Continuum radio emission as well as line radiation due to H I and O H at 21 and 18 cm, respectively, have been measured from the galactic nuclear region. The continuum source designated Sag.A has been localized in a region of dimension ~ 10 pc (1 pc = 3×10^{18} cm) near the galactic centre. Investigations using radio lines have revealed an exceedingly complex velocity field of the gas. There is evidence of shells of gas being ejected from the galactic centre, similar to the case of Seyfert nuclei.

Infra-red measurements of the galactic centre in the 1·5–1500-μm range have been made by several recent investigators. Its spectrum also shown in Fig. 2 bears a striking resemblance to that of Seyfert galaxies. The nucleus of our galaxy appears to be acting like a miniature Seyfert galaxy, at a power level of $10^8 L_\odot$, $\sim 10^{-4}$ times the energy output of NGC 1068.

QSO's and compact galaxies. In addition to Seyfert galaxies (including the galactic centre) some QSO's and a group of "compact galaxies" have been found to emit infra-red radiation. The spectrum of the QSO 3C273 for which the most complete data is available is also shown in Fig. 2. Its spectrum at wavelengths $\lambda \lesssim 300$ μm is found to be similar to that of Seyfert galaxies, but at radio wavelengths the spectrum is inverted with respect to that for Seyferts. Data available for a number of Seyfert galaxies at wavelengths $\lambda \geq 3 \cdot 5$ μm indicate a general similarity with the properties of Seyfert galaxies. There is a suggestion that infra-red emission from Seyfert galaxies, compact galaxies and QSO's may be similar phenomena involving the same type of physical processes.

Mechanisms proposed for infra-red radiation. The optical radiation from Seyfert galaxies appears to be separable into two distinct components—a thermal component due to hot gas from which the emission lines originate, and a non-thermal component

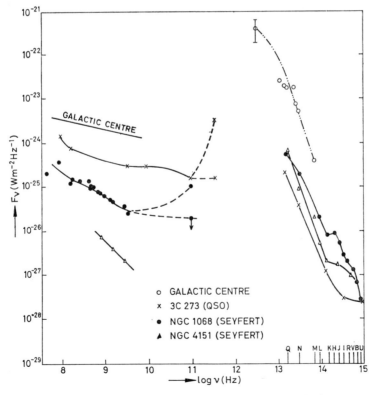

Fig. 2. Spectra (*flux densities*) of the galactic centre and of extragalactic sources in the frequency range 10^{15}–10^8 Hz. Units of ordinates: ($Wm^{-2}Hz^{-2}$ + const.).

which is most likely due to electron synchrotron emission. The non-thermal components are often strongly polarized and are also localized in fairly small volumes of space near the galactic nuclei. It is also generally accepted that the radio spectra can be explained on the assumption of electron synchrotron radiation. Attempts to account for the infra-red spectra of Seyfert type objects have given rise to a dichotomy of opinion, however.

Electron synchrotron radiation. One class of theory is based on the premiss that the entire spectrum of a Seyfert nucleus—optical, infra-red and radio wavebands—must arise from electron synchrotron emission. Relativistic electrons ejected from a source or sources of an unspecified character emit electromagnetic radiation by gyrating in magnetic fields. In order to produce the observed spectrum

$$F_\nu \propto \nu^{-2}$$

in the 2·2–22-μm waveband we require an electron energy spectrum

$$N(E) \propto E^{-5}$$

which must remain invariant from source to source ($N(E)$ is the electron number density per unit energy interval). It is normally supposed that the turn-over in the far infra-red spectrum at $\lambda \approx 100$ μm occurs due to synchrotron self-absorption. Excessively large magnetic field strengths $\gtrsim 10^{20}$ G are required if all the energy output of the brightest Seyferts originates in a single source at the centre. (This assumes that the turn-over at $\lambda \approx 100$ μm occurs due to synchrotron self-absorption.) Much smaller fields $B \approx 1 - 100$ G would suffice if a large number $\sim 10^2$ of smaller sources are involved. The main objection to the electron-synchrotron model for infra-red emission is that it requires the electron energy distribution which is chosen in a rather *ad hoc* manner to remain invariant from source to source. Furthermore, there does not appear to be any natural explanation for the sharp high-frequency turn-over of Seyfert spectra at a wavelength $\lambda \approx 2$ μm.

Other non-thermal processes discussed in the present context are Compton scattering from lower frequency (radio) photons, and proton synchrotron radiation. Both these processes are considerably less efficient than the electron-synchrotron process.

The second important category of explanation of the infra-red spectra of Seyferts involves a central primary source of ultra-violet radiation, soft X-rays or low energy cosmic rays surrounded by an extended shell of dust particles. Dust particles at varying distances from the primary energy source take up different temperatures, particles nearer the source being systematically hotter than the more distant ones (see Fig. 3). Thus different parts of the dust shell contribute to radiation at different infra-red wavebands. The energy output from the primary source is absorbed more or less completely by the dust shell and re-emitted in the infra-red. The ob-

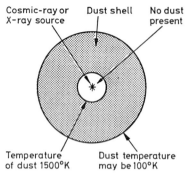

Fig. 3. *Schematic illustration of dust shell around primary energy source.*

served high-frequency cut-off in the Seyfert spectra at $\lambda \approx 2$ μm is readily accounted for by the fact that dust grains cannot survive at temperatures exceeding 1500–2000 K—that is at distances from the primary source interior to a critical value. Hotter dust grains, which if present could contribute to shorter-wave infra-red radiation, quickly sublime under conditions prevailing in the galactic nuclei. The stability of the dust shell against dispersal by radiation pressure has also been discussed and it may be shown that the photoelectric charge induced on dust grains produces a strong coupling between dust and gas which prevents the rapid escape of grains. Computed infra-red spectra for dust models have been shown to be in agreement with the observational data (see Fig. 4). The observed peak in the Seyfert spectra at $\nu \approx 10^{12}$ Hz and the high-frequency cut-off at $\lambda \approx 2$ μm are well reproduced by the model calculations. The best number density distribution of solid particles is found to be

$$\varrho \propto R^{-0.6}$$

where R is the radial distance from the central source. The several curves in Fig. 4 correspond to different values of the total optical depth in the shell to the primary radiation. (Note that optical depth τ implies attenuation by factor $e^{-\tau}$.) Although the results shown here correspond to graphite particles, the general trends are similar for other likely materials. The above remarks apply equally well to models where the primary energy source is ultra-violet light, X-rays, or cosmic rays.

Infra-red variability. A crucial issue in deciding between the various radiation mechanisms proposed for strong infra-red galaxies relates to their infra-red variability. There are reports—yet tentative—of rapid fluctuations in the infra-red output of Seyfert galaxies at 2 and 10 μm. At 2·2 μm, the flux from NGC 4151 is reported to vary on a timescale of months, whilst that of NGC 1068 is reported to vary on timescales $\lesssim 1$ day. At longer wavelengths data at 10 μm for NGC 4151 indicate variations

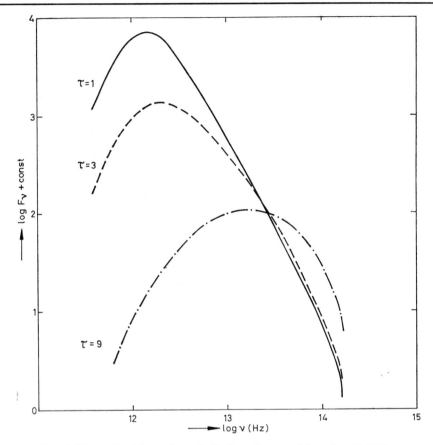

Fig. 4. Normalized fluxes for a shell of graphite particles of radii 0·05 μm with a number density law $\varrho \propto R^{-0.6}$ for various values of the total optical depth τ to the primary radiation. The inner and outer boundaries of the shell are chosen so that grain temperatures are 1500 K and 12 K respectively. The spectra are found to be insensitive to the assumed values of the total source luminosity and the absorption efficiency of grains with respect to the primary radiation.

of $\lesssim 50\%$ in the flux over timescales $\lesssim 1$ year. The 10-μm flux from NGC 1068 is reported to vary by $\lesssim 25\%$ over a timescale of 1 year.

It is possible to attribute fluctuations in the energy output of Seyfert galaxies to oscillations in the energy output of the primary central source on the basis of dust models. But an upper limit is set by the light travel time across the radiating region. A shorter-period variation of the central energy source, it it exists, cannot clearly be reflected in the infra-red emission. For a prescribed primary source luminosity the several types of dust model predict different minimum timescales for infra-red variability. The results of calculations for a source of luminosity $10^{12} L_\odot$ (appropriate for NGC 4151, 1068) are set out in Table 1.

Although the preliminary reports of infra-red variability of the Seyfert galaxies NGC 1068, 4151 are

Table 1. Predicted minimum period of variability for dust models

Wavelength (μm)	Primary UV source	Primary source 10-MeV nucleons or 30-keV X-rays
2	2 yr	2 days
10	50 yr	1 month

in discord with the predictions for a dust shell heated by ultra-violet radiation, it is consistent with those for heating by 10-MeV nucleons or 30-keV X-rays. A cosmic-ray or X-ray spectrum peaking at these energies is characteristic of some types of objects within the galaxy, e.g. supernovae remnants. The interstellar cosmic-ray spectrum is also characterized by a maximum flux at an energy close to 10 MeV. The spectral properties demanded of the primary energy source in the present model are thus not exceptional.

The main difference in the two types of primary source arises from the fact that dust grains are very much more efficient at absorbing energy from an ultra-violet source than from a source of cosmic rays or X-rays. Thus, for a given source luminosity the distance at which dust grains acquire a particular temperature (say, 500 K) is larger for the case of heating by ultra-violet radiation; hence the light travel time, and consequently the minimum timescale for variability, is longer in the former case.

It should be stressed that much shorter periods of variability than those reported at present would rule out dust models for explaining the infra-red emission of Seyferts. The electron synchrotron model, on the other hand, could involve radiating regions of much smaller dimensions and thus provide an explanation for very rapid fluctuations in the infra-red flux.

The question remains how the nuclei of Seyfert galaxies could provide such vast quantities of energy in the form of relativistic electrons, ultra-violet photons, X-rays or cosmic ray nucleons and also dust particles. One possibility is that extremely massive star-like objects of the type originally discussed by F. Hoyle and W. A. Fowler are present in the centres of galaxies, and that the explosions of such objects could give rise to the required high-energy particles and photons. The situation could be similar to that occurring in supernovae explosions. Sporadic explosions of "massive objects" in the nuclei of Seyfert galaxies could lead to infra-red variability of the type actually observed in NGC 1068 and NGC 4151.

As the exploded material expands outwards from a central "massive object" it gets progressively cooler until, eventually, temperatures ~2000 K low enough for solid particle formation could occur. Solid particles which grow to sizes of a few tenths of micron may gather to form shells around the central object and account for the infra-red emission from Seyfert-type galaxies.

Bibliography

BURBIDGE, G. R. and STEIN, W. A. (1970) *Astrophys. J.* **160**, 573.
HOYLE, F. and WICKRAMASINGHE, N. C. (1968) *Nature*, **218**, 1126.
LOW, F. J. and KLEINMANN, D. E. (1968) *Astron. J.* **73**, 868.
NEUGEBAUER, G. and BECKLIN, E. (1971) *Ann. Rev. Astron. Astrophys.* **9**, 67.
REES, M. J., SILK, J. I., WERNER, M. W. and WICKRAMASINGHE, N. C. (1969) *Nature*, **223**, 788.
WICKRAMASINGHE, N. C. (1971) *Nature*, **230**, 166.

N. C. WICKRAMASINGHE

INSULATION OF ELECTRIC POWER CABLES

Introduction

Cables may be divided into three main categories: high-voltage power, low-voltage power (distribution) and telecommunication. The insulation of each of these categories possesses some characteristics peculiar to the type and others which are mutual. For example, the value of permittivity is of vital importance for both high-voltage and high-frequency cables. Amongst other important characteristics are permissible temperature rises set up by both steady-state and transient conductor currents, ability to withstand the mechanical stresses set up by bending, dielectric loss, breakdown characteristics on a.c., d.c. and impulse, flame-retarding abilities.

Development of Power Cables

From the earliest developments paper tapes impregnated with insulating oil have been extensively used. The historic rigid-conductor cable used by Ferranti employed such insulation at 10 kV and the electrical and mechanical properties are such that this insulation is still widely used in a more refined form.

At voltages below 19 kV to earth oil-impregnated paper-insulated cables (solid type, i.e. non-draining) are used often with the three conductors contained in a single sheath. The three conductors are stranded and insulated separately and then laid up helically together. The space between and around the conductors is packed with paper or jute to form a circular surface which is then further wrapped with insulation. This is called the belted type of cable (see Fig. 1). With the three-core cable high electric stresses are set up tangentially to the paper insulation surfaces in which direction the insulation strength is weakest. To overcome this each core is wrapped with a conducting layer of metallized paper which

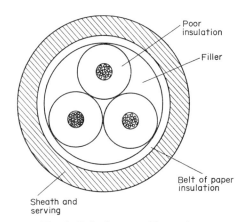

Fig. 1. Belted type cable insulation.

converts the cable electrically into three single-core cables with the electric stress completely in the radial direction. This form of construction was originated by Hochstadter and is known as the "H" type. As power supply system line-to-line voltages increased above 33 kV the solid-type cable became prone to breakdown because of the voids formed (small pockets of air or gas) in the insulation when constituent parts of the cable expanded and contracted to different extents with the heat evolved on load current cycles. The stress across these voids is relatively high and local discharges occur, which through complex processes such as heat evolution and erosion of the neighbouring insulation often caused complete failure of the insulation.

In many countries, for voltages above 33 kV, the type of cable system in most common use is the oil-filled cable with paper/oil insulation, due to Emanueli. In the oil-filled cable the hollow centre of the conductor is filled with insulating oil maintained under pressure by reservoirs feeding the cable along the route. As the cable heats on load, oil is driven from the cable into the reservoirs and vice versa, hence the creation of voids is avoided. An alternative is the gas-pressure installations where nitrogen at several atmospheres maintains a constant pressure on the inner sheath, compressing the dielectric, and so preventing the formation of voids. In the U.S.A. the paper/oil-insulated cable is often installed in a rigid pipe containing the insulating oil and a conductor oil-duct is not required. At the highest line to line voltages, at present of the order of 400 kV, the oil-filled cable has a design working stress of 15 kV/mm. The design of the insulation is based on transient (impulse 1/50 μsec waveform) voltages rather than those of power frequency.

In spite of earlier hopes for plastic insulations, the combination of paper and oil still forms the most effective dielectric at the higher voltages. The difficulty of manufacturing Polythene (polyethylene) cables without the creation of voids in the insulation has limited their use to voltages below 70 kV to earth. The temperature limitation of these cables is 70°C, above which the conductor tends to sink in the hot plastic, becoming offset and creating higher voltage stresses. This limit has been raised to 90°C by the use of cross-linked polyethylene.

Parameters of Insulation

The dielectric loss, due to leakage and hysteresis effects in the dielectric, is usually expressed in terms of the loss angle δ; $\delta = 90 - \varphi_d$, where φ_d is the dielectric power-factor angle. The dielectric loss is given by $\omega C V^2 \tan \delta$, where C = capacitance to neutral, and V = phase voltage. For modern high-voltage power cables $\tan \delta$ lies between 0·002 and 0·003, but this value increases with temperature above 60°C in oil-filled cables. In low-voltage cables this loss is negligible, but becomes very appreciable in cables at 275 kV and above. The relatively high value of $\tan \delta$ for paper/oil insulation compared with some plastics is due to rotation of the molecular dipoles in the a.c. field. Of great importance is the thermal stability of insulation which is largely dependent on the power factor–temperature relationship for insulation. For modern high-voltage paper/oil insulation a typical relationship is shown in Fig. 2.

The maximum allowable temperature rise in cable insulation is determined by a mixture of scientific analysis and practical experience. Failure usually occurs where some undue stress is applied due to a fault condition, when high mechanical and thermal stress are set up.

Factors which influence the maximum permissible temperature rise in cable insulation are as follows:

(a) The differential expansion between the insulation, the surrounding sheath and the conductor sets up mechanical stresses and subsequent abrasion of the insulation. This is due to the different temperature coefficients and temperature rises of these materials.
(b) Changes in the electrical properties of the insulation especially in the dielectric loss.
(c) Changes in mechanical and chemical properties which result in electrical changes.

The recommended steady-state working conductor temperature of oil-filled or gas pressure cables is 85°C.

Cable Insulation Materials

In paper/oil insulation cellulose paper tape is wound helically around the cable core. The tapes are usually wound with a gap (butt gap) between turns to allow movement on bending. Overlapped helices are also possible, but never helices which are butted exactly to each other. As the butt gap represents a weak point in the insulation (the breakdown strength is dependent on the properties of the impregnating oil), the gaps are staggered between layers. The bending radius of a cable is approximately expressed as follows:

$$\left(\frac{\text{radius of one complete helix}}{\text{bending radius of cable}} \right) \div \left(\frac{\text{butt space}}{\text{tape width}} \right)$$

which in practice is in the order of 0·1.

The conductor radius in power cables is usually determined by heat transfer considerations and insulation thickness determined from this and the voltage to be withstood. At the higher voltage the conductor radius may be determined by voltage rather than temperature rise. The breakdown strength of paper/oil insulation depends on the tape thickness and it is usual to vary the tape thickness through the insulation of a cable such that the required breakdown stress at any radius is matched by a tape of the appropriate thickness. Because of the smaller depth of oil entrapped, the breakdown voltage for thin tapes is higher than for thicker

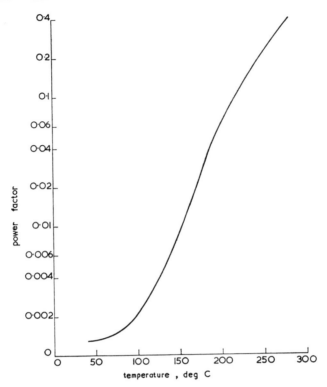

Fig. 2. Power factor—temperature curve for paper-oil insulation.

tapes. Tape width depends on the bending radius and is usually 8 mm upwards.

The use of plastics for lower voltage cables (11 kV and below) is well established. Extruded polyethylene solid insulation has been used successfully for voltages to earth up to 70 kV, but difficulties have been encountered in extruding polyethylene without voids on a commercial scale. The discharges occurring in these voids have so far limited the operating stress, but there are indications that this is being overcome. A summary of modern insulation materials that are now in use for voltages at or below 76 kV (conductor to earth) is given in Table 1.

Polyvinyl chloride is used extensively for low-voltage mains distribution cables. More recently ethylene propylene rubber (EPR) compounds have been used for high-voltage cables and show good electrical and mechanical properties. The power factor–temperature curves for various solid insulants are shown in Fig. 3 and the resistance of such materials to partial discharge is illustrated in Fig. 4.

Insulation for High-Voltage Direct Current Cables

Owing to the absence of capacitance effects and dielectric losses with direct voltage, high-voltage

Table 1

Material	Dielectric resistance (Ω cm)	ε_r	Power factor tan δ	Flame retardance	Continuous operating temp. (°C)
Chemically cross-linked polythene	10^{19}	2·3	0·0002	Burns	90+
Polyvinyl chloride	10^{14}	4·5–7	0·08	Self-extinguishing	70
Butyl rubber	10^{16}	3·4	0·008–0·015	Burns	85
Ethylene propylene	10^{16}	2·8	0·003–0·015	Burns	90

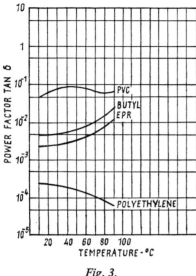

Fig. 3.

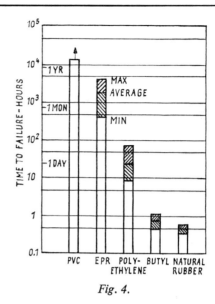

Fig. 4.

cables could play an increasingly important role in direct-current transmission links. The loss in the dielectric on direct current is only about 3% of that on alternating current. Whereas the electric stress distribution in alternating-current cable insulation is determined by the capacitance, in direct-current cables it is determined by the electric resistance of the dielectric. The electrical resistivity of conventional dielectrics is very temperature dependent; for oil-impregnated paper, for example, the resistivity at 20°C is one hundred times that at 60°C. In direct-current cables thermal considerations not only determine the current rating but also influence the electric stress distribution in the dielectric. The electrical resistivity also varies with the electric stress. Instead of the electric stress decreasing through the dielectric from the conductor to the sheath, in direct-current cables the stress increases and can be larger at the sheath.

Future Developments in High-Voltage Power Cable Insulation

Work has been in progress for some years on the replacement of paper tapes by plastic material in conjunction with a liquid insulant, e.g. oil. A list of commercially available materials with properties is given in Table 2, although new materials are becoming available. Of particular interest in these

Table 2

	Permittivity ε_r	Dielectric loss tangent $\tan \delta$	$\varepsilon \tan \delta$	Density kg/m³ $\times 10^3$	Thermal resistivity in tape construction $K - m/W$	Maximum operating temperature degC
Polystyrene	2·5	0·0003	0·00075	1·05	12·00	60
Polyethylene (high-density)	2·3	0·0003	0·00069	0·96	4·00	85
Polypropylene	2·15	0·0010	0·00215	0·90	10·00	85
Polytetrafluoroethylene (p.t.f.e.)	2·0	0·0002	0·00040	2·20	6·00	150
Polyethyleneterephthalate (p.e.t.e.)	2·96	0·0020	0·00590	1·40	8·00	125
Polycarbonate	3·2	0·0010	0·00320	1·20	7·25	85
Oil-impregnated paper	3·3	0·0025	0·00825	0·85	5·00	85

Taken from *Proc. I.E.E.*, **112**, 1965, page 111.

materials are the low values of power factor and permittivity which determine dielectric loss and capacitance respectively. These quantities increase with voltage and above a voltage to earth in the region of 76 kV become of extreme importance.

Compressed gas such as sulphur hexafluoride (SF_6) is also being developed as insulation, the conductor being supported by spacers within a containing sheath. An active research programme is also being pursued for the development of cryogenic cables both superconducting and resistive. For this a number of possibilities for insulation are being considered as follows: plastic tapes impregnated with liquid helium or nitrogen, high vacuum, and the cryogenic fluids themselves which are good electrical insulators.

Bibliography

BARNES, C.C. (1964) *Electric Cables*. Pitman: London.

B. M. WEEDY

ION IMPLANTATION. Ion implantation is concerned with the controlled injection of energetic ionized atoms (or molecules) into solid targets. The effects of this process are:
(1) to introduce, into the solid, a different atomic constituent which can influence the physical and chemical properties of the solid to a depth similar to that to which the ions penetrate before coming to rest, and
(2) by displacing atoms of the target as a result of collisions between the ions and target atoms and subsequently between target atoms themselves, to create a disorder in the atomic arrangement of the target and again influence both chemical and physical properties.

Ions of most atomic species are produced generally by electron impact ionization in a variety of source configurations, the most efficient containing a high density plasma such as confined arc. The species to be ionized is introduced as a gas or appropriate vapour to the source and, in the case of a solid source charge material, the vapour is achieved by thermal evaporation or by ion sputtering of the material. Ions are extracted from the source by a well defined electrostatic field and are then electrostatically accelerated to the required energy, focused into the required beam configuration by electrostatic or magnetic lenses and transported, through high vacuum to minimize ion losses from the beam via gas atom collisions, to the solid target. It is frequently required to inject a single atomic, or even isotopic component into the solid and in these circumstances mass separation of the components in the ion beam can be effected by a number of different electrostatic and/or magnetic field configurations.

In two important areas of ion implantation, however, ions are accelerated directly from the plasma, through the gas supporting the plasma and immediately strike the target. The first application in which this occurs is ion pumping, where ions of the gaseous species in a vacuum system are formed by electron impact and are accelerated to energies of 2–5 keV into a cathode. The ions penetrate the cathode and are trapped within the atomic matrix, thus constituting a gaseous removal or pumping action. In the second application ions of an appropriate species are formed in a gas supported discharge and these ions together with a high flux of neutral atoms of the same atomic species fall upon a target or substrate. Again some ion removal or burial occurs and this appears to act as a key or catalyst to promote the improved attachment of the high flux of neutrals to the substrate, thus allowing growth of films of the required species on the substrate. This process of ion plating appears to produce very uniform and highly adherent film coatings.

In the more conventional meaning of ion implantation, ion beams of well defined energy, usually in the range of some hundreds of eV to several MeV and most frequently between 20 keV and 200 keV, are focused on to well defined targets. It is possible to achieve, under favourable circumstances, beam densities in the region of 10 ma/cm^2, which represents an ion flux to the target of the order of 10^{17} ions/cm^2/sec and defined beam spot sizes to the order of 10000 Å. Deflection of the ion beam by electrostatic or magnetic fields allows beam sweeping of a target to provide for larger area implantations, whilst this can also be achieved by defocusing of the beam or mechanical oscillation of the target.

The ability to define the area of implantation accurately is a positive advantage of ion beam techniques, particularly in certain semiconductor applications, since it allows accurate spatial delineation of the target where property changes are achieved. A further advantage is that rather accurate control of the quantity of implant can be obtained by controlling the ion flux and time of implant. It is also possible to specify, relatively accurately, the resulting concentration of implant as a function of depth in the target, since the ion penetration depth can be calculated and measured to reasonable precision (e.g. to within several hundreds of Angstroms in most cases but to within several Angstroms in certain circumstances). If the target is amorphous (no long range atomic symmetry) then the incident ions slow down in the solid via a succession of collisions with target atoms which are spatially uncorrelated and statistically distributed. When the ion energy degrades to the order of about 25 eV it can no longer overcome the interatomic forces in the lattice and comes to rest. Since the collisions are statistically distributed, the spatial rest points of the ions are also statistically distributed, generally assuming a Gaussian form as a function of depth into the target parallel to the direction of ion incidence with a mean depth R_p, and spread δR_p determined by the ion and target species and the ion energy. Up to an incident energy E approximately equal to M_1 keV (where M_1 is the atomic weight of the ion) the mean ion penetration depth increases roughly proportional to ion energy but at higher energies the depth is proportional to the square root of energy. The spread of the depth distribution also increases with increasing energy, as does the transverse distribution of stopping points perpendicular to the direction of ion incidence. This transverse spread is always less than R_p but for light ions in heavy targets the transverse distribution of stopping points may be almost as broad as the depth distribution.

Table 1 indicates theoretical values of R_p and δR_p for a number of ion-solid combinations and experimental measurements have largely confirmed these values. As a rough guide, the mean penetration depth is between 5 Å and 100 Å per keV of ion energy at energies below M_1 keV. Although these results apply to amorphous or randomized atomic arrangements, this situation can be simulated, with crystalline solids by arranging that the ion beam enters the solid in an incidence direction well removed from a major crystalline axis or plane. The ion then "sees" the lattice as a relatively opaque collection of atoms. If, however, an ion enters a crystal close to or parallel to a major axis or plane it sees the lattice as a collection of parallel strings or planes of atoms running into the crystal, and the effect of this symmetrical arrangement is to provide for those ions, incident on the surface at a distance greater than a few tenths of an Angstrom from a surface atomic position, a gentle steering action. This steering or channelling action causes the ions to oscillate between neighbouring atomic rows or planes whilst continuing into the crystal, prevents close collisions between the ions and target atoms, thus minimizing energy loss and allowing deeper penetration of the ions into the lattice. In such a channelled trajectory the major source of ion energy loss is via excitation of atomic electrons and since this process is known to lead to a rate of energy loss proportional to ion velocity, the maximum ion penetration depth in a channel is proportional to the square root of the incident ion energy. The fraction of ions which undergo such channelling processes and the depths reached by these ions depend upon ion type, energy and direction of incidence and crystal type, structure, orientation and temperature. In particular the fraction channelled and the penetration depth increase with increasing ion energy and increasing lattice transparency (the number of atomic rows or planes per unit area parallel to the ion beam direction). Figure 1 illustrates the channelling effect by comparing the ion penetration depth profile for K ions in amorphous tungsten oxide and parallel to the major axes in single crystal tungsten. This figure clearly indicates that a considerable fraction of incident ions can be channelled under favourable circumstances and that penetration depths can be enhanced by more than an order of magnitude by this process.

In the absence of channelling the average concentration of implanted ions, following injection of a known ion dose, can be estimated from the R_p and δR_p values from the equation

$$N_c = \frac{0.4 \times \text{ion dose per unit Area}}{\delta R_p}$$

Table 1. Projected Ion Ranges (R_p) and Range Standard Deviations (δR_p) for a Number of Ions in Various Target Materials

Target and ion species		Ion energy (keV) (R_p and δR_p measured in micrometres)						
		10		50		100		200	
		R_p	δR_p	R_p	δR_p	R_p	δR_p	R_p	δR_p
Diamond	Al	0.0079	0.0020	0.0392	0.0081	0.0850	0.0151	0.1860	0.0272
Diamond	Sb	0.0049	0.0004	0.0151	0.0012	0.0259	0.0021	0.0462	0.0035
Si	He	0.1015	0.0479	0.4696	0.1042	0.8129	0.1263	1.3304	0.1435
Si	B	0.0383	0.0189	0.2024	0.0627	0.3977	0.0939	0.7253	0.1259
Si	P	0.0144	0.0055	0.0610	0.0198	0.1233	0.0354	0.2538	0.0610
Si	Sb	0.0087	0.0015	0.0269	0.0045	0.0457	0.0074	0.0809	0.0125
Ge	He	0.0609	0.0255	0.3175	0.1262	0.5998	0.1657	1.0593	0.2010
Ge	B	0.0253	—	0.1259	0.0645	0.2552	0.1012	0.5010	0.1449
Ge	As	0.0068	0.0029	0.0224	0.0088	0.0397	0.0150	0.0733	0.0258
Ge	Sb	0.0057	0.0019	0.0175	0.0056	0.0295	0.0091	0.00517	0.0153

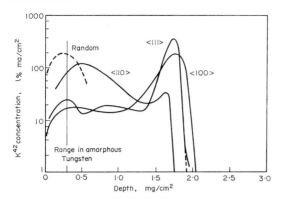

Fig. 1. Range distributions of 500 keV K^{42} ions in single crystal tungsten.

but it should be remembered that the concentration actually follows a Gaussian type depth profile. At low ion energies this profile intersects the surface so that there may be a relatively uniform concentration near the surface and then a relatively rapid decrease of concentration with increasing depth. Since the characteristics of the depth profile are energy dependent, however, it is possible to build up, rather accurately, a desired depth profile by injecting controlled numbers of ions at various energies and this is a very useful feature for semiconductor implantation for accurately locating a junction with a known impurity or dopant concentration profile on the surface side of the junction.

From the above, and particularly in view of the comments on ion plating, it might be anticipated that it is possible to continue to increase the concentration of implant in the solid indefinitely. This is not the case since, during implantation, the ion beam continuously displaces atoms from the target surface by a process known as sputtering. The effect of this is to erode the surface and leads to removal of both target material and implanted species. Consequently, there is always a competition between ion entrapment in the solid and the removal of earlier implant and this generally leads to a saturation concentration of implant of 10–20% atomic fraction of the target. If the sputtering rate is low (which occurs for low energy ions) it is possible to greatly enhance this concentration and indeed build up superficial layers of implant alone.

It is thus seen that implantation can result in composition changes in solids and this has particular value in introducing electrically active dopants into semiconductors, passivation of surfaces to corrosion, surface hardening, optical and magnetic changes in surface layers and numerous other as yet unexplored or briefly investigated areas. However, we have seen that ion implantation also causes changes in lattice perfection since, in slowing down, the ion transfers to lattice atoms energy which may, if it exceeds the energy binding the atom to its lattice site, be sufficient to displace the atom, creating a vacancy and a moving interstitial. If the energy transfer is large enough, the displaced atom itself may carry away enough energy to cause further displacement and the growth of a cascade of displaced atoms is readily envisaged. An order of magnitude estimate of the number of displaced atoms per incident ion is given by the equation

$$\gamma(E) = \frac{E_{\text{elastic}}}{2E_d}$$

where E_{elastic} is the fraction of energy lost in elastic processes. The spatial region over which these displacements occur is very similar to that in which the ions come to rest and there is both an extension of this displaced atom cascade parallel to the beam direction and transverse to this direction. However, the mean displaced atom or disorder depth is smaller than the corresponding ion penetration depth (i.e. the peak of disorder density occurs closer to the surface than the peak of the ion penetration profile) whilst the disorder depth distribution function is always broader than the Gaussian ion range profile. To a first approximation the volume of the disordered region is proportional to the square of the incident ion energy. Following the production of these defect or "Frenkel" pairs (vacancy and interstitial) however a rearrangement will generally ensue via strain and thermally activated defect migration so that larger aggregates of defects, which create less lattice strain, form. In metals the final form of disorder is generally of extended defects, such as dislocations, with (depending on the temperature) a background of point defects, whilst in semiconductors, the generation of initially individual and highly disordered (quasi amorphous) zones is observed. Increased ion fluence leads to an overlap of these zones and production of an amorphous layer centred on or about the mean ion range. In metals the final defect concentration never exceeds 0·1% whilst in semiconductors the level may approach 100%. However, in semiconductors, implantation at temperatures in excess of ~400°C prevents the formation of this gross disorder so that damage is manifest in dislocation entanglements as in metals. Clearly these high disorder levels can and will affect the physical properties of solids and in semiconductors can seriously militate against the successful achievement of only known energy levels required to be introduced by the implanted dopant ions.

The major application of ion implantation has been in the production of electronic devices in semiconducting materials and we will now look at some of the devices that have been produced.

The more conventional technique in device fabrication is diffusion, in which the dopant is usually deposited from a vapour onto the surface of the semiconductor slice and then driven into the required depth in a high temperature diffusion stage. Figure 2 shows a photograph of both a diffused and an ion implanted junction in silicon. The junctions are exposed by a technique of sectioning through the doped substrate and then staining them to expose by contrast the change from p- to n-type material. The ion

Fig. 2. Photographs of sectioned and stained p–n junctions in silicon: left: a phosphorus-diffused junction; right: a phosphorus-implanted junction, showing that implantation produces a more uniform penetration. The horizontal mark indicates the line of the junction in each case. The host silicon material was epitaxially grown on sapphire. (Photograph supplied by courtesy of R. W. Lawson, Post Office Research Department, Martlesham, Suffolk.)

implanted junction can be seen to be more uniform due to the well defined ion range. Diffusion does, of course, produce a junction at a well-defined depth but since any faults or irregularities in the substrate often provide faster diffusion paths for the dopant, the final junction edge is often irregular. The silicon substrate used here was epitaxially grown on sapphire and this type of material is well known to be heavily faulted. The depth to which an ion penetrates a substrate is not influenced to any great extent by the substrate material provided one neglects channelling effects. Consequently, it is expected that junction depth of ion implanted devices will be extremely uniform. This is a great advantage particularly for making large area, shallow junctions as are required for example for nuclear particle detectors. Ion beams can be easily scanned over large areas to provide extremely uniform doping and this advantage together with a low energy ion beam to produce shallow junction depths makes the implantation technique ideal for this type of device.

A conventional device is manufactured by diffusing the impurity into the substrate through windows cut in a protective oxide which defines the area over which the junction is formed. Ion implanted devices are doped by accelerating the impurity as an ion beam into the substrate, using similar windows to define the required junction areas. This is illus-

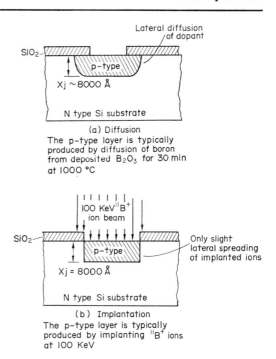

Fig. 3. Comparison of implanted and diffused junction shapes.

trated in Fig. 3a. The diffusion takes place in three dimensions and so the extremities of the diffusion profile (i.e. at the position where the pn junction is formed) are smooth and rounded. The three dimensional movement also takes the dopant sideways under the oxide so that the pn junction at the substrate surface is protected by the oxide layer. In a typical diffusion step, boron would be diffused into the substrate in an oven for 20 min at a temperature of $\simeq 1000°C$. If the junction depth x_j was 8000 Å then the lateral spread under the oxide might be approximately equal to x_j. In an implanted junction shown in Fig. 3b, an energy of $\simeq 100$ keV of B^+ ions would be required to produce a junction at the same depth but in this case the lateral spread of the ions is so small (<1000 Å) that there is a very abrupt edge to the pn junction under the edges defined by the ends of the oxide windows. The radius of curvature of the junction edge is an important parameter in predicting junction breakdown voltage and so the sharp edges, in an implanted device, produced by the mask lead to high electric fields and reduced breakdown voltages compared to the diffused junctions with their smoothly varying edges. The mask edges shown in Fig. 4 are shown vertical and are of course "ideal", but in reality window edges have quite different shapes. In many cases the mask used to define the implant area is a metal, as shown in Fig. 4, which is used for contacts to electrodes. The metal, such as Al, is deposited onto

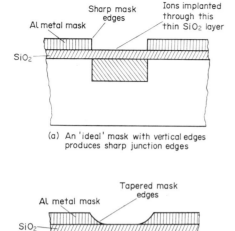

Fig. 4. Ideal and real mask edges used in ion implantation.

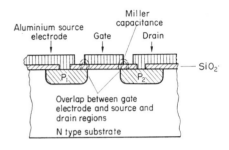

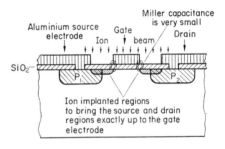

Fig. 5. Conventional and ion implanted MOST.

a thin oxide layer through which the implanted ions are injected. In this way the whole of the substrate and junction can be protected or passivated. The edge of the aluminium is shown sloping which is typical for etched aluminium. This means that the implanted ions are penetrating different thicknesses so that the junction underneath is also tapered as shown. The radius of curvature and hence the electric field at the junction are now reduced with a subsequent raising of the breakdown stage.

The small lateral spread of implanted ions under the oxide forming the window edges has played a major role in promoting the use of ion implantation in device fabrication. Let us consider this with reference to the metal-oxide-semiconductor-transistor (MOST) which is extensively used in large scale integrated circuit arrays where its low power consumption is an attractive feature. Figure 5 shows a cross-section through two MOSTs that are basically identical except that an implantation stage has been employed in producing the second device. In the conventional diffused device (Fig. 5a) the sideways diffusion of the implant under the oxide leads to some uncertainty in the exact position of the edges of the source and drain *pn* junctions. Consequently when the aluminium gate electrode is deposited there is an uncertain amount of overlap between the gate and the source and drain regions. The aluminium must be placed over the whole of the channel and

in between source and drain for the device to operate satisfactorily. In order to achieve this, because of the uncertainty of the extremities of the source and drain regions, a certain amount of overlap is inevitable. Unfortunately this overlap introduces parasitic capacitance effects (or Miller effects) that reduce the performance of the device.

Figure 5b shows how this overlap can be eliminated by using an ion implantation stage during production of the MOST. The source and drain regions are formed by conventional diffusion and then electrode contacts are laid down for source, drain and gate. The gate electrode is made deliberately narrow, however, so that it does not cover the whole of the channel and certainly does not overlap the source and drain. An implantation stage is next carried out using the gate metal electrode as a mask. Ions penetrate the oxide above the channel and carry the source and drain regions exactly up to the gate electrode. The thickness of metal prevents the remaining part of the channel from being doped during this implantation stage. Consequently, a device is produced with a very low parasitic capacitance so that it can switch at much faster speeds and operate up to higher frequencies.

Another example of the use of on implantation in device fabrication is found in the production of varactor diodes. These diodes have a very small junction area and are designed to utilize the basic property of the dependence of junction capacitance on reverse bias voltage. The width of the space charge region at the junction varies with applied bias voltage and if the carrier concentration is uniform then the junction capacitance varies as the inverse square of the applied voltage, i.e. as $v^{-1/2}$. An example of this type of junction is the Schottky barrier between a metal and semiconductor. If the carrier concentration falls off with increasing depth the capacitance of the junction varies more rapidly than $v^{-1/2}$ and such diodes are know as hyperabrupt. Ion implantation is well suited to producing a doping profile that falls off rapidly with distance into the semiconductor whilst at the same time the surface concentration can be kept level (by variation of the implant ion energy) when good Schottky barriers are possible. By varying the shape of the implant profile a wide range of capacitance-voltage characteristics are made possible whilst the precision and uniformity of the implantation technique is also advantageous in obtaining closely matched or identical varactors for various device applications.

Whilst, in the above, we have detailed some of the more important current applications of ion implantation to the semiconductor device field, it should be reiterated that there is a growing interest in such techniques to modification of physical and chemical properties of solids in general. This is well illustrated in the proceedings of a recent conference, entitled "Ion Implantation in Semiconductors and Other Materials", in which applications to corrosion, passivation and surface wear resistance of metals, electrical conductivity, changes in metallic oxides, refractive index changes in glasses, changes in the properties of magnetic bubbles in garnets etc., were all subjects of discussion.

Bibliography

MAYER, J. W., ERIKSSON, L. and DAVIES, J. A. (1970) *Ion Implantation in Semiconductors*, Academic Press.

Proceeding of the International Conference on Application of Ion Beams to Semiconductor Technology, Grenoble, France, 1967.

Proceedings of the International Conference on Atomic Collision Phenomena in Solids, Sussex, England, 1969.

Proceedings of the First International Conference on Ion Implantation in Semiconductors, Thousand Oaks, California, 1970.

Proceedings of the Second International Conference on Ion Implantation in Semiconductors, Garmisch-Partenkirchen, Germany, 1971.

Proceedings of the International Conference on Ion Implantation in Semiconductors and Other Materials, Yorktown Heights, New York, 1972.

G. CARTER and W. A. GRANT

L

LASERS IN BIOLOGY AND MEDICINE.
Studies of the biological and physical effects of laser radiation began shortly after the production of the first ruby laser by Maiman in 1960. Ophthalmologists recognized the laser not only as a potential hazard to the eye but as a possible means of therapy for certain eye diseases. Ordinary incoherent light sources such as the Sun, the carbon arc, and the xenon arc lamp had already been exploited for this purpose and it was logical to evaluate the potential therapeutic use of this new light source. These initial studies have gradually broadened out into other fields of medicine using many of the numerous types of laser which have now become available. Here, however, it is proposed to limit the discussion to the main types of laser which are commercially available and their effect on the eye, the skin, certain other organs, neoplastic tissue, and cell preparation.

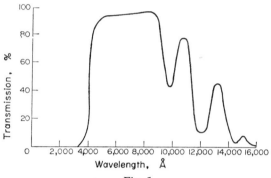

Fig. 1

Eye

The transmission curve of the refractive media of the human eye (Fig. 1) immediately gives some idea of how they will be affected by various types of laser. In the visible and near infra-red region of the spectrum most of the light is transmitted to the retina and it is here, as the energy or power of the laser is increased, that the first lesions are produced. Most pulsed lasers produce full thickness retinal damage but with fairly low-power gas lasers selective retinal damage can be produced. Using a helium–neon laser (6328 Å) the damage is confined to the pigment epithelium and receptor cell layers whereas with argon (5145 Å or 4880 Å) or helium cadmium (4416 Å) the initial damage occurs in the molecular cell layer and the nerve-fibre layer which will result in a visual field defect. A helium–neon laser is therefore an ideal instrument for treating detachments because this difficulty can be avoided. Laser lesions which are produced in the retina scar in about 4 to 7 days and these scars are a means of reattaching retina which has been detached or of preventing the detachment from spreading further. Up to the present ruby lasers have been mainly used for detachment surgery but the histological evidence is that with suitable development the helium–neon system should supersede it. The advantage of lasers in this context is that no anaesthesia or fixation of the eye is necessary, as it is with incoherent light sources, so that the patient can co-operate and look in the direction required by the surgeon. The way in which the laser is operated makes a marked difference to its ocular effects. Q-switched lasers, for instance, cause extensive damage not only in the retina but also in the clear media of the eye and thus are not considered for ocular therapy.

Other work with gas lasers (Smart et al., 1970) has also shown that, when mode locking occurs, of two nominally identical lasers having the same power output, say 5 mW, as measured using a conventional power meter and the same beam divergence, say 2 mrad, the one that is mode locking will produce an immediate lesion in the pigment epithelium and receptor cell layers, while the other with a true C.W. output will not produce a lesion even with exposure times of 20 minutes. Other lasers operating in the ultra-violet (nitrogen) or far infra-red (carbon dioxide) simply cause damage to the corneal surface, as below 4000 Å and above 16,000 Å the refractive media of the eye are opaque.

Skin

The next target organ which is usually considered from a hazard or therapeutic point of view is the skin. Being much less sensitive than ocular tissue comparatively high energies are required to produce macroscopically observable changes in normal caucasian skin while negroid skin has thresholds ranging from 3 to 10 times lower. Depending on the mode of operation of the laser and the energy used the effect on the skin can vary from a mild erythema which persists for a few hours to full thickness damage with eventual sloughing. Using a ruby laser for example a Q-switched pulse of about 1 joule, with a peak power of 50–70 MW, only transient reddening of the skin occurs whereas when operated in the passive mode a 20-joule pulse will give full thickness damage. When the skin is pigmented or tattooed severe damage occurs at lower energies and lasers have been used for the surgical removal of tattoos. The treated area usually heals normally and although somewhat less pigmented than formerly the scar is cosmetically acceptable. Since the skin is the target organ most likely to be chronically exposed to laser radiation a great detail of speculation has occurred regarding the possible carcinogenic effects of such radiation.

Up to the present there is no evidence for such carcinogenic effects although abnormal mitoses (cell divisions) have been seen by Goldman and there is some evidence that sensitization can occur. There is also evidence that the number of cells in the stage between divisions is increased, in some instances doubled (Gordon) possibly indicating an inhibition cell division. In other instances Gordon has found evidence that the rate of cell division in tissue-cultured melanomata is increased following exposure to a ruby laser, but it is very difficult to postulate the exact mechanism by which these changes are produced. At present it can only be argued that unnecessary exposure of the skin to laser radiation should be avoided since there is an increased incidence of tumours in skin which is repeatedly traumatized by any physical or chemical means. Particular attention should be paid in this respect to lasers working in or near the ultra-violet region of the spectrum since there is a well-recognized association between chronic exposure to this part of the spectrum and an increased incidence of skin cancers.

Cranium

Another area in which there is also some uncertainty is the exposure of the cranial contents to laser radiation either via a craniotomy or via the intact skull. Experiments with small animals such as mice, rats and guinea pigs have shown that irradiation of the intact skull using high-energy pulsed lasers can produce severe brain damage frequently leading to the death of the animal (Conference on laser safety, 1966), the energy necessary to produce this effect being roughly proportional to the amount of absorbing material between the site of application and the brain. The question of damage to the human central nervous system is obviously extremely important, since small pulsed ruby lasers are routinely used to treat retinal detachments. In this sort of application it has been shown that a film placed on the back of the skull records laser light when the laser beam is directed into the eye, but the indications are that it is at too low a level to be important. Even at very much higher levels rabbits whose eyes have been exposed to energies in excess of 70 joules have shown no signs of neurological involvement over a subsequent period of 2 years. Similarly human choroidal melanosarcoma and melanosarcomata of the forehead have been exposed to high-energy lasers without any neurological sequelae. Consequently it would seem that the possibility of producing brain lesions in the larger mammals, especially in the intact animal, using high-energy lasers without focusing is remote. In this area in particular long-term studies may prove to be extremely difficult to carry out since it must be shown that any brain lesion which subsequently develops is related to the laser exposure and not to intercurrent cranial trauma or disease.

Other Uses

The use of lasers in relation to vascular disease or surgical procedures involving the vascular system has been frequently mentioned but an extensive series of cases does not yet seem to have been described. Lasers have been applied to research problems such as testing the effectiveness of various drugs as anticoagulants. At present such work is bedevilled by the uncertainty of inflicting a reproducible injury on a small blood vessel so that a clot would form. The time which it takes for an anticoagulant to cause the dispersal of the clot, or its effectiveness in preventing the formation of a clot, can be measured much more accurately using a laser since a very precisely controlled and reproducible injury can be inflicted on the vessel wall. Also in relation to the vascular system a high-power CO_2 laser system has been suggested for use as a "light knife". When a highly vascular organ such as the liver is incised the resulting haemorrhage can cause a great deal of difficulty by obscuring the operative field until the bleeding vessels are located and sealed off. High power CO_2 laser systems, however, are capable not only of incising the organ but of sealing off most of the blood vessels as they are transected. This offers some hope of bloodless surgery but other damage at the edges of the incision may lead to some difficulties with wound healing. It has been suggested that a CO_2 laser might also be used to resect malignant tumors: since there is no contact with the growth the risk of spreading tumour cells by manipulating the growth would be reduced. Again, however, an extensive series of cases or animal experiments does not seem to have been reported.

A pulsed ruby or neodymium laser attached to a microscope so that it can be focused onto a specimen on the microscope stage is beginning to be an extremely useful research tool since a single cell or part of a cell can be irradiated and the effect of this treatment on the subsequent metabolism and reproductive behaviour of the cell can be studied.

Some workers have used pulsed or C.W. argon lasers attached to microscopes where, because of their better beam characteristics they can be focused down to a spot small enough to be capable of damaging a portion of one arm of a chromosome. By studying the structural form and function of the daughter cells it is possible to begin to identify which parts of various chromosomes are responsible for particular cell features and cell functions.

Probably the most important question to many medical workers is the question of whether or not to use lasers to treat neoplastic disease and it is probably in this field that the most varied results have been reported. It has been observed that irradiation of only small parts of a tumour using a ruby laser can lead to a delayed regression of the tumour after about 20 days. This could, for instance, be suggestive of the provocation of an immune response from the host. Some workers, however, consider it mandatory to irradiate the tumour with sufficient energy to result in its immediate destruction. There is also evidence of an alteration in the behaviour of the malignant growth. Both enhancement and retardation of growth rates have been reported by Goldman and Gordon respectively. It was shown that when a tumour which was uniformly fatal to the host in 60 days was irradiated with a high-energy ruby laser, material collected from the plume ejected from the tumour and subsequently transplanted gave rise to tumours which produced no fatalities over a period of 6 months. It is difficult to reconcile many of the results repeated because of differences between the lasers used and their mode of operation, wavelength-specific effects, differences in tissue culture methods, and the different immunological status of the various tumour hosts. It is probable that many of these questions will only be answered by an extensive study of the effects of lasers on malignant tissue at the level of the single cell. For further reading the most comprehensive account of the application of lasers in Biology and Medicine is given by Wohlbarsht et al., (1972).

See also: Radiation, non-ionizing, hazards of

Bibliography

GOLDMAN, L., *Biomedical Aspects of the Laser*, Berlin–Heidelberg–New York: Springer Verlag.
GORDON, T. E. Orange Memorial Hospital, 550 North Bumby Avenue, Orlando, Florida. Personal communication.
Proceedings of the First Conference on Laser Safety 19–20 May 1966 at Orlando, Florida. Sponsored by Martin Company, Orlando, Florida, in co-operation with the Office of the Surgeon General, U.S. Army.
SMART, D., MASON, N., MARSHALL, J. and MELLERIO, J. (1970) New ocular hazard of Mode Locking in C.W. lasers, *Nature* 227 (5263) 1149–1150.
THEWLIS, J. (1969) Laser safety. In *Encyclopaedic Dictionary of Physics* (J. Thewlis, Ed.), Suppl. Vol. 3, 185, Oxford: Pergamon Press.
WOHLBARSHT, M. L. (ed.) (1972) *Laser Applications in Medicine and Biology*, Vol. 1, Plenum.

D. SMART

LOCATION AND REDUCTION OF NOISE

Introduction

Noise is an environmental pollutant and as such should be eliminated or minimized wherever possible. Its effect varies widely in degree ranging from noise that interferes with rest or sleep, noise that annoys, noise that interferes with communication, to noise that produces hearing damage and is harmful to health. It can arise from a wide variety of sources such as a dry bearing in an alarm clock, the tyres of a car, or from a jet aircraft on take-off, and its effect on the recipient is dependent upon a number of parameters. In addition to the ear responding to sound pressure and frequency, other factors such as the noise intermittency, the environment where the noise is heard and effects such as the psychological relationship between the noise source and the observer determine the seriousness of the noise problem. There are many criteria for assessing these factors which are fully discussed elsewhere (Noise Advisory Council, 1973; Burns, 1968) and it is first of all necessary to assess whether noise reduction is necessary by using the appropriate criterion.

Noise Description

In order to take any action towards reducing noise, the noise must first be described in a physically meaningful manner. A single measurement using a sound-level meter giving the sound pressure level (S.P.L.) of the noise in dB is a basic definition, but since the frequency of the components constituting the noise are equally important, then this factor must also be assessed and this can be done with varying complexity and usefulness. The simplest measurement allowing for effect of frequency is the A level weighting of the dB scale. The low-frequency components of the noise are suppressed by using a standard filter and the S.P.L. of the filtered signal is quoted as dB(A), this being a number which is more indicative of the effect of the noise on the ear than a straight S.P.L. reading. By passing the noise through a set of filters, each of which has a width equal to an octave and by measuring the S.P.L. of the noise contained in each of these bandwidths, an octave-band analysis can be obtained and represented graphically. There are $\frac{1}{3}$-octave filters which further subdivide the audio-frequency range but the most useful diagnostic tool is the narrow-band analysis. This is obtained by using a filter whose centre frequency is gradually varied over the whole

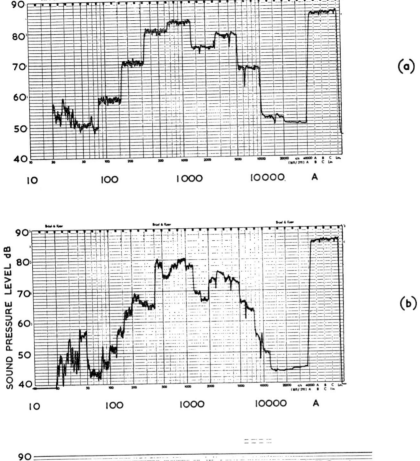
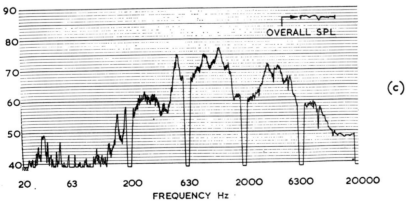

Fig. 1. Frequency analysis of noise. (a) Octave analysis. (b) $^1/_3$ octave analysis. (c) Narrow band analysis (6% band width).

range of frequency investigated, the output of the filter being plotted against the frequency, this process usually being done automatically by a pen recorder. The octave, third octave and a typical narrow-band analysis of noise from an engine are shown in Fig. 1. The narrow-band filter may be one of two types, a constant percentage bandwidth which has a constant Q-value and is most widely used, or the constant bandwidth which is becoming increasingly used, particularly in the field of random vibrations. Typical bandwiths would be, say, 6% of the centre frequency in one case and 5 Hz in the other.

Narrow-band analysis provides us with all the necessary information to identify the amplitudes and frequencies of all significant harmonics precisely. It enables the tracing of the source of noise by finding the source of individual harmonics and one can assess the effect of the noise upon the human ear by reference to the equal loudness contours. The way in which the noise will be attenuated by walls, distance and incidental absorbing surfaces can be determined and the analysis provides a basis for the design of mountings, enclosures, absorbing ducts and silencers.

The pressure-time history of noise can easily be obtained in the form of an oscillogram and is particularly useful when the noise is repetitive and localized bursts of noise can be related to particular occurrences.

But depending upon the resources available, any one or combination of the above measurements can be used as a basis for the following discussion.

Location

The procedure adopted to locate the noise will depend upon the nature of the problem which could be to reduce noise generally in some factory installation where there are several noisy machines, or the focusing of attention upon one particular type of noise emanating from a complex machine. When noise from a multi-source system is mainly from one source, then it is pointless to attempt reduction of the noise contributed by the minor sources since their contribution is insignificant. This is illustrated by considering two noise sources, one of which is 10 dB greater than the other; the elimination of the lesser source will have no effect on the measured S.P.L. It is therefore important to concentrate attention only on the predominant noise source.

If the machine is large and contains many individual noise sources, i.e. a large diesel engine, or if the noise is generated from several sources in a workshop, then a useful first approach is to systematically take a series of readings of S.P.L. and plot contours of equal sound-pressure level (Fig. 2). These contours will immediately indicate the location of those surfaces of the machine or the items of equipment which require noise suppression. The noise can be produced directly by the source, as in the case of fans, jets, road vehicle tyre noise, etc., but more commonly it is produced indirectly by radiation of vibrating surfaces excited by a source of energy remote from the offending surface. In this latter condition it is equally important to identify the excitation source as it is to locate the noise-radiating surface.

The surface vibration can be monitored using either a hand-held probe for quick diagnosis or by attaching to the surface an accelerometer or similar transducer (Harris and Crede, 1961) and comparing the transducer output with that from the microphone. The comparison can be made by relating the frequency spectra and checking for coincidence of harmonics in both signals or by examination of the oscillograms and checking for correlation of bursts of surface vibration with equivalent disturbances in output of the microphone (Fig. 3). Oscillograms of noise are also useful in determining the noise sources

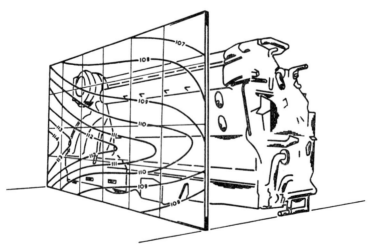

Fig. 2. Isobars of acoustic pressure (dB) (J. Skorecki, 1963).

from machinery which is cyclic in nature since the various bursts of noise recorded on the oscillograms can be related to particular mechanical occurrences within the cycle, i.e. the operation of valves in the internal combustion engine.

Stethoscope kits are available which enable the ear to listen to localized areas and where appropriate suspected surfaces can be blanked-off by the attachment of a damped lead plate (Priede, 1960) which masks the noise radiated from the surface.

The precise source of indirect noise is often difficult to define and it is sometimes necessary to resort to a trial-and-error method (Fielding, 1972) which entails changing those parameters likely to affect the noise produced. The elimination of certain machine components, the variation of clearance between moving surfaces, the insertion of damping material in the suspected vibration path, and the change of mass and stiffness of the structure, are examples of the parameters which can be changed. A well-planned programme of experiments where careful control is kept of all parameters can confirm or eliminate possible vibration effects.

Artificial excitation of the machine or structure can be produced by means of an electrodynamic vibrator attached at a suitable point and the response of surfaces determined over a range of frequencies whilst the force or acceleration of the input point is kept at a predetermined level thus revealing any unwanted resonances in the structure. It is the combination of the vibration response and the sound radiation qualities which determine the basic character of the structure in radiating noise when exposed to varying forms of excitation force (Clarkson, 1972).

Methods of noise and vibration detection have been developed recently which use relatively complex and expensive equipment. One example is the impulse-testing technique where impulses are applied to the structure with a suitable hammer. The pulse input and the system response of either the radiated noise or surface vibration are recorded simultaneously, their Fourier transforms computed and ratio taken and the resulting frequency response data obtained almost immediately. The equipment (Roth, 1971) involves a two-channel analogue to digital converter (A.D.C.), a small dedicated computer and a display unit. Another example is the use of the hologram (Powell and Stetson, 1965) whereby using a laser, a photographic image is produced of the vibrating object showing a system of interference fringes which map contours of constant vibration amplitude (Fig. 4) enabling a detailed analysis of the modes of vibration of complex structures to be made (Satter et al., 1972).

Lastly it is worth while remembering that the origin of the noise can be very remote from the surfaces radiating the noise, as is the case of the central-heating system where the radiator panels in a bedroom are excited by vibrations transmitted along the pipes from the excitation source which may be the fuel pump attached to the boiler.

Reduction of Noise

Before embarking upon any method of noise reduction it is first of all necessary to realize that objectionable noises can be produced by surfaces whose movements are very small indeed. An efficient radiating surface vibrating at 1000 Hz with an amplitude of approximately 0.1 μm will generate a S.P.L. of 80 dB immediately adjacent to the surface. It would seem at first sight to be an almost impossible task to reduce further this minute surface movement, but fortunately it can usually be accomplished to some extent by relatively simple methods.

(i) *Reduction of the excitation at source.* If the energy source of the noise is known, then initial efforts should be directed at reducing the available energy, or modifying it so that the eventual noise is reduced correspondingly. There is little that can be done at source to reduce noise generated directly by fans or air jets since they are by their nature acoustic phenomena and to alter character of energy level would alter their function. However, where noise-radiating surfaces are excited indirectly, reduction or modification of the noise-source energy is effective. Typical examples are the dynamic balancing of rotating and reciprocating members, the reduction of clearance between moving parts which will reduce the impacts that occur due to the relative movement of the parts and the smoothing of cam profiles so

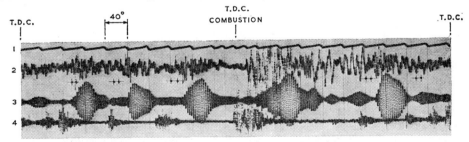

Fig. 3. Oscillogram of a diesel engine showing outputs from: (1) camshaft angle marker; (2) microphone (noise signal); (3) the noise signal after it has been passed through the wave analyser tuned to 3100 Hz; (4) accelerometer mounted on engine surface.

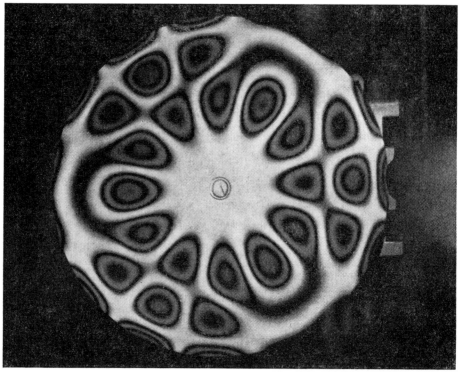
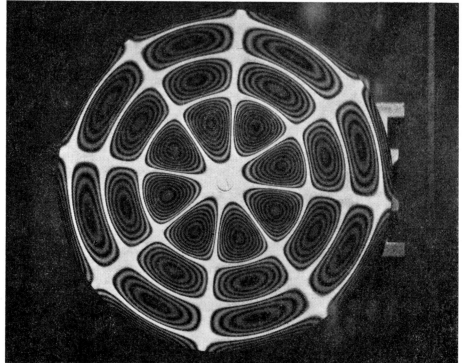

Fig. 4. Photographs of a time-averaged hologram of a vibrating metal disc using He-Ne laser light. (Photographs by K. W. Topley, M.M.P.A., taken during work carried out in Department of Mechanical Engineering, Loughborough University of Technology.)

that the impact between the follower and its stationary support is reduced.

A modification to the shape of the force pulse which excites the radiating surface can be beneficial. By smoothing the pulse, thus creating a less rapid build-up, the magnitude of the higher frequency components in the force spectrum is reduced, making the pulse less effective in exciting movement in surfaces whose natural frequencies tend to lie in the higher frequency range (Clarkson, 1972). This effect can sometimes be achieved by improvement of lubrication. If impacting surfaces are well lubricated, the oil layer between the surfaces will cushion the impact and the force impulse. This can be emphasized by a suitable design of components (Fearon, 1964).

(ii) *Isolation*. Large surfaces on a machine are particularly efficient at radiating noise and surfaces such as walls, floors and windows are no exception to this rule. It is therefore necessary to prevent vibration from the source reaching these surfaces by providing some form of barrier. This is accomplished by suitably designed resilient connections between the machine and the floor (Crede, 1951), and between the structure of the machine and its external surfaces. These connections, which usually increase flexibility between the constituent parts, reduce the force transmitted to the radiating surface, and thus its displacements and radiated noise.

If the load-carrying structure of the machine is isolated from the outside panels, a considerable reduction of noise can be achieved, but unfortunately many conventional machines incorporate components which combine the function of load carrying with that of enclosure, i.e. the engine sump and crankcase, with the result that all forces are transmitted through the large outside surfaces of the machine which are the efficient noise radiators.

(iii) *Alteration of structural response*. In the same way that the harmonic content of the excitation force can be altered to mismatch the natural frequencies of the noise-radiating surface, it is obviously possible to alter the response of the surface in relation to the excitation force and so reduce its effectiveness in converting the excitation energy into acoustic energy. This can be achieved by altering the stiffness/mass ratio of the structure, thereby moving the natural frequencies of the structure either above or below those of the exciting force. An increase of stiffness for the same mass of structure is possible either by using a light alloy material (Priede *et al.*, 1964–5) instead of steel or cast iron and increasing the thickness or by redistributing the mass more effectively as ribs and stiffeners.

Structural response can be altered by the introduction of damping into the structure by employing laminar or sandwich construction for the panels (Ruzicka, 1961). Panels fabricated of alternate layers of sheet metal and viscous material designed to dissipate energy by shear deformation can also have considerable structural strength. If the surfaces are relatively light-weight, such as car door panels, the application of a mastic paint or mastic impregnated felt will provide considerable reduction of surface vibration, with beneficial effect on the radiated noise.

(iv) *Sound absorption*. If the noise generated by a source cannot be reduced to a desirable level, the environment in which the noise is heard can be modified by adding acoustically absorbent material to the walls and ceiling thus reducing the reflected noise. This treatment will be most effective at some distance away from the source of noise where the S.P.L. of the reflected noise is comparable with that radiating directly from the source. Aerodynamic sources, such as fans and exhausts, can be ventilated along ducts lined with absorbent material (King, 1965) to reduce the noise that reaches the outside.

(v) *Attenuating structures*. This most radical but effective measure is accomplished by the use of a mass barrier between the noise and the ear. It can be in the form of an ear defender (ear plug) or a brick wall or an enclosure. A wall of single-brick thickness plastered on both sides will produce a noise attenuation of approximately 45 dB, which is considerably greater than that attained by most of the methods already discussed. If the machine is totally enclosed, it is often necessary to provide ventilation which must be carefully designed to prevent noise leakage. Similarly, doors giving access to the enclosure must be equally as massive as the enclosure and have an airtight seal if the effectiveness of the enclosure is to be maintained.

The noise reduction methods just described are represented diagrammatically in Fig. 5, together with the associated noise spectra recorded by a microphone positioned at the point marked ⊗. Figure 5a shows the machine when all the attempts to reduce the noise at source have been made. The effect achieved by mounting the machine on flexible supports (isolation) is illustrated in Fig. 5b. Most of the higher-frequency noise is radiated directly by the machine surfaces and it is only the lower frequencies of the noise spectrum which are reduced. Figure 5c shows how structural modification by damping treatment applied to a machine surface can reduce isolated peaks on the spectrum. The reduction of reflected noise from a wall surface by the application of noise absorbent material is illustrated in Fig. 5d. A large reduction in the noise spectrum level occurs after total enclosure of the machine, by means of a solid partition, Fig. 5e, but the misuse of absorbent material for this purpose only results in a very small attenuation of the high-frequency portion of the spectrum, as is shown in Fig. 5f.

Figure 5g shows the importance of isolating the enclosed machine from the floor and how the low-frequency components are reduced, and Fig. 5h shows how absorbent material lining the interior surfaces of the enclosure will reduce the build-up of reflected noise within the enclosure and hence the noise measured at the microphone.

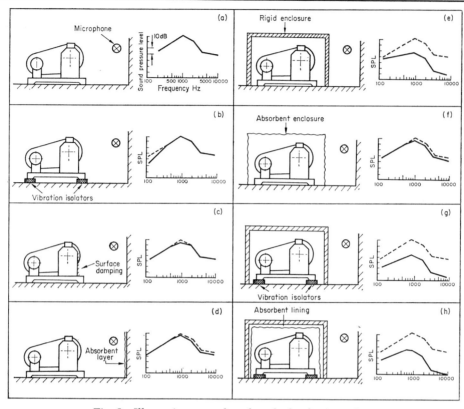

Fig. 5. Illustrative examples of methods of noise reduction.

Bibliography

BURNS, W. (1968) *Noise and man*, John Murray.
CLARKSON, B. L. (1972) The social consequences of noise, *Proc. Inst. Mech. Engrs.* **186**, 8/72.
CREDE, C. E. (1951) *Vibration and Shock Isolation*, Wiley.
FEARON, W. (1964) Waterside attack of diesel engine cylinder liners, *J.R.N.S.S.* **19**, 1.
FIELDING, B. J. (1972) The investigation of mechanism noise in a diesel engine, *Proc. 3rd World Congress, I.F.T.O.M.M.*, Vol. B, Paper B1.
HARRIS, C. M. and CREDE, C. E. (1961) *Shock and Vibration Handbook*, Vol. 1, Chap. 13, McGraw Hill.
JOHNSON, A. C. J. (1962) Sound analysis, in *Encyclopaedic Dictionary of Physics* (J. Thewlis, Ed.), **6**, 560, Oxford: Pergamon Press.
KAY, R. H. (1962) Wave analyser, sound, in *Encyclopaedic Dictionary of Physics* (J. Thewlis, Ed.), **7**, 684, Oxford: Pergamon Press.
KING, A. J. (1961) Loudness and its measurement, in *Encyclopaedic Dictionary of Physics* (J. Thewlis, Ed.), **4**, 354, Oxford: Pergamon Press.
KING, A. J. (1962) Noise and its abatement, in *Encyclopaedic Dictionary of Physics* (J. Thewlis, Ed.), **5**, 30, Oxford: Pergamon Press.
KING, A. J. (1965) *The Measurement and Suppression of Noise*, Chapman & Hall.
NOISE ADVISORY COUNCIL (1973) *A Guide to Noise Units*, Department of the Environment.
POWELL, R. L. and STETSON, K. A. (1965) Interferometric vibration analysis by wavefront reconstruction, *J. Opt. Soc. Amer.* **55**, 12.
PRIEDE, T. (1960) Relation between form of cylinder pressure diagram and noise in diesel engines, *Proc. Auto. Div. Inst. Mech. Engr.* **175**, 63.
PRIEDE, T., AUSTEN, A. E. W. and GROVER, E. C. (1964–5) Effect of engine structure on noise of diesel engines, *Proc. Auto. Div. Inst. Mech. Engrs.* **179**, Pt. 2A, No. 4.
ROTH, P. (1971) Detecting sources of vibrations and noise using H. P. Fourier analyser, Application Note 140-1, Hewlett-Packard.
RUZICKA, J. E. (1961) Damping structural resonances using visco-elastic shear-damping mechanisms, Pt. I, II, *Trans. A.S.M.E.*, **83**, Series B, No. 4.
SATTER, M. A., DOWNS, B. and WRAY, G. R. (1972) Reduction of noise at the design stage: a specific case study, *Proc. Inst. Mech. Engrs.* **186**, 26–72.
SKORECKI, J. (May 1963) Vibration and noise of diesel engines, A.S.M.E. Paper No. 63-OGP-2.

B. J. FIELDING

M

MACH EFFECT (VISUAL). In the visual Mach effect a varying luminance distribution in a scene is perceived by the eye with enhanced gradients to the luminance and in some cases with apparent "bands" or "fringes" near a contour. For example, in Fig. 1 the full line represents the luminance distribution across a division in a physical sense, i.e. as it would be measured by a device with linear response such as a photoelectric cell. The broken line represents what the eye perceives, i.e. the relative sensation. This is in fact difficult to measure quantitatively but the effect is very large and it is not difficult to see.

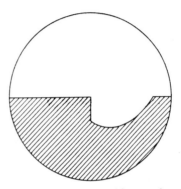

Fig. 2. Rotating disc as used by Mach to show the bands.

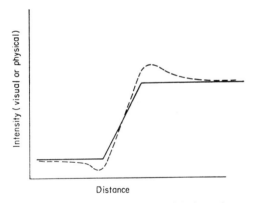

Fig. 1. The Mach bands. The full line shows the physical intensity distribution and the broken line the visually perceived distribution.

This effect was first noted by Ernst Mach (1838–1916) in 1865 in the course of experiments with rotating discs. A disc with a black sector as in Fig. 2 will produce a luminance contour as in Fig. 1 by persistence of vision, if it is set in rapid rotation. The resulting *Mach bands* are easily seen and are very striking. It was shown by Mach that the effect has nothing to do with the rotation of the disc; any suitable luminance distribution will serve and the scale of angular subtense of the detail can vary over a wide range.

The mechanism of formation of Mach bands is not well understood. Several models have been proposed, one group of which uses the concept of inhibition. According to this concept a stimulus at a given receptor (rod or cone in the retina) inhibits to some extent the responses of neighbouring receptors to light; this is shown schematically in Fig. 3 where the inhibitory effect is proportional to the ordinate. Such an effect can easily be seen qualitatively to produce Mach bands. Mach himself showed that the effect could be represented by a model in which an increment proportional to the second derivation of the luminance distribution was added to the luminance to give the subjective effect. In a third group of models the response of the eye is Fourier analysed and a transfer function which exceeds unity for certain spatial frequencies is set up. However, it seems clear that the Mach effect is very non-linear, whatever may be its mechanism, and a transfer function treatment would be inappropriate.

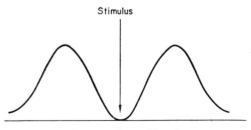

Fig. 3. Inhibition of neighbouring receptors postulated as a mechanism of Mach band formation.

The Mach effect serves a useful purpose in vision in intensifying contours and edges. In this respect it may be regarded as complementary to the Weber–Fechner law, according to which the brightness of large uniform areas is perceived on a logarithmic scale. Thus the non-linearity of the eye enhances important detail such as contours but compresses unimportant detail such as absolute intensities into a shorter range.

The Mach effect plays a somewhat different part in vision through optical instruments, in particular in such instruments as microscopes where frequently enough magnification is used to make diffraction structure in the image clearly visible. The diffraction structure of microscope images is complicated and it depends in detail on the mode of illumination. We restrict ourselves first to the *physical image*, i.e. that which would be measured by a linear detector. If the condenser numerical aperture is very much smaller than the objective NA the illumination is substantially coherent and the physical image of an edge between light and dark regions contains much diffraction structure. If on the other hand the condenser NA is much greater than that of the objective, a condition which can only be achieved in low-resolution microscopy, the illumination is effectively incoherent and there is little diffraction structure. We denote the ratio, condenser NA divided by objective NA, by S; Fig. 4 shows the physical images of edges for three cases, the value $S = 0.8$, partially coherent illumination, corresponding roughly to the optimum for general microscopy. The visual images of such edges show clearly how the Mach effect emphasizes even the slight inflections in the nearly incoherent images. Figure 5 shows some comparisons of physical and visual images, both experimental, which illustrate this point. Mach's own model in terms of the second derivative leads to the result that the Mach bands should have extrema at the stationary points of the second derivative of the physical image; these points are indicated on the graphs and there is some measure of confirmation.

The practical outcome for microscopy, however, is that there is in the Mach bands a further source of artefacts which can falsify visual images. The

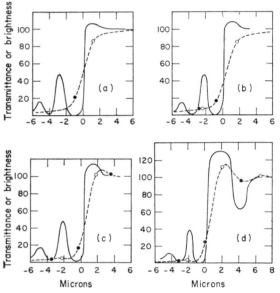

Fig. 5. A comparison of physical (broken line) and visual (full lime) images of an edge. The coherence parameter S takes the values (a) 5·7, (b) 1·0, (c) 0·6 and (d) 0·32. The circles mark points where the second derivative of the physical image is stationary. (After W. N. Charman and B. M. Watrasiewicz, J. Opt. Soc. Am. **54**, *771 (1964).)*

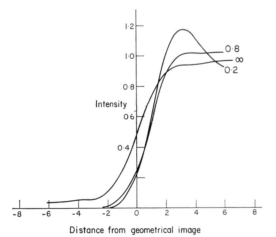

Fig. 4. Physical images of an opaque edge, showing diffraction structure. The three graphs, $S = 0.2, 0.8$ and ∞ correspond to substantially coherent, partially coherent and incoherent illumination respectively. The abscissa scale is in diffraction units such that 3·83 is the radius of the first dark ring of the Airy pattern.

sophisticated microscopist is familiar with the effects of geometrical aberrations and of diffraction but it is now seen that Mach bands can produce much more marked spurious effects than diffraction under some conditions. The effects appear to all observers but with individual variations in the band structure. Stabilization of the retinal image causes the bands to disappear. Also they are much reduced at scotopic levels of retinal illumination. No definite chromatic effects have been observed; however, if the bands are produced by enhancement of diffraction effects, as in the microscope, their spacing changes if the wavelength is changed, so that they are undoubtedly linked to inflections in the diffraction structure.

Bibliography

RATCLIFF, F. (1965) *Mach Bands: Quantitative Studies on Neural Networks in the Retina*, San Francisco: Holden-Day, Inc. (This book includes translations of Mach's original papers on the topic.)

WELFORD, W. T. (1969) The Mach effect and the microscope, pp. 41–76 of *Advances in Optical and Electron Microscopy*, Vol. II, eds. R. Barer and V. E. Coslett, New York: Academic Press.

W. T. WELFORD

MAGNETISM, MEDICAL APPLICATIONS OF.

Historical.

Medical applications of magnetism have probably been experimented with ever since magnetism was discovered. Hali Abbas wrote in the tenth century that a loadstone (lodestone) held in the hand would cure gout and spasms. Paracelsus in the early sixteenth century also believed in the remedial powers of magnets in the cases of hernia, dropsy and jaundice. However, William Gilbert, who was the first Queen Elizabeth's personal physician, as well as one of the leading researchers in magnetism, decried such medical uses of magnets. Mesmer, around 1800, was aware of the well-known fact that the faculty of exerting a force on iron and loadstone could be transmitted from one piece to another by a stroking action, and he claimed that similar stroking with magnetized objects had a beneficial effect on his patients. He called this "animal magnetism". Later he dropped the use of the iron or loadstone but kept the stroking action, and practiced his "animal magnetism" in order to cure and otherwise influence humans. Elisha Perkins, a physician, patented his Perkins Metallic Tractors in 1776 as a cure for illness or pain, and it is said that George Washington used the device. Later Gaylord Wilshire patented his Ionico, a magnetic collar that was supposed to magnetize the blood and cure many ailments (Wilshire Boulevard in Los Angeles is named after him). Even today "magnetic belts" and the like, good for whatever ails you, are offered all over the world. A few years ago, in spite of the fact that many physicists had experienced no obvious adverse effects from long exposure to fields of several thousand oersted, a patent was granted in several countries for a magnetic shield designed to protect the hands of a physician when applying a rather small magnet to his hapless patient as a cure for heart disease. There are also patents on clothing containing curative magnets, a magnetic chair, headband, couch and the like. Magnets are even offered as a remedy for insomnia, and, when put into shoes, to relieve tiredness of the feet. It has also been claimed that magnetic fields could retard the ageing process. Needless to say, none of these claims has been proved by rigorous investigations.

Direct Influence of Magnetic Fields

Although the curative power of magnets remains within the realm of superstition, certain biological effects of magnetic fields have been established. The first scientific experiments on the influence of magnetic fields were performed in Edison's laboratory in the year 1892, where dogs and experimenters themselves were exposed to fields of several thousand oersted without suffering any ill effect.

Time-varying magnetic fields produce two known effects, namely the sensation of light and that of muscle stimulation. D'Arsonval was the first to describe the light effect in 1892. Other famous physicists like Thompson also investigated the effect and compared it with other light-producing stimuli like pressure and electric pulses. The magnetic light-sensations, known as *phosphenes*, are described as colourless or occasionally light-blue shimmering luminosites appearing on the borders of the visual field. They are reported to be produced when alternating fields of 10 to 100 Hz are applied to the temporal areas of the human head. It is generally assumed, though not scientifically demonstrated, that the effect is produced in the retina.

Stimulation of muscles has been reported by a number of investigators. It seems that nearly every muscle, when subjected to a changing magnetic field, contracts. Montgomery employed the sartorius muscle of the frog to demonstrate the equivalence of this effect and of that produced by electrical methods. This effect of time-varying magnetic fields could have important applications, for instance in stimulating the heart muscle for pacemaking, at least in emergencies until electrodes can be inserted. In such time-varying magnetic fields the mechanism is obviously connected with the electric potentials produced.

Turning to fields where the inhomogeneity or the field itself is the main factor (and not induction due to time varying fields) the phenomena become much less understood. In the case of constant magnetic fields, a direct influence could theoretically be exerted via the orientation of materials whose magnetic susceptibility tensor has different components, or by the orientation of acicular or anisotropic structures. Chemical processes could undergo a shift in equilibrium in a magnetic field if the substances involved had different susceptibilities. Any process involving movement of ions could be influenced as a result of changes in the paths of the ions in a magnetic field, and this might in turn become important for ions passing through biological membranes. Inhomogeneous magnetic fields can, of course, exert additional forces on materials of different susceptibility.

No rigorous theories concerning these or other possible mechanisms exist, but the effects produced should be small in fields up to several thousand oersted. Magnetic forces are small for all but ferromagnetic materials, and there are practically none of the latter present in living organisms. The only known biological material of very high magnetic susceptibility is the coating of the radulae (or "teeth") of chitons, a class of lower marine animals. However, in biological systems even very small effects could be cumulative and cause significant changes.

It has been suggested by Labes that magnetic fields can influence life processes through *liquid crystals*. It is well known that liquid crystals orient in fields of the order of 100 oersted, and it has been shown by Svedberg that in such oriented liquid-crystal systems there can be marked changes in diffusion coefficients and rates of chemical reactions. In living bodies, liquid crystals and materials that are close to liquid crystals exist and through these the rate of life processes might be influenced by magnetic fields.

Normal life on earth is subjected to the geomagnetic field, which has a strength of 0·3–0·6 oersted and an orientation varying from horizontal to vertical. Humans have been subjected to fields of several thousand oersted in the course of constructing cyclotrons, for example, and also in studies on the effects of the long-term application of external magnetic fields in the thousand-oersted range. In none of these cases has any effect on humans been reported, except for the magnetic phosphenes and muscle stimulation discussed earlier.

A large number of studies of different biological aspects have been carried out on various organisms. Unfortunately, some of the results were found to contain discrepancies. However, there are indications that the action of enzymes can be modified by a field.

An investigation at the University of California, Berkeley, showed that there is synergism between the effects of several environmental variables and the influence of the magnetic field appears when these variables are chosen so that the system is near the limit of stability. In the study of the flour beetle, for instance, it was found that if the survival rate became small, due to relatively high temperature, the application of a constant magnetic field of several thousand oersted increases this rate and reduces the development of malformations. There were also indications that the agglutination time of erythrocytes and other blood parameters could be influenced by magnetic fields.

Most experiments were carried out below 10 k oersted. As higher fields become more readily available, any biological influence should become more pronounced (most effects should increase more than proportionally with increasing field strength) and incidentally, this would have to be taken into account in any instrumental applications of magnetic fields such as those discussed below.

A special case is that of strongly inhomogeneous fields. These are easily produced in small volumes and Levengood, for instance, has demonstrated pronounced morphological abnormalities when such fields were applied to developing amphibian eggs and embryos.

Magnetically Operated Devices

As magnetic fields interact very weakly with the diamagnetic or paramagnetic substances of living bodies, but very strongly with ferromagnetic materials, one can design simple devices in which a ferromagnetic object is inserted into the body and controlled from the outside by a magnetic field. The object in the body can be quite small and simple, while the equipment outside can, in principle, be as complex as necessary.

Magnetic instruments have been designed for various clinical purposes. Several such devices extract iron splinters from the eyes and other parts of the body. Another makes it possible to recover, without surgery, objects such as hairpins, nails, and safety pins swallowed accidentally. A simple application of magnetic forces is in the alleviation of eyelid drooping (ptosis), which may result from myasthenia gravis (muscle paralysis) and a number of other causes. Here small magnets are attached to the eyebrow and a strip of ferromagnetic material to the eyelashes. This keeps the eyelid open. Alternatively, one can improve eyelid closure if necessary. For these purposes $SmCO_5$ magnets embedded in polyurethane are used.

Magnetic catheters can be used to reach certain locations in the body which are otherwise inaccessible without surgery. The Pod, one version of such a catheter, having a permanent magnet as its tip has been developed recently. The main purpose of the device is to transport and release at a desired location drugs or other chemicals for therapeutic, diagnostic or research purposes. The Pod could also be used to introduce microtransducers into arteries or veins in order to obtain pressure curves or oxygen saturation values and other important data.

The efficiency of propulsion of a magnetic device such as the Pod inside the body can be greatly increased in the following way. Any magnetized body with no centre of symmetry will move under the influence of an alternating magnetic field, due to frictional or viscous forces. If one attaches a flexible plastic tail to a small magnet in a fluid and applies an alternating magnetic field, the magnet will oscillate transversely and the tail will lash, propelling the magnet through the surrounding medium by hydrodynamic forces. A cylindrical magnet 1 mm in diameter and 3–4 mm long, plus a tail of similar length, can travel 30 cm/sec in water under the influence of a 50 c/s alternating field of about 100 oersted. Such a device can be used to guide a soft catheter. An advantage of this method is that the driving force depends on the field strength and not on the gradient of the field; the driving force, therefore, does not drop off so much with distance from the magnet. In practice, a d.c. field is applied as well, in order to keep the magnet moving in the right direction. If the magnet progresses in a tube, like an artery, no d.c. field is required for its guidance unless a branch is reached. The vibrations of the magnet and the plastic catheter tube also facilitate passage around obstacles and reduce contact with the inner surface of the blood vessel. In most cases the catheter is inserted so that it moves in the direction of the blood flow, but it is also possible to propel it against the blood flow in veins.

Using such a device the obliteration of arteriovenous malformations by high-frequency cauterization and selective angiography has been reported. A similar device has been used to deliver contrast material for selective bronchography and to remove excess contrast material in the bronchial tree. It can also be applied to recover saline-washed bacteriological and cytological specimens from the bronchiae. However, in all these applications only a few patients have been treated so far and the method has yet to be developed further before it can be adopted for clinical use.

The introduction of carbonyl-iron powder into the bloodstream, as contrast material which can be controlled by magnetic fields, has been proposed. It has been shown that such ferromagnetic particles which are not captured by the reticuloendothelial system can be removed by an extracorporeal bypass.

Other potential applications of magnetic control concern the moving of materials in the gastrointestinal system. For example, it becomes possible to obtain better control of contrast material for X-ray diagnosis. Replacing the barium sulphate, now generally used in such a procedure by a ferromagnetic material, would open the way to control by magnetic fields. The whole procedure could thus be streamlined, possibly resulting in less X-ray exposure to the patient and a reduction in the amount of material to be taken. In addition, intestinal loops containing the material could be moved and in this way the so-called mobility could be tested.

The possibility of moving intestines by filling them with a ferromagnetic material and applying an external magnetic field could open the way to greater efficacy of radical radiation therapy of intra-abdominal, intra-peritoneal and pelvic malignant tumors, such as carcinoma of the cervix, ovaries or colon. Because of possible radiation damage to the intestinal tract there is a limit to the amount of irradiation which may be administered to other organs in the area, and this falls short of the tumoricidal dose in most cases. By moving the intestines magnetically they could be kept out of the path of the radiation, thus permitting more effective radiation therapy.

Measurements

The electrical activity of the human body supplies very important data on the functioning of various organs. Electrocardiography, electroencephalography and electromyography are all widely used in medicine. However, in all cases electrodes must be connected to the skin, and what is actually measured is influenced by the electrical properties of the skin, thus losing spacial information. Neither can d.c. currents inside the body be measured. If, however, the magnetic field associated with the electrical activity is measured outside the body, it should theoretically be possible to obtain more detailed information on the functioning of the organs.

Techniques for measuring weak alternating low-frequency magnetic fields have been developed to the point where, under certain conditions, fields as low as 10^{-9} oersted can be detected. The activity of the heart is about 10^{-7} oersted (peak value) near the body, and it has been found possible to draw magnetic maps which give important information on the electrical activity of the heart; d. c. currents have also been found. The significance of these currents is not understood as yet, but it may be that they are related to certain heart conditions. The most advanced system exists at M.I.T.'s National Magnet Laboratory, where Cohen has performed experiments using one coil in a shielded room to increase the sensitivity considerably and thereby obtain more precise data from these magnetic fields. It was also found possible to measure the magnetic field related to the alpha-waves of electroencephalograms. With this facility it is also possible to measure ferromagnetic material in the lung; e.g. asbestos, iron dust, if inhaled, are retained and can be diagnosed.

A promising application of electromagnetic technique is in the measurement of *blood flow*. The clinical methods used for this purpose at present are far from satisfactory and electromagnetic flowmeters are potentially more reliable.

Electromagnetic flowmeters are based on electromagnetic induction, i.e. the production of an electric potential between two stationary electrodes by the movement of a conductor in a magnetic field. The potential is proportional to the vector product of the field vector and the velocity of the conductor. If the conductor is a flowing liquid, the velocity is not uniform over the cross-section normal to the direction of flow. It has been shown, however, that at least in simple non-turbulent flow, an instrument can be designed to integrate and measure the total volume passing between the electrodes. An advantage of this type of instrument is that it measures volume flow directly. In fact, Faraday measured the flow of the River Thames by this method, the magnetic field being the intrinsic field of the earth.

In applying this principle to the measurement of blood flow, a serious difficulty is that the electrical conductivity of blood is rather low. Moreover, the blood vessels themselves are also electric conductors and though their conductivity is much lower than that of blood, they act as a shunt and lower the electric potential generated. A practical difficulty in the design of blood flowmeters is the undesirability of piercing the blood vessel at the point of measurement in order to insert the electrodes. One solution is to place the electrodes outside the blood vessel and measure the potential capacitatively by applying an alternating magnetic field. The a.c. technique has the advantage of being insensitive to interference from DC potentials which can be of the same order of magnitude as the effect being measured. The first flowmeters based on this principle called for implantation of an AC magnet around the blood vessel. The difficulty of doing this was the reason why these flowmeters were not widely used. The answer seems to have been found in the recent development of catheter flowmeters. In these, the a.c. magnet is outside the body and the electrodes are inside the blood vessel but are introduced at a convenient point by means of a catheter and guided to the point of measurement. The catheter is introduced percutaneously through a guide wire, the latter in turn being introduced through a needle. The electrodes, which are packed inside a small tube, are pushed out when the point of measurement is reached and their spring-like support expands until they are pushed against the walls on opposite sides of the bood vessel. Once in position, the electrodes can also be used to measure the lumen of the blood vessel, by taking an X-ray picture.

Transmission Through the Skin

Prosthetic devices located internally or externally have been designed for many purposes. Operation of many internal devices requires transmission of driving energy into the body and operation of an external device requires transmission of signals from the body to the outside. Wires cannot be used for this purpose since it is not possible, at present, to pierce the skin of humans permanently. However, electromagnetic methods can be used. The following are some of the difficulties involved in its design; there must be high efficiency, the implanted part must be small, positioning of internal and external parts must be exact, there must not be too much pressure on the skin, and so on.

A *transcutaneous power transformer* working at 3·7 kHz designed to drive an artificial heart, is capable of transmitting over 30 watts of power through intact skin. Modified models of the transformer which have an efficiency of 80–90% can also be used at higher power levels to charge implanted power-storage devices; up to 100 watts can be transmitted in this way. The transformer consists of two circular "cup cores" 6 cm in diameter, one placed subcutaneously and the other externally.

A rather unconventional way of increasing the coupling efficiency of a transcutaneous transformer is to shape the skin and underlying tissue into a multiple connected space, held inside the core of a torrodial transformer. This involves an unusual but rather simple plastic operation to form the connected space. (An ear pierced for ear-rings represents such a formation, but it is too small for practical applications.) Due to the continuous magnetic path that is provided by the toroid, low and unregulated frequencies can be used without the need for tuned coils and oscillators but the rather unusual shape of the skin might not be acceptable in the majority of cases.

Myoelectric signal transmission from the body to the outside would be the best control to operate prosthetic limbs. The advantage of this method being that no additional power supply is necessary.

To sum up, it can be seen that there are as yet no routine clinical applications using magnetic fields directly and only limited instrumental applications. Despite the fact that the development of new magnetic materials and more powerful magnets have brought great advances in magnetic devices in many fields, as far as medicine is concerned, the use of magnetism is only in its infancy.

Definitions of Some Medical Terms Used

Angiography—radiography of the cardio-vascular system.

Bronchography—radiography of the trachea, bronchi or the bronchial tree.

Reticuloendothelial system—a system of phagocytic special cells which show a special affinity for certain colloidal dyes, finely suspended particles and lipoid matter introduced into the circulation.

Bibliography

ACETO, H., TOBIAS, C. A. and SILVER, I. L. (1970) Some studies on the biological effects of magnetic fields, *IEEE Trans. Magnetics MAG* **6**, 368.

BARNOTHY, M. F. (1964 and 1969) *Biological Effects of Magnetic Fields*, New York: Plenum Press, vol. 1 and vol. 2.

BEISCHER, D. E. (1962) Human tolerance to magnetic fields, *Astronautics*, **7**, 24.

COHEN, D., EDELSACK, E. A. and ZIMMERMAN, J. (1970) Magnetocardiograms taken inside a shielded room with a superconducting point-contact magnetometer, *Appl. Phys. Lett.* **16**, 278.

DAVIS, L. D. and PAPPAJOHN, K. (1962) Bibliography of the biological effects of magnetic fields, *Fed. Am. Soc. Exp. Biol. Proc.*, Suppl. 12.

DRILLER, J., CASARELLA, W., ASCH, T. and HILAL, S. K. (1970) The Pod bronchial catheter and associated biopsy wire, *Med. Biol. Engn.* **8**, 15.

DRILLER, J., HILAL, S. K., MICHELSEN, W. J., SOLLISH, B., KATZ, L. and KONIG, W., JR. (1969) Development and use of the Pod catheter in the cerebral vascular system, *Med. Res. Engn.* **8**, 11.

FREEMAN, M. W., ARROTT, A. and WATSON, J. H. L. (1960) Magnetism in medicine, *J. Appl. Phys.* **31**, 404 S.

FREI, E. H. (1972) Biomagnetics, *IEEE Trans. Magnetics, MAG* **8** 407.
GILBERT, W. (1600) De magnete Magneticisque Corporibus et de Magno Tellure, *Physiol. Nova, London.*
GROB, D. and STEIN, P. (1971) Samarium-cobalt magnetic prostheses, *J. Appl. Phys.* **42**, 1318.
KOLIN, A. (1970) Evolution of electromagnetic blood flowmeter, *UCLA Forum Med. Sci.* No. 10, 383. Los Angeles: Univ. of Calif. Press.
LEVENGOOD, W. C. (1969) *J. Embryol. exp. Morph.* **21**, 23.
LUBORSKY, F. E., DRUMMOND, B. S. and PENTA, A. Q. (1964) Recent advances in the removal of magnetic foreign bodies from the esophagus, stomach and duodenum with controllable permanent magnets, *Am. J. Roentgen* **92**, 1021.
MONTGOMERY, D. and WEGGEL, R. (1969) Magnetic forces for medical applications, *J. Appl. Phys.* **40**, 1039.
NAKAMURA, T., KONNO, T., TSUYA, N. and HATANO, M. (1971) Magneto-medicine: magnetism in medical uses, *J. Appl. Phys.* **42**, 1320.
SAWYER, K. C. (1967) *Management of Foreign Bodies in the Food and Air Passage*, p. 191, Springfield: Ill.: Charles C. Thomas.
TAREN, J. A. and GABRIELSEN, T. O. (1970) Radiofrequency thrombosis of vascular malformations with a transvascular magnetic catheter, *Science* **168**, 138.

E. H. FREI

MEDICAL ULTRASONICS. Ultrasonic techniques are important in medicine because their use provides the best solution to certain problems in diagnosis, therapy and surgery. This is because ultrasound has unique interactions with living systems; and, in diagnostic examinations, any hazard seems to be much less than that associated with ionizing radiation.

Physical Fundamentals

Most medical applications of ultrasound employ frequencies in the range 1–15 MHz. The usual transducer is a disc of polarized lead zirconate titanate. For continuous-wave applications, this transducer may be air-backed and so mounted that the restriction of its vibration is minimized. However, for applications involving the use of short pulses, the bandwidth of the transducer is increased by the attachment of an absorbent loading material to its rear face.

In the steady state, a circular piston of radius r vibrating cophasally generates a field in which the central axial distribution is given by

$$\frac{I_x}{I_0} = \sin^2\left[\frac{k}{2}\{(r^2 + x^2)^{1/2} - x\}\right]$$

where I_0 is the intensity at the source, I_x is the intensity at any distance x from the source, and $k = 2\pi/\lambda$, where λ is the wavelength ($\lambda = c/f$, where c is the propagation velocity and f is the frequency). Hence, if $r^2 \gg \lambda^2$,

$$x'_{max} = r^2/\lambda$$

where x'_{max} is the position of the last axial maximum moving away from the source. The last axial maximum marks the transition between the near and far fields of the ultrasonic beam. In the near field, the energy is mainly confined within a cylinder; but in the far field, the beam diverges conically in such a way that the central lobe is reduced to zero at angles of $\pm\theta$ given by

$$\sin\theta = 0.61\lambda/r.$$

A transient ultrasonic beam is close to the corresponding steady state conditions, becoming more uniform with decreasing pulse duration.

The ultrasonic field may be focused by means of a lens, a mirror, or a curved transducer. Ray optics may be used as a first approximation in predicting the position of the focus; more accurate calculations involve consideration of the wavelength.

The power P of an ultrasonic beam is usually measured by determination of the vector force $\mathbf{F}$ associated with radiation pressure. In the case of complete absorption, the relationship is

$$\mathbf{F} = P/\mathbf{c}$$

Another, less important, method of power measurement is by calorimetry.

Ultrasonic Characteristics of Biological Tissues

The propagation velocities, characteristic impedances (equal to ϱc, where ϱ is the density), and absorption coefficients for some materials commonly encountered in medical ultrasonics are given in Table 1. The propagation velocity is almost independent of the frequency. The absorption in air and water is due chiefly to viscosity, and is proportional to the square of the frequency. However, in soft tissues, the absorption occurs mainly as a result of

Table 1

Material	Propagation velocity (ms^{-1})	Characteristic impedance (10^6 kg m^{-2} s^{-1})	Absorption coefficient at 1 MHz (dB cm^{-1})
Air	330	0·0004	12
Blood	1570	1·61	0·18
Bone	4080	7·80	13
Brain	1540	1·58	0·85
Fat	1450	1·38	0·63
Liver	1550	1·65	0·94
Water	1480	1·48	0·0022

relaxation, and is approximately proportional to the frequency.

Biological Effects of Ultrasound

(a) Thermal effects

The heating of tissue due to absorption may be sufficient to cause damage. Many tissues are irreversibly damaged by exposure to temperatures in excess of 50°C; at lower temperatures, the effects of heat are usually reversible and non-injurious.

(b) Mechanical effects

A medium supporting an ultrasonic wave is subjected to mechanical forces. In biomedicine, the most important is a shearing force due to a velocity gradient. This force may be sufficiently large to break macromolecules. The velocity gradient is amplified by the presence of gas-filled bubbles, due to either stable or transient cavitation.

(c) Hazards in diagnostic procedures

There is no thermal hazard in contemporary diagnostic techniques. Diagnostic intensities appear to be below the thresholds for mechanically induced damage. There is no statistical evidence of an increase in foetal abnormality following ultrasonic investigations.

Diagnostic Applications

(a) Pulse-echo techniques

(i) *Basic principles.* Most diagnostic applications are based on the pulse-echo method which gives information about the positions of relecting surfaces within the body, and, to a lesser extent, clues as to the natures of these reflectors. Other methods of investigation, such as X-radiology, do not readily give such information about soft-tissue structures.

A continuous train of short pulses of ultrasonic energy, each of a few microseconds duration, is generated by a suitable transducer. These pulses are directed along a narrow beam into the patient. Echoes which return from interfaces within the patient may be detected by the transducer, delayed by times corresponding to depth along the ultrasonic beam. The amplitude of each echo depends upon the characteristic impedances of the media on each side of the corresponding interface, and the geometry and orientation of the boundary. In the ideal case of normal incidence on a plane interface,

$$\frac{I_r}{I_i} = \left(\frac{Z_t - Z_i}{Z_t + Z_i}\right)^2$$

where I_i and I_r are the incident and reflected intensities, and Z_i and Z_t are the characteristic impedances on the incident and transmitting sides of the interface. The echo amplitude is also controlled by the absorption which occurs in the media through which the ultrasonic pulse is propagated.

Pulse-echo techniques are thus based on three principles: (1) ultrasonic energy can be confined into short pulses within a narrow beam; (2) ultrasound travels at almost the same velocity in all soft tissues; and (3) ultrasound is partially reflected at most tissue interfaces.

Pulse-echo techniques depend upon the satisfactory detection of ultrasonic signals. This requirement seriously restricts the application of the method to the examination of structures surrounded by, or containing, either gas or bone. This is because both gas and bone present large characteristic impedance mismatches (and hence large reflections) at interfaces with soft tissues, and they also have relatively high absorptions. In purely soft tissues, compensation for absorption may be provided by swept (time-varied) gain in the receiver amplifier.

(ii) *Methods of display.* In the *A-scope* echoes are displayed (after electrical amplification) on a cathode-ray oscilloscope as vertical deflections of a horizontal time-base. The depth along the ultrasonic beam is proportional to the distance along the time-base. The same information can be presented (on a *B-scope*) by brightness-modulation of the time-base. A two-dimensional scan of a plane within the patient can be generated by integrating a B-scope display whilst the transducer is moved around the patient's body, if the direction and position of the time-base are linked to those of the ultrasonic beam. Scanners to perform this function may be manually operated with the transducer in contact with the patient's skin, or mechanically driven, with the transducer moving in water which acts as a coupling medium. An improved image (a *compound B-scan*) is obtained if the probe is oscillated as it is moved around the patient, because this increases the likelihood of normal incidence with specular reflectors.

The movements of structures may be studied by *time-position recordings* in which a B-scan is deflected at constant velocity normal to the ultrasonic time-base across a photographic film or an electronic storage tube. Alternatively, an analogue converter may sometimes be used to derive a signal proportional to the instantaneous position of an isolated moving structure.

Modified forms of the *C-scope* (which is a brightness-modulated display in which the *x*-deflection corresponds to the bearing of the target, and the *y*-deflection, to its elevation) have also been used to display medical diagnostic pulse-echo information.

(iii) *Resolution.* The resolution of a pulse-echo system is conveniently defined as being equal to the distance which appears on the display to be occupied by a point target in the ultrasonic beam. The resolution is determined by the geometry of the beam and the dynamic range of the system. At any given distance from the transducer, the dynamic range is the difference between the maximum echo amplitude which is received as the ultrasonic beam is swept across the target, and the minimum echo amplitude which produces a registration on the display, both referred to the input of the receiver. In the case of a typical flat disc transducer, and a dynamic range of 10 dB, the range resolution is about $1\cdot5\,\lambda$. With a disc diameter of $25\,\lambda$, the lateral resolution is about $10\,\lambda$ at a range of $120\,\lambda$. Some improvement in lateral resolution over a limited range can be achieved by focusing the ultrasonic beam.

(b) *Continuous-wave techniques*

The earliest attempts to use ultrasound in medical diagnosis were based on transmission methods analogous to those used in X-radiology. These attempts faied, chiefly because of the large variations in the absorption in the normal. The performance of continuouswave imaging is also limited by standing waves, although dynamic studies have been made using the ultrasound image camera. Preliminary results of holographic imaging have been encouraging, but many practical problems remain to be solved. Some attempts to overcome these difficulties have involved the pulsing of the ultrasonic field.

(c) *Doppler techniques*

If two transducers are arranged so that one receives ultrasonic energy transmitted from the other, the transmitted and received frequencies are equal provided that the path length is fixed. However, if the path length changes with time, the received frequency is shifted by the Doppler effect. For example, if the transducers are co-planar, the change in path length may be due to the movement of a reflector which directs the transmitted beam back to the receiver. In clinical practice, such a system gives shift frequencies which fall in the audible range. Consequently, in many applications it is sufficient to listen to the Doppler signal: but in some specialized investigations, electronic methods of analysis can give additional data.

Simple Doppler systems indicate only target velocity. Directional information can be obtained by the use of a phase-sensitive detector. It is also possible to sample target movements according to range by electronic gating.

Applications in Clinical Diagnosis

The scope of the applications of ultrasonic techniques in clinical diagnosis is indicated in Table 2. The pulse-echo techniques are generally useful because they provide information about the positions of targets. The amplitude information can sometimes be used in the differentiation between normal and abnormal tissues. The information available from other characteristics of the echo, such as its energy-frequency spectrum, is largely unexplored.

Therapeutic Applications

Ultrasonic irradiation is used in the physiotherapy of a wide variety of soft-tissue ailments. Circular transducers of about 3 cm diameter are common,

Table 2

Technique	Diagnostic site	Clinical application
A-scope	Brain	Localization of mid-line structures
	Skull	Measurement of foetal biparietal diameter
	Liver	Localization of fluid-filled cavities
	Kidney	Differentiation of cystic and solid lesions
	Lung	Diagnosis of pleural effusion
	Eye	Length measurements
B-scope	Uterus	Diagnosis of multiple pregnancy; localization of placenta
	Abdomen	Diagnosis of gynaecological and other abnormalities, including those of liver, kidney and spleen
	Breast	Diagnosis of abnormalities, especially cysts
	Eye	Tumour diagnosis
Time-position recording	Cardiovascular	Assessment of heart (particularly mitral) valve function; diagnosis of pericardial effusion
Doppler	Cardiovascular	Detection of foetal heart movement; localization of blood vessel occlusions; determination of blood-flow profiles

with intensities in the range $0\cdot25$–3 W cm^{-2}. The frequency is generally either 1 or 3 MHz. Each application is for 3–10 min, and these may be repeated perhaps ten times during a course of treatment.

The biological mechanisms which are involved have not been systematically investigated. It seems likely that the treatment is chiefly valuable because of the particular distributions of heat which can be obtained.

Surgical Applications

(a) *Neurosurgery*

At the focus of a suitable system, such as that of a plane disc transducer and a plastic lens, the ultrasonic intensity may be sufficiently high ($0\cdot2$–$2\cdot5$ kW cm^{-2}) to damage tissue, with irradiation times of 2–$0\cdot1$ sec. By an appropriate choice of frequency and angular convergence, the focal damage can be achieved without injury to the tissues lying between the source and the lesion. Thus, it is possible to induce a *trackless* lesion. At intensities above about 0·7 kW cm⁻² at 3 MHz, the heat due to irradiation is inadequate to account for the biological damage, and the mechanism is at least partly non-thermal. In the brain, the method has been used in the treatment of Parkinson's disease and for the destruction of the pituitary gland in hormone-dependent cancer. However, the technique has not enjoyed unqualified success, partly because it is necessary first to remove a portion of the skull on account of the high absorption of ultrasound which would otherwise occur in the bone. The method has some value in experimental neuro-anatomy.

(b) *Vestibular surgery*

The ultrasonic irradiation of the vestibular end organ (part of the inner ear) can destroy the balance function, whilst the hearing generally remains unimpaired. The method is of value in the treatment of the small proportion of patients with Meniere's disease who require surgical intervention. Alternative operations are either dangerous or involve a total loss of hearing. The ultrasonic energy isusually applied to the lateral semicircular canal following some preliminary surgery; less commonly, the ultrasound may be applied to the round window. An intensity of around 20 W cm⁻² is required for the lateral canal approach (or around 3 W cm⁻², for the round window); the irradiation time is about 10–30 min. There is some evidence that the biological mechanisms are purely thermal.

See also: Radiation, non-ionizing, hazards of Ultrasonics in medial diagnosis

Bibliography

BOCK, J. and OSSOINIG, K. (Eds.) (1971) *Ultrasonographia medica*, I, II, and III, Vienna: Verlag der Wiener Medizinischen Akademie.

GROSSMAN, C. C., HOLMES, J. H., JOYNER, C. and PURNELL, E. W. (Eds.) (1966) *Diagnostic Ultrasound*, New York: Plenum Press.

WELLS, P. N. T. (1969) *Physical Principles of Ultrasonic Diagnosis*, London and New York: Academic Press.

WELLS, P. N. T. (1970) *Rep. Prog. Phys.* **33**, 45.

WELLS, P. N. T. (Ed.) (1972) *Ultrasonics in Clinical Diagnosis*, Edinburgh and London: Churchill Livingstone.

P. N. T. WELLS

MESONIC ATOMS. In the broad sense of the word, any system of negative and positive particles which are bound together through electromagnetic interaction and show somewhat similar characteristics as those of ordinary atoms, i.e., electronic atoms, may be called "atoms". Thus, in addition to electrons, any of the following particles may form atoms: $\mu^-, \pi^-, K^-, \Sigma^-, \bar{p}$ (anti-proton),... The existence of these exotic atoms, i.e. muonic, mesonic, hyperonic, and anti-protonic atoms, is now experimentally well established. Very often all of these exotic atoms are simply called mesonic (mesic) atoms.

The first predictions of the possibility of forming such atoms were made by Wheeler and by Fermi and Teller, both in 1947. After inconclusive observations of muonic and pionic atoms by various researchers, Camac obtained more conclusive evidence of pionic atoms in 1952. The first comprehensive study of exotic atoms, however, was carried out in 1953 by Fitch and Rainwater who, through the systematic measurement of muonic X-ray spectra, showed that the charge radii of nuclei are given by $R = r_0 A^{1/3}$ where $r_0 \simeq 1 \cdot 2$ and A is the mass number. This value of nuclear radius constant r_0 was smaller than had been previously assumed and is in better agreement with the values which are now accepted. (In most cases of interest, the transition energies of exotic atoms fall in the range of gamma-rays. It is customary, however, to call them X-rays as in the case of ordinary atoms.)

The study of exotic atoms occupies a unique place in physics since it encompasses atomic, molecular, nuclear, and elementary particle physics. In practice, however, mesonic atoms have been most extensively studied as a source of information on nuclear structure. There has been a rapidly increasing interest in this subject in recent years and already about 400 papers on mesic atoms have been published. The usefulness of negative mesons in mesic atoms as nuclear probes was recognized in the early days of research on mesic atoms and many interesting features of the mesic X-ray spectra have been anticipated from theoretical considerations. With NaI scintillation detectors, however, the verification of many theoretical predictions was out of reach. The development of high-resolution lithium-drifted germanium gamma-ray spectrometers in 1964 initiated a new era of active research.

When a negative meson is stopped in matter, it first loses energy by ionization. It is then further slowed down by collisions with electrons and nuclei until finally captured by an atom into a high Bohr orbit. The meson then cascades down first by Auger transitions and, as it reaches lower orbits, increasingly by radiative transitions, emitting mesic X-rays. Radiationless transitions inducing fissions in some nuclei are possible, although these are not the important modes of cascade in mesic atoms.

The large mass of mesons compared to that of electrons allows mesic atoms to exhibit quite different characteristics as compared to the ordinary atoms. For example, as one can see from eq. (2) below, the meson orbits are closer to the nucleus so that the effects of the finite extension of nuclear charge and of vacuum polarization produce significant energy level shifts. (The Bohr radius of muonic atoms is only $\sim 2 \cdot 56 \times 10^{-11}$ cm.) For heavy elements the average radius of mesic atoms in the ground state lies inside the nucleus. For a muonic lead atom, for example, the probability of finding the ground state muon inside the nucleus is approximately one-half. For mesons which spend part of their time inside the nucleus, the charge seen by the mesons is decreased, thus raising the energy levels of the mesic atoms. The gross features of the energy levels of a mesic atom can be obtained by solving the Schrödinger equation for the meson-nucleus system assuming only the electromagnetic interaction. The energy E_n of levels with principal quantum number n, radius r_n of the nth orbit, and classical velocity v_n in that orbit, can be expressed by the usual Bohr equations with the mass replaced by the meson-nucleus reduced mass μ:

$$E_n = -\mu c^2 \frac{(\alpha Z)^2}{2n^2} \quad (1)$$

$$r_n = \frac{\hbar^2}{\mu c^2} \frac{n^2}{Z} \quad (2)$$

$$v_n = \alpha c \frac{Z}{n}, \quad (3)$$

where α is the fine structure constant. To get a more accurate expression for the energy, however, one must solve the Dirac equation for muonic atoms and the Klein–Gordon equations in the case of pionic and kaonic atoms.

I. Muonic Atoms

Of all exotic atoms, muonic atoms have been most extensively investigated and the results obtained have reached a high degree of sophistication. With the completion of high-intensity proton accelerators, meson factories, around the world (LAMPF, Los Alamos, U.S.A.; TRIUMF, Canada; SIN, Switzerland; JINR, Dubna, U.S.S.R.), it will soon be possible to produce muonic atoms and molecules copiously, thus allowing use of the bent-crystal spectrometer.

The finite nuclear size effect on energy levels is obtained by numerically solving the Dirac equation for a potential calculated from some type of assumed charge distribution of the nucleus. In addition, we have to consider further corrections, the most significant arising from vacuum polarization, Lamb shift, anomalous magnetic moment, and nuclear polarization. For the ordinary hydrogen atom, the radiative corrections are mostly due to electron self-energy effects and the contribution from vacuum polarization is much smaller. The situation is reversed for the muonic atoms, owing to the proximity of muons to the nucleus. For example, the vacuum polarization effect for the $1S_{1/2}$ level in ^{238}U is about 74·5 keV, whereas the muon self-energy effect (Lamb shift) is only $-3·2$ keV. It is noteworthy that the vacuum polarization effect was confirmed in the measurement of the muonic phosphorus atom X-ray spectra.

For highly deformed nuclei, the electric quadrupole potential is of the same order as the excitation energies of the rotational levels and the fine structure of the muonic K X-rays. The electric quadrupole (E2) interaction mixes the nuclear ground state with various excited states and also with the various muonic states. As a result of resonances between various states, additional structure appears in the spectra of muonic atoms with highly deformed nuclei. This phenomenon is usually referred to as the dynamic E2 effect. This effect, first predicted by Wilets in 1954, has been experimentally observed and the results show that the muonic X-ray spectra of deformed nuclei can be described adequately by the collective model. The magnetic dipole hyperfine structure has also been observed for muonic atoms of a number of nuclides. Extensive measurements of isotope shifts have been made for many elements. Wu and her collaborators observed for the first time in 1968 the isomer shifts in muonic atoms. Recently various authors have made accurate measurements of the isotone shifts in the Ca, Ba, and tin regions. The isotone shift differs from the isotope shifts in that the former measures the energy difference of atoms when protons are added to the nucleus with the number of neutrons remaining the same, whereas for the latter the role of protons and neutrons are reversed. The study of isotope as well as isotone shifts reveal the nuclear shell structure. For example, the rms radius of ^{48}Ca (neutron magic number) is found to be smaller than that of a lighter isotope ^{40}Ca.

The determination of the charge distribution of spherical as well as deformed nuclei traditionally has been one of the main objects in the study of muonic X-ray spectra. Recently very extensive investigation of nuclear charge distributions in terms of two or even more parameters has been undertaken through an analysis of muonic X-ray spectra. In general the results are in good agreement with values obtained from electron scattering experiments, although small discrepancies persist. Muonic atoms, together with electron scattering experiments, have become two of the most important sources of our knowledge of the nuclear charge distribution. Although it has long been recognized that different transition energies in muonic atoms determine different parameters of nuclear charge distribution, it has not been clear exactly what parameter is determined from the measurement of a particular transition energy. According to the recent work of Ford and Wills, a particular transition energy determines, with a certain degree of model independence, a particular moment of nuclear charge density $\langle r^k \rangle$ and thereby the equivalent radius R_k defined by $R_k = [(1/3)(k+3)\langle r^k \rangle]^{1/k}$, where k is *not* in general an integer. Note that, for $k = 2$, $\langle r^k \rangle$ and R_k reduce to rms radius and the usual definition of the radius of equivalent uniform charge distribution, respectively. Contrary to common belief, even $2p \to 1s$ transitions do not determine exactly rms radius except for the lightest atoms. For example, $2p_{1/2} \to 1s_{1/2}$ transitions in muonic ^{206}Pb atoms determine $\langle r^{0.80} \rangle$ while $2p_{3/2} \to 1s_{1/2}$ transitions determine $\langle r^{0.83} \rangle$ instead of $\langle r^2 \rangle$.

Actually the model independence of the moment of nuclear charge density $\langle r^k \rangle$ has limited validity and a modified form of the moment, such as $\langle r^k \exp(-r) \rangle$ shows better model independence. The advantage of the analysis of muonic atom data in terms of generalized equivalent radii is that it gives a useful insight and guidance for detailed analysis of data. Since the measurements of electron scattering and muonic X-rays have reached the stage of considerable sophistication, it would be interesting to see whether a simultaneous fit to these two sets of data can be achieved from a common static nuclear charge distribution. All analyses so far, however, indicate small but possibly important inconsistencies between these two sets of experiments. Most of the calculations have been made using the so-called Fermi charge distributions with or without centre depression. Actually there exists no reason why the nuclear charge distribution should be represented by a Fermi distribution. For the better understanding of muonic atom and electron scattering data, further refinement in both theory and experiment is desirable. On the side of muonic atoms, more precise evaluation of various corrections including, perhaps, the finite size effect of the muons would be necessary. At the same time, more accurate analysis of the electron scattering data with such corrections as those due to nuclear polarization and absorption would be necessary for simultaneous analysis to be more meaningful.

II. *Muonic Molecules*

Muonic molecules are systems in which a negative muon binds together two nuclei. Some examples are $(p\mu^-p)^+$, $(p\mu^-d)^+$, and $(d\mu^-d)^+$, and in practice only such ionic states are of interest. Undoubtedly the most important discovery about muonic molecules is that of the catalysis of fusion in hydrogen by muons. Owing to the large mass of the muons, the two nuclei in $(p\mu^-d)^+$, for example, can approach to within a distance of the order of the range of nuclear forces. Even when the kinetic energy of the nuclei is much lower than the Coulomb barrier, the following nuclear fusion reaction can occur as the result of tunnelling:

$$p + d \to \text{He}^3 + \gamma + 5.5 \text{ MeV}. \quad (4)$$

On the other hand, in an ordinary HD molecule the probability of a nuclear reaction is virtually nill; it would be necessary to wait on the average $\simeq 10^{20}$ years to observe one such reaction in 250 gallons of liquid HD. It is tempting to speculate on the possibility of sustaining nuclear fusion in hydrogen with muons. Subsequent theoretical study, however, has shown that this is not possible.

III. *Muonium*

Instead of replacing electrons in ordinary atoms by one of the other kind of negative particles, we can also consider atomic and molecular systems in which various positively charged particles form bound systems with electrons. For example, in muonium a positive muon replaces the proton in an ordinary hydrogen atom.

Precision measurements of muonium ground-state hyperfine-structure intervals were made by Telegdi's group and by Hughes'. The determination of hfs splitting and the muon magnetic moment can yield a value of the fine structure constant α. The recent value of $1/\alpha$ thus determined is 137·03654 (30) (2·0 ppm) which agrees well with the currently recommended value, 137·03602 (21).

The chemical interaction of muonium, which may be regarded as a light isotope of hydrogen, with atoms and molecules is expected to resemble the interactions of hydrogen atoms. Inasmuch as the hydrogen reactions are simulated by the muonium reactions, the investigation of the latter may open a new physical method for studying the former. Muonium chemistry is still in its early stage of development. The meson factories will certainly facilitate its study.

IV. *Pionic and Kaonic Atoms*

In muonic atoms the interaction between orbital particles and the nucleus is almost entirely electromagnetic. In hadronic atoms (all kinds of atoms except leptonic atoms, the latter being ordinary atoms and muonic atoms), however, the strong interaction in addition to the electromagnetic interaction plays an important role. For example, the strong interaction affects the cascade scheme for the lower values of orbital angular momentum and causes energy level shifts and level width broadening. It is to be noted that the nuclear polarization effect is much greater in pionic atoms than in muonic atoms. A nucleus is not a homogeneous and uniform distribution of matter and the pion is sensitive to the granularity of a nucleus. Since such a structure can be seen in the states of virtual excitation of the system, this effect is equivalent to the nuclear polarization.

One of the interesting applications of pionic and kaonic atoms is the study of the surfaces of heavy nuclei. Johnson and Teller suggested in 1954, that for a nucleus stable against beta decay, the neutron distribution would extend beyond the proton distribution and that in an experiment involving the negative pion-nucleus interaction, one might find a nuclear surface region populated mainly by neutrons. It has been also emphasized by Wilkinson, that the study of kaonic atoms may provide us with information about the extreme surface region of the nucleus. Since K$^-$-proton and K$^-$-neutron interactions can easily be distinguished from one another, one can even estimate the neutron-to-proton ratio in the extreme peripheral region of nuclei. There have been suggestions again in recent years that the nuclear surface of a heavy nucleus is predominantly populated by neutrons. At present, however, the evidence for the neutron "atmosphere" is not conclusive.

V. Sigmic and Anti-protonic Atoms

At the time of this writing, Xionic atoms (Ξ^--p systems) have not been observed. Backenstoss and his collaborators at CERN recently obtained the most convincing evidence for the observation of sigma hyperonic (sigmic) atoms as well as the first measurement of anti-protonic atom spectra.

The low energy K^- mesons were produced by the CERN Proton Synchrotron and about 100 K^-/burst could be stopped in a target of 8 g/cm² thickness. Σ^- hyperons which give rise to Σ^- hyperonic atoms are predominately produced through the reactions:

$$K^- + p \rightarrow \Sigma^- + \pi^+$$
$$K^- + n \rightarrow \Sigma^- + \pi^0. \quad (5)$$

In sigmic atoms nuclear capture occurs at a remote region of the nuclear surface and therefore the study of sigmic atoms may lead to exploration of the extreme periphery of nuclei as was pointed out by Burhop. Sigmic atom X-ray spectra should show a fine structure due to the magnetic moment of negative sigma hyperons. Therefore the study of sigmic atoms may also lead to a measurement of the Σ^- magnetic moment.

In a low momentum separated beam produced by the CERN Proton Synchrotron, about 300 anti-protons/burst could be stopped in a target of 4 g/cm² thickness. Figure 1 shows the anti-protonic X-ray spectrum in Tl. All strong lines in this spectrum are identified as anti-protonic atom transition lines except for two groups of lines at 74 and 84 keV. These lines are due to the electronic K_α and K_β transitions in Pb and Tl. No clear pionic transition lines appear, indicating that pions are effectively rejected in this experiment.

VI. Biomedical and Industrial Uses of Mesonic Atoms

In a paper in *Nature* in 1961, Fowler and Perkins suggested the radiological application of negative pions. For the advantages and disadvantages of negative pions for the purpose of radiotherapy, the reader is referred to a book edited by Yuan

Several years ago, a very ingenious idea was conceived by Rosen who showed that negative muons might be useful for medical diagnostic purposes by permitting *in vivo* analysis of the chemical composition of tissues and organs. Figure 2 shows the simulated muonic X-ray spectrum from "standard man". We see in this figure the actual spectrum from X-ray sources embedded in tissue-equivalent material having the composition of "standard man". The X-ray energies and intensities were calculated to correspond to what would be observed from stopped muons in the above system. It has been further suggested that negative muons might be used as a sensitive means for elemental analysis of inorganic materials. The work of Malanify and of Kim to explore a quantitative analytical technique for measuring elemental and isotopic constituents

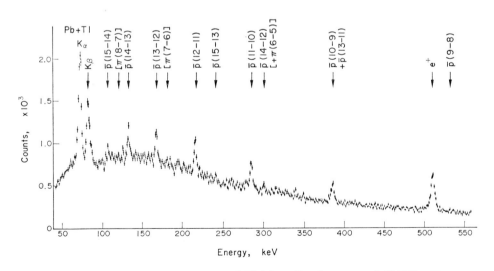

Fig. 1. Anti-protonic X-ray spectrum of Tl (from Bamberger et al. (1970) with permission of North-Holland Publishing Co.).

in certain materials utilizing the muonic X-rays emitted following muon capture may be regarded as the first serious attempt of an industrial application of the study of muonic atoms.

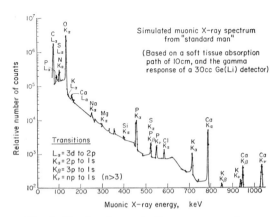

Fig. 2. Simulated muonic X-ray spectrum from "standard man". [Rosen (1971)].

Bibliography

ANDERSON, H. L. et al. (1969) *Phys. Rev.* **187**, 1565.
ALVAREZ, L. W. et al. (1957) *Phys. Rev.* **105**, 1127.
BACKENSTOSS, G. (1970) *Ann. Rev. Nuclear Sci.* **20**, 467.
BACKENSTOSS, G. et al. (1970) *Phys. Lett.* B **33**, 230.
BAILEY, J. M. et al. (1971) *Phys. Rev.* A **3**, 871.
BAMBERGER, A. et al. (1970) *Phys. Lett.* B **33**, 233.
BARRETT, R. C. (1970) *Phys. Lett.* B **33**, 388.
BURHOP, E. H. S. (1967) *Nucl. Phys.* B **1**, 438.
BURHOP, E. H. S. (1969) *High Energy Physics* vol. 3, p. 109. New York: Academic Press.
ELTON, L. R. B. (1967) *Proc. Int. Conf. on Electromagnetic Sizes of Nuclei*, p. 267. Ottawa: Carleton University.
ERICSON, T. E. O. (1970) in *High Energy Physics and Nuclear Structure*, p. 488. New York: Plenum Press.
FORD, K. W. and WILLS, J. G. (1969) *Phys. Rev.* **185**, 1429.
FOWLER, P. H. and PERKINS, D. H. (1961) *Nature*, **189**, 524.
HUGHES, V. W. (1966) *Ann. Rev. Nucl. Sci.* **16**, 445.
JOHNSON, M. H. and TELLER, E. (1954) *Phys. Rev.* **93**, 357.
KIM, Y. N. (1971) *Mesic Atoms and Nuclear Structure*. London: North-Holland.
MALANIFY, UMBARGER, AUGUSTSON, WALTON and MORGAN (1971) Private communication.
ROSEN, L. (1971) invited paper, Fourth Int. Conf. on High Energy Physics and Nuclear Structure, Dubna, USSR, 7–1 September.
TELEGDI, V. L. (1969) In: Bederson et al. (Ed.) *Atomic Physics*. p. 551. New York: Plenum Press.

WILKINSON, D. H. (1967) *Comments on Nuclear and Particle Phys.* **1**, 112.
WU, C. S. and WILETS, L. (1969) *Ann. Rev. Nuclear Sci.* **19**, 527.
WU, C. S. and WILETS, L. (1971) *Comments on Atom. Molec. Phys.* **2**, 195, and papers cited there.
YUAN, L. C. L. (1971) *Elementary Particles*. London: Academic Press.

Y. N. KIM

METEOROLOGICAL OBSERVATIONS FROM SATELLITES. Measurements from satellites provide the best hope of obtaining the vast amount of data required as a starting point for accurate global scale numerical weather forecasts currently made possible by the rapid growth in the size and speed of computers. The minimal requirements formulated for the Global Atmospheric Research Program (GARP) are set out in Table 1 (W.M.O., 1971). The present routine observing network of synoptic radiosondes falls considerably short of the requirements, especially over ocean areas, furthermore, extension of the system to the required density is a costly affair. Satellite measurements are compara-

Table 1. Observational Requirements for GARP

All elements to be measured with a horizontal resolution of 500 km

	No. of levels in troposphere	No. of levels in stratosphere	Time between observations (hours)	Accuracy
Wind	4	3	12	±3m*sec⁻¹
Temperature	4	3	12	±1°C
Relative humidity	2	0	12	±30%
Pressure	1 (at least)	0	12	±0·3%
Sea surface temperature	1	0	72	±1°C

* 2 m sec⁻¹ in tropical regions.

tively cheap, are global in scale, are free from the difficulties associated with differing instrument types and are relatively easy to collate. Table 2 gives a complete list of American meteorological satellites to date. These have carried experiments for mapping cloud positions using television photography (this also yields wind estimates from changes between successive pictures), for obtaining temperature as a function of height and geographical position, and

Table 2. Weather Satellite Launch Dates to November 1972

Satellite	Launch date
TIROS-1	1 April 1960
TIROS-2	23 Nov. 1960
TIROS-3	12 July 1961
TIROS-4	8 Feb. 1962
TIROS-5	19 June 1962
TIROS-6	19 June 1963
TIROS-7	21 Dec. 1963
TIROS-8	21 Dec. 1963
TIROS-9	22 Jan. 1965
TIROS-10	1 July 1965
ESSA-1	3 Feb. 1966
ESSA-2	28 Feb. 1966
ESSA-3	2 Oct. 1966
ESSA-4	26 Jan. 1967
ESSA-5	20 April 1967
ESSA-6	10 Nov. 1967
ESSA-7	16 Aug. 1968
ESSA-8	16 Dec. 1968
ESSA-9	26 Feb. 1968
NIMBUS-1	28 Aug. 1964
NIMBUS-2	15 May 1966
NIMBUS-3	14 April 1969
NIMBUS-4	8 April 1970
NIMBUS-E	Dec. 1972
ATS-1	7 Dec. 1966
ATS-3	5 Nov. 1967
ITOS-1	23 Jan. 1970
ITOS-A	11 Dec. 1970
ITOS-C	July 1973
ITOS-D	15 Oct. 1972

determining the distribution of atmospheric constituents such as ozone and water vapour.

Satellite Orbits

Two orbital configurations are particularly suitable for meteorological use.

(i) *Geostationary orbit*

A satellite in equatorial orbit at an altitude of 36,000 km has a period of 24 hr and so always occupies the same point in the sky relative to an observer on the ground. A geostationary satellite can observe approximately one-third of the earth continuously, although horizontal resolution is necessarily poor away from the sub-satellite point.

(ii) *Sun-synchronous orbit*

Complete global coverage can be obtained with a satellite in polar orbit. An altitude of 1000 km gives 15 orbits per day while the earth rotates underneath, so that the sub-satellite point passes within 12° longitude of any given position twice per day, once travelling north, once travelling south. An orbit inclined to the earth's axis experiences a torque from the earth's equatorial bulge which produces a gradual precession relative to inertial space. With suitable choice of orbital parameters the orbit may be made to rotate through 360° in a year. Under these circumstances a given latitude is always crossed at the same local time and the orbit is said to be sun-synchronous.

Vehicles

The first series of satellites designed specifically for atmospheric studies was TIROS (Television and Infrared Observation Satellite), a cylindrical eighteen-sided polyhedron about 60 cm high and 1 meter across, weighing about 170 kg. TIROS was spin-stabilized so that its axis was fixed in space rather than relative to the earth's surface, a fairly serious limitation arising from early technology. TIROS wheel satellites are improved versions which have the spin axis perpendicular to the orbital plane, so that instruments mounted on the rim view the earth perpendicularly once per rotation of the satellite instead of once per orbit as previously (Rados, 1967). The later TIROS series satellites, renamed ESSA (Environmental Survey Satellite) are wheel-type satellites in sun-synchronous orbits.

The Nimbus series are experimental meteorological satellites used as test beds for new techniques, the most useful of which become standard equipment on operational satellites. The Nimbus spacecraft is larger (3·5 mm high), heavier (500 kg), provides more power (450 watts as against 30 watts) and more advanced than TIROS. In particular it is earth stabilized, using horizon sensors to keep its axis perpendicular to the earth surface. All Nimbus satellites occupy polar sun-synchronous orbits at altitudes around 1000 km.

ITOS (Improved TIROS Operational System) incorporates features based on experience with Nimbus. In particular three axis stabilization is used as in Nimbus, although the basic construction is different, consisting of a cube of 1 m side, weighing 350 kg. The solar paddles provide about 150 W mean power. They are launched into sun-synchronous orbits at 1460 km altitude (Schwalb, 1972).

The ATS (Advanced Technology Satellite) series, launched into geostationary orbit is used mainly for cloud photography, the orbit being too distant for detailed infra-red work. ATS satellites are spin-stabilized cylinders about 140 cm long and 150 cm in diameter (Warnecke and Sunderlin, 1968).

The Soviet satellite program revolves around the Cosmos series. Those which are equipped for atmospheric studies with television cameras and infra-red and microwave sensors are given the name Meteor. These satellites use inertial wheel stabilization to keep the instruments pointing downwards and they occupy polar retrograde orbits (Ganopol'skiy et al., 1969).

Visible Imagery. Satellite Cloud Photography

Cloud photography from satellites using television techniques has proved a relatively straightforward but beneficial application of remote sounding: there is good agreement between the cloud configuration and a weather map, which is particularly valuable over the oceans where conventional upper air observations are few. In many cases experienced observers can recognize the onset of potentially catastrophic systems such as hurricanes. Some meso-scale phenomena have been revealed for the first time, including cloud clusters, certain lee wave systems and eddies downwind of islands. Movement of clouds between successive pictures provides a method of determining the wind distribution, although there are difficulties in identifying cloud movement with the local wind speed and in fixing the cloud height (Fujita, 1968). Areas of extremely high brightness on cloud photographs have been shown to be well correlated with deep, precipitating clouds (Suomi, 1969).

The first television pictures of the earth were taken with a 500-line vidicon camera mounted on TIROS-1. The Nimbus I and II satellites carried an improved system called the Advanced Vidicon Camera System (AVCS) (Rados, 1967). This consists of three identical cameras mounted at 35° to each other, arranged so that the three pictures have an overlap with each other and with the next orbit. Each camera has a 1 inch vidicon with a 40 msec exposure time. This is scanned by an 800-line raster of 800 points. A triplet of photographs is taken each 91 sec, while the satellite is on the illuminated side of the earth. The system is switched off on the other side. Compensation for the variable illumination of the surface is provided by an iris-type aperture stop linked directly to the solar paddle shaft, the position of which is related to the solar zenith angle. The pictures are stored on a tape recorder until a command from a ground station causes them to be transmitted to the ground station via the 5 watt S-band transmitter. Here mosaics are built up on to various map projections, land outlines and geographical grids added and analyses made for transmission to interested meteorological centres.

The usefulness of cloud photographs in forecasting depends on the speed with which they are made available. Accordingly an additional system known as Automatic Picture Transmission (APT) is provided, which allows pictures of lower spatial resolution to be received almost instantaneously by operators of low cost receiving stations anywhere from where the satellite is visible. A storage type vidicon is used with a scanning rate of 4 lines per second. The picture, transmitted at 136 MHz with a 20 kHz bandwidth, can be received on comparatively simple equipment (Portune and Owen, 1970).

On Nimbus III and IV the AVCS was replaced by a single image dissector camera system (IDCS) which couples superior signal to noise ratio with the ability to sense a greater dynamic range. Pictures from IDCS are compatible with the APT system.

The ATS satellites carry cameras which produce photographs of the earth every 24 min. These were in colour for ATS III. The satellite spins about an axis parallel to the earth's axis at 100 rpm. This motion is used to scan lines of constant latitude through a telescope, while a precision step mechanism moves the scan slightly in latitude, covering the entire disc in 2400 steps. Fibre optics carry light from the focal plane of the telescope to three photomultipliers which have different spectral responses centred in the blue, green and red. Horizontal resolution at the ground is about 4 km (Warnecke and Sunderlin, 1968).

Infra-red Imagery

Visible imagery considered thus far cannot be employed on the dark side of the earth. However, cloud positions here may be determined by infra-red techniques which are able, in addition, to furnish estimates of the temperature of the surface or cloud top. Use is made of spectral regions in which the atmosphere is substantially transparent, principally 3·5–4·1 μm and 10·5–12·5 μm. The first of these spectral intervals was used on all satellites prior to Nimbus III because detector performance is superior. Some scattered sunlight is detected, however, so that on the illuminated hemisphere the pictures obtained are qualitative only.

The intensity of radiation I_r observed in one of these spectral intervals centred at wave number r under cloud free conditions is

$$I_r = \varepsilon \tau B_r(T)$$

where ε is the emissivity of the surface; τ the transmission between surface and satellite, and $B_r(T)$ the Planck function corresponding to the surface temperature T.

Emissivity of the land varies considerably; sand and rock surfaces can have emissivities as small as 0·5 whereas areas well covered with vegetation have emissivities near unity. The emissivity is unity over the ocean for these wave-numbers. Consequently use of the observations for surface temperature measurement is in practice restricted to ocean areas. This is not a serious restriction since land temperature is too variable in time and space for easy application.

The absorption in both intervals is due to carbon dioxide and water-vapour. Of these CO_2 is most easily corrected for since its distribution is constant and known, whereas water-vapour amounts are rather variable. However, over the oceans water-vapour amounts are well correlated with temperature and it has been found possible in the 3·5–4·1 μm region to allow for the transmission factor τ by adding to the brightness temperature a correction which is a function of brightness temperature and viewing angle only. This correction is always less than 3°C (Smith et al., 1970). In the 10·5–12·5 μm band, correction is more difficult since there is a contribution from water dimer molecules, at present poorly understood. Corrections up to 10°C may be necessary. A further correction for haze, again poorly understood, is necessary over land but this is negligible over ocean areas.

Nimbus I, II and III carried the high (spatial) resolution interferometers (HRIR) for infra-red imaging. These were single channel instruments using the 3·5–4·1 μm "window" with a narrow field of view equivalent to a resolution of 3 km at the surface. Pictures are built up by scanning from horizon to horizon using a rotating object mirror. Advancement of the scan is obtained from the movement of the satellite.

On Nimbus IV the longer wavelength channel 10–12 μm is used, giving a coverage on both day and night side of the earth, with a resolution of about 7 km. An additional channel 6·5–7·0 μm was included to observe emission in the water-vapour band. The resulting instrument is known as the Temperature and Humidity Infrared Radiometer (THIR).

The ITOS series carry an instrument known as the scanning radiometer which is similar to THIR except that the 6 μm channel is replaced by a visible channel of 0·5–0·7 μm. ITOS-D will carry a further scanning radiometer with much higher horizontal resolution. The improved performance is obtained by using the superior cadmium–mercury–telluride photoelectric detector.

Measurement of Vertical Distribution of Temperature

At any frequency in the infra-red for which an atmospheric constituent possesses strong absorption, the radiation leaving the top of the atmosphere is a function of the distribution of absorbing gas and the distribution of temperature throughout the atmosphere. Consequently if the temperature is known, a measurement of intensity yields some information about the distribution of constituent. This forms the basis of several instruments for determining atmospheric composition not pursued further here. Conversely, if the distribution of constituent is known, intensity measurements yield information about the vertical temperature distribution. Carbon dioxide is a suitable working constituent, for it is uniformly mixed in the lower 100 km of atmosphere and has absorption bands at 15 μm and 4·3 μm which are not substantially overlapped by those of other atmospheric constituents. Several instruments have used measurements at these wavelengths to derive the vertical distribution of temperature below the satellite by methods described below. The process is frequently called "temperature sounding".

Theory of temperature sounding

Consider an infinitely deep atmosphere which is horizontally stratified. The intensity of radiation of frequency r emitted upwards by a thin slice of atmosphere containing a path length of du of absorber at temperature T is $k_r du B_r(T)$, where $B_r(T)$ is the Planck function at frequency r and temperature T and k_r is the absorption coefficient. Of this a proportion τ_r will reach the top of the atmosphere where

$$\tau_r = \exp\left(-\int k_r du\right)$$

The integral being taken over the region of atmosphere between the slice and the top of the atmosphere. Integrating over all such slices to find the total intensity I_r at the top of the atmosphere,

$$I_r = \int B_r(T) k_r du \left(\exp \int -k_r du\right)$$
$$= \int_{\tau_r=0}^{1} B_r(T) d\tau_r$$

A more convenient vertical component is $y = -\ln p$, where p is the atmospheric pressure in atmospheres. (For an isothermal hydrostatic atmosphere, y is a linear function of height.)

$$I_r = \int B_r(T) \frac{d\tau_r}{dy} dy$$
$$= \int B_r(T) K_r(y) dy$$

In other words, the intensity is a weighted average of the black body function, the weighting function $K_r(y)$ being $d\tau_r/dy$.

As the general shape of K_r is similar to the curves C to E of Fig. 2a, little radiation reaches satellite levels from the lower portions because it is absorbed on the way. On the other hand at very high levels there are few molecules to emit. Consequently the largest contribution comes from some middle level. The precise shape of the weighting function curve depends upon the distribution of absorber and the wavelengths chosen. In general the maximum value of K_r occurs at higher levels for highly absorbed wavelengths than for weakly absorbed wavelengths.

For a wavelength in the wings of a single line of a molecular absorption band of a uniformly mixed absorber such as CO_2 in the lower atmosphere (e.g. below about 50 km so that collision broadening is dominant and Doppler broadening negligible) it can be shown that the "half-width" of the weighting function in units of $\ln p$ is independent of wavelength, corresponding approximately to 10 km (Houghton and Smith, 1970).

Since the width of lines, for instance in the 15 μm CO_2 band, varies from 0·1 to 0·001 cm^{-1} over the range of atmospheric pressures involved, a conventional spectroscopic system possessing the monochromatic response described above would need a resolution better than 0·1 cm^{-1} in the wings of a line and as high as 0·001 cm^{-1} near the line centres; otherwise the effective weighting function is rather broad, being an average of curves with maxima at different heights.

Many of the radiometers designed for temperature sounding possess a spectral bandwidth nearer to 5 cm^{-1} which is wide enough to include several lines. This is not too serious for the lower atmosphere because most of the radiation in a region containing several lines comes from the line wings where the variation of absorption with wavelength is small. The absorption in the line wings is relatively weak, so the bulk of the energy accepted by the instrument comes from the lower atmosphere.

Variations in the central frequency and bandwidth of the instrument produce variation in the height of the maximum weight. If several wavelength regions can be found for which the weighting functions peak at different heights, remote sounding becomes possible.

Retrieving the Temperature Profile

In practice a temperature profile is not uniquely determined by measurements of intensity in a small number of spectral intervals; there are an infinite number of temperature distributions which would give the same intensities. Allowance must be made also for noise in the observations and furthermore account must be taken of the effects of cloud on the measurement. Cloud is usually opaque to infrared radiation but within the field of view cloud cover may be only partial and the tops at varying levels. In short, additional constraints must be sought to permit a unique solution.

For the cloudless case a number of arbitrary constraints are possible. For example, if the instrument has p channels the solutions could be made to be straight lines between p points or the sum of p orthogonal functions. The foregoing may be applied even with no prior knowledge of atmospheric structure. However, since atmospheric statistics are available they may be used to constrain the solution to something more realistic. Examples of this kind of constraint are that the difference between the solution and the mean profile from the statistics should be a minimum, or that the difference between the solution profile and some first guess (say a computer forecast) should be a minimum.

A more direct approach using atmospheric statistics is to regard the solution profile, **b**, as a linear function of the observed radiances, **I**. Thus

$$\mathbf{b} = D\mathbf{I}$$

where the estimator D is chosen to give a best fit to some statistical ensemble for which the temperatures are known, i.e. choose D such that

$$\sum_{s \in S} (\mathbf{b}_s^* - D\mathbf{I}_s)^2 \text{ is a minimum,}$$

where s denotes membership of a statistical ensemble, S; $\mathbf{b}_s^*$ is a profile measured by conventional means, i.e. balloon sondes and rockets, and $\mathbf{I}_s$ is the observed radiance arising from that profile.

The linear approximation is very poor in the presence of clouds, consequently the method can only be applied to clear situations. The principal advantage is that the statistical information is built up by the satellite itself from measurements coincident with the conventional measurements and no knowledge of the shape of the weighting functions or the absolute calibration of the instrument are required. This method has been used by Smith, Woolf and Jacob (1970) using the SIRS-A instrument on Nimbus III (see below).

Retrieval of the temperature profile in the presence of cloud is complicated by the non-linearity of the transfer equations. Furthermore, the statistical distribution of clouds is not adequately known, nor are all the necessary optical characteristics properly established.

One method of obtaining profiles in cloudy conditions used by Smith, Woolf and Jacob assumes the presence of a single broken layer of cloud. On the basis of a first guess of the temperature profile, measurements from two radiance channels may be used to give estimates of the cloud height and amount in this single layer. This information can be used to modify the radiances of the other channels to a value appropriate to clear conditions. A profile is derived for the clear conditions by the method described above. The latest profile now provides a new first guess and the process is repeated until a stable solution is obtained. Two or three iterations usually suffice.

Rodgers (1970) has devised a different technique for use with the SCR measurements on Nimbus IV (see below). For each profile in a collection obtained by conventional means the radiance which would have been observed by the SCR is computed. On the basis of statistics so derived, the retrieval process selects the temperature profile and cloud distribution with the greatest probability of occurrence for the given measurement and noise level. Maximizing the probability is a non-linear problem for which a library-search method is used. Accurate knowledge of the weighting functions is required in pre-computing the statistics. The calibration of the instrument must also be known for the retrieval stage (cf. the method of Smith, Woolf and Jacob above).

Remote Sounding Instruments

(i) Satellite infra-red spectrometer (SIRS)

Nimbus III carried the first sounding instrument with extended vertical coverage—the Satellite Infrared Spectrometer. This is known as SIRS-A in distinction to SIRS-B, a later version on Nimbus-IV. The region from the ground to around 35 km is scanned by eight channels of 5 cm^{-1} resolution in the 15 micron band (r_2) of CO_2. Figure 1 shows the eight weighting functions and a sample retrieval obtained by the method outlined in the previous section.

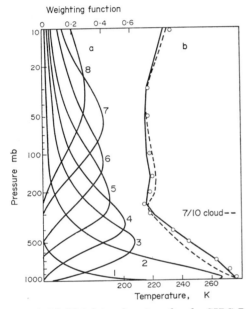

Fig. 1. (a) Weighting functions for the SIRS-B instrument. (b) A comparison of satellite deduced temperatures (continuous line) with radiosonde measurements (circles). The first guess in the declouding process is shown as a dashed line (after Smith, Woolf and Jacob, 1970).

The instrument is a grating spectrometer achieving channel selection by placing eight slits at the correct orientation in the image plane. The field of view is alternated between space and the target by a mirror chopper (a highly polished rotating blade) placed in front of the grating. Alternating signals are produced at the thermistor bolometer detectors from which the radiance for each particular channel is obtained. The target is usually earth but occasionally an internal black-body of known temperature is viewed as a means of calibration. The horizontal resolution at the surface is about 200 km (Dreyfus and Hilleary, 1962).

SIRS data have been routinely incorporated into operational meteorological analysis, providing a significant improvement in forecasting (Smith, Woolf and Fleming, 1972).

(ii) Selective chopper radiometer (SCR)

The selective chopper radiometer (SCR) in Nimbus IV is a six-channel instrument (see Fig. 2) with six independent radiometers with interference filters of spectral band pass of a few cm^{-1} in the 15 micron γ_2 band of CO_2. In the four lower channels a cell of CO_2 is introduced into the light path. This removes the radiation at highly absorbed wavelengths, i.e. the line centres. Since such radiation originates in the high atmosphere, the vertical resolution of the lower channels is much sharper, approaching the ideal monochromatic response. The detector is a thermistor bolometer and as in the SIRS, the signal is produced by chopping the light path between space and earth view, or space and black body for calibration. The remaining channels measure the radiation absorbed in the CO_2 cell. Consequently they have maximum weighting functions rather high in the stratosphere where this radiation originates.

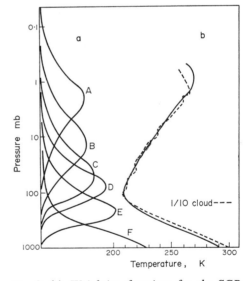

Fig. 2. (a) Weighting functions for the SCR. (b) Comparison of satellite deduced temperatures (solid line) with rocket sonde measurements (dashed line) made at Wallops Island on 27 August 1970.

For these channels (A and B) the light path is alternated between a cell containing CO_2 and an empty cell. An alternating signal is produced in the detector due to the absorption in the CO_2 cell. As a precaution against spurious signals arising from differences in the two paths other than those due to absorption in the CO_2, calibration cycles are occasionally made in a wavelength at which CO_2 does not absorb. Lack of balance of the two paths is corrected using some of the lower channel measurements. Good temperature profiles up to 45 km can be retrieved (see Fig. 2). This has provided for the first time a way of studying the global temperature structure of the high stratosphere on a daily basis (Barnett et al., 1972). Previously temperature measurements above the burst height of conventional radiosonde balloons (30 km) were provided only by meteorological rockets. The stations are situated predominantly in the United States and only a handful of firings are made throughout the world on any one day.

A more elaborate S.C.R. will be flown on Nimbus-E. Improved detectors (pyroelectric) permit a narrower field of view, equivalent to about 25 km at the surface and the addition of more channels will give more information about clouds. The extra channels include one at 520 cm^{-1} for sounding water vapour, four channels specifically designed to monitor cirrus cloud, a window channel at 900 cm^{-1} and two shorter wavelength window channels at 2·3 μm and 3·5 μm which can be used to determine low cloud cover on the night side where sunligh is not a problem.

(iii) *Pressure modulator radiometer*

A major difficulty to be overcome in selecting radiation from the very high atmosphere, i.e. from the line centres, is in balancing the two light paths so that the signal is produced solely from absorption in CO_2 and not from say, uneven chopping or non-identical cells. The pressure modulator radiometer (PMR) proposed for Nimbus-F, will use only one light path per channel employing a single cell of CO_2. An oscillating piston will vary the pressure of the CO_2, thus altering the width of the absorption lines in the cell. The alternating signal produced at the detector then arises only from radiation very close to the line centres. A balloon-borne test of a PMR is reported by Taylor (1972).

The satellite instrument will monitor atmospheric temperatures between 45 and 90 km.

Other Sounding Techniques

(i) *Infra-red interferometer spectrometer* (IRIS)

IRIS is a Michelson interferometer capable of obtaining the entire spectrum of the atmosphere from 5 to 25 microns. Nimbus III and IV carried versions of this instrument capable of spectral resolution of 5 cm^{-1} and 2·8 cm^{-1} respectively.

Data are retrieved at the ground as Fourier interferograms. After processing to obtain the inverse Fourier transform a spectrum is produced of which the portion near 667 cm^{-1} may be used for temperature retrieval. The spectra also contain emission bands of water vapour and ozone from which some information of their distribution may be obtained (Hanel et al., 1972).

(ii) *Nimbus-E microwave spectrometer* (NEMS)

Oxygen, which is uniformly distributed in the atmosphere up to about 100 km has strong absorption lines near 5 mm wavelength. Nimbus-E will carry an experimental microwave sounding instrument with five channels known as Nimbus E Microwave Sounder (NEMS). Three of the channels are in the oxygen absorption bands (53·65, 54·90 and 58·80 GHz), one is a window channel (31·40 GHz) and the remaining one is in a water-vapour absorption band (22·24 GHz) (Rosenkrentz et al., 1972). The principal advantage of microwave sounding is that the measurements are virtually unaffected by clouds. Disadvantages include the weight and power requirements, difficulties in defining the field of view and inferior accuracy of radiance determination compared with infra-red techniques.

(iii) *Limb scanning*

A different approach to measuring stratospheric temperatures uses an instrument with a very narrow field of view looking at the atmosphere tangentially instead of straight down. A very sharp weighting function is obtained without too much loss of horizontal resolution. Measurements at different heights are made by scanning the limb rather than spectroscopically so a fairly large spectral range can be used, partially offsetting the effect of a narrow field of view on the amount of energy collected (Gille and House, 1971).

Nimbus E will carry an instrument, the limb scanning radiometer (LMR) with a 1 mrad field of view capable of a height resolution of about 2 km at 70 km altitude.

The science of mapping the atmosphere from space is still in its infancy; great improvements are to be expected in both instrument design and the techniques for deriving the atmospheric state from the observations. Nonetheless, the achievements to date are impressive. Within a few years the activity of the past decade should culminate in a better understanding of the environment and improved weather forecasting with all the potential economic advantages to mankind which that implies.

Bibliography

BARNETT, J. J., CROSS, M. J., HARWOOD, R. S., HOUGHTON, J. T., MORGAN, C. G., PECKHAM, G. E., RODGERS, C. D., SMITH, S. D. and WILLIAMSON, E. J. (1972) *Quart. J. R. Met. Soc.* **94**, 17.

DREYFUS, M. G. and HILLEARY, D. T. (1962) *Aerospace Eng.* **21**, 42.
FUJITA, T. (1968) *COSPAR: Space Research*, No. 9, p. 557.
GANOPOL'SKIY, V. A., GORODETSKIY, A. K., KESATIN, A. M., MALKEVICH, M. S., ROZENBERG, G. V., SYACHINOV, V. I. and FARAPONOVA, G. P. (1969) *Atmos. and Ocean. Phys.* **5**, 258.
GILLE, J. C. and HOUSE, F. B. (1971) *J. Atmos. Sci.* **28**, 1427.
HANEL, R. A., CONRATH, B. J., KUNDE, V. G., PRABHAKARA, C., REVAH, I., SALOMONSON, V. V. and WOLFORD, G. (1972) *J. Geophys. Res.* **77**, 2629.
HOUGHTON, J. T. and SMITH, S. D. (1970) *Proc. Roy. Soc. Lond.* A**320**, 23.
PORTUNE, J. E. and OWEN, C. M. (1970) *Weather*, **25**, 104.
RADOS, R. M. (1967) *Bull. Amer. Met. Soc.* **48**, 326.
RODGERS, C. D. (1970) *Quart. J. R. Met. Soc.* **96**, 654.
ROSENKRENTZ, P. W., BARATH, F. T., BLINNS, J. C. III, JOHNSTON, E. J., LENOIR, W. B., STAELIN, D. H. and WATERS, J. W. (1972) *J. Geophys. Res.* To be published.
SCHWALB, A. (1972) *N.O.A.A. Tech. Memo.* NESS 35. Washington, D.C.
SMITH, W. L., RAO, P. K., KOFFLER, R. and CURTIS, W. R. (1970) *Mon. Wea. Rev.* **98**, 604.
SMITH, W. L., WOOLF, H. M. and FLEMING, H. E. (1972) *J. App. Met.* **11**, 113.
SMITH, W. L., WOOLF, H. M. and JACOB, W. J. (1970) *Mon. Wea. Rev.* **98**, 582.
SUOMI, V. E. (1969) *Global Circulation of the Atmosphere* (Ed. Corby), p. 222. London: Royal Meteorological Society.
TAYLOR, F. W. (1972) *App. Optics*, **11**, 135.
WARNECKE, G. and SUNDERLIN W. S. (1968) *Bull. Amer. Met. Soc.* **49**, 75.
WORLD METEOROLOGICAL ORGANIZATION (1971) *Report of the sixth session of the joint GARP organizing committee*. Geneva: World Meteorological Organization with I.U.G.G.

R. S. HARWOOD

MICROWAVE PHONONS. The field of research which is commonly described as "microwave phonons" is nothing like as exotic as its name implies, for it is little more than an extension of the technique of ultrasonics to much higher frequencies. The name has arisen as follows. The usual way of generating high-frequency ultrasonic waves is to use a transducer to convert electromagnetic waves into sonic waves, without change of frequency. There is then a considerable reduction in wavelength because the velocities of sonic waves are much less than the velocities of electromagnetic waves, the factor being of the order of 10^{-5}. The term microwave is commonly applied to electromagnetic waves of frequencies in the range 10^9 to 10^{11} Hz. When sonic waves of these frequencies are created they are also described as microwaves. This is a more accurate description, for an electromagnetic wave of frequency 10^{10} Hz has a wavelength in free space of 0·03 m, whereas a sonic wave of the same frequency, travelling through a solid, typically has a wavelength of approximately that of visible light, say 5000 Å. The term phonon is used because the quantum nature of sonic waves becomes more apparent as their frequency rises. The properties of ordinary acoustic waves, which strictly should also be described by quantum mechanics, can usually be given in a classical framework because it is comparatively difficult to do experiments which demonstrate that a classical description is inadequate. As the frequency rises it becomes easier to demonstrate the inadequacies of classical descriptions, and the need for a quantum description is readily apparent at frequencies of the order of 10^9 Hz and above. The termination "-on" is generally a good indication that a quantum description is being used. (The further implication that a quasi-particle description is also being used is not always intended.) Microwave phonons, then, is the field concerned with ultra-high-frequency acoustic types of vibrations. They are microwaves because they have very short wavelengths, and phonons because a quantum description is required. (Microwave phonons, sonics, ultrasonics and acoustics are used more or less interchangeably.)

In the above definitions it has seemed desirable to emphasize the similarities between microwave phonons and the more familiar ultrasonics. From the experimental point of view there are some striking differences, which is one of the reasons why microwave phonons is a separate field. If one takes an ultrasonic wave propagating through an electrically insulating solid and steadily increases its frequency, it is usually found that the attenuation rises so much that, by extrapolation, virtually nothing would get through a reasonable thickness of material at frequencies near 10^9 Hz. However, if the solid is cooled it is found that the attenuation decreases, and in many substances there is hardly any attenuation at helium temperatures, even though there may be complete absorption at room temperature. So, experiments which are virtually impossible at room temperature become a real possibility at helium temperatures. Microwave phonons is therefore often regarded as a low-temperature research field. This is not entirely true because one of the purposes of the study is to learn about attenuation mechanisms (e.g. phonon–phonon and spin–phonon scattering) with a view to exploiting the knowledge to make less absorbing solids for room temperature use. Some success has been achieved in this direction.

Another difference arises from the method of generating the waves. The usual way of producing ultrasonic waves is to use a resonant transducer, which typically consists of a piezoelectric crystal which will just contain a whole number of half-wavelengths of the ultrasonic wave, and which is excited electrically. As the frequency increases there is usually no significant decrease in the efficiency of the piezo-electric transduction mechanism, but there is a real problem in that one of the dimensions of the resonant transducer is becoming very small. It can therefore be expected to be mechanically fragile and difficult to use. In the early days of microwave phonons this problem was overcome by using a non-resonant transducer, in which the generator essentially launched an ultrasonic wave into the acoustic equivalent of free space (a non-attenuating crystal) in much the same way that a radar set launches an electromagnetic wave into free space. However, with improvements in thin film technology there has been a return to resonant, but highly loaded, transducers, crystalline or partially crystalline films of cadmium sulphide being commonly used. These films are deposited directly on to a convenient substrate. (Zinc oxide, zinc sulphide and lithium niobate are also used.)

A problem with ultrasonic waves in general is that of detecting them when they are in a medium which is not piezo-electric. The simplest procedure is to pass them into another material, which is piezo-electric, and there detect them using the associated electric fields. This procedure introduces surfaces at which the waves can be reflected and scattered, and again as the frequency increases the wavelength decreases and scattering off small irregularities becomes a relatively more serious problem. So, a good deal of care has to be taken with surface preparation in microwave phonons. It is usual to have surfaces flat to less than the wavelength of sodium light and successive surfaces as parallel as possible. A rather similar difficulty occurs in bulk propagation, for the wavefront typically extends for many thousands of wavelengths in the direction perpendicular to the direction of propagation. As the wave progresses, the front will encounter inhomogeneities and differential attenuation and scattering may occur across it. This can lead to loss of detection, particularly where the detector responds to the mean phase across the wavefront. There is, in consequence, a good deal of interest in detectors which respond to the acoustic power, by exciting transitions between pairs of energy levels. Such transitions are very similar to those induced by microwave electromagnetic waves (with possibly different selection rules), and it is therefore natural to use a quantum description and introduce a "phonon" as the quantum of acoustic energy in analogy with the "photon", the quantum of electromagnetic energy.

Most of the above account has been given with propagation through electrical insulators in mind. The attenuation of microwave ultrasonics in normal metals at low temperatures is quite high and this restricts the range of possible experiments. Even so, the attenuation may be less than that of the corresponding electromagnetic waves and there is therefore an advantage in using phonons if it is necessary to inject quanta deep into a conductor. The lower frequency range of microwave phonons has been used extensively for the study of Fermi surfaces and the electron–phonon interaction. In superconductors microwave phonons show little attenuation if the quantum is less than the energy gap. A variety of superconductor-insulator junctions are being investigated, as possible generators and detectors of microwave phonons. There has been considerable interest in propagation in semiconductors, because of the possibility of transferring energy from the charge carriers into the acoustic wave, via the acoustoelectric effect. The general idea is similar to that of the travelling wave tube amplifier. The carriers are given a drift velocity by the application of an external electric field. If the drift velocity is close to the velocity of the acoustic wave energy transfer can occur, particularly if the medium is piezo-electric. A given charge carrier will appear to experience a time independent electric field, due to the piezo-electricity, against which it either does work or has work done on it. The resulting force on the charge carriers leads to bunching and in general energy can be transferred from the carriers to the acoustic wave. This idea attracted much attention at one time, for it seemed to promise that very considerable amplification could be obtained in a small sample. One of the problems with existing microwave electromagnetic systems is that the size and weight of the system is to some extent determined by the electromagnetic wavelength. The possibility that by converting electromagnetic waves to acoustic waves much smaller geometries could be used was very attractive, for it seemed that in principle most of the manipulations, amplification, detection, mixing, etc., which are commonly performed on electromagnetic waves could also be done with acoustic waves. So far this programme is only in its infancy, and the technological problems which have been encountered suggest that it may be some time before maturity is reached. Also, the use of surfaces waves seems more promising.

A basically similar type of effect occurs in magnetic materials, which can propagate wave motions in the magnetic moments (spin waves). A magneto-elastic coupling between microwave phonons and spin wave modes will allow energy exchange when the propagation velocities are similar. A number of devices have been constructed which rely on this phenomenon. They are not primarily amplifiers of microwave phonons, but variable delay devices. In a number of receiving systems it is desirable to delay an incoming electromagnetic signal by a time of the order of microseconds. An electromagnetic wave travels approximately 300 m in 1 μsec, so one possibility is to put the signal into one end of a 300 m cable and take it out of the other end 1 μsec later. It is attractive to convert the signal into acoustical energy, before delaying. Then the length needed for the delay is reduced by a factor of approximately

10^{-5}, to approximately 3 mm. If the acoustic delay line also has suitable magnetic properties, variable delay can be achieved by means of an externally applied magnetic field.

We have emphasized the connection between electromagnetic waves and microwave phonons, which is obvious enough when one considers electromagnetic/sonic transducers. There is, however, another connection, which involves incoherent waves. Black-body electromagnetic waves are generally regarded as random in frequency, amplitude and phase, and are emitted by hot sources according to the laws of black-body radiation, with some allowances for the possibility that they may not be entirely black-body radiators, but grey bodies. A hot source in a material medium also emits black- (or grey-) body acoustic radiation. In fact, in electrically insulating materials the heat energy transported by the thermal conductance process travels as sonic waves—black-body acoustic radiation. (In electrical conductors the bulk of the heat is transported by charge carriers.) In electrical insulators at room temperature the heat is carried mainly at frequencies near 10^{13} Hz. But, in the helium temperature range the dominant frequency for transport falls to approximately 10^{11} Hz. There is no essential difference between these thermally generated waves and the microwave phonons obtained from transducers. It is just that they have a wider frequency spread and are very much less coherent. Indeed, if one can work with a wide spectrum of incoherent waves, as distinct from a line spectrum of coherent oscillations, thermal sources are much simpler and cheaper to use. (It is perhaps relevant to note that until the advent of the laser practically all optical spectroscopy had used incoherent sources!) So far the analogues of diffraction gratings, lenses, etc., have not been developed for these very high-frequency sonic waves. Nevertheless, the understanding of the attenuation mechanisms of microwave phonons has already been used to modify materials and so alter their thermal conduction properties at room temperature.

We have not mentioned propagation in liquids and gases as these are relatively unexplored fields. No propagation is expected in gases. Propagation in liquid helium is an interesting field and laser radiation in ordinary liquids may convert to acoustic waves, of short mean free path. Nor have we discussed surface waves, because at present the frequencies in use are below those associated with microwave phonons. Surface waves are, however, sometimes described as "microwaves" (microwave ultrasonics).

Few comprehensive reviews have been written, but a full discussion will be found in *Microwave Ultrasonics in Solid State Physics* by J. W. Tucker and V. W. Rampton, published by North-Holland Press, 1972.

A comprehensive list of references is to be found in *A Bibliography of Microwave Ultrasonics* by A. B. Smith and R. W. Damon, *I.E.E.E. Transactions on Sonics and Ultrasonics*, SU-17, 86–111 (1970), and a number of specialized reviews are to be found in the series "Physical Acoustics", edited by W. P. Mason and R. N. Thurston, particularly in volumes IVa, V and VIII.

K. W. H. STEVENS

MIRROR ELECTRON MICROSCOPE. A special type of electron microscope used for the direct observation of the form of the microfields of a solid surface. The mirror electron microscope is unique in that the specimen is neither struck by electrons nor does it emit them. The electron beam, whose axis in the specimen region is normal to the specimen, is reflected at equipotential surfaces very close to the physical surface of the specimen. This is achieved by holding the specimen at a small negative voltage relative to the cathode from which the beam originates. While reversing their direction of motion, the electrons are travelling very slowly and therefore are strongly influenced by the surface fields to which they are exposed. The form of the surface microfields depends on the geometric shape and the electric and magnetic properties of the surface. The reflected beam carries back information about these microfields. The interest of the technique derives from its ability to depict the form of fields directly and from the fact that the specimen remains largely undisturbed by the electron beam. Though the principle of the technique of mirror electron microscopy was discovered in Germany before the Second World War, it is only since 1955 that any serious attempt has been made to exploit its possibilities.

In the mirror electron microscope, the incident and reflected electron beams can either share a common axis over the whole of their length, in which case one or more electron optical elements act simultaneously on both forward and reverse beams, or the beams may be separated by, for example, a magnetic prism. Even in a separated beam system, however, it is impracticable to depart from axial symmetry in the immediate vicinity of the specimen. Thus the object stage of the microscope must contain components which are common to both incident and reflected beams. Image formation is therefore quite different from that in a conventional transmission electron microscope, and in fact the resolution of most instruments is no better than that of the optical microscope (2000 Å). The magnification range is usually about 200× to 4000×. However, it is uniquely sensitive to surface field variations and for this reason has been applied to a variety of problems, e.g. observation of magnetic domain patterns, piezoelectric, ferroelectric and photoconductive surfaces and in semiconductor surface physics.

The low resolution of most instruments is due to the shadow projection mode of imaging. This also gives rise to characteristic distortions which can make interpretation of the micrographs difficult. Attention has been given recently to a more satisfactory form of imaging in which there is proper point-to-point correlation. This offers the prospect of improved resolution (300 Å) and greater ease of interpretation; the sensitivity is, however, lower than in a shadow projection instrument. Details of the various designs of instrument will be found in the references quoted below.

The mirror electron microscope has not as yet achieved any commercial success, and most instruments are to be found in universities and research institutes.

Bibliography

BOK, A. B. *et al.* (1971) *Advances in Optical and Electron Microscopy.* Ed. R. Barer and V. E. Cosslett, Vol. 4, pp. 161–261.

HAWKES, P. W. (1972) *Electron Optics and Electron Microscopy*, pp. 190. Taylor and Francis.

M. E. BARNETT

MODULAR ELECTRONIC EQUIPMENT

Introduction

A long-accepted practice of electronic engineers has been to break down large electronic measurement and control systems into smaller self-contained units for ease of specification, development, production, system fault-diagnostics and general physical convenience. The concept is recent (1955), however, of designing these units as unifunctional modules that physically and operationally are compatible and interchangeable and can be used therefore as common parts to provide the basis of both simple and complex systems. The widespread introduction of transistors enabled the first effective modular schemes to become a practical reality (1960). Since then the art of modularity has been taken to the more advanced levels of data-processing, such that, using integrated circuits, hardware modularity has begun (1970) to have strong interactions with the modularity of software, or at least its transportability between computer-based systems.

The instrumentation associated with the nuclear sciences has, more perhaps than most others, the opportunity and need to take advantage of the changing operational practices created by the data processing revolution of the 1970s. This progressive situation has been assisted by the introduction of the modular concept at a very early stage in the first place, and secondly, by a continuous development of the concept for real-time computer-aided measurement and control of systems.

The Modular Concept

As in many other areas (building construction, furniture, etc.) the term "modular" was applied in the electronics world initially to equipment units whose physical sizes were, in one respect or another, multiples of a basic dimension and any selection of these modular units could be combined in a standard framework for either operational convenience, economy of costs or just general tidiness. A classical example is the 19-in. international standard instrument-panel and rack (I.E.C. publication 297). Rectangular panels, all of horizontal dimensions 19-in. but of height $(n_0 U)$†, can be fitted across the front aperture of the rack in any order and one above the other. Boxes of given maximum horizontal and vertical dimensions can be fitted to and behind these front-panels to contain the electronic circuit components. Thus a range of modular 19-in. front-panel units may be built and inserted in these "Post Office" racks—so called because of their first recorded use in the British Post Office.

However, the modern concept of "modular" now describes units which are not only physically compatible but also operationally compatible and in this respect the term is also applied to computer programs. This concept began to develop during the late 1950s/early 1960s, when transistors made possible the design and construction of relatively small and unifunctional modules which could be fitted into a framework contained within a 19-in. front-panel unit.

With a wide selection of these unifunctional modules (amplifiers, filters, discriminators, registers, etc.) to choose from, practically any size or variety of measurement/control system could be assembled in these frameworks, in racks (Bisby, 1965; Costrell, 1970; Fabre and Viguie, 1955).

Such a framework (shelf, bin, crate, drawer) (Fig. 1) provides a volume whose height is less than $n_0 U$ (to allow for horizontal tie-bars), whose width (W) is less than the clearance in the standard 19-in.

Fig. 1. Typical CAMAC crate and modules. (Neg. H. 161,861.)

---

† $U = 1 \cdot 75$ inches.

rack (to allow for vertical tie-bars) and whose in-rack depth can be any dimension limited only by what is physically reasonable. This volume (V) can be subdivided into equal subvolumes of $V/(n_h n_d n_w)$, where n_h, n_d, n_w are the dividing factors, and each subvolume can be occupied by an electronic unit. However, for ease of access, etc., these subvolumes are usually created by subdivision of the horizontal dimension (W) only, and the modules thereby have front aspects of horizontal width W/n_w, height $n_0 U$ and any depth (D). For any one self-compatible set of modules, the values of W, D, n_0 and n_w have to be fixed and a module of one set will be physically incompatible with a module of another set if any one of the corresponding parameters are different in the two sets, providing that the corresponding parameters are not related factorially.

Physical compatibility is therefore an essential requirement of any modular range of equipment and other characteristics must also be compatible if the full modular concept is to apply. Just as the selected values of W, D, n_0, n_w specify the basis of physical compatibility, compatibility rules must apply to these other characteristics and all equipment units must satisfy them. Furthermore, the rules must not be modified, except in such a way that early units remain compatible with later units.

The following aspects of compatibility are listed in order of the diminishing frequency of their occurrence in the many modular schemes now available. A fully comprehensive modular-unit scheme will possess specifications or rules for all these characteristics:

Mechanical specification

This defines the physical dimensions of the subunits and the framework into which these units can be assembled to form systems. Every modular scheme has this level of compatibility. For plug-in modular units, this specification will also include the physical location and features of electrical mating parts, the guidance and fixing of modules into the framework.

Power supply specification

This defines the voltage, ripple, impedance and other characteristics of the source of power which a unit can expect to find available to it in a system. It implies the allocation of pins, on mating connectors of plug-in units, to specific power lines and also the means of interconnection. Given mechanical and power supply compatibility, any unit will not only fit with any other unit, but also be operational when the power is on.

Signals and connectors specification

This aspect of compatibility assumes particular importance in a modular scheme. By subdividing a total system into its specific functions, as subunits, these physically separate functions need to be interconnected in a much more standardized way than would be the case if they were designed as an integral part of a purpose-designed system. The analogue and digital signal outputs from one unit should drive inputs of another unit without the need for buffering and via an interconnecting cable or network having defined characteristics. Otherwise the flexibility of a modular system can be much degraded. Not all modular schemes possess this level of compatibility and therefore interconnections must be done on each assembly of units by *ad hoc* and highly specific buffering and cabling.

Data transfer specification

When data, in digital signal form, have to be transferred from one unit to another in an assembly of units, there must be defined procedures for doing this, if the highest degree of compatibility is to be achieved (Bisby, Apr. 1971, Dec. 1971).

Ergonomic specification

The front-panel of most modular units is an operational interface between the operator and the measurement/control system. The panels are fitted with input/output connectors, control knobs, switches, display lamps/meters/indicators, etc. Each of such items needs to be labelled as to its function, and positioned for easy, infallible operation or viewing. This can be achieved and an assembly of units can have a good appearance by defining a glossary of terms and abbreviations and by specific rules covering front-panel layout. The glossary prevents the use of the same word on two different units to mean two different things, or two different words on different units to mean the same thing. It completely eliminates the jargon peculiar to an individual designer. Good layout rules prevent interconnecting cables impeding displays or controls, which thereby have good visual and manual access.

A classification of modular unit schemes according to the levels of compatibility achieved can be a guide to the degree to which a particular scheme can compare with a comprehensively compatible scheme in claiming the total advantages and disadvantages, given briefly below.

Advantages

1. Flexibility in system size and configuration.
2. Expansion and up-dating of systems at marginal costs.
3. Rapid replacement of faulty units leading to short down-times.
4. Reduction of stock-holding of equipment and spares thus providing high utilization of capital investment.

5. Rationalization of documentation and test procedures giving low-cost diagnostics and maintenance.
6. Quality control over batch production without substantially increased costs.
7. Addition of new facilities at low development costs.
8. Release of design effort for more appropriate tasks.
9. Rapid evaluation of possible solutions to a particular measurement or control problem.
10. Interchange of operating and technical/maintenance staff.

Disadvantages

1. Costly and prolonged development of the compatibility rules for a scheme.
2. Extreme difficulty of maintaining or modifying the compatibility rules as particular technologies change or develop.
3. Additional cost and sources of unreliability for features which, though relevant in the total applications context, may not be necessary for a specific application.
4. Acceptance of a particular scheme involves commitment to that scheme thereafter.

Clearly, if a scheme has mechanical compatibility only, that scheme may not lay claim to many of the advantages nor suffer the disadvantages 1 and 3. On the other hand, a scheme with mechanical, power and signal compatibility could claim advantages 1 to 9 inclusive and have disadvantages 1 to 4. The extent to which the advantages and disadvantages apply, of course, depends on whether they are being considered by a user of equipment or a supplier/manufacturer of equipment.

Bibliography

BISBY, H. (Mar. 1965) The design principles and role of comprehensive unit system of electronic equipment with particular reference to the Harwell 2000 System, *J.I.E.R.E.* **29**, 3, pp. 185–195.
BISBY, H. (April 1971) Modular instrumentation in nuclear applications, *Nuclear Engineering International*.
BISBY, H. (Dec. 1971) The CAMAC interface and some applications, *Radio and Electronic Engineer*, **41**, 527–37.
COSTRELL, L. (Oct. 1970) Development and current status of the standard NIM system, N.B.S. Technical Note 556. Washington: U.S. Government Printing Office, DC. (1955) 20402.
FABRE, R. and VIGUIE, R. (1955) *L'Onde Electrique*, **35**, (344), 1108–15.
I.E.C. Publication 297.

H. BISBY

MODULATION SPECTROSCOPY

Abstract

The definition, scope, general techniques, and objectives of modulation spectroscopy are discussed. An outline of the theory and experimental techniques are given for wavelength modulation spectroscopy, wavelength derivative spectroscopy, spatial modulation, and modulation by temperature, hydrostatic pressure, uniaxial stress, electric and magnetic fields, secondary light beams, and electrochemical deposition of thin films, among others.

I. Introduction

Since the development of a quantum theory of atoms fifty years ago, electromagnetic radiation has become a standard probe of quantized electron energy levels in all phases of matter. Optical spectroscopy deals specifically with the measurement of the interaction of electromagnetic radiation with matter, via phenomena such as reflection, absorption, luminescence, photoconduction, Raman scat-

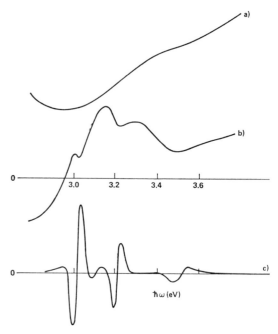

Fig. 1. Spectra illustrating the structural enhancement obtainable with modulation techniques. (a) Direct reflectance spectrum of crystalline Ge from 2·8 to 3·8 eV; (b) first derivative of (a), obtained by means of wavelength-derivative reflectance; (c) third derivative of (a), obtained by means of electro-reflectance [spectra in (a) and (b) after Sell and Kane, (1972), spectrum in (c) after Aspnes (1972)].

tering, etc., and the theoretical interpretation of the optical spectra thus obtained in terms of the electronic properties of materials. Some transitions, especially in complex many-atom systems such as large molecules or solids, are difficult to observe and identify in optical spectra, either because they occur with low probability or because they only produce weak singularities (see Fig. 1). This situation should be contrasted to that which occurs in isolated atoms, where optical absorption spectra consist essentially of sharp lines. Modulation spectroscopy attacks the problems associated with the optical spectroscopy of many-electron systems by measuring and interpreting *changes* in optical spectra resulting from various perturbations applied to the material under study. The simplest aspect of modulation spectroscopy, the use of derivative techniques to enhance weak structure and improve resolution, is not a recent development. The novel feature was introduced in a paper (Seraphin, 1964) that described the effect of an electric field on the reflectivity of a crystalline solid. Seraphin showed that the modulation of some property of the specimen yielded spectra containing information beyond that obtainable from simple derivative techniques, and that the measurement of these spectra by reflection was feasible. These results stimulated the rapid growth of modulation techniques in the latter half of the 1960s, which was fueled by a convergence of interests between high-resolution spectroscopy and band structure calculations in solid-state physics, and also by the commercial availability of relatively inexpensive phase-sensitive detectors. An enormous amount of experimental and theoretical material on modulation spectroscopy is now available in the literature, much of which is discussed in the general references at the end of this article.

The ability of modulation techniques to enhance weak structure in optical spectra has resulted in their recent extensive development and application in solid-state physics. Consequently, we shall emphasize this aspect most heavily. Applications to atomic systems have been discussed by Gordy (Gordy, 1967). As currently, used, the term modulation spectroscopy carries the connotation of perturbation-induced relative changes, $\Delta I/I$, in the intensity, I, of a reflected or transmitted monochromatic beam of light which are small, typically of the order of a few per cent down to experimental detection limits of parts-per-million. To achieve the necessary signal-to-noise ratios, some form of data processing or signal averaging is mandatory, and this usually takes the form of a modulation of the reflectance, R, or transmittance, T, of a sample (which may be a solid, liquid, or gas) by a periodic application and removal of some form of perturbation. The resultant periodic, perturbation-induced change in light intensity reflected or transmitted by the sample is phase-sensitively detected and displayed as a function of wavelength or energy of the monochromatic incident beam of light. The fact that the perturbation-induced changes are small permits the standard optical reflection and transmission equations to be expanded to first order to relate the experimentally-measured relative changes, $\Delta R/R$ or $\Delta T/T$, in the reflectance or transmittance of the sample to the theoretically-calculated perturbation-induced change, $\Delta \varepsilon$, in its complex dielectric function, ε. This provides a simple and direct interface between experiment and theory.

A particularly attractive and powerful feature of modulation spectroscopy is the great flexibility afforded by the wide range of perturbations which may be used and the consequent wide variety of physical interactions which may be studied, either for their own intrinsic interest or as spectroscopic tools. Some of the more common perturbations, with the generic terms which describe them when used in the reflection mode, include modulation by temperature variations (thermoreflectance), modulation by a secondary or pump light beam (photoreflectance), modulation by an electric field (electroreflectance or ER), modulation by a magnetic field (magnetoreflectance or MR), and modulation by an applied stress (piezoreflectance or PR). The first is an example of a scalar or isotropic modulation. The latter three are examples of vector and tensor perturbations which introduce preferred directions in the perturbed system and therefore generally lower its symmetry. Two other (scalar) techniques commonly used are termed wavelength modulation spectroscopy (WMS; wavelength-modulated reflectance or WMR in the reflection mode), where the wavelength of the monochromatic probe beam is itself modulated about some centre value, and wavelength derivative spectroscopy (WDS; wavelength-derivative reflectance or WDR in the reflection mode), where the relative reflectance is measured to high accuracy and the resulting spectrum differentiated numerically with respect to wavelength. Neither WMS nor WDS involve the direct perturbation of the sample itself, but both techniques employ the same general experimental approaches and theoretical analyses as the direct perturbation methods. Other modulation techniques which have been used include spatial modulation, a form of comparative or double-beam spectroscopy where perturbed and reference samples are periodically interchanged in a measurement system, modulation by electron beams or electric currents, and thin-film modulation, where a third, thin-film phase is periodically introduced and removed between a substrate and ambient. The latter technique is particularly useful in studying the optical properties of the film itself, and has seen considerable recent development in the field of electrochemistry for the study of adsorbed species which are stable only in a specific chemical environment or in limited ranges of interface potential. It is also possible to apply one steady-state perturbation to a sample while modulating with another,

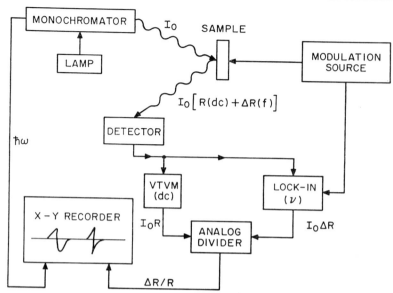

Fig. 2. Block diagram of a typical modulation spectroscopy system, shown operating in the reflectance mode.

in order to enhance certain transitions or to achieve other specific objectives such as symmetry analysis or studies of energy band nonparabolicities in solids. Magnetoelectroreflectance and piezoelectroreflectance are two examples of combined steady-state and modulated perturbation techniques.

II. *Experimental*

Regardless of the perturbation technique involved, the experimental configuration outlined in Fig. 2 is used in nearly all modulation spectroscopy measurements (Cardona, 1969). The optical components shown in the upper left are the same as those used in any optical spectroscopy system. The lamp and monochromator provide a tunable, quasimonochromatic beam of light of intensity I_0, which is reflected or transmitted by the perturbed sample and converted into an electrical signal by the detector. Useful sources include Nernst glowers or Globars for wavelengths in excess of several microns, quartz-halogen tungsten-filament lamps for the near-IR through the visible, Xe arc lamps from 400 to 200 mm (3 to 6 eV), and deuterium arc discharge lamps at higher energies into the near vacuum UV. Either prism or grating monochromators may be used. Detectors include PbS or PbSe photoconductors in the near IR, silicon photodiodes or photomultipliers with GaAs cathodes from the near IR through the visible, and bialkali or trialkali photomultipliers at higher energies. The use of mirror optics avoids defocusing problems caused by dispersion in the index of refraction of lenses for spectral measurements made over wide ranges in photon energy.

The modulation source in Fig. 2 consists of a frequency reference at some frequency, v (typically in the low audio range), and a transducer to apply the desired perturbation to the sample. The frequency reference provides the phase reference for the system. Transducer details depend on the perturbation applied and will be discussed in conjunction with specific techniques. The periodic perturbation of the sample causes a periodic variation, $\Delta I = I_0 \Delta R$, at the frequency v, in the intensity of the reflected light reaching the detector. The detector output voltage or current therefore consists of an a.c. component proportional to $\Delta I = I_0 \Delta R$ and having a fixed phase relationship with respect to the frequency reference, a d.c. component proportional to the total average reflected intensity $I = I_0 R$, and noise.

The demodulation system at the bottom of Fig. 2 consists of the phase-sensitive detector or lock-in amplifier tuned to the modulation frequency v, a divider, and a recorder. The phase-sensitive detector synchronously rectifies the detector output in phase with the frequency reference, and averages the resulting output signal to extract the a.c. component proportional to $I_0 \Delta R$ in the detector output. The noise component bears no fixed phase relationship to the frequency reference and hence averages to zero. Division of the demodulated signal by $I_0 R$ to obtain $\Delta R/R$ may be done in a number of ways depending on the detector used. The simplest method is that commonly used with photomultiplier tubes. The average anode voltage is compared to a reference voltage in a feedback loop which varies the cathode potential, and therefore the photomultiplier gain, in order to maintain the average anode voltage constant. Thus the signal corresponding to $I_0 R$ is held fixed, and the phase-sensitive detector yields directly an output proportional to $\Delta R/R$. In silicon detectors with fixed gain but with negligible values of dark current, the d.c. output current is proportional to $I_0 R$ and the division can be performed electrically by means of analogue dividers or logarithmic converters, or the lamp current may be controlled by feedback techniques. In PbS and PbSe cells the gain is fixed and the dark current is often not negligible. In this case, the combination of an optical chopper between source and monochromator and a second phase-sensitive detector operating synchronously with the chopper (at a frequency other than the modulation frequency) may be necessary to extract a signal proportional to $I_0 R$. Note that with any of the above methods, it is not necessary to measure absolute intensities. The spectral dependences of lamp output, monochromator throughput, atmospheric attenuation and detector response cancel in division, hence the optical design of a modulation spectrometer is much simpler than that of an absolute instrument. The above discussion for the reflection mode also applies to modulated transmission and other modulation measurements.

When some care is used in regulating light sources to minimize lamp flicker, and in securely mounting optical components to minimize vibrational effects, it is possible to achieve a signal-to-noise ratio in $\Delta R/R$ (or $\Delta T/T$) which approaches the shot-noise limit. This is given approximately by $\Delta R/R \cong N^{-1/2}$, where N is the total number of photons to which the detector will respond in a period equal to the time constant of the signal averaging system. As an example, a system with a 10^{-6} watt flux of 2 eV photons impinging on a detector of quantum efficiency of 0.2 will result in

$$(10^{-6} \text{ watt}/2 \text{ eV})(\text{eV}/1.6 \times 10^{-19} \text{ joule})(0.2)$$
$$\cong 10^{12} \text{ events/sec}$$

being registered, whence a detectability limit of $\Delta R/R \sim 10^{-6}$ can theoretically be achieved for a 1 sec averaging time. Signal-to-noise ratios of the order of 5×10^{-7} have been reported.

III. Scalar Perturbations

Scalar perturbations are isotropic, introducing no preferred directions in a sample. Thus, modulation spectra obtained with scalar perturbations have the symmetry of the unperturbed material. This class of perturbations includes wavelength-modulation and wavelength-derivative spectroscopy, spatial or substitutional modulation, modulation by hydrostatic pressure, temperature modulation, and thin-film modulation. Scalar modulation spectra, except for spatial, substitutional, or thin-film modulation, are generally related to the first derivative of the reflectance, as shown in Fig. 1b. While not as powerful as vector perturbations for analytic purposes, scalar perturbations have the distinct advantage of being applicable to virtually any type of sample.

(a) General theory

The experimentally-measured modulated relative intensity change, $\Delta I/I$, in the reflected or transmitted light is determined by the perturbation-induced change, $\Delta \hat{\varepsilon} = \Delta \varepsilon_1 + i \Delta \varepsilon_2$, in the complex dielectric function, $\hat{\varepsilon} = \varepsilon_1 + i \varepsilon_2$, of the sample through the optical properties of the system. For simplicity, let the sample be isotropic so $\hat{\varepsilon}$ is a scalar. Although the detailed form of $\Delta \hat{\varepsilon}$ depends on the perturbation, modulation spectra are dominant in spectral regions containing singularities in $\hat{\varepsilon}(E)$, where $E = \hbar \omega$ is the energy of the incident light. These singularities can be approximated by the functional form

$$\hat{\varepsilon}(E) = F(E) + G(E)(E - E_g + i\Gamma)^\eta, \quad (3.1)$$

where F and G are slowly varying complex functions of E. For a weak singularity such as a three-dimensional critical point in the joint density of states of a crystalline solid for direct allowed transitions, $\eta = \frac{1}{2}$, E_g is the critical point or threshold energy, and Γ is the lifetime broadening parameter. For a strong singularity such as an exciton or atomic absorption line, $\eta = -1$, E_g is the transition energy, and the full-width half-maximum energy is 2Γ.

For a transmission experiment, in the simplest approximation where reflection losses at the surface and multiple-reflection effects are neglected, $I \cong I_0 \exp(-\alpha d)$ where I_0 is the incident intensity, d is the sample thickness, and $\alpha = \text{Im}(4\pi \hat{n}/\lambda)$ is the absorption coefficient. Here, λ is the wavelength of light in vacuum, and $\hat{n}$ is the complex index of refraction of the sample ($\hat{n}^2 = \hat{\varepsilon}$). If $\hat{\varepsilon} \to \hat{\varepsilon} + \Delta \hat{\varepsilon}$ as a result of a scalar perturbation then

$$\Delta I/I = \Delta T/T \cong \text{Im}\{(2\pi d/\lambda)(\Delta \hat{\varepsilon}/\hat{n})\}. \quad (3.2)$$

Due to the orders-of-magnitude variation of α with wavelength, transmission modulation spectroscopy has seen only limited application in solids, usually near fundamental absorption edges where $\hat{n}$ is essentially real and $\Delta I/I$ arises from $\Delta \varepsilon_2$. Modulated reflectance measurements are more important. Here $\Delta I/I$ and $\Delta \hat{\varepsilon}$ are related by the Fresnel reflectivity equations, which for normal incidence reduce to $R = \varrho^2$; $\varrho \exp(i\theta) = (\hat{n} - \hat{n}_a)/(\hat{n} + \hat{n}_a)$; $\hat{n}_a$ is the (complex) index of refraction of the ambient (Cardona, 1969). The differential form yields

$$\Delta I/I = \Delta R/R \cong \text{Re}\{2\hat{n}_a \Delta \hat{\varepsilon}/[\hat{n}(\hat{\varepsilon} - \hat{\varepsilon}_a)]\}. \quad (3.3)$$

The complex quantity $\Delta \hat{\varepsilon}$ can usually be obtained directly from modulated reflectance data by means of a Kramers–Kronig transform. The necessary equations for a modulated reflectance spectrum are:

$$\Delta\theta(E) = \frac{E}{\pi} \mathscr{P} \int_0^\infty \frac{dE'}{(E'^2 - E^2)} \left[\frac{\Delta R(E')}{R(E')} - \frac{\Delta R(E)}{R(E)} \right];$$
(3.4a)

$$\Delta\hat{\varepsilon} = [\hat{n}(\hat{\varepsilon} - \hat{\varepsilon}_a)/\hat{n}_a] \left[\tfrac{1}{2}(\Delta R/R) + i\Delta\theta \right]. \quad (3.4b)$$

The second term in brackets in eq. (3.4a), $\Delta R(E)/R(E)$, contributes nothing to the value of the integral but aids by removing the singularity at $E' = E$ for computational purposes. $\Delta\hat{\varepsilon}$ can usually be computed more easily and accurately from modulated reflectance data as compared to $\hat{\varepsilon}$ from direct reflectance data since $\Delta R/R$ is a more localized function in energy than R.

(b) *Wavelength-modulation spectroscopy (WMS)*

In this technique, the wavelength of the incident light-beam is modulated about an average value by a transducer in the form of a rotationally-vibrating refractor plate inserted in the monochromator just behind the entrance slit, or by vibrating the slits themselves. This causes the image of the entrance slit to oscillate, which in turn causes the diffracted beam to oscillate relative to the exit slit, resulting in a periodic variation of the wavelength of the output beam. The sample itself remains in the unperturbed condition. In principle, WMS gives rise to a first-derivative modulation spectrum described by

$$\Delta\hat{\varepsilon} = (d\hat{\varepsilon}/dE) \, \Delta(\hbar\omega) \qquad (3.5)$$

where $\Delta(\hbar\omega)$ is the photon energy modulation resulting from the action of the refractor plate. However, the technique is sensitive to variations of intensity with photon energy arising from all sources, including absorption lines in air, spectral emission lines in lamps, prism or grating anomalies, etc., in addition to the desired variation of dielectric function with energy, and thus cannot be used over wide spectral ranges without some form of double-beam compensation. A number of double-beam techniques have been developed to minimize these spurious effects (Batz, 1972; Shen, 1973).

(c) *Wavelength-derivative spectroscopy (WDS)*

In wavelength-derivative spectroscopy, the reflectance or transmittance of a sample is measured to high accuracy relative to a reference, and its derivative as a function of energy is calculated numerically. As in WMS, the sample is not perturbed. The measurement is accomplished by a double-beam comparative technique, where the monochromatic beam is periodically deflected by the transducer. The transducer may be a vibrating mirror or a transparent chopper, which deflects the beam between, for example, a reflecting sample and a reference such as an aluminium mirror. Thus, WDS spectra cannot be described directly in terms of a change in dielectric function in eqns. (3.2) and (3.3), as with WMS, but rather must be expressed as the ratio of two intensities. In a reflectance (WDR) configuration, the spectrum measured is R/R_{ref}, which upon differentiating with respect to energy becomes

$$\frac{d}{dE} \left(\frac{R}{R_{\text{ref}}} \right) = \frac{R}{R_{\text{ref}}} \left[\frac{1}{R} \frac{dR}{dE} - \frac{1}{R_{\text{ref}}} \frac{dR_{\text{ref}}}{dE} \right]. \quad (3.6)$$

If R_{ref} is nearly independent of photon energy, then eqn. (3.6) can be approximated by

$$\frac{d}{dE} \left(\frac{R}{R_{\text{ref}}} \right) \cong \frac{1}{R_{\text{ref}}} \frac{dR}{dE}. \quad (3.7)$$

If this in turn is divided by the original data, R/R_{ref}, the ideal first-derivative WMR spectrum, $R^{-1}(dR/dE)$, is obtained but without the sensitivity to spectral variations in light intensity arising from sources other than the sample itself. Signal-to-noise ratios in WDS spectra are similar to those encountered by direct perturbation methods even though the relative reflection or transmission intensities are determined by different electronic detection techniques (Sell, 1970, 1973). The precision of absolute reflectances determined by the double-beam measurement prior to differentiation depends on the precision to which the reflectance of the reference surface is known. The spectrum in Fig. 1b was taken by WDR measurements on a germanium crystal using an aluminium reference mirror.

(d) *Spatial or substitutional modulation*

In spatial or substitutional modulation, the difference in reflectance or transmittance between two samples which are identical except for a small difference in, for instance, impurity concentration, is measured by periodically interchanging the two samples. Each sample remains in the unperturbed condition. The periodic interchange may be accomplished either by using the WDS configuration described in the preceding paragraph with the second sample replacing the reference mirror, or by using a fixed beam and mounting the samples on a rotating disc so that they alternately intercept the light. The latter technique is restricted primarily to room-temperature measurements. Since both samples typically have similar optical properties, spatial modulation spectra can be described by means of eqns. (3.2) and (3.3) as a difference of first-derivative form:

$$\Delta\hat{\varepsilon} = \hat{\varepsilon} - \hat{\varepsilon}_{\text{ref}} \cong (d\hat{\varepsilon}/dQ) \Delta Q \qquad (3.8)$$

where ΔQ represents the difference between samples in the parameter under study. Spatial modulation has been used to study the effect of grain boundaries and impurity concentrations in the crystals, and the effects of dilute alloying on noble metals (Beaglehole, 1973).

(e) Modulation by hydrostatic pressure

Hydrostatic pressure, P, produces negligible changes in the optical properties of gases, but may produce substantial changes in liquids and solids due to volume changes which influence the overlap of electronic wave functions and therefore broaden absorption lines and shift critical point thresholds. Thus from eqn. (3.1), these changes produce first-derivative spectra described by:

$$\Delta \hat{\varepsilon} \cong \eta G(E)(E - E_g + i\Gamma)^{n-1} \left(-\frac{dE_g}{dP} + i\frac{d\Gamma}{dP} \right) \Delta P. \quad (3.9)$$

In solids, the derivative (dE/dP) dominates. In cubic crystals the shift of the energy gap, $\Delta E = (dE_g/dP)\Delta P$, can be written

$$\Delta E_g = a \operatorname{Tr}(\mathbf{e}) = -aP(S_{11} + 2S_{12}) \quad (3.10)$$

where $\mathbf{e}$ is the second-rank strain tensor, $\operatorname{Tr}(\mathbf{e}) = \Delta V/V$ is the relative volume change of the crystal, and S_{11} and S_{12} are independent components of the fourth-rank compliance tensor, $\mathbf{S}$. The coefficient, a, is a deformation potential which describes how much a specific critical point energy will shift under a volume change, and is typically of the order of 1–10 eV. Critical points have characteristic deformation potentials and therefore respond differently to hydrostatic pressure. Due to the difficulty of periodically modulating pressures in the kilobar range, hydrostatic pressure modulation measurements to date have been performed under steady-state conditions in high-pressure cells, in conjunction with other modulation techniques, or have been obtained as a component of uniaxial stress modulation (Pollak, 1973), as discussed later.

(f) Temperature modulation

Temperature modulation experiments are performed with the sample mounted in partial thermal isolation on a heat sink. Periodic temperature variation can be obtained by resistive heating, or by passing current pulses through the sample itself or through a heater to which the sample is attached. Periodic illumination with intense light beams has also been used. Since heat conduction is a relatively slow process, thermal modulation works best on very thin samples. Epitaxial films on alkali halide substrates lend themselves well to this technique. By eqn. (3.1), it follows that thermal modulation spectra can be described by

$$\Delta \hat{\varepsilon} \cong \eta G(E)(E - E_g + i\Gamma)^{n-1} \left[-\frac{dE_g}{dT} + i\frac{d\Gamma}{dT} \right] \Delta T \quad (3.11)$$

The spectral lineshape is determined by the relative contributions of dE_g/dT and $d\Gamma/dT$, which are usually comparable in magnitude. In metals or heavily-doped semiconductors, the temperature shift of the plasma frequency may also be important. The quantity dE_g/dT is negative in most materials (the lead salts being a notable exception), and is determined by two principal effects: the volume change due to thermal expansion of the lattice, and the shift in effective electron potential due to the fluctuations of the ion cores about their equilibrium positions. The former effect is completely equivalent to modulation by hydrostatic stress, and the latter is described theoretically in terms of Debye–Waller factors. Although thermal modulation yields first-derivative spectra, the extra contributions from potential fluctuations and the lifetime broadening term result in thermal modulation spectra which may appear quite different from those obtained by other first-derivative modulation techniques (Cardona, 1969; Batz, 1972).

(g) Thin-film modulation

The change in reflection or transmission caused by the deposition of a thin film on a substrate can be studied by modulation techniques if the film can be readily deposited and dissolved. Many electrochemical reaction products form adsorbed species which are stable only in limited ranges of the interface potential and fall into this category. The reflection change at normal incidence caused by a thin film of thickness $d \ll \lambda$ and dielectric function $\hat{\varepsilon}_f$ adsorbed on a substrate is given by eqn. (3.3), if

$$\Delta \hat{\varepsilon} = -4\pi i d\hat{n}(\hat{\varepsilon}_f - \hat{\varepsilon}_a)/\lambda. \quad (3.12)$$

Expressions valid for non-normal incidence with the incident light polarized either parallel or perpendicular to the plane of incidence are more complicated (McIntyre, 1973). It is possible to invert the non-normal incidence equations to calculate $\hat{\varepsilon}_f$ in terms of d and the experimentally-measured quantities $(\Delta R/R)_\parallel$ and $(\Delta R/R)_\perp$, the relative reflectance changes parallel and perpendicular, respectively, to the plane of incidence, without performing a Kramers–Kronig analysis. This is important, for in contrast to other modulation techniques, reflectivity changes induced by thin films are not derivative-like and therefore show little tendency to be localized in energy.

Thin-film modulation has been used for studies of thin electrochemically-deposited layers of Ag on Pt, for oxidation-reduction studies of Ni electrodes, and other systems. The same formalism applies to the description of the electroreflectance effect in metals (McIntyre, 1973).

IV. Vector and tensor perturbations

Vector and tensor perturbations contain an inherent preferred direction and therefore differ from scalar perturbations by lowering the symmetry of the system to which they are applied. The most commonly used perturbations of these types are electric and magnetic fields, and uniaxial stress. The resulting perturbation-induced change in the dielectric function is in general tensorial, even in isotropic systems where the unperturbed dielectric function is scalar. This results in the appearance of additional polarization dependences in modulation spectra produced by vector and tensor perturbations, which are useful in correlating the spectra to specific transitions. The necessary modifications to eqns. (3.2) and (3.3) in the special case of an isotropic unperturbed system can be written

$$\Delta T/T \cong \text{Re } \{(2\pi d/\lambda)(e^i e^j \Delta \hat{\varepsilon}^{ij}/\hat{n})\} \quad (4.1)$$

$$\Delta R/R \cong \text{Re } \{2\hat{n}_a e^i e^j \Delta \hat{\varepsilon}^{ij})/[\hat{n}(\hat{\varepsilon} - \varepsilon_a)]\} \quad (4.2)$$

where $\Delta \hat{\varepsilon}^{ij}$ is the ijth component of the second-rank tensor $\Delta \hat{\varepsilon}$, and e^i and e^j are components of the unit polarization vector, $\mathbf{e}$, along the i and j axes. Summation over $i, j = x, y, z$ is implied throughout. The possibility of selecting different relative orientations of the polarization, the perturbation, and the crystallographic axes in solid-state applications makes modulation spectroscopy with vector and tensor perturbations an especially powerful technique for symmetry analysis in the band structure of solids.

(a) *Modulation by uniaxial stress*

Modulation by application of a uniaxial stress is obviously restricted to solids, and is most successful when applied to materials which can undergo relatively large ($>$0.1%) deformations without fracturing. Figure 3 shows some typical arrangements for applying a.c. stress to an optical specimen. The mechanical method shown at the top involves a magnetically- or pneumatically-driven compression frame or a pair of PZT transducers in a yoke configuration. In the other configurations shown at the bottom, the transducers drive the sample directly, either by bonding the transducer to a specimen of appropriate length and operating the system at mechanical resonance, or by evaporating a thin film sample directly onto a polished surface of the transducer. The a.c. techniques are limited to relative deformations of the order of 10^{-5}. The recent trend in piezoreflectance has been to apply a larger d.c. stress as a steady-state perturbation and to rely on a second modulation technique to resolve the stress-induced changes in optical spectra which result (Sell, 1973).

A uniaxial stress is a second-rank tensor which can be represented by the dyadic product $X = \hat{X}\hat{X}X$, where $\hat{X}$ is the unit vector in the stress direction and X is the magnitude of the applied stress. Since the stress influences optical properties by distorting the unperturbed crystal and thereby changing the overlap of atomic wave functions, stress modulation spectra are more closely related to the second-rank strain tensor, $\mathbf{e}$. The strain tensor has six independent components: e^{xx}, e^{yy}, e^{zz}, $e^{xy} = e^{yx}$, $e^{yz} = e^{zy}$, and $e^{zx} = e^{xz}$. The stress and strain tensors are related by the fourth-rank compliance tensor $\mathbf{S}$ according to $e^{ij} = S^{ijkl}X^{kl}$. In cubic crystals, $\mathbf{S}$ has only three

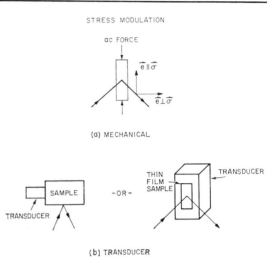

Fig. 3. Diagrams of commonly-used configurations for stress modulation. In the mechanical configuration shown at the top, the sample size is typically of the order of $1 \times 1 \times 10\,mm^3$, and the stress is applied by end loading as shown. In the transducer configuration shown at bottom left, stress modulation is accomplished by operating the transducer-sample configuration at mechanical resonance. At bottom right, the transducer directly stresses the thin-film sample which is either rigidly mounted or deposited directly on the transducer.

nonzero components of the forms $S^{xxxx} = S_{11}$, $S^{xxyy} = S_{12}$, and $S^{xyxy} = S_{44}$. For a given stress, it is advantageous to take linear combinations of the six independent strain tensor components to generate three types of terms, each having an independent deformation potential. The three types and linear combinations are: hydrostatic pressure, $\text{Tr}(\mathbf{e}) = (e^{xx} + e^{yy} + e^{zz})/\sqrt{3}$; tetragonal shear, $(e^{xx} - e^{yy})/\sqrt{2}$ and $(2e^{zz} - e^{xx} - e^{yy})/\sqrt{6}$; and trigonal shear, e^{xy}, e^{yz}, and e^{zx}. The normalization is that employed by Kane (Kane, 1969). Uniaxial stress always results in a nonvanishing hydrostatic pressure term and one or more shear terms.

The dominant effect of the hydrostatic pressure term on the optical properties is to shift threshold energies, $\mathbf{E}_g$, according to a deformation potential as discussed previously. The shear terms act similarly but may differ in sign as well as in the value of a deformation potential, depending on the symmetry of the transitions involved. By utilizing the optical selection rules through polarization measurements, the same structure in modulation spectra may appear to shift to higher or lower energies with stress, if the structure is due to transitions which are equivalent or degenerate in an unperturbed crystal. Symmetry analysis is based on a systematic study of these effects. Degenerate bands, such as the uppermost valence bands of semiconducting materials at the centre of the Brillouin zone, can often be split sufficiently under the action of the shear strain terms to eliminate the warping effect of the degeneracy and to approach simple parabolic behaviour.

If a scalar modulation technique such as WMR or WDR is used to provide the resolution necessary to follow small stress-induced changes in optical spectra under a static uniaxial stress, the polarization dependences are characteristic of the stress. When vectorial modulation techniques such as electroreflectance are combined with stress perturbations, the symmetry analysis must be based on tensorial representations of a higher order.

(b) *Electric field modulation*

Electric field modulation (electroreflectance in the reflection mode; electroabsorptance in the transmission mode) is unique among modulation techniques for a number of reasons. Even though electric-field-induced changes in $\hat{\varepsilon}$ are sufficiently small so that eqns. (4.1) and (4.2) remain valid, electric field modulation exhibits well-defined low-field, intermediate-field, and high-field regions of response, wherein both the lineshape and the functional dependence of $\Delta\hat{\varepsilon}$ changes substantially with field. Moreover, the precise nature of the field interaction itself can take on a variety of forms depending on the sample. In solids, electroreflectance spectra range from being similar to piezoreflectance spectra in ferroelectrics, to loss-function spectra in metals, and to third energy-derivative spectra when taken at sufficiently low fields in semiconductors. Figure 1c gives an example of the latter type of spectrum. Still another type of response, resonant linear electrooptic, is obtained under certain conditions in semiconductors lacking inversion symmetry. In addition, the problem of obtaining uniform fields is not trivial, and the fact that nonuniformities in the perturbing field can seriously distort experimental spectra has led to much confusion in the past. Thus, electromodulation spectra tend to be much more difficult to measure and analyse than those obtained by other forms of modulation spectroscopy. Nevertheless, these spectra are often more highly structured and contain more information than those obtained by other techniques. Consequently, the electroreflectance technique has been applied to insulators, semiconductors, and metals, and also to low-conductivity liquids. The most easily interpreted spectra have been obtained from semiconductors and insulators, where the theory is well known and uniform fields can be obtained without undue difficulty (Aspnes, 1973; Fischer and Aspnes, 1973).

Several configurations are commonly used to obtain electromodulation spectra, as shown in Fig. 4. On insulators ($\varrho > 10^8$ ohm cm), the preferred configuration is the transverse electroreflectance (TER)

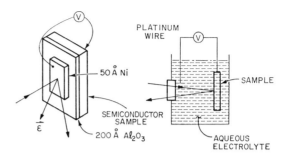

(a) LONGITUDINAL

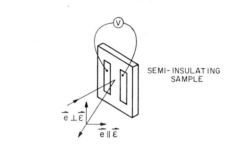

(b) TRANSVERSE

Fig. 4. Diagrams of commonly-used configurations for electric field modulation. In the longitudinal or surface-barrier geometry, the modulating field is perpendicular to the surface. The MOS (field-effect) and electrolyte (surface-barrier) configurations shown at the top illustrate two commonly-used longitudinal techniques. In the transverse geometry shown at the bottom, the modulating field is in the plane of the surface, and is obtained by applying a high voltage between two contacts as shown. The possibility of obtaining spectra with light polarized either parallel or perpendicular to the electric field is indicated explicitly in the transverse geometry.

geometry, where the field is developed in the plane of the surface by applying a high voltage between two contacts evaporated on the surface and separated by a narrow (~1 mm) gap. This technique enables spectra to be taken with the light polarization oriented both parallel and perpendicular to the modulating electric field. Somewhat less general but more versatile is the surface-barrier geometry, where the field is generated perpendicular to the surface by forming a boundary with a transparent, non-ohmic contact. Surface-barrier electroreflectance (SBER) spectra have been obtained by using metal–semiconductor or metal–electrolyte interfaces, *p–n* junctions, heterojunctions, thin-film MOS structures, and field-effect configurations. In semiconductors, it is desirable to modulate to nonequilibrium depletion conditions to obtain spectra which are simplest to analyse.

At low values of the modulation field in insulators and semiconductors, ER spectra can be described by means of third-rank linear and fourth-rank quadratic resonant non-linear optical susceptibility tensors. The linear response occurs only in crystals which lack inversion symmetry. For crystals of zincblende symmetry there is one nonvanishing coefficient, and the linear field-induced change, $\Delta\hat{\varepsilon}_{lJ}$, has the form $\Delta\hat{\varepsilon}_l^{ij} = X^{ijk}\mathscr{E}^k$, where $X^{ijk} = X_{123}$ if $i \neq j \neq k \neq i$. No theory has yet been given for the lineshape. The quadratic field-induced change, $\Delta\hat{\varepsilon}_q^{ij}$, has the form (Aspnes, 1973)

$$\Delta\hat{\varepsilon}_q^{ij}(E) = \frac{(\hbar\Omega)^3}{3E^2} \frac{\partial^3}{\partial E^3} \{E^2 \hat{\varepsilon}^{iJ}(E)\} \quad (4.3)$$

which is the third derivative of the unperturbed dielectric function. The quantity $(\hbar\Omega)$ is called the characteristic energy and is given by

$$(\hbar\Omega)^3 = e^2 \mathscr{E}^k \mathscr{E}^l \hbar^2/(8\mu^{kl}) \quad (4.4)$$

where $(\mu^{kl})^{-1} = \hbar^{-2} \partial^2 E_{cv}/\partial k_i \partial k^k$ is the second-rank interband reduced-mass tensor. The low-field limit applies provided $|\hbar\Omega| < \Gamma/3$. If $\mathscr{E} = 10^5$ V cm^{-1} and $\mu = 0.1\, m_e$, then $\hbar\Omega = 21$ meV. Thus most higher-interband ER spectra where, $\Gamma \sim 50\text{--}100$ meV, can be obtained in this limit.

For excitonic or atomic line spectra, the term "low-field" also means that perturbation theory is applicable. This occurs whenever $e\mathscr{E}a < E_8$, i.e. the field-induced potential drop across the exciton is appreciably less than the binding energy. In this situation, the lineshape is derivative-like and is determined by the competition between the shift of the exciton binding energy with field and the field-induced broadening.

At intermediate fields, perturbation theory breaks down, and the quadratic low-field effect evolves into a more complicated response. The field-induced change in the dielectric function can be represented approximately as a convolution integral over the energy band structure with an Airy function kernel (Aspnes et al., 1968):

$$\Delta\hat{\varepsilon}^{iJ} \cong \frac{4\pi^2 e^2 P_{cv}^i P_{vc}^j \hbar^2}{m^2 E^2} \int_{BZ} d^3k \left\{ \frac{1}{\hbar\Omega} Gi\left(\frac{E - E_g + i\Gamma}{\hbar\Omega}\right) \right.$$
$$\left. + i\frac{1}{|\hbar\Omega|} Ai\left(\frac{E - E_g + i\Gamma}{\hbar\Omega}\right) - \delta(E - E_g + i\Gamma) \right\}. \quad (4.5)$$

Here, P_{cv} is the momentum matrix element and Ai and Gi are Airy functions. The lineshapes are characterized by a large number of slowly-decreasing oscillations called subsidiary or Franz–Keldysh oscillations, which are a direct manifestation of the quantum interference of the one-electron wave functions in the crystal. The period of these oscillations can be used to determine the magnitude of the applied field or the interband reduced mass. In the parabolic band approximation, the intermediate-field response is known as the Franz–Keldysh effect.

For excitons, the intermediate-field region corresponds to a complete dissociation of the unperturbed exciton. Thus modulation spectra of excitons under intermediate-field conditions are simply the negative of the unperturbed exciton lineshape.

At sufficiently high fields, the field-induced potential drop, $e\mathscr{E}a_0$, across a unit cell in the crystal becomes comparable to interband energy separations. The entire energy band structure must then be recalculated to take the field perturbation term into account. High-field effects have been observed as polarization effects in the fundamental edge transitions of semiconductors, where, due to the band degeneracy existing in the unperturbed crystal, any field is literally a high field (Fischer and Bottka, 1970). The polarization behaviour is adequately explained by accounting for the increased volume of momentum space spanned by the field-accelerated electron. For example, the highly symmetric state at the centre of the Brillouin zone takes on the symmetry character of lower symmetry states off the zone centre parallel to the applied field.

Although most of the emphasis in electromodulation spectroscopy has been on interband transitions in crystalline solids, other materials and other types of optical transitions may be profitably studied by this technique. Modulated luminescence in complex molecules has been used to determined excited state dipole moments via the Stark shift. Electroabsorption in the charge transfer complex of 2,4,7-trinitro-q-fluoranone in poly-n-vinylcarbazole offers potential in elucidating the details of the electronic interactions between donor and host molecules (Weiser, 1973). Impurity and defect spectra in semiconductors and insulators also respond to electric field modulation, although the physical mechanisms have not been completely described.

(c) *Magnetic field modulation*

Magnetic field modulation has several advantages over the vectorial techniques previously discussed: magnetic fields may be applied to any system, regardless of its state of phase or mechanical or electrical characteristics, field uniformity is not difficult to achieve, and dramatic sharpening of the lower absorption thresholds in solids into a series of discrete Landau levels enables highly precise values of effective masses and absorption thresholds to be obtained. Zeeman effects on discrete line spectra also enable the atomic symmetry of the atomic or excitonic spectra to be determined. In general, Zeeman shifts are sufficiently strong that modulation techniques are not necessary for their measurement. On the other hand, magnetic field modulation is not well suited to studies of higher interband transitions because excited-state lifetimes are too short to permit strong magnetic field effects to be observed with currently-available techniques, although some success has recently been achieved by combining magnetic field modulation with circular-polarization measurements in metals. Also, it is difficult to modulate intense magnetic fields directly, and the information content in small modulations about large average values is limited. Consequently, magnetic fields, as with uniaxial stress measurements, are commonly used as a static perturbation, with the actual resolution of the resulting structure being performed by other techniques such as electroreflectance (Aggarwal, 1973).

The square-root density of states of solids near the fundamental absorption edge breaks up into a series of Landau levels in a strong magnetic field. These levels have the form of a summation of one-dimensional density-of-states functions in the effective-mass approximation:

$$\varepsilon^{ij}(E, H) = \frac{e^2 P_{cv}^i P_{vc}^j \hbar \omega_c}{m^2 E^2} \sum_{n=1}^{\infty} (E - E_n + i\Gamma)^{-1/2} \quad (4.6)$$

where $\omega_c = \dfrac{eH}{\mu c}$ is the cyclotron frequency, and

$$E_n = E_g + (n + \tfrac{1}{2}) \hbar \omega_c.$$

For nonparabolic bands, the interband reduced mass, μ, increases and the separation between the Landau levels decreases as the energy increases. Magnetoreflectance measurements provide a unique opportunity for mapping the nonparabolicity of the low lying bands.

The change in dielectric function with magnetic field is usually quite large whenever $\hbar\omega_c > \Gamma$, so that it is more accurate not to consider the change in $\hat{\varepsilon}$ as a small quantity, $\Delta\hat{\varepsilon}$, but rather to consider the magnetically-perturbed crystal as having a new dielectric function $\hat{\varepsilon}'$, with $\Delta\hat{\varepsilon}$ taking the meaning of a small change in $\hat{\varepsilon}'$ measured by means of a secondary technique such as WDR, WMR, or ER. If ER is used, care must be taken to distinguish the case of parallel fields, where eqn. (4.6) applies, from the crossed-field configuration, where all transitions shift to lower energy by an amount $\Delta E = (m_c + m_v) \varepsilon^2 c^2 / 2H^2$, and the selection rule $\Delta n = 0$ for the pure magnetic field case breaks down to allow valence-conduction transitions between Landau levels of different n. The resulting magneto-electroreflectance spectra are exceedingly rich in structure and difficult to analyse (Aggarwal, 1972).

IV. Symmetry analysis

The major difficulty in the direct spectroscopy of solids is the tentative and uncertain procedure by which structure is correlated with transitions from specific regions in the Brillouin zone. Vector and tensor modulation techniques have largely alleviated this problem by allowing the transition symmetry

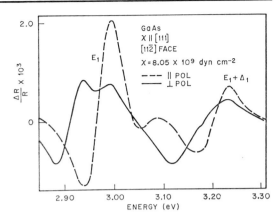

Fig. 5. Piezoreflectance spectrum of the E_1 and $E_1 + \Delta_1$ transitions of GaAs for unmodulated stress along the [111] direction. The spectra show the difference obtained for light polarized parallel and perpendicular to the stress axis, as indicated in the figure. A surface-barrier electrolyte electroreflectance configuration was used in this measurement [after Pollak and Cardona (1968)].

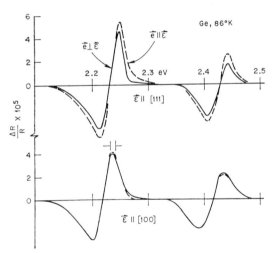

Fig. 6. Transverse electroreflectance spectrum of the E_1 and $E_1 + \Delta_1$ transitions of Ge. Upper curves: field parallel to [111]; lower curves: field parallel to [100]. The spectra correspond to light polarized parallel and perpendicular to the applied field as indicated in the figure [after Fischer and Bottka (1970)].

to be determined via perturbation-induced optical anisotropies.

As an example, we consider the doublet known as E_1, $E_1 + \Delta_1$ which occurs in all semiconductors having diamond or zincblende crystal structures. This doublet arises from transitions near the L point in the Brillouin zone; that is, a set of eight equivalent points with axes of $\langle 111 \rangle$ symmetry are involved. A uniaxial stress or an electric field parallel to any $\langle 111 \rangle$ axis removes the eight-fold degeneracy; two of the eight are parallel to the perturbation axis, while the other six are intersected at an angle of 70°32'. This yields a stress-induced doublet for appropriately polarized light, which is revealed in electroreflectance (Fig. 5). A perturbation along the [100] direction does not produce splitting because all $\langle 111 \rangle$ axes remain equivalent under the perturbation. Similarly, the electric field vector itself can be used for symmetry diagnostics in the transverse configuration; optical selection rules coupled with removal of degeneracies lead to polarization-dependent spectra (Fig. 6) which clearly indicate the $\langle 111 \rangle$ symmetry of this pair of transitions.

V. Other perturbations

(a) Modulation by secondary light beams

An intense beam of light may also modify optical properties. In the reflection mode, modulation by a secondary light beam is known as photoreflectance. Several different physical mechanisms may be involved in the modulation process. The excess free carriers resulting from excitation of electron-hole pairs across the fundamental energy gap may screen excitons, broadening sharp absorption lines and decreasing the exciton binding energy, resulting in a broadening and shift of the exciton line to higher energy. This results in a typical first-derivative change in $\bar{\varepsilon}$, similar to other scalar perturbations, except that the broadening and energy shifts are now related to screening parameters of the line itself. It is theoretically possible to optically pump a solid to the extent that the bands become degenerate and absorption edges shift to higher energy (Burstein shift), or to initiate optical transitions from an originally nondegenerate conduction band minimum. Alternatively, the optically-excited carriers may screen existing surface electric fields arising from specific distributions of surface states and/or surface traps, and thus modulate the optical properties via an electric field mechanism. High photon fluxes may also cause thermoreflectance effects. Generally, a photomodulation spectrum contains contributions from several of the above. Since there are less ambiguous methods of investigating the effects of the various mechanisms described above than by exciting them with a light beam, photoreflectance has not been widely used.

(b) Modulation of carrier concentration (solids)

Optical properties of degenerate bands differ from those of equivalent but nondegenerate bands due to the Pauli exclusion principle. Transitions normally allowed between valence and conduction bands may not take place if either band becomes degenerate; interconduction band transitions not possible in the absence of conduction band degeneracies may appear if the conduction band becomes degenerate. If the location of the band extrema are known, the observation of this type of modulation spectrum provides a powerful alternative method of symmetry analysis in crystals. The most conclusive observations of this type have occurred with the onset of interconduction band transitions where degeneracy was achieved by biasing the semiconductor strongly positive in the surface region in a surface-barrier electroreflectance configuration. Spectra obtained in this way are also influenced by electroreflectance effects. Substitutional or spatial modulation between nondegenerate and degenerate samples has also been used to study these effects.

(c) Modulation by current flow

A steady-state current applied to a metal or semiconductor can be represented as a redistribution of filled electron states in momentum space. In principle, this redistribution can be observed optically as a distortion of the shape of the absorption edge, and the lineshape for the effect has been calculated. However, experiments to date have been inconclusive due to the thermal heating caused by the large current flows necessary to achieve sufficient modification to be observed. Thus, current modulation spectra are characteristic of thermal modulation.

(d) Modulation by electron beam bombardment

In this technique, the sample under study is made the anode of a high-vacuum discharge tube. Modulation is achieved in principle by filling normally empty conduction band states, and thus inhibiting optical transitions via a Pauli exclusion approach. The extremely short lifetimes of electrons in excited states do not allow significant concentrations to be built-up with available techniques, however, and the sample heating which results from dissipation of the beam power leads instead to thermal modulation. Thus these spectra are equivalent thermal modulation spectra.

VI. Raman scattering

Raman scattering, while itself a complete branch of spectroscopy, is also related to modulation spectroscopy in the general sense that Raman spectra also result from perturbations of the solid. However, in contrast to conventional modulation techniques, the relevant perturbations here are not applied externally but rather are due to phonons, the intrinsic elementary excitations of the lattice. A second difference lies in the fact that phonon energies tend to be relatively large with respect to attainable resolution of typical monochromators. Therefore, the energies of the scattered photons are shifted sufficiently from the energy of the incident beam so that the scattered photons may be discriminated by ordinary spectroscopic techniques (Cardona, 1973).

Modulation techniques have been applied as an aid to interpretation of Raman spectra and as a means of extending Raman spectroscopy to normally-inactive modes by lowering the symmetry of the specimen by means of an externally-applied vectorial or tensorial perturbation. In the first case, wavelength-modulation or wavelength-derivative techniques have been used to enhance weak structure in Raman spectra. This technique has proven to be especially fruitful in second-order Raman spectra, which arise from two-phonon processes. By the usual optical selection rules, the two phonons involved in the scattering event must have equal and opposite momentum (neglecting the small momenta of the photons), whence second-order Raman spectra yield directly the joint density of phonon states. By performing wavelength modulation or wavelength differentiation, critical points in the phonon joint density of states may be enhanced. Thus, second-order Raman spectroscopy places phonon spectroscopy in a position completely equivalent to the ordinary optical spectroscopy of electronic states (Yacoby and Yust 1973).

The reduction of symmetry to observe normally-inactive Raman modes has been done mainly for high-symmetry crystals (cubic and hexagonal) by means of externally-applied electric fields. As in modulated reflectivity measurements, the electric field usually influences the observed Raman spectrum by several mechanisms, including the Franz-Keldysh effect on the electronic states involved as well as symmetry-lowering effects which act on optical selection rules. Both effects have been calculated and observed. The Franz-Keldysh mechanism is most important if the incident and scattered photon energies lie close to a critical point in the joint density of electron states (Pinczuk and Burstein 1973).

Bibliography

General references on modulation spectroscopy:
CARDONA, M. (1969) *Modulation Spectroscopy*, New York: Academic.
Proceedings of the First International Conference on Modulation Spectroscopy (B. O. Seraphin Ed.), Amsterdam: North-Holland (1973).
Semiconductors and Semimetals (R. K. Willardson and A. C. Beer Eds.), Vol. 9: *Modulation Techniques*, New York: Academic (1972).
Specific references:
AGGARWAL, R. L. (1972) In *Semiconductors and Semimetals, op. cit.*, p. 457.
AGGARWAL, R. L. (1973) In *Proceedings ..., op. cit.*, p. 263.
ASPNES, D. E. (1972) *Phys. Rev. Letters*, **28**, 168.
ASPNES, D. E. (1973) In *Proceedings ..., op. cit.*, p. 676.
ASPNES, D. E., HANDLER, P. and BLOSSEY, D. F. (1968) *Phys. Rev.* **166**, 921.
BATZ, B. (1972) In *Semiconductors and Semimetals, op. cit.*, p. 316.
BEAGLEHOLE, D. (1973) In *Proceedings ..., op. cit.*, p. 442.
CARDONA, M. (1973) In *Proceedings ..., op. cit.*, p. 470.
FISCHER, J. E. and ASPNES, D. E. (1973) *Phys. Stat. Sol.* (b) **55**, 9.
FISCHER, J. E. and BOTTKA, N. (1970) *Phys. Rev. Letters*, **24**, 1292.
FISCHER, J. E., KYSER, D. S. and BOTTKA, N. (1969) *Solid State Commun.* **7**, 1821.
GORDY, W. (1967) In *Handbook of Physics* (E. U. Condon and H. Odishaw Eds.), 2nd ed., pp. 7–92, New York: McGraw-Hill.
KANE, E. O. (1969) *Phys. Rev.* **178**, 1368.
MCINTYRE, J. D. E. (1973) In *Advances in Electrochemistry and Electrochemical Engineering* (R. H. Muller Ed.), Vol. **9**, p. 61, New York: Wiley-Interscience.
PINCZUK, A. and BURSTEIN, E. (1973) In *Proceedings ..., op. cit.*, p. 131.
POLLAK, F. H. (1973) In *Proceedings..., op. cit.*, p. 892.
POLLAK, F. H. and CARDONA, M. (1968) *Phys. Rev.* **172**, 816.
SELL, D. D. (1970) *Appl. Opt.* **9**, 1926.
SELL, D. D. (1973) In *Proceedings..., op. cit.*, p. 942.
SELL, D. D. and KANE, E. O. (1972) *Phys. Rev.* **B5**, 417.
SERAPHIN, B. O. (1964) In *Physics of Semiconductors: Proceedings of the Seventh International Conference* (M. Hulin Ed.), p. 165, Paris: Dunod.
SHEN, Y. R. (1973) In *Proceedings..., op. cit.*, p. 718.
WEISER, G. (1973) (To be published).
YACOBY, Y. and YUST, S. (1973) In *Proceedings..., op. cit.*, p. 34.

D. E. ASPNES and JOHN E. FISCHER

MONSOON METEOROLOGY

Definition and Extent of the Monsoons

The monsoons, which blow in response to the annual change in the difference in atmospheric pressure over land and sea, result from the difference in temperature between land and sea, and where great continents border an ocean, temperature differences are large. When the Sun moves north of the equator in the Northern Hemisphere summer, the land mass of Asia because of its relatively low heat capacity is rapidly warmed. On the other hand, the northern Indian Ocean and the western Pacific Ocean store the Sun's heat within their deep surface layer. Consequently, the land gives off heat more readily than the sea, and the air over the land becomes warmer and the air pressure lower than over the neighbouring ocean.

Thus during summer, air flows from the Indian Ocean toward lower pressure over southern Asia, ascending as it is heated over the land until it reaches a level at which the pressure gradient is reversed, whereupon it flows on a return trajectory from land to sea, where, descending, it is once more taken up by the landward-directed pressure gradient (Fig. 1). As long as the land is significantly warmer than the sea this great circulation persists.

In winter the reverse occurs. The low heat capacity of Asia relative to the northern Indian and western Pacific Oceans insures that air over the land is colder than over the sea. We observe then the typical winter monsoon in which, at low levels, air flows out from the continent over the sea where it rises and returns in the middle and higher layers of the troposphere to the land, sinks to the surface, and resumes the cycle.

Over Africa the monsoons have a somewhat different character from those of Asia. During the Northern Hemisphere summer the desert areas of North Africa heat rapidly and pressure there falls in the same way as over Asia. South of the equator

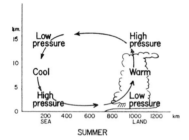

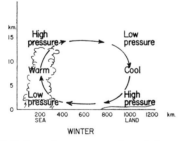

Fig. 1. Schematic representation of the vertical circulations associated with the summer and winter monsoons.

over Africa, during the Southern Hemisphere winter, cooling occurs, thus establishing a pressure gradient across Africa from south to north, which, in turn, sets up a massive flow of air, also from south to north, across the equator.

Because the deflecting force due to the earth's rotation reverses its direction at the equator and is weak in equatorial regions, air flow is more directly from high to low pressure than is the case in higher latitudes. The influence of Africa on the atmospheric circulation extends 800 km east of the continent and merges with the influence of Asia farther north. In the Northern Hemisphere summer, wind circulates in a huge gyre from the south-east around the northern edge of the South Indian Ocean anticyclone and toward the coast of Africa near the equator, swinging into the south and then south-west to parallel the African, Arabian and Asian coasts and, finally, sweeping across India, Burma and the Indo-China–Thailand peninsula as the southwest or summer monsoon.

Six months later a complete reversal takes place. Northern Africa is cold and southern Africa is warm and so the winds blow from the north across the equator in the western Indian Ocean.

Climate in the vicinity of Australia is monsoonal but with less intensity, while monsoon "tendencies" have been identified in many other regions—for example, Texas, the Caspian Sea and even parts of Europe. However, since an annual shift in circulation direction is fundamental to the monsoon concept, I have defined the monsoon area as encompassing regions with January and July surface circulations in which

1. the prevailing wind direction shifts by at least 120 degrees between January and July;
2. the average frequency of prevailing wind directions in January and July exceeds 40%;
3. the mean resultant winds in at least one of the months exceed 3 m sec^{-1};
4. Fewer than one cyclone–anticyclone alternation occurs every 2 years in either month in a 5-degree latitude–longitude rectangle.

These criteria are satisfied in a contiguous region extending from western Africa to Indonesia with southward protrusions to Madagascar and northern Australia (Fig. 2).

I have deliberately avoided mentioning weather as a monsoon criterion, although many writers have attempted to graft onto their circulation criteria a

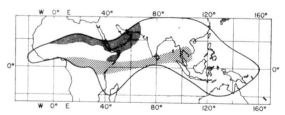

Fig. 2. The area satisfying monsoon criteria is enclosed by the full line. Within this area, deserts (annual rainfall < 250 mm) are stippled and the region with a bimodal annual rainfall variation is hatched. Except for a small portion of eastern peninsular India the remainder of the monsoon area experiences a summer or autumn rainfall maximum.

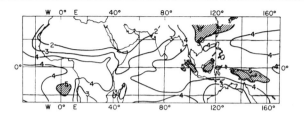

Fig. 3. Mean annual cloudiness (in eights of sky covered) based on 3 years of weather satellite photographs.

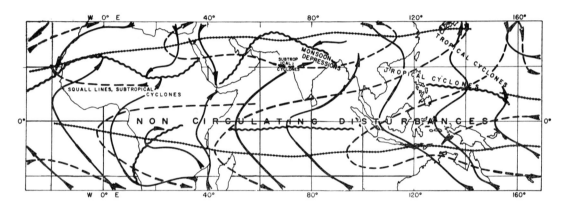

Fig. 4. July. 1. Mean resultant winds at the surface (full lines, thickened where speed exceeds 7 m sec^{-1}). Major pressure troughs denoted by wriggly lines are termed "heat troughs" over the continents and "near-equatorial troughs" elsewhere. 2. Mean resultant winds at about 12·4 km above the surface (dashed lines, thickened where speed exceeds 20 m sec^{-1}). Major pressure ridges are denoted by joined-diamond lines. Labelling indicates preferred regions for synoptic systems.

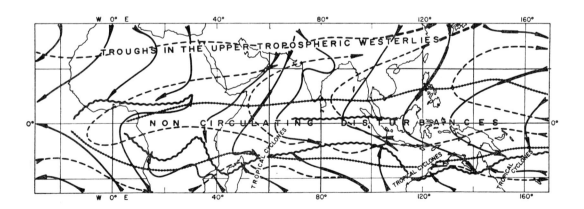

Fig. 5. January. As for July except that thickened dashed lines denote speeds >60 m sec^{-1}.

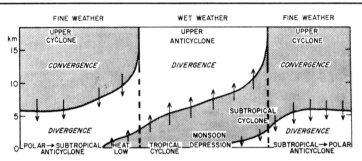

Fig. 6. Circulation components of the monsoons arranged schematically according to weather, divergence, and vertical motion. Levels of non-divergence are denoted by heavy lines.

"wet summer/dry winter" requirement. In fact no simple correlation exists between the direction of the surface pressure gradient and rainfall.

In the monsoon area most *rain* falls in summer or in autumn, except for a near-equatorial band possessing a double maximum (Fig. 2). The summer maximum is associated with heat-trough intensification, although the heat troughs themselves are dry (Fig. 6). The double maximum is probably associated with temporary intensification of near-equatorial troughs in the transition seasons.

Role of the Himalayan–Tibetan Massif in the Monsoons

The chain of mountains extending from Turkey to West China produces a sharp discontinuity in surface monsoon characteristics along about 100°E by protecting the land to the south from cold polar outbreaks. Thus in summer the South Asian monsoon is stronger (more intense heat trough) than the East Asian whereas in winter the East Asian monsoon is the stronger (more intense polar anticyclone).

Figure 3 reveals a large cloudiness gradient between north-west India and western China, a distribution which prevails throughout the year and which is also reflected in the rainfall.

This weather discontinuity derives from the mountain-plateau mass of the Himalayas and Tibet which pervasively affects the entire Northern Hemisphere monsoon area throughout the year, and at times even influences climate in the Southern Hemisphere monsoon area.

In the broadest sense, the Northern Hemisphere monsoon area comprises three parts: east of Tibet, where spring and summer are wet and significant winter precipitation falls; west of Tibet, where deserts dominate; and south of Tibet, where summers are wet and winters are desert-like. The 12·4-km mean resultant winds (Figs. 4 and 5) reflect this distribution. East of Tibet they are divergent, and west and south-west of Tibet convergent throughout the year. South and south-east of Tibet, the winds are divergent in summer and convergent in winter. These distributions derive from the combined mechanical-thermal effect of Tibet.

During *summer*, central and south-eastern Tibet is a powerful elevated radiational heat source with condensation along the Himalayas aiding the effects of radiation. The subtropical ridge aloft overlies southern Tibet (Fig. 4) and compressional heating produced by mechanical subsidence is negligible. The heating increases the N.–S. temperature gradient of the summer monsoon and produces a speed maximum in the upper tropospheric easterlies south of India. Thus upstream to the east of 70°E, rising motion beneath divergent easterlies favours clouds and rain. Downstream to the west of 70°E subsidence beneath convergent easterlies keeps skies clear.

Central and south-east Tibet are almost snowfree during *autumn, winter* and *spring*, and so continue to act as a high-level radiational heat source. Pressure surfaces are raised there, diminishing the S.–N. temperature gradient and the strength of the subtropical jet to the south. Air subsides immediately east of the plateau and so is further warmed by compressional heating. Over central China it flows alongside very cold air which has swung around the northern edge of the massif (Fig. 5). Here the extremely large temperature gradient produces an exceptionally strong jet stream. With speeds increasing downstream eastward of 100°E, upper tropospheric divergence favours large-scale upward motion and increased cloudiness.

West of 100°E, the generally convergent upper tropospheric westerlies over the Middle East favour sinking and decreased cloudiness.

Year-long subsidence persists over the great deserts of southwest Asia and north Africa, extending to the surface in winter anticyclones and to the lower troposphere in summer heat lows (Fig. 6).

Rain During the Monsoons

The monsoons, as modified by Himalaya–Tibet, set the stage for weather. Over and near the continents the cold dry air of the winter monsoon usually predisposes weather to be fine, whereas the warm moist air of the summer monsoon usually predisposes weather to be unsettled. Nevertheless, despite the fact that during the monsoons moving *surface* disturbances are rare and surface winds are notably steady, rainfall, on scales from days to weeks, is surprisingly variable. The cause lies in synoptic changes effected by moving disturbances and by intensification and decay of quasi-stationary bad-weather systems. Resultant weather is locally modified by the diurnal cycle and by meso-scale distortions and is also constrained within the envelopes of important more extensive fluctuations with periods similar to those of the middle-latitude Grosswetterlagen.

The most common synoptic components of the monsoons are conveniently subsumed within three genera:†

1. circulations in which downward motion and fine weather predominate, i.e. *polar and subtropical anticyclones*, and **heat lows**;
2. circulations in which upward motion and wet weather predominate, i.e. **tropical cyclones, monsoons depressions**, and **subtropical cyclones** (Fig. 5);
3. other systems with marked weather gradients, i.e. *troughs in the upper tropospheric westerlies, the Polar Front*, near-equatorial troughs, quasi-stationary non-circulating disturbances, surface transequatorial flow, and **squall lines**.

Figures 4 and 5 show the regional distribution of some of these synoptic components.

Tropical cyclones, monsoon depressions and squall lines significantly change the direction of the surface monsoon circulation as they develop and move. But the other synoptic systems either move very little or do not significantly change surface wind directions. Thus quite variable weather may occur in the regime of a superficially steady monsoon. Most rain falls in one of these two circumstances, either:

1. As "*rains*" from deep nimbostratus with embedded cumulonimbus when vertical wind shear and lower tropospheric convergence both are large (although rain intensity may fluctuate considerably, skies remain predominantly overcast). *Rains* fall in association with accelerating surface westerly winds equatorward of tropical cyclones, monsoon depressions or subtropical cyclones (Fig. 6) or near the early summer Polar Front of South-east Asia.
2. As "*showers*" from scattered towering cumulus or cumulonimbus, when vertical wind shear and lower tropospheric convergence both are small. Overland afternoon thunder*showers* occur before and after the summer monsoon, near the heat trough, during "breaks" in the *rains* and in the shadow of mountain ranges during the *rains*. Over the sea, however, thundershowers are rare in these situations because insolational heating is insufficient to trigger them.

---

† Predominantly summer systems are shown in **bold face**, predominantly winter systems in *italics*.

March of the seasons

During spring and autumn, the transition seasons between winter and summer monsoons, complex changes take place both in the large-scale circulation and in the mixtures of synoptic systems. The changes differ within the monsoon area and also from one year to the next. The spring transition lasts about twice as long as the autumn transition because in spring the normal equator-pole global temperature gradient is being reversed, while autumn marks return to the normal gradient. As might be expected, since the monsoons arise from the differing heat capacities of land and ocean, circulation in a transition season changes first in the surface layers. After several weeks, the change extends throughout the troposphere and the transition season is over.

The march of the seasons may be treated briefly and somewhat superficially by following the annual variations of major surface low pressure troughs (Figs. 4 and 5).

In the *Indian Ocean–South Asia* region, an east–west oriented trough exists throughout the year in the tropics of each hemisphere. In the Northern Hemisphere summer the trough lies well away from the equator over southern Asia and there may be a very weak secondary trough close to the equator (Fig. 7). As the year progresses from summer to winter, the primary trough weakens and dissipates and the secondary trough becomes predominant. The sequence reverses from winter to summer. This model satisfactorily removes two difficulties inherent in the old concept of a single trough undergoing an annual oscillation between the summer tropics of both hemispheres—apparently discontinuous latitudinal jumps, and the usual absence of double rainfall maxima between 10° and 25°N. The transition season "jumps" are caused by rapid cooling in autumn and heating in spring of the deserts of north-west India and West Pakistan.

South of the equator, despite considerable day-to-day fluctuations, a single trough follows the annual march of the sun.

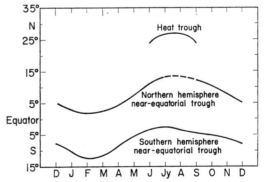

Fig. 7. Annual latitudiual variation of lower tropospheric pressure trough over the Indian Ocean.

Over *Africa*, a weak heat trough is sometimes found between 5–10° latitude in each hemisphere during the transition seasons. Then as the year advances, the trough in the winter hemisphere disappears and the trough in the summer hemisphere intensifies and follows the sun poleward. The sequence then reverses through the following transition season.

The north Indian Ocean–South Asian regime is weakly duplicated in the northern Australian region while over East Asia, absence of a vigorous *heat* trough leads to a relatively regular annual movement of the trough.

Over the ocean, troughs near the equator spawn weakly cyclonic bad-weather systems; farther from the equator, tropical cyclones may develop in the troughs. When the midsummer monsoon trough of northern India extends into the northern Bay of Bengal, monsoon depressions develop there. *Over land*, west of 70°E, fine weather prevails in the troughs (heat troughs), but on their equatorward sides subtropical cyclones develop.

Concluding Remarks

During summer and winter, although surface disturbances are rare, middle- or upper-tropospheric disturbances are not. In the transition seasons differences between the monsoon area and surrounding areas are least. Thus latitude for latitude, the range and complexity of annual variations are greater in the monsoon area than beyond.

The general character of the monsoons and their interregional variations reflect juxtaposition of continents and oceans. However, without the great mechanical and thermal distortions produced by the Himalayas and the Tibetan Plateau, the vast Northern Hemisphere deserts would be less desert-like, and central China would be much drier and no colder in winter than India.

Within the monsoon area annual variations are seldom spatially or temporally in phase. The climatological cycles merely determine *necessary* conditions for certain weather regimes; synoptic changes then control where and when the rain will fall and how heavily, and whether winds will be destructive.

The great vertical circulations comprising the monsoon (Fig. 1) often undergo wide-ranging nearly simultaneous accelerations or decelerations, apparently triggered by prior changes in the cold parts of the circulations.

That synoptic-scale disturbances often appear to develop and to weaken in response to changes in the major vertical circulations might explain why many of the disturbances are quasi-stationary. In turn, synoptic-scale vertical motion determines the character of convection and the efficiency with which energy is transported upward from the heat source.

Synoptic-scale lifting, by spreading moisture deeply through the troposphere, reduces the temperature lapse rate and increases the heat content in mid-troposphere. Thus, though it diminishes the intensity of small-scale convection and the frequency of thunderstorms, it increases rainfall and upward heat transport. Conversely, synoptic-scale *sinking*, by drying the mid-troposphere, creates a heat minimum there, hinders upward transport of heat and diminishes rainfall. However, the increased lapse rate favours scattered intense small-scale convection and thunderstorms.

In the monsoon area the character of the weather, on the scale of individual clouds, seems to be determined by changes occurring successively on the macro- and synoptic scales. *Rains* set in—not when cumulonimbus gradually merge—but when a synoptic disturbance develops, perhaps in response to change in a major vertical circulation. *Showers* too are part of the synoptic cycle. Individually intense, but collectively less wet, they succeed or precede *rains* as general upward motion diminishes.

Bibliography

ARAKAWA, H. (ed.) (1969) *Climates of Northern and Eastern Asia*, Amsterdam, London, New York: Elsevier Publishing Company.

CHANG, J.-h. (1967) *Geogrl. Rev.* **57**, 373.

CHANG, J.-h. (1971) *Geogrl. Rev.* **61**, 370.

GENTILLI, J. (ed.) (1971) *Climates of Australia and New Zealand*, Amsterdam, London, New York: Elsevier Publishing Company.

GRIFFITHS, J. F. (ed.) (1972) *Climates of Africa*, Amsterdam, London, New York: Elsevier Publishing Company.

INDIA METEOROLOGICAL DEPARTMENT (1967–) *Forecasting Manual*, Poona: Deputy Director General of Observations (Forecasting).

RAMAGE, C. S. (1969) *Oceanogr. Mar. Biol. Ann. Rev.* **7**, 11.

RAMAGE, C. S. (1971) *Monsoon Meteorology*, New York, London: Academic Press.

RAMAGE, C. S., MILLER, F. R. and JEFFERIES, C. (1972) *International Indian Ocean Expedition Meteorological Atlas*, Vol. 1, *The Surface Climate of 1963 and 1964*, Washington, D.C.: National Science Foundation.

RAMAGE, C. S. and RAMAN, C. R. V. (1972) *International Indian Ocean Expedition Meteorological Atlas*, Vol. 2, *Upper Air*, Washington, D.C.: National Science Foundation.

THOMPSON, B. W. (1965) *The Climate of Africa*, London, New York: Oxford University Press.

WALKER, J. M. (1972) *Weather*, **27**, 178.

YOSHINO, M. M. (ed.) (1971) *Water Balance of Monsoon Asia*, University of Tokyo Press.

Contribution No. 72-8 of the Department of Meteorology, University of Hawaii.

C. S. RAMAGE

MYOELECTRIC CONTROL. Myoelectric (muscle-electric) control seeks to use the superior ability of the brain and the central nervous system

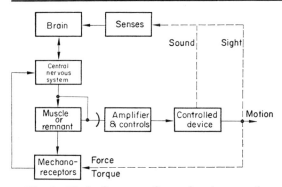

Fig. 1. Block diagram of myoelectric control system.

in decisional, optimizing, adaptive, multivariable control of complex peripheral limb tasks. The output of the central processor is a train of electric pulses to the appropriate muscles. It is these signals which are detected and used to effect control of artificial limbs, of powered braces for paralysed limbs, or of limb-like actions on machines. A block diagram showing the elements of a myoelectric control system is shown in Fig. 1.

The Source of Myoelectric Signals

The excitation signals to the muscles (efferent) and the feedback signals from various mechano-receptors which provide information about the position, velocity and force of the resulting limb action (afferent) are passed along neural transmission lines (axons) as local changes of sodium and potassium ion concentrations (Katz, 1966). These changes are detectable as small variations of axonal potentials relative to the surrounding body fluids as indicated in Fig. 2a. The axons are not continuous tubes but are short lengths separated at intervals of about 1 mm by membranes which enable a kind of relay action to take place, each cell in turn initiating the next down the line.

The wave shape at any point is similar, but differs in its basic time constants which depend on the axon diameter. This leads to a propagation of these potentials with a speed which depends on the axon diameter, the smaller fibres implying slower speeds. These speeds are of the order 2–100 m/s. The frequency at which the efferent and afferent pulses are generated depends on the rate of action and force exerted by the peripheral muscle, and these can reach 200–300 pulses per second.

A number of muscles fibres are served by one nerve axon, according to the precision and coordination required of the muscle. Thus the eye, middle ear, larynx and pharynx will have one axon to excite

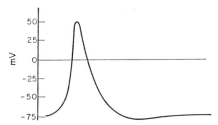

(a) Typical nerve pulse potential axon core relative to adjacent extracellular fluid

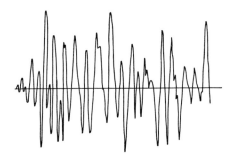

(b) Typical myogram recorded with surface electrodes

Fig. 2. Typical myoelectric signals.

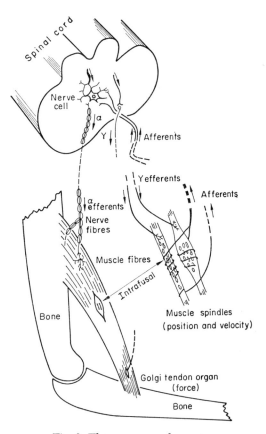

Fig. 3. The neuromuscular system.

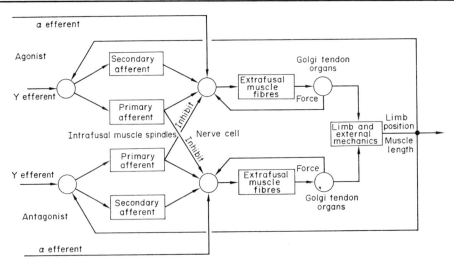

Fig. 4. Block diagram of the neuromuscular control system.

fewer than 10 muscle fibres, whereas a lower limb axon might serve more than 600 muscle fibres within a diameter of 1·5 mm.

Figure 3 illustrates the anatomical components of the neuromuscular system and Fig. 4 its corresponding control loop.

The shape and Spectrum of Observed Myoelectric Signals

Viewed spatially, the transmission of the impulse could be the movement of a short column of charge along a tubular filament. The pulse duration of about 1–2 ms and the velocity of about 50 mm/ms imply a column of some 25 mm.

From the viewpoint of the sensitive electrodes, the nerves and muscle fibres form an array of more or less parallel transmission lines. There are fewer paths connecting the muscles and the nerve cells in the spinal cord than in the bulk of muscles where there is branching to engage many muscle filaments acting in parallel. The form of the waveform will depend on the disposition of the electrodes relative to these transmission lines. For example, two electrodes placed along an axon will register a diphasic signal due to the progression of the pulse first under one electrode, then under the other. The delay between the positive and negative portions of the pulse will depend upon the distance between the electrodes and the propagation velocity.

Similarly, if the electrodes are placed astride the nerve, say, relying on conductive pickup, the detected wave will be monophasic. It may even result in a zero signal if the nerve is always equidistant from the two electrodes.

In practice, apart from electrodes that locate in the core and sheath of an individual axon, all detected myographic signals will be of the basic form outlined above. The complexity of the waveform will depend on the number, phase and orientation of the nerve and muscle fibres that fall within the range of the electrodes. This will be obviously greater in the region of bulk muscle where the spatial integration is highest, than elsewhere.

The basic diphasic wave of which the detected myoelectric signal is composed approximates to the first derivative of the Gaussian distribution curve. The Fourier transforms of Gaussian functions have the same functions as the originals. The frequency spectrum of such a function with the pulse width $\tau = 2$ ms is compared with various quoted spectra in Fig. 5. The spectrum reaches a maximum at about 200 Hz and for all practical purposes all the information will pass in a band 10 Hz to 2 kHz.

The actual repetition frequency of myoelectric signals varies with tension, position and speed of

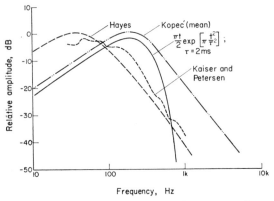

Fig. 5. The spectrum of myoelectric signals.

shortening of the muscle group. However, this does not affect the spectral envelope but only the number of spectral frequencies.

The effect of collecting several axonal and muscle signals will be to cause instantaneous summations and cancellations. This generally gives the wave the appearance of being a random impulsive noise with limited bandwidth. However, if the pulse width and shape from all sources is constant, the frequency spectrum is unchanged by the number of sources or their different rates. In practice, the fact that there is a higher frequency "tail" to the spectrum suggests that there are some signal sources with shorter pulse widths.

Methods of Detecting Myoelectric Signals

Basically, there are three ways of applying electrodes to detect myoelectric signals: under, through or on the surface of the skin. These methods are called respectively: subcutaneous or implants; hypodermic or percutaneous, and surface contact.

Implants

The signals can be picked up by implants that are intimately attached to the muscle group one wishes to use. The implants divide naturally into those that are self-energized and those that are externally energized.

The self-energized units employ a miniature battery or a biological battery converting enzymes into electric current in a fuel cell. They could also use the bioelectric potentials themselves, or use a radio-isotope or a piezoelectric source. Batteries have a finite life and require periodic recharging and replacement.

The externally energized units can be considered as passive circuits that are interrogated by the external transmitter-receiver. The muscle signals are used to modulate or modify the natural frequency of the implanted circuit, so that when it is electromagnetically coupled to the transmitter, it directly alters the latter's timing.

Crosstalk is the determining factor when implants are used at several sites. There are also several problems involving materials and surgical procedures.

The waveform detected by implanted units is the aggregation of many fibre signals from the region of the exposed electrode tips. The range will be approximately equal to the electrode spatial separation. In most cases, protuding flexible wire electrodes are used with several millimetres separating them.

Suggestions have been made periodically that implants could be used to monitor individual motor units. This implies accurate location of the electrodes separated by 50–200 μm lying within a distance of this dimension from the nerve or muscle unit chosen.

Hypodermic and percutaneous electrodes

These two methods are dealt with together since they both utilize electrodes that pass through the skin. The essential difference between the two systems is in whether the wires are unsupported or are guided through the skin by a hypodermic tube. In practice, the tube-supported type facilitates placing accuracy, whereas the flexible wire electrodes often make shallow angles with the skin surface. The virtue of these electrode systems is that they penetrate the insulating layer of skin, and make closer contact with the activated nerve and muscle. Moreover, these electrodes are easily inserted without elaborate surgical procedures; they require only a simple local sterilizing technique.

The waveform from these electrode systems is generally of the complex diphasic type. However, selectivity can be increased by manipulating the wires to optimum sites and by using short uninsulated tips. It has been claimed that one motor unit or sub-unit can be located in this manner.

Surface electrodes

From the point of view of procedure and preparation, the surface electrodes are the simplest. The myoelectric signals from the muscle and nerves are conducted through the extra-cellular fluids and body tissues to the skin surface. The conductivity through the skin is poor and is subject to wide variations even within small areas, due to the pores, hairs, etc. This often leads to a variability of the signal.

One can regard the skin as a membrane which is slightly permeable to ions. Contact with the metal electrodes produces a metal-ionic solution interface having all the usual characteristics of a chemical half-cell. There is set up a potential which depends on the metal, its surface and the concentration of ions in solution. This is governed by the Nernst equation.

It is usual to employ chlorided silver electrodes in conjunction with a conductive paste or saline solution to ensure a saturated concentration of halide ions and a stable half-cell potential. Then the potential of each electrode is about 200 mV.

Electronic amplifiers for myoelectric signals

Because the sources of nerve and muscle signals convert so little energy, they are unable to supply much current. The conductive paths between the sources and the electrodes have high resistance. Therefore, the detecting amplifiers must present at least an order higher input resistance, say, greater than 1 MΩ.

Although the electrodes reach equilibrium, differences of temperature, abrasion of the electrodes by the skin, interposing of fine hairs, etc. cause small fluctuations of the contact potential. When direct current amplifiers are used, this presents a difficulty with the zero stability. These small direct voltage changes are often enough to saturate high gain amplifiers. The body, being essentially a bag of conductive fluid, is subject to mains induction and aerial effects. Any interference collected by the electrodes is amplified along with the true signal and, although it is possible to filter this from the output, the magnitude is sometimes sufficient to saturate the amplifier before the output stage. Therefore, the reduction of these effects is best attempted in the first stage of the amplifier by providing the greatest possible rejection of common-mode signals. Also some consideration must be given to grounding of the body in the region of the electrodes. Figure 6a represents the body side of the E.M.G. detection system.

An example of an amplifier system is used for exploring myographic signals (Godden, 1968) with a surface electrode system is given in Fig. 6b. The input impedance is greater than 200 MΩ at 50 Hz and the common mode rejection ratio is better than 80 dB. The direct voltage gain is about unity compared with the mid-band gain of 80 dB and the low frequency cut-off point occurs at 10^{-3} Hz. The noise at the output is of the same order as the residual mains induction and referred to the input is about 3 μV.

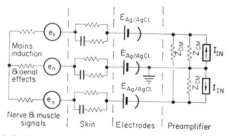

(a) Equivalent circuit of myoelectric detection circuit

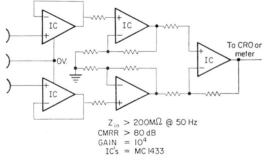

(b) Block diagram of myoelectric detection amplifier

Fig. 6. Myoelectric detection system.

Signal Processing

The type of signal processing that is applied to the raw electromyogram depends on the characteristics of that signal. In deciding which is used one must also take into account the physiological nature of the signals and the type of control that is to be effected. It is essential that the nature and reliability of the particular signal be tested and understood. The influence of changes of muscle tone, inherent hysteresis and fatigue must be reconciled with the control task that is envisaged. In some cases, such as in prehension, the physiological deficiences of the signal can be taken up in the control system, but similar backlash, inertia or threshold dead-band built into a position-controlled limb would introduce imprecision and the possibility of instability.

Three other factors are invariably associated with the detected myoelectric signal and therefore can affect its processing. They are noise, artefact and power line induction.

The noise is created within the amplifier, being thermal noise due to the discrete nature of the electron current in the electronic components. Generally, noise is not a serious problem except if it is a large part of the total signal, which should not be true of modern amplifiers, or if threshold levels are set at or below the noise peaks.

Artefacts of an amount sufficient to interfere with the myogram need not be tolerated with existing techniques.

Power line induction can be dealt with by adequate common-mode techniques, or by selective filtering.

Modes of Control

Two main types of control are possible:
(a) on–off or threshold,
(b) linear or proportional.

In the first, one accepts that, for one reason or another, the signal parameters are variable or are not reproducible enough to relate the conscious effort exerted by the subject with the position or force of the controlled device in a one-to-one correspondence. Usually, there is certainty only in whether the signal is present or absent. Then, one often resorts to sensing whether or not the signal amplitude exceeds a threshold level that is set sufficiently high to prevent spurious operation by noise, induction or resting signal levels.

Linear control assumes that the signals are perfectly reproducible and that they are not subject to fatigue, hysteresis or artefact, or at most only by such an amount that can be accommodated by in-built backlash arrangements in the controlled circuits or device.

With an on-off type of input signal there is no possibility of continuous direct control of position; the obvious parameter to control in this case is velocity. This is simply achieved by making the signal operate spring-returned hydraulic or pneumatic valves, or biased electric or electronic switches.

Velocity control is employed in most types of crane, and it is a characteristic of these machines that there must be considerable training to acquire operating skill. The same characteristic can be observed in velocity controlled prostheses. The angular rate of such units must compromise between the large delays implied by low velocities, and the prediction and overshoot problems involved at high velocities.

For position determined systems a signal proportional to the desired displacement is required. Any variability of the signal due to artefact, fatigue, etc., will effect a position change unless provision is made to accept a degree of dispersion. Generally, any artifices employed to mask spurious signals will themselves produce problems of delay and imprecision. However, when a linear system is achieved, this is known to require least training for successful operation.

Medical Applications of Myoelectric Control

The application of myoelectric control to medical problems such as the control of artificial limbs (prosthetics) and powered splints and braces (orthotics), is limited by three interface factors:

(1) the man-machine control and interface problems,
(2) the shortage of myoelectric sources for adequate control,
(3) the nature of the myoelectric signals.

When separate controls are used for each joint rotation there are presented to the user decisional and computational problems, successive reaction and external delays and system lags. For these reasons no devices with more than two such uncoupled controls have been accepted by patients. To use a coupled or "end point" control system implies a linear position control strategy which then relies on the nature of the myoelectric signals.

Myoelectric signals, due to the nature of the neuromuscular servo system from which they come, are subject to non-linearity, hysteresis and fatigue. Also, the way in which these signals are extracted can introduce artefact signals, distortion and pick-up. These imperfections can be tolerated by an incrementally adjusted device, such as in a velocity-controlled system, but they are incompatible with desirable position-controlled systems.

Formidable problems arise with prosthetic devices intended for higher levels of amputation. The number of prospective controlled degrees of freedom increase as the number of available, accessible neuromuscular systems decrease. Even when the sources are present, such as in orthotic systems, the need for these to be free of cross-talk means that the muscles must either be well isolated or that an attempt be made to use individual motor units within the same muscle. There is no evidence that several single motor units can be simultaneously exploited from the same muscle. It can be shown that in the absence of single motor control, there are no more independent myoelectric controls available than there are alternative mechanical (bone) movements which are generally more manageable or acceptable for control.

At present, devices successfully using myoelectric controls include:

(a) a number of hooks or hands employing either open-loop velocity- or closed-loop force-controlled systems,
(b) a position-controlled elbow,
(c) various hand, wrist and elbow prosthetic developments using sequential controls,
(d) various powered orthotic arm developments, usually associated with wheelchair structures.

Bibliography

KATZ, B. (1966) *Nerve, Muscle and Synapse*, McGraw-Hill.

MATTHEWS, P. B. C. (1964) Muscle spindles and their motor control, *Physiol. Reviews*, **44**, 2, 219–288.

GODDEN, A. K. (1968) *The Techniques of Myoelectric Control of Prostheses and the Prospects of this Type of Control for Thalidomide Casualties*. Oxford University Engineering Laboratory Report 1,048,68, March 1968.

A. K. GODDEN

N

NEUTRON RADIOGRAPHY—THE PRESENT POSITION.† Radiography with neutrons was described in the *Encyclopaedic Dictionary of Physics* in 1962 by J. Thewlis. The present article is meant to supplement the previous description by providing information on the current position in this growing field.

Neutron radiography produces images of opaque objects by the use of neutrons; the method is essentially the same as X-radiography except that neutrons are used as probing radiation. Neutrons are attenuated in a different manner from X-rays, and these differences lead to a number of advantages for the use of neutron radiography. Table 1 lists the energies of neutrons used for radiography and gives some characteristics of radiography in each energy range.

Although radiography has been performed with neutrons in all these energy ranges, most of the effort has involved thermal neutrons. Therefore this discussion will involve primarily thermal neutrons.

As compared with X-radiography, thermal-neutron radiography presents an improved capability to discriminate between some neighbouring materials in the periodic table. Also, thermal neutrons tend to present a reversal of X-ray attenuation in that most heavy elements are relatively transparent to neutrons, whereas several light elements (notably hydrogen, lithium and boron) are highly attenuating. In addition to these attenuation differences, a significant advantage for neutron radiography follows from the fact that neutron attenuation varies for individual isotopes. Thus, neutron radiography permits differentiation between ^{235}U and ^{238}U, or between ^{113}Cd and other isotopes of cadmium. A final demonstrated advantage of neutron radiography is that inspections of highly radioactive material can easily be performed; this advantage follows because several neutron detection methods show little or no sensitivity to gamma radiation.

Table 1. Energies of neutrons used for radiography

Type	Energy range	Comments
Cold	Below 0·01 eV	High cross-sections decrease the transparency of most materials but also increase the efficiency of detection. A particular advantage is the reduced scatter in materials at energies below the Bragg cutoff, for example, iron at energies below 0·005 eV.
Thermal	0·01–0·3 eV	Good discriminatory capability between different materials, and ready availability of sources.
Epithermal	0·3–10,000 eV	Excellent discrimination for particular materials by working at energy of resonance. Greater transmission and less scatter in samples that contain materials such as hydrogen and enriched reactor fuel materials.
Fast	10 keV–20 MeV	Good point sources of fast neutrons are available. At the lower energy end of the spectrum, fast-neutron radiography may be able to perform many inspections done with thermal neutrons, but with a panoramic technique. Good penetration capability because of low cross-sections. Poor material discrimination

---

† Work performed under the auspices of the U.S. Atomic Energy Commission.

Neutron Sources

Neutrons for radiography have been obtained from reactors, accelerators and radioactive sources. The fast neutrons typically available from these sources have been slowed in energy by surrounding moderators for thermal-neutron radiography. After moderation it is necessary to bring out a well-defined thermal-neutron beam to perform radiography. Several types of collimators have been made for this purpose, straight, multisectioned (Soller Slit type) and diverging. Most work has been done with diverging beams because of the common need for large radiographic coverage. Collimator quality has been defined by the ratio of the length (L) to the diameter of the inlet aperture (D). Ratios of L/D on the order of 7 to 100 are common for radioactive and accelerator sources. For reactor sources, a value of 50–200 is common and values as high as 1000–2000 have been used.

The geometric unsharpness of radiographs improves for large values of L/D; good resolution comes with a high L/D ratio. On the other hand, the neutron intensity available for radiography decreases rapidly for large L/D values; intensity is inversely proportional to $(L/D)^2$. Typical neutron intensities from various sources are as shown in Table 2 after taking into account the loss in neutron flux from moderation of the fast neutrons (typically a loss of 100 to 1000 for the ratio between fast-neutron yield and the peak thermal-neutron flux in the moderator), the loss due to the insertion of the neutron-absorbing collimator (typically a factor of 5 to 10), and the loss due to collimation (~2000 for an L/D of 10 to ~10^5 for an L/D of 100).

Many types of reactor have been used for thermal-neutron radiography. The high neutron flux generally available provides the capability for high-quality radiographs and short exposure times. Although truck-mounted reactors are technically feasible, one must normally consider a reactor as a fixed-site installation with the requirement that the objects be brought to the reactor for radiography. Investment costs are generally high, but small, medium-cost reactors can provide good results. The actual cost per neutron is usually favourable for the reactor source.

The accelerators used for thermal-neutron radiography have been primarily the low-voltage type that employ the reaction $H^3(d, n)He$ (normally designated a (d-T) generator), high-energy X-ray machines in which (x, n) reactions have been used, and Van de Graaff accelerators that use reactions such as $^9Be(d, n)^{10}B$. The (d-T) accelerators presently provide total 4π fast-neutron (energy ~14 MeV) yields in the range of 10^{10} to 10^{12} n/sec. Target lives in sealed neutron tubes are reasonable (100 to 1000 hr depending upon yield), and the sealed-tube system presents a source similar to an X-ray machine. A significant amount of neutron radiographic work has been performed with the (d-T) neutron generators.

The neutron radiographic work with (x, n) reaction neutrons is important because organizations will have the capability of using the same machine for high-energy X-radiography (the threshold X-ray energy for neutron production in beryllium is 1·66 MeV) or neutron radiography. Work with a 5·5-MeV linear accelerator showed that useful neutron radiography could be performed and that the changeover time from one radiation to another could be accomplished in 1 hr. Beam intensities for thermal-neutron radiography with the 650 R/min at 1-m X-ray output machine were on the order of 5×10^4 n/cm^2-sec for fair beam collimation.

Much higher beam intensities have been obtained with 2·0-MeV Van de Graaff acceleration of deuterons onto a beryllium target. An intensity of $1·2 \times 10^6$ n/cm^2-sec was achieved for medium collimation, and it was estimated that an improvement factor of 6 could be realized if an acceleration voltage

Table 2. Characteristics of thermal neutron radiographic sources

Source type	Typical radiographic intensity (n/cm^2-sec)	Application criteria	Comments
Radioisotope	10^1–10^4	Poor-to-medium resolution, long exposure times	Stable operation, low-to-medium investment cost; can be portable; low intensity
Accelerator	10^3–10^6	Medium resolution, average exposure times	On–off operation; medium investment cost; can be portable; medium intensity
Reactor	10^5–10^8	Excellent resolution, short exposure times	High intensity; medium-to-high investment cost; difficult to transport

Table 3. Some radioactive sources for neutron radiography

Source	Reaction	Half-life	Cost[a] in thousands of dollars	Average neutron energy (MeV)	Neutron yield (n/sec-g)	Gamma dose[b] (rads/hr at 1 m)	Gamma-ray energy (MeV)	Comments
^{124}Sb-Be	(γ, n)	60 days	25	0·024	$2·7 \times 10^9$	$4·5 \times 10^4$	1·7	Short half-life and high γ background, available in high-intensity sources; low neutron energy is an advantage for thermalization
^{210}Po-Be	(α, n)	138 days	20	4·3	$1·28 \times 10^{10}$	2	0·8	Short half-life; low γ background
^{238}Pu-Be	(α, n)	89 years	310	~4	$4·7 \times 10^7$	0·4	0·1	High cost; long half-life
^{241}Am-Be	(α, n)	458 years	1500	~4	1×10^7	2·5	0·06	Easily shielded γ output; long half-life; high cost
^{241}Am-^{242}Cm-Be	(α, n)	163 days	—	~4	$1·2 \times 10^9$ (80% Am) (20% Cm)		0·04 0·06	Increased yield over ^{241}Am-Be for relatively little more cost but with a short half-life
^{242}Cm-Be	(α, n)	163 days	—	~4	$1·46 \times 10^{10}$	0·3	0·04	High yield source, but half-life is short
^{244}Cm-Be	(α, n)	18·1 years	35[c]	~4	$2·4 \times 10^8$	0·2	0·04	Long half-life, low γ background are attractive. Source can also be used as a spontaneous fission source, with about half the neutron yield; potentially a readily available inexpensive source
^{252}Cf	Spontaneous fission	2·65 years	200[d]	2·3	3×10^{12}	2·9	0·04 0·1	Very high yield source, present cost is high but projected future cost makes it attractive, small size and low energy are advantages for moderation

[a] Only the cost of the radionuclide is given and is normalized to a source total yield of 5×10^{10} n/sec. See W. C. Reinig, *Nucl. Appl.* **5**, 24 (1968).
[b] The γ-ray dose is normalized to a neutron yield of 5×10^{10} n/sec.
[c] The cost is based on a proposed cost of $170/g, as quoted by Stewart, Horwitz and Youngquist, *Nucl. Appl. Tech.* **9**, 987 (1970).
[d] The cost is based on the present price of $10/μg. Future costs may be on the order of $1–2/μg.

of 3 MeV were used. Total fast-neutron yields of the order of 10^{12} n/sec are realistic for several high-voltage acceleration reactions.

Of the many radioactive sources, those that have been most widely used are summarized in Table 3. Although radioactive sources offer the best prospect for portable neutron radiographic equipment, one should recognize that the 10^5 or more decrease between radiographic thermal-neutron intensity and the total fast-neutron yield from the source dictate the need for long exposure times. A transportable radioactive source installation for neutron radiography is certainly feasible.

Detection Methods

The primary detection methods for thermal-neutron radiography have used photographic or X-ray film. This film can be exposed directly to the neutron beam with a conversion or intensifying screen in the so-called direct-exposure method. Alternately, the film can be used to record an autoradiographic image from a radioactive, image-carrying screen in a technique called the transfer method. Useful materials for conversion or intensifying screens are listed in Table 4.

Conversion screens of thin gadolinium metal foil and ^{6}Li or ^{10}B loaded scintillators have been the most widely used direct-exposure methods. The scintillators provide useful images with total exposures of about 5×10^5 n/cm^2. The high speed and favourable relative response to neutrons and γ-rays make the scintillator method attractive for use with non-reactor neutron sources. For higher intensity beams, gadolinium screens are extensively used. They provide better uniformity and image sharpness (a high-contrast resolution of 10 μ has been reported), but a fast-film exposure at least 30 times greater than that of the scintillator is required. For non-reactor neutron source work, the use of double metal screens (gadolinium and gadolinium, or rhodium and gadolinium) offers a speed advantage. For typical reactor or high-intensity work, an exposure of about 10^9 n/cm^2 will provide a good quality result with a slow film such as Eastman Kodak Single R film and a single gadolinium screen. The single emulsion film presents some advantage in reduced response to γ-radiation.

For the transfer method, the most widely used materials have been indium and dysprosium. The half-lives of the activities produced are 54 min and 2·35 hr, respectively. This method is especially valuable for situations that involve high γ backgrounds, as in the case of radiography of a radioactive object such as irradiated reactor fuel. The γ-radiation has little effect on the activity of the image-conversion screen and, therefore, no effect on the resultant radiograph. On the other hand, a disadvantage of the transfer method is that activation saturation imposes a lower useful neutron-intensity limit. For exposures of several half-lives with fast films, intensities as low as 6×10^3 n/cm^2-sec have been used with dysprosium.

When the two primary detection approaches are compared, the direct-exposure method offers high speed, infinite image-integration time, and the best observed spatial resolution. The transfer method offers γ insensitivity and somewhat improved contrast because of the lack of sensitivity to some scattered and secondary radiation.

Another approach to neutron-image detection is a dynamic one in which light from a scintillator is observed by a television camera. The light can be amplified by a separate or integral image-intensifier tube. The major advantage of this method is dynamic response, which thereby permits such applications as the study of fluid flow in a closed metal system, or the study of metal casting.

Although the major technical advantage of a neutron television system is that object motion can be followed, another attraction exists that has appeal for industrial inspection. One can picture a production-line approach in which parts could be continuously inspected by a neutron radiographic television system. Such a system could certainly be used to perform simple inspections. However, it must be realized that the television approach produces images in a frame time (typically $1/25$ to $1/30$ sec), so relatively high thermal-neutron intensities are required if one is to obtain reasonable image statistics. A thermal-neutron intensity of 5×10^6 n/cm^2-sec is needed by present equipment.

Table 4. *Properties of some thermal-neutron radiography detection converters*

Material	Useful reactions	Cross-section for thermal neutrons (barns)	Half-life
Lithium	^{6}Li(n,α)^{3}H	910	Prompt
Boron	^{10}B(n,α)^{7}Li	3830	Prompt
Rhodium	^{103}Rh(n)^{104m}Rh	11	4·5 min
	^{103}Rh(n)^{104}Rh	139	42 sec
Silver	^{107}Ag(n)^{108}Ag	35	2.3 min
	^{109}Ag(n)^{110}Ag	91	24 sec
Cadmium	^{113}Cd(n,γ)^{114}Cd	20,000	Prompt
Indium	^{115}In(n)^{116m}In	157	54 min
	^{115}In(n)^{116}In	42	14 sec
Samarium	^{149}Sm(n,γ)^{150}Sm	41,000	Prompt
	^{152}Sm(n)^{153}Sm	210	47 hr
Gadolinium	^{155}Gd(n,γ)^{156}Gd	61,000	Prompt
	^{157}Gd(n,γ)^{158}Gd	254,000	Prompt
Dysprosium	^{164}Dy(n)^{165m}Dy	2200	1·25 min
	^{164}Dy(n)^{165}Dy	800	140 min
Gold	^{197}Au(n)^{198}Au	98·8	2·7 days

That number could be improved by the use of more highly absorbing detectors or longer integration times.

Applications

A wide variety of neutron radiographic applications are now in use; about ten worldwide organizations offer commercial neutron radiographic service. Two major types of application account for much of this activity. One involves the neutron radiographic inspection of highly radioactive material, primarily irradiated reactor fuels; see Fig. 1. The other involves the inspection of a variety of small, metal-jacketed explosive devices; see Fig. 2.

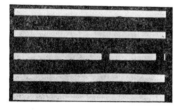

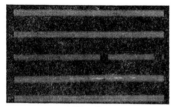

Fig. 1.

Fig. 2.

In both cases, the radiographs provide a variety of information. For the fuel work, the radiograph not only shows the condition and position of the fuel, jacket and components within the assembly, but also gives information about the change in the dimensions of the fuel, shows deposits in the void regions, and could be used to indicate the isotopic distribution within the fuel. In the case of explosives, the presence of the explosive in a metal device can be determined, and observations about explosive density, voids, and foreign material, such as oil, can be made.

Some additional neutron radiography applications include inspections of adhesive-bonded-metal honeycomb assemblies, location of fluids and lubricants in metal systems, inspection of boron-filament composites, studies of fluid migration in metal systems, determination of uniformity of neutron attenuation in reactor control rods or fuels, inspection of brazed joints, determination of hydrogen content in metals and organics, determinations of moisture content in ceramics, and inspections of turbine blades, castings, electronic components, and welds.

Some interesting recent neutron radiographic application work has included studies of diffusion in a D_2O/H_2O system, observations of hydride† in welded titanium hardware, the dynamic distribution of water in a metal heat-pipe assembly, detection of icing in the carburettor of a gasoline engine, the uniformity of a radioactive ^{244}Cm–Be neutron source, and several investigations that used neutron radiography in biological work. Neutron radiography could be applied to four primary biological areas. These include: (a) observations of tissue ordinarily masked by bone in an X-radiograph; (b) study of tumors in bone (bone must have a minimum of surrounding tissue); (c) observation of variations in hydrogen density, as is often the case in tumorous tissue; and (d) the possible use of a variety of contrast agents to examine biological systems. Recent work in this field has resulted in a useful application that involves tumor visualization in bone excised from human jaws.

The use of thermal neutron radiography for practical inspection work is growing. Future developments should lead to improved methods and practical applications of radiography with neutrons beyond the thermal-energy range.

Bibliography

BARTON, J. P. (1970) Radioisotope sources and neutron radiography, *Chem. Engng. Prog. Symp. Series*, **66**, 106, 91.

BARTON, J. P. (1972) A Visual Image Quality Indicator (VISQI) for neutron radiography, *J. Materials*, **7**, 18.

BARTON, J. P. and PERVES, J. P. (1966) Underwater neutron radiography with conical collimator, *Brit. J. Nondestructive Testing*, **8**, 74.

BERGER, H. (1969) Neutron image intensifier, *Encyclopaedic Dictionary of Physics* (J. Thewlis, Ed.), **3**, 222, Oxford: Pergamon Press.

BERGER, H. (1971) Neutron radiography, in *Ann. Rev. Nucl. Sci.* (E. Segrè, Ed.), **21**, 335, Palo Alto: Annual Reviews.

BERGER, H. (1972) The present state of neutron radiography and its potential, *Mat. Eval.*, **30**, 55.

BROWN, M. and PARKS, P. B. (1969) Neutron radiography in biological media-techniques, observations, and implications, *Am. J. Roent. Rad. Ther. Nucl. Med.* **163**, 254.

† Hydride concentrations ~500 ppm were observed; more recent work has revealed concentrations as low as 50 ppm.

HAWKESWORTH, M. R. and WALKER, J. (1969) Review: Radiography with neutrons, *J. Mat. Sci.* **4**, 817.

KALLMANN, H. (1948) Neutron radiography, *Research*, **1**, 254.

MATFIELD, R. (1971) Neutron radiography, *The Atom*, No. **174**, 84.

Neutron Sources and Applications, Proceedings of an American Nuclear Society National Topical Meeting, 1971, Report CONF-710,402, 3 vol., National Technical Information Service, Springfield, Virginia 22151.

RAY, J. W. (1969) Neutron radiography, *Res. and Dev.*, **20**, 7, 18.

SPOWART, A. R. (1972) Neutron radiography, *J. Physics E: Scientific Instruments*, **5**, 497.

THEWLIS, J. (1956) Neutron radiography, *Brit. J. Appl. Phys.*, **7**, 345.

THEWLIS, J. (1962) Radiography by neutrons, in *Encyclopaedic Dictionary of Physics*, **6**, 104, Oxford: Pergamon Press.

H. BERGER

NEUTRON STARS. A neutron star is a highly condensed stellar remnant with a mean density comparable with that of the atomic nucleus. The possibility of the existence of such stars was first suggested in 1939 when Oppenheimer and Volkoff showed that they might be the end product of the evolution of stars too massive to form white dwarfs, but it is only since the discovery of pulsars in 1967 that experimental evidence for their existence has become available; it is now generally accepted that pulsars are rapidly rotating neutron stars.

A star is originally formed by the onset of gravitational contraction in a diffuse cloud of gas and dust of the order of 1 light year (10^{13} km) across. As such a cloud contracts, it converts its gravitational potential energy into thermal energy and eventually becomes hot enough to start "burning" its hydrogen by means of thermonuclear reactions in its core. When this occurs, the resulting build-up of internal thermal pressure soon brings the collapse to a temporary halt, producing a stable main-sequence star with a radius of several million km and a mean density of the order of 10^2–10^4 kg m^{-3}. When the star has exhausted its various thermonuclear fuels, however (after going through a complicated evolutionary process whose exact nature depends on its original mass and composition), its gravitational contraction is resumed and can end in one of three ways depending on the mass of the star. In the case of a star of less than about 1·5 solar masses, Chandrasekhar has shown that the collapse is eventually brought to a permanent halt by the progressive build-up of so-called *degeneracy pressure* in the free electron gas that permeates the material of the star. (As an electron gas is compressed to progressively higher densities it eventually forms a degenerate Fermi gas in which the energies of the individual electrons are determined not by the temperature of the gas but by the requirements of the Pauli exclusion principle, which does not allow all the electron velocities to be small even when the temperature is low; as the density of such a gas increases, the mean velocity of the electrons increases, thus causing the pressure to rise.) Such a star becomes a *white dwarf* with a radius comparable with that of the Earth and a mean density of the order of 10^9–10^{10} kg m^{-3}. In the case of a heavier star, on the other hand, the gravitational forces are too powerful for the collapse to be halted in this way, and it is generally believed that such a star, after undergoing one or more catastrophic explosions in which it loses a large part of its mass, ends up either as a *neutron star* or as a *black hole* depending on its final mass. In the case of a stellar remnant of less than a certain critical mass (originally given by Oppenheimer and Volkoff as 0·7 solar masses but now thought to be nearer 3 solar masses), the collapse is eventually brought to a permanent halt by the build-up of degeneracy pressure in the free neutron fluid of which the star becomes increasingly composed as its mean density rises above about 3×10^{14} kg m^{-3}. Such a body becomes a neutron star with a radius of the order of 10 km and a mean density of the order of 10^{17} kg m^{-3}. If the mass exceeds the Oppenheimer–Volkoff limit, however, the gravitational forces are able to overcome even this new degeneracy pressure and the body collapses beyond the neutron star stage, becoming a black hole (a region from within whose boundaries no signal of any sort can escape) when its radius falls below its Schwarzchild radius (*see* BLACK HOLES IN SPACE). Oppenheimer and Snyder have shown that the collapse of such a body should continue indefinitely, since there are no known forces capable of bringing it to a halt.

Although the theoretical possibility of the existence of neutron stars has been known since 1939, it was only during the late 1960s that positive experimental evidence for the existence of such objects was obtained. The breakthrough occurred in 1967 when the Cambridge radioastronomers Hewish and Bell discovered the first pulsar, an event that triggered off an extensive and highly successful search for other similar objects. (Pulsars are faint, highly compact, intra-galactic radio sources which emit pulses of electromagnetic radiation at regular intervals, their periods being of the order of 1 second or less and being constant to better than 1 part in 10^8.) By a process of elimination, it was found that the only astronomical objects capable of producing periodic signals of the type emitted by pulsars were neutron stars, and, in 1968, Gold suggested that pulsars were in fact rapidly rotating neutron stars, their pulses being produced by some form of "lighthouse" effect as they rotated. In the same year, Staelin and Reifenstein discovered a very-short-period pulsar at the centre of the Crab Nebula, a pulsar that was subsequently positively identified as a faint blue star near the centre of the Nebula. This discovery led to the general acceptance of Gold's theory, since it was known that the Crab Nebula was the remains of a supernova explosion that occurred in 1054—the sort of event that had been suggested as a possible birthplace of neutron stars. (Indeed, Wheeler (in 1966) and Pacini (in 1967) had previously put forward the apparently far-fetched suggestion that the source of the enormous amount of energy being emitted by the Crab Nebula was in fact a rotating neutron star, but, until 1968, their theory could not be tested since no one knew what a neutron star looked like.) A pulsar similar to that in the Crab Nebula has been discovered at the centre of another supernova remnant (in Vela) and this has also been identified as a rotating neutron star.

Since 1968 a great deal of information about the nature of neutron stars has been obtained, partly by the theoretical investigation of the properties of ultra-dense matter and partly by the study of pulsars—particularly those in the Crab Nebula and in the Vela supernova remnant. With regard to the origin of neutron stars, the theory that they are the end product of the evolution of massive stars (probably O and B stars) has proved to be in good agreement with the experimental facts, since the galactic distribution of pulsars has been found to be almost identical to that of stars of this type and it has also been shown that the observed periods of pulsars are consistent with their having been produced by the gravitational collapse of such stars. O and B stars have rotational periods of the order of 10 days or less, and, if angular momentum were conserved during their gravitational collapse, the initial rotational periods of the resulting neutron stars would be of the order of 10^{-5} second. In practice, however, it is believed that most of the original angular momentum is dissipated in the form of gravitational radiation during the course of the collapse, thus increasing the initial rotational periods of the neutron stars to roughly 10^{-2} second, periods that subsequently become progressively larger as the neutron stars dissipate their remaining angular momentum through the medium of their powerful magnetic fields. This picture is consistent with the fact that all the pulsars so far discovered have periods that fall in the range from 0·03 to 4 seconds and are steadily increasing.

Astrophysicists have also been able to build up a fairly convincing picture of the internal structure of neutron stars. It is now generally believed that they consist of a solid outer crust a few kilometres thick containing a superfluid core that may or may not possess a solid centre. The solid outer crust is believed to consist of a regular lattice of stable nuclei with atomic numbers in the range from 42 to 140 embedded in a highly degenerate free electron gas, the density of the material increasing from roughly 10^{11} kg m^{-3} just below the surface of the star to roughly 3×10^{17} kg m^{-3} at the inner boundary of the crust. The nuclei in this solid lattice are kept firmly in place by the repulsive Coulomb forces that act between them, thus giving the material of the crust a rigidity roughly 10^{18} times greater than that of steel. The superfluid core is believed to be completely devoid of nuclei (which break up when the density rises above 3×10^{17} kg m^{-3}) and to consist mainly of free neutrons plus some free protons and electrons. At its centre, the pressure is believed to be so great (over 10^{28} atmospheres) that even neutrons, protons and electrons become unstable, turning into hyperons and mesons; the density at the centre of the heaviest neutron stars is believed to be in excess of 10^{18} kg m^{-3}. Some astrophysicists believe that the central regions of neutron stars may consist of solid matter formed by the crystallization under intense pressure of the various fluids of which the core is composed.

The above picture of the internal structure of neutron stars has proved to be capable of explaining many of the observed properties of pulsars. In particular, it accounts for the occurrence of pulsar "glitches"—sudden discontinuous decreases in period first observed in the case of the Crab pulsar in 1969. These glitches are believed to be brought about by the sudden relaxation of stresses in the solid crust or core of the rotating neutron star that constitutes the energy source of the pulsar, stresses that are produced by the fact that the equilibrium configuration of the star becomes progressively more spherical as it slows down. (The events that give rise to glitches are sometimes referred to as "starquakes".) Strong evidence for the existence of a superfluid core in neutron stars is provided by the fact that the period of a pulsar takes a comparatively long time to settle down after such a glitch, indicating that the coupling between the various layers of the star is very weak.

See also: Black holes in space. Pulsars. Recent advances in radio and X-ray astronomy.

Bibliography

CLARK, J. W. and CHAO, N. C. (1972) The Crystallization of neutronic matter, *Nature*, **236**, 37.

GORENSTEIN, P. and TUCKER, W. (1971) Supernova remnants, *Scientific American*, **225** (1), 74.

LUYTEN W. J. (1962) White dwarfs, in *Encyclopaedic Dictionary of Physics* (J. Thewlis, Ed.), **7**, 759, Oxford: Pergamon Press.

MESTEL, L. (1961) Degeneracy in astrophysics, in *Encyclopaedic Dictionary of Physics* (J. Thewlis, Ed.), **2**, 274, Oxford: Pergamon Press.

OPPENHEIMER, J. R. and VOLKOFF, G. M. (1939) On massive neutron cores, *Phys. Rev.* **55**, 374.

OSTRIKER, J. P. (1971) The nature of pulsars, *Scientific American*, **224** (1), 48.

PINES, D., SHAHAM, J. and RUDERMAN, M. A. (1972) Corequakes and the Vela Pulsar, *Nat. Phys. Sci.* **237**, 83.

RUDERMAN, M. A. (1971) Solid stars, *Scientific American*, **224** (2), 24.
SCARGLE, J. D. and PACINI, F. (1971) On the mechanism of the glitches in the Crab Nebula Pulsar, *Nat. Phys. Sci.*, **232**, 145.
WRUBEL, M. H. (1962) Stellar evolutions, in *Encyclopaedic Dictionary of Physics* (J. Thewlis, Ed.), 6, 880, Oxford: Pergamon Press.

H. I. ELLINGTON

NEUTRONS, CHEMICAL BINDING OF. In recent years considerable effort has been devoted to the study of the internal electromagnetic structure of the neutron and, in particular, to the weak, short-range, spin-independent, velocity-independent interaction between the neutron and the electron. One contribution to such an interaction, the Foldy term, is related to the anomalous magnetic moments and is expected to arise if there is a charge separation within the neutron. A further contribution, the so called intrinsic neutron–electron interaction, may be associated with current theoretical models in which a virtual dissociation of the neutron into a proton and a negative π-meson is believed to occur for part of the time. However, the maximum estimated value of the intrinsic term is far less than expected on theoretical grounds.

It is conceivable that interactions of this type could lead to the formation of a bond between a neutron and an electron. In consequence, intense interest was generated when, in 1969, Grant and Cobble presented evidence for the existence of a weakly bound n^- species in LiF at 4 K. Consideration of the experimental results suggests that bonds are formed between neutrons and electrons trapped in F-centres in the crystal with a bond energy of about 0·1 eV and that the species decomposes with a half-life of some 30 s.

Basically, the experiment consisted of introducing 10^{17} to 10^{18} trapped electrons per cm^3 into small sections of LiF crystal by irradiation with gamma-rays and then irradiating the crystal, maintained at a temperature of 4 K, with thermal neutrons produced by a 14 MeV generator and moderated in paraffin wax. The specimens were rapidly transferred to a shielded neutron counter and allowed to warm up while counting was in progress. The neutrons were observed as a difference between pulse-height spectra accumulated for a specimen irradiated at 4 K and one resulting from runs in which the specimen had been subjected to neutron irradiation at room temperature, consequently, neutrons resulting from other reactions and background counts from other types of radiation should have been eliminated. The spectra attributed to the neutrons liberated from the crystal after the end of the irradiation were very similar in form to those arising from a moderated ^{252}Cf calibration source. The count observed was of the order of 43 ± 8 events accumulated over 23 runs for irradiation at a flux of 2×10^6 neutrons cm^{-2} s^{-1}.

In a series of subsidiary experiments, it was shown that removal of the trapped electrons by heating the crystal reduced the observed count rate by a factor of 4, while interposing a cadmium shield between the specimen and the detector, resulted in a far greater reduction. Other reactions which may have resulted in the production of delayed neutrons or spurious counts were examined and rejected as alternative explanations of the observed results. These results were interpreted on the basis that the neutrons had been trapped by the F-centre electrons and retained in the crystal for at least 40 s at a temperature of 4 K.

From the initial results it was tentatively suggested that the number of neutrons trapped increased with the intensity of the neutron flux but that saturation effects limited the increase of concentration with time for irradiations longer than 60 s. Some modifications were introduced in the light of later experiments which suggested that the effect decreased with both increasing neutron flux, being no longer observed for fluxes greater than 10^7 neutron cm^{-2} s^{-1}, and with increasing neutron exposure, falling to 0·4 neutrons per run after an exposure of $1·7 \times 10^{11}$ neutrons cm^{-2}; such crystals could be revitalized by thermal annealing followed by gamma-ray irradiation. Retention of neutrons was not observed for other materials examined nor in a number of other LiF crystals similarly treated.

Workers from other laboratories have attempted to confirm these results using neutrons produced in both generators and nuclear reactors at fluxes of up to 3×10^9 neutrons cm^{-2} s^{-1}, all without success. While a number of these experiments were undertaken under somewhat different conditions, the lack of any supporting evidence must cast serious doubts as to whether the observed effect can be attributable to the formation of a chemically bound species of the type proposed. The fact that such effects have only been observed in a material in which the high capture cross-section would be expected to limit both the rate of formation and the lifetime of a loosely bound species of this type supports this view.

While, qualitatively, the picture of an interaction between an electron and the extended, intrinsic electromagnetic field of the neutron may appear attractive, theoretical calculations desprove the existence of a bound state resulting from such interaction.

In the case of a neutron interacting with an F-centre electron through the attractive spin-independent electrostatic interaction determined by the neutron electromagnetic form factor a variational calculation may be used to estimate the binding energy. The interaction is assumed to be sufficiently weak as to produce no change in the F-centre electron wave function and the electrostatic potential is taken as being determined mainly by the F-centre electron. If the intrinsic charge density of the neutron is calculated by taking the non-relativistic limit of the neutron electromagnetic current it may be shown that the integral of the effective potential $V(r)$, $-\int_0^\infty dr V(r) r$, is approximately $10^{-8} (\hbar^2/M)$, where M is the mass of the neutron, orders of magnitude below the requirement for a bound state. Further, the potential energy of the neutron charge distribution in the external electrostatic potential may be estimated to be of the order of 10^{-10} eV. Rough estimates of the magnitude of the interaction energy arising from the spin-spin interaction indicate a value of about 10^{-8} eV. It is, therefore, very unlikely that either of these interactions could lead to the formation of a bound state.

Trapping of neutrons due to the whole, or to parts, of a block of material forming a potential well and their subsequent release could produce effects very similar to those likely to be observed from the decay of a bound species of the type postulated and future experiments in this field would need to be very specific on this point. To formulate such an effect mechanisms must be defined to cover the processes of trapping, retention and release of the neutrons.

If a material is assigned a refractive index $n^2 = 1 - N\lambda^2 b/\pi$, where b is the bound coherent scattering length, N is the number of scattering atoms per cm^{-3} and λ is the wavelength of the neutron $= h/(2ME)^{1/2}$, total reflection of neutrons with appropriate wavelengths will be experienced if the angle of incidence is less than the critical angle. This critical angle can be as high as $\pi/2$ in a variety of materials for neutrons with wavelengths in the range 500–1000 Å (energies: $3 \times 10^{-7} - 8 \times 10^{-8}$ eV). Recent experiments have demonstrated that such "ultra-cold" neutrons can be retained in large metal containers for the order of 15s; the fact that the neutron "lifetime" under such conditions is far less than the theoretical value, governed mainly by the natural decay of the neutron, has not been satisfactorily explained at the present time. If this effect is postulated as the retention mechanism, regular crystals of materials with a negative scattering length could form traps for neutrons in an appropriate energy state. For the majority of materials, which exhibit positive scattering lengths, the presence of gross irregularities such as large pores, grain boundaries or precipitated phases would be required. The size of traps of this type must be greater than a critical value, of the order of 2×10^{-6} cm for an energy less than 10^{-7} eV. Transition to, and from, the trapped state would be dependent on collective effects in the crystal with a phonon being emitted during an inelastic scattering collision. Using the Born approximation Yu Kagan has produced expressions for the density of bound states and the trapping rate under such conditions. He concludes that for a binding energy of about 10^{-7} eV

the "lifetime" of the "bound" neutron may amount to tens or hundreds of seconds at the temperature of liquid helium and that the trapping rate would be of the order of 0·1–1 neutrons s^{-1} for an incident flux of 10^{10} neutrons cm^{-2} s^{-1}; in this formulation the lifetime is found to be independent of the binding energy. However, experiments on ultra-cold neutrons have suggested that the role played by surface effects and inelastic scattering is not well understood and it is likely that such calculations of the lifetime of trapped neutrons represent upper limits rather than values likely to be observed under practical conditions.

At the present time authoritative opinion would appear to be strongly against the view that the formation of a neutron–electron bond is possible.

Bibliography

BEALL, FOWLER, W. and MARTIN, T. P. (1970) *Phys. Rev. Lett.* **24**, 557.
DRILL, S. D. and ZACHARIASEN, S. D. (1961) *Electromagnetic Structure of Nucleons*. Oxford: University Press.
FOLDY, L. L. (1958) *Rev. Modern Phys.* **30**, 471.
GRANT, T. J. and COBBLE, J. W. (1969) *Phys. Rev. Lett.* **23**, 741.
GRANT, T. J. (1971) Ph. D. Thesis, University of Purdue.
ISAAC, G. R. *et al.* (1969) *Phys. Lett.* **31**B, 63.
KAGAN, YU (1970) *JETP. Lett.* **11**, 147.
KROHN, V. E. *et al.* (1969) *Phys. Rev. Lett.* **23**, 1475.
PALEVSKY, H. (1962) Neutron, in *Encyclopaedic Dictionary of Physics*. (J. Thewlis, Ed.), **4**, 810, Oxford: Pergamon Press.
RIETSCHEL, H. *et al.* (1970) *Phys. Lett.* **31**A, 83.
SCHLETT, D. W. (1970) *Phys. Rev. B.* **2**, 3437.

H. BRIDGE

NONDESTRUCTIVE TESTING, AUTOMATION OF. Nondestructive testing is the term which is used to cover a wide range of techniques that have been developed to monitor product quality during or after manufacture, or provide maintenance surveillance of structural integrity during service. The essence of the techniques is that they are based on visual or physical methods of assessment which do not render the product in any way unserviceable as a result of performing the test. For the inspection of surfaces the tests usually used are primarily designed to improve the visual contrast of the defect or provide some more effective means of drawing an operator's attention to a defective region; coloured or fluorescent dyes, magnetic particles, light or infra-red beams are the most generally used aids. For confirming structural soundness within a sample some probing beam is necessary which will penetrate into the sample but be affected, in some easily identifiable way, by internal defects in its path; X-rays, γ-rays, nuclear particle beams, ultrasonic energy, and eddy currents are used and variations in transmissivity, reflectivity or scatter monitored and the data interpreted, directly or by inference, in terms of structural features within the product.

Most of these techniques have been developed over the years as passive inspection procedures with an in-built reliance on an operator and his intuitive skill and experience to interpret the test data so that a proper judgement can be made on its significance or importance. With growing emphasis towards a more positive approach to nondestructive testing, the techniques are being adapted to provide continuous monitoring of quality, utilizing feed-back data to identify trends and in certain situations to control the significant process variables themselves. Moving into these fields of application calls for a fresh appraisal of the test techniques themselves to assess their suitability for this new role, greater emphasis on methods and objectives of signal processing and data presentation, and the introduction of automation concepts to do away altogether with the need for operator control at any stage in the procedure.

Automated nondestructive testing is now well established for the ultrasonic testing of welds and special mechanical "crawlers" have been designed to move in a controlled fashion and accurately position the transducer arrays along the weld line to provide a fully automatic system of inspection. An automated "crawler" of a different type has been designed with a sensing device to locate a pipeline weld from the inside and automatically align an X-ray tube so that completely reproducible radiographs of girth welds can be taken rapidly, with an assurance of optimum radiographic sensitivity under the difficult and onerous conditions associated with pipeline construction. At the primary product stage automatic means of monitoring steel bars and billets for centre-line inclusions are now regularly installed and the test signals are used to operate the cropping mechanism and minimize material wastage. At a later stage automatic means of inspection and gauging fast moving sheet are also well established using both ultrasonic and X-ray techniques.

In manufacturing processes where a high level of sophisticated quality control is demanded and where there is a high and continuous throughput, much effort has been directed towards automating nondestructive testing techniques. Sorting bridges are regularly used as on-line equipment to carry out a comparative test of quality or heat treatment although many of them have been developed empirically and so need to be applied under conditions of strict control within narrow limits of product variability. Eddy current methods have also formed the basis of many satisfactory on-line installations for ensuring product uniformity by monitoring electrical resistivity. This is a property which very sensitively reflects variations in a manufacturing process by the effect they produce on microstructure. Again the greatest care must be exercised not to extrapolate such tests beyond the limits for which they have been properly evaluated and calibrated. Similar on-line eddy current tests are increasingly being applied to detect foreign material (e.g. metal fragments) in food manufacturing or processing plants. When set with a suitable threshold level of detectability, these systems are completely automatic and can usually be used in such a way that they can be made to actuate automatic pass–reject gates so that suspect products are immediately removed from the line.

Since its late arrival on the scene, the nuclear industry has tended to be a pace-setter in inspection requirements and the high level of testing required, the demanding materials specification and the higher throughputs as nuclear power programmes extend, has caused considerable attention to be directed to automated procedures. In the case of AGR stainless steel fuel cans, for example, a fully automated testing station has been developed at Harwell, and installed at Springfields, which is capable of accepting a production batch of tubes, inspecting each one along its entire length for bore uniformity, wall thickness variations and cooling fin profile uniformity. The rig is then designed to analyse the many thousands of digitized readings obtained on each tube from each of the three tests against a predetermined specification in a small on-line computer and produce a go/no-go decision instantaneously, without any intervention or decision making by an operator. This could be a forerunner of similar installations where inspection costs are of secondary importance to product integrity.

To meet the needs of automated nondestructive testing and anticipate the new situations that its more widespread application will introduce, there is considerable activity to design techniques and systems to match up to the more exacting demands which are inevitably associated with automated operation.

With radiography, automatic inspection has become more of a reality with the considerable improvement in direct-view imaging methods that do not rely on film recording, with its attendant problems of delay and expense. Fluoroscopy, television presentation, image intensifiers and image amplifier panels can now give adequate sensitivity for many product quality monitoring requirements and can form the basis of in-line continuous inspection when combined with suitable mechanical handling devices to orientate the components for optimum radiographic examination. The more recent evaluation of X-ray sources with focal spots as small as 10 microns diameter makes projection fluoroscopy even more attractive for certain applications and opens up the possibility of automated inspection of miniature components with far greater detail sensitivity than is possiple with conventional X-ray equipment. Direct-view presentation of radiographic images is only part of the way towards true automation, and pattern recognition techniques based on computer analysis of digitized image line scans is the ultimate goal which is now being sought in a number of centres. The inference, of course, is that one is able to programme a computer to carry out the complex and often rather subjective analysis of image detail now performed by the eye and brain of an experienced radiologist. Recent work has shown that relatively simple features such as porosity and depth of weld penetration can be monitored automatically with complete objectivity in this way from radiographic films of fuel element end welds. This is but one step from fully automated in-line inspection linked directly to the welding operation where a scanning radiation sensor replaces the film and provides direct data input to the computer.

Since X-rays and γ-rays can be so easily detected and their signal strengths analysed by radiation-sensitive detectors, they also form a useful starting point for a number of automated inspection procedures for in-line measurement of sheet thickness, coating uniformity, composition and level.

In the ultrasonics field the realization of fully automated systems is more remote. Automatic systems have already been referred to, but the data from such tests still has to be recorded and subsequently analysed. Rail testing cars are now routinely in use in many countries. These again are fully automatic in their ultrasonic inspection procedures but present significant problems of interpretation, which in themselves form an interesting subject for an automated system of data reduction and analysis. Ultrasonic holography is a technique of particular significance in this context since the result achieved is, in many ways, analogous to a radiographic film presentation. As with X-ray scanning, the output data from ultrasonic holography can be digitized and could readily be computer-processed to provide an objective assessment of the location and characteristics of internal defects. Ultrasonic thickness measuring systems are already being used in-line, and have effectively provided feed-back data during a mechanical or electrochemical machining operation to either inform or control as the situation demands.

In every technique of this type it is, of course, essential to ensure that none of the instrument or test variables can influence the decision-making of the computer and because of this, parallel emphasis must be directed to the performance of the transducer. This is usually an essential link both in the sending out of interrogating signals and in receiving the signals modulated by the information on which the defect analysis has to be based. Its performance must be clearly understood and its characteristics allowed for in the analysis.

Automated nondestructive testing is now a reality in many situations. However, the technical advancements in recent years in the test procedures themselves and in methods of signal processing and data handling should make their application an easier step, at that time in the future when a wider management appreciation of their economic benefits exists.

Bibliography

Automated inspection for defects and dimensions. *SIRA Conference Proceedings.* Adam Hilger Ltd., 1969.

Automated non-destructive testing. *Ultrasonics for Industry, 1971 Conference Proceedings,* pp. 76–85.

Cableless X-ray pipeline crawler. *Non-destructive Testing,* 3 (2), 83–87, 1970.

COTTERELL, K. (1969) Have digital computers a place in NDT? *Non-destructive Testing,* 2 (4), 269–273.

HALMSHAW, R. (1967) Automation in Radiology, *Proc. 5th Int. Conf. on NDT,* pp. 398–402.

HITT, W. C. (1970) Automated ultrasonic testing and facsimile recording systems in use in the United States, *Proc. 6th Int. Conf. on NDT,* Hannover, Report B15.

GOULD, W. S. (1972) The new economics of electromagnetic methods of NDT. *Brit. J. NDT* **14** (1), 1–5.

SAILER, W. (1970) Electromagnetic test and sorting systems and their application in the automotive industry. *Proc. 6th Int. Conf. on NDT,* Hannover, Report C11.

SHIRAWA, T. and HIROSHIMA, T. (1970) Automatic magnetic inspection of hot-rolled round steel bars. *Proc. 6th Int. Conf. on NDT,* Hannover, Report F12.

Ultrasonic "eyes" aboard British testing train. *Materials Evaluation,* **29** (6), 16A, 1971.

R. S. SHARPE

O

OPTICAL COMMUNICATION SYSTEMS

Introduction

The current world-wide interest in optical communication systems stems from man's apparently insatiable appetite for more telecommunications capacity. This need might be met by overlaying more and more of the existing systems but economic studies show that as the information flow along a given route increases, new techniques become more economic and in particular systems using light (visible or near infra-red) begin to look attractive. The exact point at which a clear economic advantage is reached is a complex function of system design, component cost and reliability and the characteristics of the particular route. But it has been recognized as being already near enough in time to stimulate a great deal of research and development.

Recent studies have fallen into two categories, those involving the transmission of light through the atmosphere (or outer space) and those in which the light is transmitted through some prepared guiding medium. The latter may consist of a series of lenses, a hollow pipe or a glass fibre which acts as a dielectric waveguide. In either category a wide variety of components have been studied for use as sources and detectors, modulators and amplifiers but in recent years interest has centred on a few for application in the most immediately attractive systems.

The ultimate possibilities for an optical system are illustrated by the figures for the lens guide. The losses per lens can be made as low as 0·08 dB. Thus for a lens every 100 m and a loss between repeaters of 50 dB, the repeater spacing is of the order of 60 km. Since the guide has extremely low dispersion, bandwidths of 10^{14} Hz might be transmitted. However, such systems attract little attention at present since the bandwidths that are likely to be used in the next two decades lie in the range 10^6 to 10^9 Hz with an absolute upper limit of 10^{10} Hz per channel.

Transmitter

(a) *General considerations*

The transmitter acts as a source of light which is modulated in order to carry information. The source is usually a laser since this offers a desirable combination of narrow linewidth, high intensity and good spatial coherence. But on grounds of simplicity and economics, incoherent sources will be used on occasion, the GaAs light-emitting diode (LED) being the most attractive. In the system environment, it is necessary to select the source bearing in mind its ease of modulation, coherence, linewidth, power efficiency, cost and output wavelength in order that the overall system shall work.

(b) *Sources*

The most commonly used are the GaAs semiconductor laser and LED, the HeNe and CO_2 gas lasers and the NdYAG solid state crystal laser. Their wavelengths are given in Table 1.

Since the GaAs double hetero-structure devices can be modulated to frequencies approaching 1 GHz directly by their drive current which is produced

Table 1

Source	Wavelength nm	Uncontrolled linewidth	Spatial mode control	External modulator needed	Power per mode	Power efficiency %
GaAs DH laser	850–900	10^{11} Hz	Difficult	No	1 mW	10
GaAs DH LED	850–900	10^{13} Hz	None	No	1 μW	10
HeNe	633	10^9 Hz	TEM_{00}	Yes	10 mW	0·01
HeNe	1152	10^{10} Hz	TEM_{00}	Yes	10 mW	0·01
NdYAG	1060	10^{10} Hz	TEM_{00}	Yes	10 mW to 1 W	1–10
CO_2	10,600	10^6 Hz	TEM_{00}	Yes	1 W +	10

by a semiconductor drive circuit, they are immediately attractive for inclusion in an operating system. They offer high-power efficiency and are very compact, rugged and potentially mass produced. However, at present difficulty is experienced in obtaining adequate lifetimes for many system applications, figures being quoted of 1000 hr for the LED and 100 hr for the laser, as against sought lifetimes of 10^4 to 10^5 hr. The linewidth of the laser is generally acceptable for currently proposed usage, but ultimately would limit the useable bandwidth in many systems because of the material dispersion of the transmission medium. The use of the LED is more restricted and is discussed under multimode guide.

The HeNe laser has been used for ground-based systems operating through free space, for which it provides an attractive source. But its low-power efficiency rules it out for most applications. The NdYAG laser is a powerful and rugged source and can be operated with a stable spatial mode and narrow spectral linewidth. Nevertheless, an external modulator is required to impress information upon its output and this with its additional mechanical complexity limits the laser's use.

For more sophisticated applications, the CO_2 laser is attractive, but to obtain highly sensitive detection at 10·6-micron wavelength heterodyne detection is necessary with the attendant complication of providing a stable local-oscillator.

(c) *Wavelength*

For atmospheric transmission, the wavelength must freely propagate through the Earth's atmosphere which means that it must lie within the range of the visible to 1200 nm, or within transmission bands centred on 2000 nm, 3500 nm, or 10000 nm. All the above lasers satisfy this requirement.

For guided wave systems, the loss of the guiding medium must be low and this is wavelength dependent. Optical glasses transmit the visible band out to about 1200 nm. At the shorter visible wavelengths (green or blue) Rayleigh scatter losses generally become large and beyond 1200 nm overtone bands of the OH vibration spectrum from included water adds serious absorption. In general transmission is restricted to the band 600 nm to 1200 nm, the exact values depending upon the medium. It follows that the GaAs (850 nm) and NdYAG (1060 nm) sources are the most suitable, since the HeNe laser is usually considered to have too low power efficiency.

(d) *Modulation*

Many forms of modulation have been studied. Intensity modulation of the carrier to carry AM information is conceptually the simplest but because of the quantum nature of light it requires large amounts of signal power to provide high-quality transmission. More commonly used is some form of pulse code modulation (PCM). Usually, a binary code is employed in which zero transmitted power corresponds to digit zero for that time interval, and full transmitted power corresponds to digit 1. Using such a code, considerable noise discrimination is possible at the receiver since only 0 or 1 has to be selected from the received power. The signals are "regenerated" in the electronics before retransmission in a multi-link transmission system to provide acceptable error rates for the long distance transmission of wideband information over a multi-repeater link. Much detailed design work has been done on the particular form of PCM that is most suited to optical transmission. For example, linear pulse position modulation might be used whereby the position of a single pulse in each group of N time elements signifies 1 out of N possibilities. Or binary PCM may be used in which each time element has an equal probability of being 0 or 1.

Frequency modulation has been studied at optical frequencies but it requires a considerable degree of sophistication both in the transmitter and the receiver to operate. In particular, it requires the laser source to oscillate in a single frequency with high stability, a requirement which rules out the great majority of existing lasers.

Transmission

(a) *Free space*

For transmission through the atmosphere, the modulated beam is launched through collimating optics, in the form of a telescope operated in reverse, the optical analogue of the micro-wave dish aerial, to send out an optical beam with extremely low divergence. For example, a diffraction limited beam from a 20 cm diameter aperture has a divergence of 10^{-5} radians so that after travelling 1 km the beam diameter has increased by 1 cm. In principle it would appear that extremely efficient energy transmission could be obtained from source to detector with such a system, but in practice thermal fluctuations in the gas path distort the beam and pollutants such as smoke, fog or rain scatter it, so that serious problems are encountered with signal fading at the receiver. But if this were not so, aligning such a link to an accuracy of 10^{-5} radians would be quite difficult. In practice atmospheric links are generally used over short single-hops, of order a km or less, to transmit medium bandwidth information. Typically, a link might be used across a conference hall or between adjacent buildings, carrying a colour TV picture or similar amount of information.

Optical systems have been proposed and tested in outer space, for use between spacecraft and from spacecraft to ground. In the latter case, many of the considerations affecting wholly ground-based systems apply since transmission remains through the atmosphere, although vertically rather than horizontally. But between spacecraft, the transmission is through a vacuum and the major problems concern the acquisition and tracking of the receiver and transmitter one to the other. The very low beam widths made possible by the optical system allow efficient transfer of power from transmitter to receiver over enormous distances, but the extreme tracking accuracy required coupled with the round-trip time delay involved in the transmitter discovering if it is accurately aligned with the receiver pose special problems.

(b) *Guided waves*

1. *Optical fibres.* Guided wave systems are the ones currently receiving the greatest attention since it is now likely that they will be used in the world's telecommunications networks in the 1980s. A key element is the guidance medium, and the most intensively studied is glass fibre. This consists of a core of glass, typically 3 to 50 microns diameter surrounded by a cladding of glass 10 to 50 microns thick and of slightly lower refractive index. If the core has refractive index n_1 and the cladding n_2, then typically $(n_1 - n_2)/n_1 = 0.01$. The light travels down this fibre by total internal reflection at the core cladding interface and if the fibre is made with sufficient care from sufficiently good materials, the loss may fall as low as 3 to 5 dB/km. Economically losses of less than 20 dB/km are generally required to compete with other transmission systems. And although these fibres are extremely small, with care they can be produced in kilometre lengths and ultimately formed into cables that can be handled by conventional means.

2. *Fibre attenuation.* Fibre attenuation arises from several mechanisms. Dominant are scatter and absorption. The most fundamental scatter loss is the Rayleigh scatter which is caused by minute fluctuations in the material density, fluctuations having a size much less than a wavelength. These fluctuations are caused by the amorphous nature of the glass material, by contrast to a crystal where the atoms are arranged in perfect order. At the wavelength of the GaAs laser, the best optical glasses show losses due to Rayleigh scattering of between 1 and 10 dB/km, the loss varying as λ^{-4}. Larger-scale fluctuations in composition, caused by imperfect dissolution of the glass constituents or entrapped gas bubbles, give rise to Mie scattering which can be extremely serious but which can be made negligible in a properly controlled fibremaking process. Finally scatter losses can also occur from changes in the core diameter or shape, effects which once again are avoidable in a well-controlled process.

Absorption loss in the red and near infra-red is caused by two mechanisms. One is due to transition metal ion impurities, the most significant being Fe^{2+} and Cu^{2+} with Mn^{3+} and Ni^{2+} as secondary contaminants. The other major absorber is the O—H bond which produces in the 800- to 950-nm band significant overtones of its fundamental stretching vibration. Typically, the transition metal impurities must be held to a level of $1:10^8$ by weight and the O—H content must be held below a part in 10^6 by weight. The achievement of such low scattering and absorption values presents difficult materials problems. Nevertheless, successful solutions have been demonstrated by several workers.

3. *Fibre manufacture.* Fibres are pulled from a structure whose dimensions are typically of millimetre or centimetre dimensions. This preform usually contains both the high index core region and the lower index cladding and is pulled into fibre while heated to softening. There are many different ways of forming the large-scale preform which may be assembled from solid prepared glass or may be formed directly from molten glass. Some approaches are outlined below.

One is an ion-diffusion technique in which the outer layers of a uniform refractive index glass rod or fibre have one metallic ion exchanged for another to yield a lower refractive index cladding. The best known fibre of this type is called Selfoc. Another technique is to form the fibre from a rod of one material which is slid inside a tube of lower index material. The composite structure is pulled to form a fibre. A similar approach is to extrude the two glasses from concentric crucibles at high temperature and to pull a fibre filament from the molten composite that emerges. A quite distinct approach is to pull a hollow tube to form a hollow fibre and then to fill the pulled tube with a liquid of appropriate refractive index to produce the liquid-core fibre. Each of these techniques offers advantages and has problems and there are in addition many variations on the above basic approaches. In all cases the fibremaking procedure is dominated by the need to achieve the very low loss of less than 20 dB/km in order that the system shall be economically viable.

4. *Single and multimode fibre-bandwidth.* The fibres fall into two distinct categories, single mode and multi-mode. The single mode fibre has a very small core diameter, typically 3 μm (3000 nm) such that only a single waveguide mode can be supported by the dielectric waveguide that it forms. Multimode fibres have larger cores and may support 10^3 or 10^4 propagating modes.

In a multimode fibre each waveguide mode propagates at a distinct group velocity. It follows that if a pulse of energy is launched uniformly distributed among the modes it will be dispersed in time upon detection and the bandwidth will be reduced. The degree of reduction depends in practice upon the refractive-index difference $(n_1 - n_2)$, the distance travelled and the details of propagation. If only a few modes are used out of the many, the loss of bandwidth will be less. Also if the propagating modes are tightly coupled so that the energy launched in a given mode couples into all other modes, and vice versa, then the observed dispersion will also be reduced because of the averaging effect of the mode mixing. The extent to which mode mixing can be obtained without incurring substantial loss by scattering out of the fibre remains to be established, and is a subject if intense current study.

The attraction of the multi-mode fibre lies in its ability to collect more power than a single mode fibre is capable of doing from many sources since with a diffuse source, such as an LED, the power launched into the fibre is roughly proportional to the product of the power per mode and the number of modes of the fibre (see Table 1). A single mode laser is capable in principle of launching all its power into either a single or a multimode fibre. We note also that the laser has a narrower spectral bandwidth than the LED (for GaAs, 10^{11} Hz compared to 10^{13} Hz) and for this reason the dispersion of the guide material is much less significant.

It follows that for low data rate transmission (say 10 Mbit/sec) over distances of a few kilometres the LED-multimode fibre is attractive. For high data rate (above 100 Mbit/sec) transmission over the same distance (between repeaters) the laser source is almost obligatory, and in between lies a range of applications where detailed study of the particular case is necessary to establish the best combination.

Other Components

(a) *Detectors*

Detectors have not in general presented a major problem for the fibre optical systems based upon GaAs sources. Efficient semiconductor photoconductive and P–N junctions, with and without avalanche gain, are available. They have high quantum efficiency and fast response so that for high data-rate detection little problem exists. In the visible region, photomultipliers offer exceptionally low noise detection and give response to 1 GHz or more. Beyond 1100 to 1200 nm, detection becomes more difficult and presents a further reason for using sources of wavelength equal to or shorter than the NdYAG laser. The most sensitive detection in this range is obtained by using optical heterodyne mixing on a photoconductive surface, but this requires a coherent stable, single-frequency transmitter source and a very stable local oscillator in the receiver. Both are extremely difficult to produce.

(b) *Modulators*

In the past, electro-optic and acousto-optic modulators have been constructed using bulk components, with impressive performances, but requiring great skill in their fabrication.

Recently, studies in integrated optics have brought forth a range of planar optical devices which promise to be more efficient than the bulk devices and potentially mass produceable. Acousto-optic modulators have been made in which the acoustic energy propagates as a surface wave to interact with the optical energy which is channelled in an evaporated thin-film waveguide. Very efficient interaction is possible between the two, allowing the acoustic energy to scatter the optical beam. In similar vein, electro-optic modulators have been constructed using an electro-optic crystal as a substrate onto which an optical waveguide is added together with electrodes to apply the switching fields.

(c) *Integrated optics*

In addition to the work just outlined on modulators, a whole range of other devices in planar geometry are under study. They have in common the desire to move away from the individual bulk devices, to a composite multi-element planar device formed by the processes of integrated circuits on a single substrate. Ultimately it seems likely that a laser source, with frequency and mode control filters, will be produced along with a single- or multichannel modulator and the necessary couplers, on a single substrate. It may even be possible to produce a complete optical repeater with amplifier and regenerator on a single substrate. In either case, such elements would be suitable for mass production and could open the way to the utilization of the full capability of the optical communication system.

Summary

Optical communication systems have been undergoing intensive development for a variety of applications for some years. The simplest applications can now be satisfied but the large-scale applications in the national telecommunications networks await further development and a proven ability to compete on straight economic grounds. The first systems are expected to be given field trials within 5 years and it will be surprising if none are operational within the decade. After that there will remain scope for improvement and development for many years more.

Bibliography

KOMPFNER, R. (1972) Optics at Bell, optical communications, *Applied Optics*, **11**, 2412.

MARCUSE, D. (1972) *Light Transmission Optics*, Van Nostrand Rheinhold.

RAMSEY, M. M. (1973) Fibre optical communications within the United Kingdom, *Optoelectronics*, **5**, 261.

Ross, M. (1966) *Laser Receivers*, Wiley Interscience.

SAITO, S. (1972) Optical communications in Japan, *IEEE Trans. Communication*, Com. 20. 725.
YARIV, A. *Quantum Electronics*, Wiley Interscience.

J. E. MIDWINTER

OPTICAL DATA PROCESSING. This expression is usually restricted to the processing of data of electrical origin by optical means. The processing of data of optical origin is usually referred to either as image formation and enhancement or as picture improvement, even when, as with the records from lunar and space probes, the optical data is digitized and fully processed in an electronic computer. Pattern or character recognition is another category of optical processing and is not discussed here.

Even in its more restricted sense optical data processing comprises the more usual functions of a general purpose electronic computer and can be considered under three headings: (1) data manipulation and transformation equivalent to central processor operation, (2) high-capacity high-speed random access data storage and retrieval equivalent to ferrite core storage, (3) very high capacity serial access data storage and retrieval equivalent to magnetic tape decks (but more so: see section 3).

1. *Data Manipulation and Transformation*

The attraction of optical methods for data manipulation lies in the possibility of working simultaneously in two dimensions, whereas conventional electronic methods operate in one dimension with time as the basic variable. The two dimensions in the optical case may be used for two independent variables or to provide a number of independent parallel computing channels. The first basic property which enables such results to be achieved arises from the multiplicative property of absorption filters, i.e. if a plane wave front of intensity $I(x, y)$ is incident on a transparency with transmission $T_1(x, y)$ the transmitted intensity is $I(x, y) T_1(x, y)$. A further transparency with transmission $T_2(x, y)$ leads to an output of $I(x, y) T_1(x, y) T_2(x, y)$. Such an output could be integrated overall by focusing onto a suitable detector. Alternatively an astigmatic optical system might be employed which focused in one dimension only onto a number of parallel detectors. The second basic property is the relationship between the amplitude distribution in a coherent plane wave at the front focal plane of a thin lens and the corresponding amplitude distribution at the back focal plane of the lens. The one is the exact two-dimensional Fourier transform of the other. This relationship is long established, (see any text on optics such as M. Born and E. Wolf, *Principles of Optics*, Pergamon Press, 1959). The use to which it can be put for spatial filtering is exemplified by Abbe in his theory of image formation in the microscope.

The power achieved by using this transformation from spatial to frequency domain is well demonstrated by considering the problem of spatially matching an input wavefront with an optical spatial filter. This process could be carried out in the spatial domain with a suitable reference transparency by physically displacing the incident signal relative to the reference in the directions transverse to the optic axis until a maximum output was achieved. This involves scanning in two directions, integrating the output for each scanning position, remembering the previous outputs, and finding the maximum. It takes time. By placing the transform of the reference at the focal plane of a transforming lens, the transform of the input signal can be convolved with the reference transform, and a further transform with a second lens will yield directly the coordinates of the optimally matched position of the input signal wavefront. All possible positions are effectively processed in parallel.

A practical example of multiple processing, but in a number of independent parallel channels, occurs in reconstruction of data from a synthetic aperture (sideways looking) airborne radar. In this case the radar returns from a coherent radar are recorded with a suitable offset as a photographic transparency with range across the film and distance travelled by the aerial along the film. The accumulated data at each slant range must be processed separately. This is done by illuminating the film with plane coherent light. "Images" of the original objects are now created directly in space with an azimuthal resolution of the same order as the diameter of the aerial. The images can be focused and rerecorded on an image film by using an astigmatic cylindrical lens to focus the range elements independently of each other, and a conical lens to bring the image plane for the different slant ranges normal to the optic axis of the system. The original record and the image film are now moved in synchronization to give a continuous record of the original radar scene. This processing is carried out in the spatial domain, but by adding further lenses optical matched filtering may be carried out in the frequency domain for all ranges simultaneously prior to recording. The filter must be chosen to select what is being looked for. A full account of this type of system is given by R. O. Harger, *Synthetic Aperture Radar Systems*, Academic Press, New York and London, 1970.

2. *High-Capacity High-Speed Random Access Digital Optical Stores*

The advantages of using light beams to access stores of this nature lie in the low inertia of a light beam whereby it may be deflected or switched on and off very rapidly, in the high resolution which may be achieved, and in the relative freedom of light beams from extraneous interference. In these high-speed stores resolutions approaching the wavelength of the light being used are not aimed at because of imperfections in the optical path distorting the wave front, and because of limitations in the accuracy of addressing at high speeds. Even so, quite high packing densities are achieved.

In its simplest form an optical store of this class comprises a mask, usually photographic, a number of light sources, usually semiconductor diodes, which can be individually addressed and operated, and a number of light detectors giving their outputs in parallel. The mask holds a rectangular array of bit positions perforated for "1"s. Each light source illuminates a row of bit positions on the mask, and each detector reads out a column of bit positions orthogonal to the illuminated rows. The capacity of a simple system of this nature is rather limited. The Optical Memory Systems OM 0102 memory has a capacity of about 10^5 bits with speeds of access of about 60 nsec. It is of course a read only memory and each mask must be prepared by the manufacturer.

The basic principle of using discrete bit positions on a mask can be elaborated. The light-sensing spot can be generated and positioned on the face of a cathode-ray tube. The spot can be multiplied by mirrors on a kaleidoscope principle to yield say 64 spots each imaged onto its own area on the storage mask. Each spot is capable of being placed in about 500×500 positions in its area in synchronization with the original CRT spot. Behind each area is a photomultiplier so this system can give 64 outputs in parallel. The overall capacity of a single mask is about 10^7 bits and the speed of access of 64 bits in parallel is determined by the settling time of the CRT. Although this could nominally be a few microseconds it proved difficult to design a servo to position the CRT spot to the requisite accuracy in that time, so the scheme failed to find a niche in the present computer hierarchy. The store, too, was of the read only type, but the basic unit could be adapted to write onto a photographic plate which could be developed and subsequently used as the data mask. The fundamental problem in the deflection positioning arises from trying to use analogue deflection systems to address a digitally positioned store at high resolution.

These problems can be overcome by using the holographic principle to store the data. In such a system a page of information containing typically 100×100 bits is stored as a single hologram on a photographic plate. As such a hologram need only be about 1 mm in diameter, it is feasible to put say 128×128 holograms in square array on a single plate of convenient size which thus holds about 10^8 bits of data. To read the data out, the chosen hologram is addressed by means of a laser beam steered by means of a digital light deflector. This latter uses the longitudinal electro-optic effect in deuterated potassium dihydrogen phosphate single crystals to change the polarization of the transmitted beam and is followed by birefringent calcite crystals to switch the beam in one of two directions depending on the polarization. Seven successive stages for each of the two orthogonal axes are used. Switching times of a few microseconds can be achieved, but accuracy of spot position is not too critical as the hologram is not being operated near the limit of its resolution. Each hologram when illuminated is arranged to reconstruct its 100×100 bit pattern over an array of 100×100 diode detectors ideally on a single chip complete with amplifiers and integral scanning. Only one such array is needed as the correct deflection is built into each hologram to form the reconstructions in the same place. The redundancy in the hologram array is distributed over all its elements and in this way the effect of dust and imperfections in the plate are shared, and not perhaps concentrated on a few bits as may arise in the simple form of masking store. The electro-optical deflection system is preferable to an acousto-optical one as it is digital and is capable of a higher (speed $\times$ resolution) figure of merit. Once again the store has been described in terms of using a photographic plate and is therefore basically a read only store, but again the basic design can be adapted to enable the hologram plates to be written under computer control. The plates are developed before use.

Many other materials have been considered and tried in these types of store in an attempt to find one suitable for fast read and write operation. Amongst those favoured at one time or another have been photochromic materials, optically transparent magnetic materials, and crystals (such as lithium niobate) susceptible to laser damage. Photochromics such as alkali halides with colour centres of various sorts have a very high resolution, are reversible and are quite sensitive. One wavelength of light is used to write information into the store by colouring or bleaching it, a second wavelength which ideally does not affect the material is used for reading, and a third is used to erase by restoring the original condition. Magnetic materials such as manganese bismuthide write in a bit of information by heating the relevant spot above its Curie point with a high-power laser beam and allowing it to cool in an appropriate magnetic field. Read out is effected with a laser beam of reduced power utilizing the Faraday or Kerr effect to change the polarization of the beam acording to the direction of magnetization. Writing in fresh information automatically erases the previous state. Compensated ferrites have also been proposed using a similar mechanism. Laser damage materials such as lithium niobate or KTN have been suggested for hologram recording. The laser beam gives rise to a small change in refractive index of the crystal and by using a suitable thickness of crystal a reasonable hologram efficiency can be obtained. Erasure usually requires thermal annealing. All these materials have not yet reached the standards for commercial use for one or more of the following reasons:

(a) They are difficult to obtain in large enough pieces with uniform and controllable characteristics.
(b) They deteriorate with use either by "staining" or by changing to a less sensitive state.
(c) They are too insensitive to achieve the high speeds of writing necessary under modern computer control.
(d) They are too delicate for commercial use.
(e) They are too expensive.

None of them come up to photographic materials in terms of sensitivity, availability, cheapness, reliability, reproducibility, and permanence.

3. Very High Capacity Sequential Access Digital Stores

Stores of this sort are designed to make full use of the resolution capabilities of light beams. The most common principle employed utilizes mechanical movement in one direction to scan the recording medium through a fixed light beam. This principle is of course the same as that used in magnetic tape decks and disc files, the light beam replacing the magnetic head(s). In order to select tracks the light beam may be moved at right angles to the scanning direction, for example by a mirror system. This now gives the additional facility of enabling the light beam to be locked onto a given track and follow it under servo control. In turn, this enables much narrower tracks to be used and much narrower spacing between tracks than is possible with conventional magnetic recording techniques. A number of competitors to magnetic disc, drum, and tape stores have been put forward using many of the new materials mentioned at the end of the last section. As they are all more expensive, more sensitive to vibration, and probably less reliable than conventional stores they are unlikely to reach commercial production. It is only if the potential speed or high-packing densities of optically read stores can be used to produce devices beyond the range of existing magnetic devices that a useful product can be achieved.

One of the more successful devices in commercial production is the Unicon store of the Precision Instrument Corporation. This was originally aimed at the archival store market and a reel of recording medium was to hold about 10^{12} bits compared with about 10^9 bits on a reel of magnetic tape of similar size and access time. The Unicon uses a high-power laser to burn micron holes in the metal layer of a metallized polyester tape. On read out the presence of such holes is detected by a laser at reduced power. In its most ambitious form the polyester tape was to be read with a helical scan as in modern video magnetic recorders, a rotating mirror taking the place of the rotating magnetic heads in such a recorder. In its most practical form a rectangle of recording medium is mounted on a rotating drum and a much simpler addressing system results. The capacity of an individual rectangle is much lower—say 10^9 bits—but speed of access is a fraction of a second and the rectangles can be changed mechanically in a fraction of a minute.

In order to obtain the highest packing densities some form of electron beam writing onto photographic or thermoplastic tape must be used despite the practical disadvantages of vacuum locks to allow passage of the tape through the electron beam system. In this way resolution better than the wavelength of light can be obtained. However, difficulties arise in reading out data at these densities and in practice an optical system is often used because of its sensitivity and relative freedom from interference, despite its lower resolution. At least one very large store is in operation using this hybrid system: electron beam recording on photographic film "chips" and optical readout. The film chips can be changed mechanically and it is of course a read only memory in the usual definition of that term.

The format of the data in sequential stores of the types described in this section tends to follow that of magnetic tape stores. The data is assembled in blocks with identifying codes at start and finish to remove any necessity for finding absolute positions on the records.

To sum up, optical data processing offers advantages over electronic processing in certain specialized areas where its ability to process in two dimensions, or its capability to achieve high packing densities, or its high speed of access can be utilized. It will not replace electronic processing in normal use.

A. C. MOORE

OPTICAL PROCESSING OF RADAR AND SONAR SIGNALS. Optical systems have the inherent property that they possess two degrees of freedom, as represented by the two independent coordinates which define a point on a wavefront. In this respect the optical system differs from the electronic system which possesses only time as the independent variable. As optical systems have the added property that a Fourier transform relation exists between light at the front and back focal planes of a lens, they offer advantages in some situations over electronic systems as a means of processing information. In addition, the Fourier transform is an essential mathematical function in filtering and sorting radar signals, and as two independent variables, range and speed, are required from a radar signal optical systems are directly applicable to their elucidation. The elegant optical techniques which have advanced with the development of a coherent light source, the laser, and the techniques of recording both the phase and amplitude of a light wave by holographic means now make optical processing of radar data possible.

With coherent laser light successive Fourier transforms can be taken with an optical lens arrangement. The system then behaves analogously to an electronic filter as a means of finding what frequencies are present in an input signal—a Fourier spectral analysis—or by the insertion of an optical filter, as a means of selecting one frequency component.

The Fourier transform property of a lens can be expressed mathematically. If coherent illumination is incident on a converging lens the light amplitudes $A_1(x_1, y_1)$ and $A_2(x_2, y_2)$ in the front and back focal planes of the lens, F_1 and F_2 (Fig. 1) are related by

$$A_2(x_2, y_2) = \iint A_1(x_1, y_1) \exp[-j\alpha(x_1 x_2 + y_1 y_2)] \times dx_1 dy_1 = F_{xy}(A_1(x_1, y_1))$$

where

$$\alpha = -\frac{2\pi}{\lambda f}$$

where λ is the wavelength of light, and f the focal length of the lens. The Fourier transfer F_{xy} is taken over the two independent variables x and y of the wavefront. The spectrum at F_2 of the light in the plane F_1 can be modified by the insertion of slits, stops, shaded transparencies or complicated optical filters acting on the phase of the light to select wanted information before further processing by other lens systems. The basic processing of data by a lens system can be understood from Fig. 2. The frequency components present in the step function (the slit) at F_1 are indicated by lines in the two directions x and y at F_2. The central spot corresponds with the average level of illumination emerging from the slit. The images in each direction on each side of the central spot correspond to the fundamental and harmonic frequencies comprising the step function.

An optical spatial filter can be used in the system where it is necessary to determine if a received signal consists solely of noise or of noise together with some signal of known form, or to determine whether a particular frequency component is present among many others. If the amount of noise is high or if the frequency components are numerous, a filtering scheme which increases the signal to noise ratio at the detection point is required. Referring to Fig. 2, a simple filter placed at F_2 with slits at the appropriate positions of the bright spots corresponding with the required frequency components will pass these components only. By use of a further lens a second Fourier transform can be accomplished producing a central bright spot due to the required signal only, Fig. 3. Should the required components be absent from the original signal (the optical object) the final signal will be of low intensity representing only the transmitted noise. Both Fourier analysis and spatial filtering by optical means can be applied to the processing of radar signals.

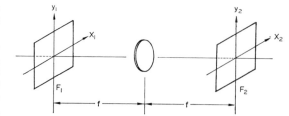

Fig. 1. Light distribution at the front and back focal planes of the lens, F_1 and F_2, are related by a Fourier transform.

The device which is essential to processing any electrical signal by optical means is a modulator of light which converts the electrical signal to a light signal. In radar there is the further essential feature that the modulator should operate in real time—the light output following faithfully the electrical signal without any delay. Many forms of light modulator have been suggested, the ultrasonic modulator having been tested to a greater extent than others in the processing of radar signals.

Amongst other devices for modulating light is the electron beam thermoplastic Lumatron tube. In this vacuated tube an electronically controlled electron

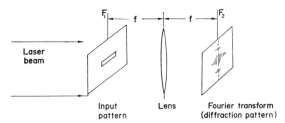

Fig. 2. Fourier transform of a wide slit has bright spots corresponding with the fundamental and harmonic frequencies contained in the slit (a rectangular pulse).

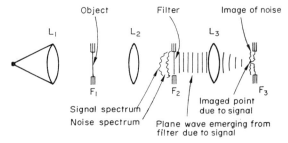

Fig. 3. An optical filter at F_2 selects the wanted object signal from noise which is detected at F_3 after a second Fourier transform.

beam scans a thermoplastic film thereby deforming it and altering its thickness. Light passing through the film is deflected through refraction at the film of varying thickness. By scanning the film at television rates the electrical signal can be converted to a light beam. The device can be further used in the same way to create a holographic filter for use in an optical-processing arrangement of the information contained in the electrical signal. Several other comparable devices are based upon the modulation of light by electro-optical crystals. Semiconductor crystals, such as potassium dihydrogen phosphate (KDP) and lithium niobate are scanned by an electron beam, as described above, and the deposition of electric charge on a face of the crystal induces local birefringence which causes rotation of the plane of polarization of light reflected from the crystal.

The ultrasonic light modulator will be discussed further. This device used in the optical processing of radar signals serves the same purpose as the electronic interface which joins measuring equipment to a computer, and turns signals into a language—in this case a light wave—which can be understood and acted upon by the data processor.

The basis of the ultrasonic light modulation is the well known effect that light is refracted at the wavefronts of an ultra-sonic wave moving in a liquid or solid. Most transparent liquids or solids experience a change in their optical refractive index as a sound wave propagates through them due to the successive compressions and rarefractions. Such changes in refractive index cause the wavefront emerging from the material to be locally advanced or retarded by the presence of the acoustical wave. The emerging light wavefront is effectively corrugated in passing through the ultrasonic modulator, the amount of corrugation being related to the intensity of the acoustical wave and to its frequency.

In its simplest form the ultrasonic light modulator would be a tube of the transparent medium with a piezo-electric crystal at one end to launch the acoustical wave which would represent the radar signal, and at the other end a good acoustical absorber to prevent reflections interfering with the wavetrain. When a plane coherent (laser) beam of light falls on the tube the wavefront of the emerging light exactly represents the radar signal. In brief, the radar signal is written across the light wavefront, every part of the signal being present in the optical aperture at one instant (Fig. 4). The amount of information which can be written on to a light wave is limited to frequencies within a bandwidth of about 100 MHz this limit being set by the acoustical losses of the medium. Materials suitable for ultrasonic light modulators are water, fused silica and lead molybdate.

To understand how the ultrasonic light modulator and the Fourier transform property of a lens can be used together to process optically the information

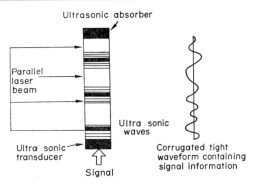

Fig. 4. Ultrasonic light modulator converts a radar signal to a corrugated light wavefront.

contained in radar echoes from targets consider the pulse-doppler type of radar in which a coded (coherent) train of radar pulses is transmitted. The echo from each pulse in the train contains information of the range through the round-trip delay of the pulse reflected from each target, and through the doppler frequency shift the speed of approach (Fig. 5). Analysis of the waveform of each pulse gives simultaneously the range and velocity of targets. By using a train of pulses the echoes from each can be processed together enhancing the resolution in both the range and velocity, so that overlapping signals from many targets can be separated.

The optical processor can be utilized to ascertain both range and speed because both of these parameters are fixed by periodicities in the radar echoes. The range of each target is fixed by the time delay between the transmitted pulse and its echo, and this can be ascertained by finding the periodicity

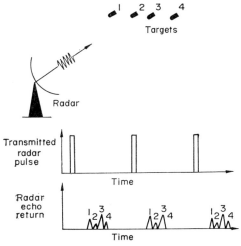

Fig. 5. Schematic diagram of pulsed Doppler radar.

of the delay throughout the train of pulses—a type of spectral analysis. Likewise the speed of approach is obtained by feeding the entire sequence of returns from the targets into a spectrum analyser and noting the doppler frequency shifts with respect to the frequency of the transmitted pulse. The two dimensional processing ability of an optical system is ideal for this computational requirement.

In an optical system, the chain of returned radar pulses together with the original transmitted pulses is presented to a simple lens system as a modulated beam of light, and the pattern produced in the back focal plane contains a Fourier analysis of the signals. Thus in one instantaneous operation, the range of the target may be obtained from the Fourier analysis of the time delay of the echoes, and the speed of approach from the Fourier analysis of the frequency shift of each echo.

One complicated way of manipulating the radar echoes would be to have a large number of acousto-optical cells stacked side by side. With a sophisticated electronic sorting and delaying arrangement, each radar pulse in turn could be directed into its own cell to be launched simultaneously with all others as an acoustical wave. Coherent light shone through all the cells would pick up the desired periodicities representing doppler shifts and delays. A more satisfactory arrangement is to use only one acousto-optical cell. As each radar pulse echo arrives it is fed into the cell and its corresponding corrugated light wave is recorded on light-sensitive material. The recordings are then compared optically.

The method of recording is important because both the phase and the intensity of the light wave must be retained for further spectral analysis. Holographic methods are ideally suited to this requirement, for by adding a coherent light beam to the radar-echo light signal, an interference pattern is created which retains both phase and intensity of the echo wave (Fig. 6).

When the entire echo of one radar pulse of the pulse train is inside the acousto-optical cell the laser light is pulsed on for the allotted hologram exposure time. The hologram occupies a vertical strip on the film, which is continously moving across the recording position. The sequence of radar returns is recorded as holograms side by side on the photographic film. Each hologram contains the required radar information—range and speed. Once this film is developed the sheet passes into the first focal plane of a lens system. It is again illuminated with coherent light and the lens system processes the information, producing in its back focal plane the Fourier analysis of the input signal. Because of the alignment of the input information on the film— a parallel series of exposed strips—a series of spots of light appear in this plane. They indicate on one axis the ranges of the targets and on the other their speeds of approach. Each spot of light is due to one of the targets and the sharpness of each is theoretically fixed by the increased resolution due to processing a whole train of pulses (Fig. 7).

A further, perhaps more significant advance might be possible by using the fact that once a spectral analysis has been made, certain frequencies can be picked out by interposing filters. A transparent pass-filter could be interposed at the back focal plane of the lens so that a particular set of frequencies present in radar returns could be swiftly recognized. Such filters representing desired information could be created by using a second acousto-optical cell, or, perhaps another type of addressable light filter driven by a computer. With a processor working in real time, a filter could even be updated or modified in order to select more exactly particularly required information.

The necessary delay between recording the echo pulses and processing the data, while the photographic film is developed, is a present disadvantage of the system. Future research in photographic ma-

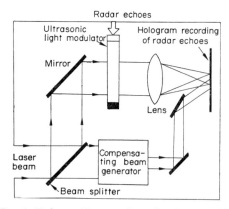

Fig. 6. Holographic recording of radar signals.

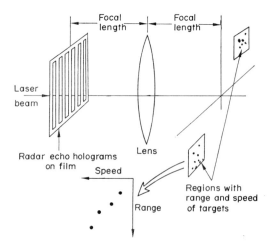

Fig. 7. Optical processing of recorded radar signals to yield range and speed of targets.

terials may reduce this delay but operation of the processor in real time may be possible by using a new class of materials in which light produces a change in the index of refraction. In such materials information can be written instantaneously, erased and rewritten. These materials consist of ferroelectric crystals such as lithium niobate and barium titanate.

The potential of the optical processor in the field of radar could be large, but it could be applied equally well to sonar and seismic work. In both of these fields the same concepts are involved as in radar—a burst of acoustical energy is released and the echoes from objects in its path are recorded. In sonar, underwater objects such as submarines, fish, or the sea bed are investigated whereas in seismic investigations subterranean geological features are explored. The interpretation of sonar and seismic signals relies on carrying out a spectral analysis of the signals.

Corrections for the known changes in velocity of sound waves with depth in sea water could be applied to sonar by interposing appropriate optical filters in the processor. Both echo-sonar or passive sonar (that is detecting underwater sources by listening devices placed in the sea, such as sonobuoys) could be processed by optical means. From an array of sono-buoys the sound wavefront emanating from a source could be reconstructed by an optical processor with suitable correction filters to investigate its position and depth.

See also: Optical data processing.

Bibliography

ARM, M. and KING, M. (1969) *Applied Optics*, **8**, 1413.

BIRCH, K. G. (1972) Spatial filtering in optical data-processing. *Reports on Progress in Physics*, **35**, 1265.

FOWLER, G. R. (1968) *Introduction to Modern Optics*. Holt, Rinehart and Winston.

JERRARD, H. G. (1971) Holography, in *Encyclopaedic Dictionary of Physics* (J. Thewlis, Ed.), Suppl. **4**, 184, Oxford: Pergamon Press.

LIPSON, H. (1972) *Optical Transforms*. Academic Press.

SLATER, K. F. (1962) Radar, in *Encyclopaedic Dictionary of Physics* (J. Thewlis, Ed.), **5**, 756, Oxford: Pergamon Press.

SWEDDOW, I. N. (1961) Fourier transforms, in *Encyclopaedic Dictionary of Physics* (J. Thewlis, Ed.), **3**, 284, Oxford: Pergamon Press.

I. FIRTH

P

PALAEOGEOPHYSICS. Geophysics is concerned with the estimation of physical parameters and forces in and near the Earth, their distribution and time-dependence. Direct geophysical observation is limited to the few hundred years of scientific activity; palaeogeophysics is concerned with extending the timespan of geophysical estimation to the geological time scale and ideally to the whole age of the Earth. This summary is itemized under the main disciplines of modern-day geophysics related to the solid Earth and its surface. Studies of natural isotopes and geochronology are not reviewed because their implications for the state of the Earth in remote geological time are chemical rather than physical. The study of the Earth's magnetic field in the past (Palaeomagnetism, q.v. for definition) takes pride of place because this is the geophysical field of force which is most commonly and unambiguously fossilized in the geological record.

Earth's Magnetic Field

Palaeomagnetic interpretation stems from the finding that magnetic iron oxides in rocks with no anisotropy of fabric commonly retain natural remanent magnetization (NRM) acquired at or soon after the time of formation of the rock, aligned in the direction of the local geomagnetic field at that time. The principal findings are:

1. The axis of NRM in young rocks (less than a few M.yr, old) at any locality averages to the axis of the field which would arise from a geocentric axial dipole (G.A.D.).

2. The axis of NRM in older rocks progressively deviates from the local G.A.D. field with increasing age. This is interpreted in terms of polar wandering (q.v.) and/or large scale systematic displacement of continents and ocean floor relative to each other and to the poles throughout geological time (continental drift, sea-floor spreading, plate tectonics). Systematic polar wandering has been shown to have been small compared with continental drift for at least the past 150 M.yr.

3. Polarity of NRM inverts frequently. Rocks of any one age worldwide have the same polarity. Hence the geomagnetic field has reversed repeatedly. Reversals have occurred, on average, more than once per M.yr for the last 100 M.yr, but perhaps less frequently in some earlier epochs. The current best estimate of time taken for the field to invert is 3000 yr. Reversals are known in rocks older than half the age of the Earth. The oldest terrestrial rocks studied to date indicate that a geomagnetic field existed at least $3 \cdot 2 \times 10^9$ yr ago.

4. Dispersions of NRM directions in rock sequences spanning a few M.yr are comparable to the dispersion of present day geomagnetic field directions around lines of present-day latitude, suggesting that secular variation has occurred in the past on a comparable scale to that in modern times.

5. When the process of magnetization is known (*see* MAGNETIZATION OF ROCKS) the magnitude of NRM can in favourable cases be related to the intensity of the local geomagnetic field in the past. Palaeointensity appears to have fluctuated in a secular fashion (cf. 4 above) but has on average been of comparable magnitude to the present geomagnetic field for at least $2 \cdot 5 \times 10^9$ yr.

Climatology

Past climates may be inferred from the occurrence of rock types and fossil floras and faunas whose modern analogues are controlled in distribution by climatic factors. Climatic zonation appears to be inevitable as long as the Earth has had an atmosphere and has orbited the Sun. It is probable that all the diverse climates evident at the present time have existed previously at various times. For example widespread glaciations are known to have occurred about 700, 500, and 300 M.yr ago, as well as in the past few M.yr; it is not certain, however, that all these glaciations were bipolar, and it seems that polar icecaps have existed for only a minor proportion of geological time. In favourable cases distinctions between mid-continental and marine climates have been demonstrated at comparable latitudes in the past. Of the climatic elements only temperature and wind direction have been estimated quantitatively, but in only isolated instances (*see* PALAEOTEMPERATURES); it is doubtful whether analogous rainfall estimates will ever be forthcoming. Nor has the subject yet advanced to the stage of assessing whether modern patterns of oceanic and atmospheric circulation are typical of the past. Although there is no reason to suppose that different factors have controlled past circulations, their relative importance may have varied. For example, it is clear that continent distribution is an important influence, and this is known to have changed drastically through time. Among other major factors, the extent to which the solar constant, the Earth's orbit, and the emissivity of the Earth surface and atmosphere may have fluctuated are not well known.

Heat Flow

The thermal history of the Earth remains a subject for theoretical consideration constrained by imprecisely known boundary conditions. Estimation of past heat flow from the geological record presently exposed at the surface may in future be achieved by comparing metamorphic facies gradients with experimentally determined $P - T$ stability fields for appropriate mineral assemblages. Some tentative attempts have already been made in this direction, and tend to indicate that when the Earth was less than half its present age, continental crust was relatively thin and geothermal gradients were relatively high.

Geoelectricity

The electrical conductivity of parts of the Earth's interior may be estimated indirectly once the relevant compositions and temperature fields are known, but estimates relating to the geological past are confined to the core based on geomagnetic dynamo theory and palaeomagnetism. The only known record of atmospheric electricity leaving a visible imprint in the geological record is in the form of palaeofulgurites. Vertical tubes of fused silicate found in desert sandstones of Permo-Carboniferous age (c 280 M.yr) in Arran, Scotland, are believed to have originated from contemporaneous lightning strikes. The magnetic effects of geologically recent lightning strikes on exposed rock outcrops are well known, and the paths of electrical discharge have been mapped over areas of the order 100 m^2. However, it is less certain whether anomalously strong remanent magnetizations in older rocks imply equally old lightning strikes, because the decay time of lightning induced magnetization is believed to be short, in geological terms.

Seismology

Evidence in the geological record regarding individual earthquakes in the past is rare and uncertain. The occurrence of submarine turbidite deposits may be associated with seismicity, but cause and effect cannot be isolated. The cumulative effect of seismic activity in the crust is preserved geologically in the form of faults, whose age, direction and amount of displacement can commonly be determined. Large strike faults involving horizontal displacement have been shown to be associated with plate motions in a fashion consistent with palaeomagnetically determined displacements. The development of fault patterns within continents can commonly be shown to arise from volcanic or intrusive igneous activity. But these studies remain essentially qualitative and belong to the geological rather than the geophysical realm.

Earth's Rotation

The only geological evidence relating to the Earth's rotation in the past derives from counts of growth lines in shallow marine fossils. The data are few, and by no means unambiguous, but if the finest growth bands are taken to be daily skeletal increments, then rhythms in banding with periods of ~30 and ~400 days may be related to the number of days in the synodical month and the year respectively. In general the counts are consistent with a decrease in number of days per year from 400 in Devonian times (c.350 M.yr ago) to the present. Mass distribution in the Earth might be a significant factor in determining such changes. It has been suggested that irregularities in rotation do coincide with accelerated plate motion which would alter the mass distribution in the crust and hence rotate the principal axes of Moment of Inertia. The data have also been related theoretically to the rotation of the Earth–Moon system and lunar tidal effects on the Earth, and lead to consistent conclusions within the (large) acceptable limits.

Geodesy

Dirac's prediction of change of G with time has been linked both with the problems of rotation rates mentioned above, and in connection with possible change in size of the Earth and the acceleration due to gravity (g) at the surface. Despite claims to the contrary no calculations yet made of ancient radius based on palaeomagnetic measurements yields a result significantly different from the present, moreover it is doubtful whether these methods are sufficiently well founded for reliance to be placed on any changes which might be postulated from future calculations by these methods. The prospects of quantitative estimation of g and ellipticity in the past seem equally remote.

Bibliography

GRAHAM, K. W. T. (1961) The remagnetization of a surface outcrop by lightning. *Geophysical Journal*, 6, 85.

McELHINNY, M. W. (1973) *Palaeomagnetism and Plate Tectonics*, Cambridge: Cambridge University Press.

NAIRN, A. E. M. (Ed.) (1961) *Descriptive Palaeoclimatology*, New York and London: Interscience.

NAIRN, A. E. M. (Ed.) (1964) *Problems in Palaeoclimatology*, New York and London: Interscience.

RUNCORN, S. K. (Ed.) (1970) *Palaeogeophysics*, London: Academic Press.

J. C. BRIDEN

PATTERN RECOGNITION. The ability to recognize patterns in input data is an essential part of the behaviour of people and animals. To a limited extent it has also been embodied in artefacts. The word "pattern" is often employed in this context with either of two distinct meanings; it may refer to a particular configuration of signals existing at any instant in the incoming pathways, or it may refer to one of the categories of such configurations to which a specific response, constituting "recognition", is appropriate. The phrase "pattern recognition" implies interpretation of "pattern" as a category. Many writers prefer the phrase "pattern classification" which is consistent with the other interpretation of "pattern". To avoid ambiguity, the task will be referred to simply as "recognition" in what follows.

It is because many widely differing input configurations may correspond to one category that the task of recognition is non-trivial. Different samples of the same handwritten word, for example, may be such that it is virtually impossible to find any precise objective statement of what they have in common. The same is true of the sound waveforms corresponding to separate vocalizations of a particular word, or the retinal images formed when the same human face is viewed from different angles while bearing different expressions.

Because people and animals perform recognition tasks with seeming ease there have been many attempts to model artificial systems on living ones. The possibility of such modelling is particularly attractive since the categories to be recognized can usually be defined only by the presentation of sample configurations, and living systems have a remarkable facility for deriving general classification rules from a succession of such presentations. This is one type of activity subsumed under the general heading of "learning". However, the operation of living systems is not sufficiently well understood that any artificial system can be said to be entirely modelled on a biological one.

In the study of both living and artificial systems particular attention has been given to visual recognition, though the tactile and auditory forms have also been considered, as have some interesting recognition tasks which are not associated with any one sensory modality. A doctor performs recognition when he considers a patient's symptom and decides they are consistent with a particular ailment. Similarly, a geophysicist may recognize that a particular configuration of data indicates some special feature such as an oil pocket, or a meteorologist that a configuration of atmospheric data presages, say, rain. Some of the techniques for automatic recognition have been applied to these multi-modal tasks.

Neurophysiological Studies

There have been two main electrophysiological studies of the processing of visual information in animals. Lettvin *et al* (1961) recorded activity in the optic nerve and optic lobes of the frog while presenting visual stimuli. They decided that, even at the retinal level, the information was processed in ways which were specially responsive to moving black dots in the visual field, which could represent flies, and to moving light–dark boundaries which could be the edges of shadows of predators. The visual system of the frog appears to be highly specialized, and not a useful prototype for a system to learn new categories during the lifetime of an individual.

The work of Hubel and Wiesel (1968) on cat and monkey has shown that in these animals the preliminary processing of visual data is of a much more general-purpose kind. Their most interesting results were obtained by recording from the visual cortex. There they found cells which responded maximally when there was a dark–light boundary of particular orientation in a particular part of the visual field, and others which were particularly sensitive to dark bars on a light background and others to light slits on a dark background. Other cells responded only to movement of one or other of these image components (edge, bar or slit) within its receptive field. Others responded only to edges, bars or slits of short length, ceasing to respond when the length was increased. A fairly clear picture is emerging of how the preliminary stages of processing of visual information are carried out, but a great deal remains to be discovered before very much can be said about the subsequent stages leading to actual recognition.

The Hubel and Wiesel results lend general support to schemes for automatic recognition in which one of the early stages of processing is the extraction of local features. The well-known "perceptron" scheme is in this category. (See Fig. 2.)

Character Recognition by Machine

Tasks of recognition which can be performed by machine fall into two fairly distinct groups, according to the degree of variability likely to occur in the input configurations belonging to one pattern category. Where the inputs are printed characters, the variability is relatively slight, though still enough to be troublesome. Even where the printing fount is standardized there are variations due to paper texture, imperfect inking (or imperfect transfer from a ribbon) and other effects.

The automatic recognition of characters is of great commercial importance since there is the need to convert a great deal of information from printed form to a form acceptable by digital computers. This can be done manually, but is clearly a tedious task which is better automated. Sometimes, but by no means always, specially designed character sets are used to facilitate automatic recognition. A familiar example is the set used for numbering cheques. This is printed in magnetic ink and read magnetically by machine, but the same principle could be used with optical sensing. These characters can be distinguished from their horizontal projections and thus need only be sensed by a vertical slit.

Whether the horizontal projection or the whole image is treated, the automatic recognition of characters is generally done by a form of template-matching. It can alternatively be described as the computation of spatial cross-correlation measures between the character presented and a set of "masks". The correlation measure will normally be much higher for one of the masks than for the others, and the character is recognized as belonging to the corresponding category.

A set of characters is particularly suited to automatic recognition by a cross-correlation method if the cross-correlation between different characters, in their perfectly formed state, is low. The "B" fount of the European Computer Manufacturers' Association (Fig. 1) has been developed to satisfy this requirement. The fount also has a pleasing appearance and is easily read by eye.

Automatic recognition with high reliability is still possible with character sets other than those specially devised. In practice characters are not perfectly formed and their background is not uniformly white. With simple equipment there is a threshold level of reflectance used to determine which areas are to be judged to be white and which black. For much of the material arising in commercial applications operation with fixed threshold is not good enough, as some of the "white" areas are in fact darker than "black" areas. Much better results are obtained by letting the decision for a particular element of area depend on the brightness of surrounding elements.

Other techniques besides cross-correlation have been used for recognition of printed characters, but it appears to be the most effective. In some schemes depending on specially devised character sets the features of the characters to which the automatic system responds are distinct from those on which recognition by eye depends. For example, where only a small set of characters is required, it is possible to let the normal shape of the character be filled with alternate black and white strips, the angle of the striping being different for each character in the set. The automatic equipment then responds to the angle of striping, the human reader to the overall shape.

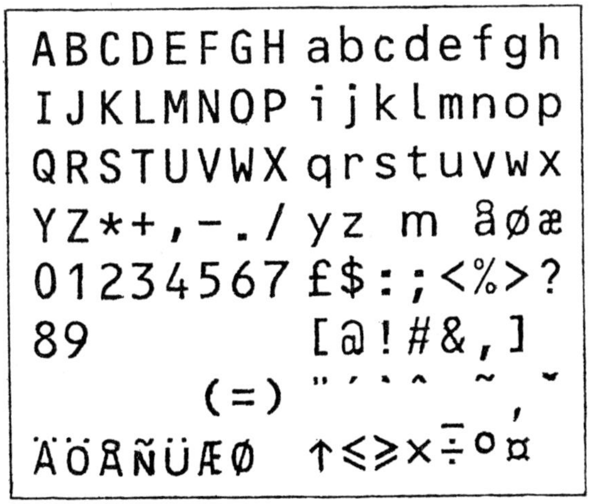

Hand-printed Characters

For hand-printed characters the cross-correlation method cannot be used. Considerable success has been achieved using other approaches, and equipment is available which can recognize such characters with the high reliability needed in commercial applications, provided the characters are reasonably well formed. Recognition is generally based on the presence or absence of special features of the character, such as a large preponderance of black below the horizontal centre line as opposed to that above it, an enclosed area above this centre line, a long vertical stroke on the left-hand edge, and so on. Attempts have been made to apply the general principle so as to allow recognition even of poorly formed characters, by taking account of a large number of special features and basing the recognition on statistical principles.

Parks (1969) describes a system which applies exactly the same procedure to the recognition of hand-printed and multi-fount machine-printed characters. It depends on identifying such local features as line terminations (such as the feet of a capital A), corners (such as the apex of an A), junctions (as at the ends of the horizontal stroke of the A) and crossovers.

The automatic recognition of cursive handwriting has also been attempted. Most of the work has depended on the assumption that information on the movement of the writing point is available, as when a telewriter is used as input. Even then the recognition has only been of single letters; the automatic system has not been required to perform the segmentation of complete words, a task which has proved to be remarkably difficult. It seems fairly certain that automatic reading of completed handwriting, which would be of great value for postal sorting, will not be achieved for some considerable time.

Speech Recognition

Devices capable of recognizing spoken words allow the interrogation of computers over ordinary telephone channels and have other potential applications. They have, for example, potential application wherever an operator must convey information to a machine but cannot conveniently use his

```
ABCDEFGH abcdefgh
IJKLMNOP ijklmnop
QRSTUVWX qrstuvwx
YZ*+,-./ yz m åøæ
01234567 £$:;<%>?
89       [@!#&,]
     (=)  " ´ ˆ  ~  ˇ
 ÄÖÅÑÜÆØ  ↑≤≥×÷°¤
```

Fig. 1. The ECMA "B" fount for optical character recognition.

hands to do so. The sorting of postal parcels is a task which places an operator in such a situation. Automatic recognition from a very limited vocabulary appears to be a practical possibility at present.

A spoken word can be represented as a "spectrogram", a graph having frequency as ordinate and time as abscissa, in which the amount of sound energy in a frequency-time domain is represented by density. Some of the methods for automatic recognition of spoken words operate in a way which is equivalent to the application of template-matching to the spectrogram. Such methods are only likely to be successful if the words are spoken in a fairly standard way. In some applications, such as the postal-sorting one, this might be no great disadvantage.

Words do, however, remain intelligible to a human listener when considerable variations are permitted in the time-durations of the component sounds. It appears that a word is characterized by a sequence of events whose precise time relations are unimportant. A recognition technique depending on the determination of such sequences is described by Hill (1969).

Perceptrons

The type of recognition system termed "perceptron" was introduced by Rosenblatt (1962). It is based to some extend on biological principles and can be "trained" by exposure to sample images. Each image is applied to a set of sensory units which could be the light-sensitive elements of a retina (Fig. 2). These are connected in randomly chosen groups to association units, which in turn are connected to response units. There is one response unit for each pattern category to be distinguished.

Both the association units and the response units have properties which appear also to be the main properties of nerve cells. That is to say, they summate effects from such of their incoming pathways as are active, and "fire" to activate their outgoing pathways if the summation exceeds a threshold. Thus the association units extract simple local features of the images. In the summations performed by the response units, each input has a "weight" associated with it, this being the amount it contributes to the sum if active. A number of "perceptron training algorithms" are known which adjust these weights while images are presented along with indications of the correct classifications.

It has been shown that the training algorithms must converge if any set of weights exists which will perform the required discrimination. However, the convergence proof gives no indication of the range of discriminations which can be performed nor of certain important characteristics such as time required for convergence. Minsky and Papert (1969) have shown that there are severe limitations on what simple perceptrons can be made to do. Nevertheless, the general principle has provided a valuable stimulus,

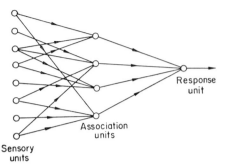

Fig. 2. Scheme of a perceptron. Only one response unit is shown. The adjustable "weights" are associated with the connections from the association units to the response unit.

and work is continuing on somewhat similar devices in which the adaptive changes are not restricted to a single functional layer.

Scene Analysis

Recognition schemes are of limited value if they can only deal with images of isolated objects against featureless backgrounds. In recent years there has been much interest in building robots capable of manipulating objects under visual control, and hence in the analysis of the pictures formed by the television cameras serving as "eyes". Such a picture may contain a number of distinct objects, some of them partly occluded by others. A system to operate under these conditions must identify elementary parts of the scene and list them with some formal representation of their relationships. The elementary parts may be lines or areas, and relationships for the latter may include "adjacent to", "above", "within" and so on.

Minsky and Papert (1969) give an account of work by Guzman on a computer programme which will analyse a scene consisting of straight lines so as to produce its most appropriate interpretation as a view of three-dimensional polyhedra. A number of papers reporting extensions of this work were presented at the Second International Joint Conference on Artificial Intelligence, of which the advance papers have been published by the British Computer Society (1971).

Barrow and Popplestone (1971) have described a computer programme which will dissect a television-camera picture into elementary areas. It produces numerical data which represent certain features of the shape and interrelationships of these areas. This data can be processed so that certain simple objects placed before the camera are recognized. That is to say, the computer programme generates a signal which usually indicates correctly the type of object presented, such operation being possible despite considerable variation in orientation of the object.

The problem of determining the significance of a scene from the formal description of its parts and their relationships is formidable if the scene is at all complex. Clowes (1969) and others have suggested that this problem has much in common with that of deducing the syntactic structure of a sentence from the sequence of words comprising it. Hence linguistic theory, which has advanced enormously in recent years under the influence of Chomsky, may have a strong bearing on recognition techniques.

Very recent studies have placed less emphasis on the linguistic aspect, regarding it as providing one method among others for specifying and manipulating tree structures. A description of an object in terms of its constituent parts is a tree structure because there is a main trunk corresponding to the object itself, which might be, for instance, a motorcar. This divides into major branches, which might be wheels, body, windows. These divide into smaller branches and finally into the twigs which are the elementary lines or areas. Recognition of an object depends on matching the tree structure specifying it to the appropriate part of the formal description of the scene. The possibilities have been discussed very fully by Guzman (1971). A major difficulty is that the classification of a branch of the tree at any level may depend strongly on context. Hence the branches cannot be treated individually before considering how they may be combined to form branches a step nearer to the trunk. This area is currently being actively studied.

Bibliography

BRITISH COMPUTER SOCIETY, Second International Joint Conference on Artificial Intelligence; advance papers, 1971.
BARROW, H. G. and POPPLESTONE, R. J. (1971) Relational descriptions in picture processing. *Machine Intelligence* 6, ed. B. Meltzer and D. Michie, p. 377. Edinburgh University Press.
CLOWES, M. (1969) Pictorial relationships—a syntactic approach. *Machine Intelligence*, vol. 4, ed. B. MELTZER and D. MICHIE, p. 361. Edinburgh University Press.
GUZMAN, A. (1971) Analysis of curved line drawings using context and global information, *Machine Intelligence* 6, p. 325.
HILL, E. R. (1969) An ESOTerIC approach to some problems in automatic speech recognition. *Int. J. Man-Machine Studies*, **1**, 101.
HUBEL, D. H. and WIESEL, T. N. (1968) Receptive fields and functional architecture of monkey striate cortex. *J. Physiol.* **195**, 215.
LETTVIN, J. Y., MATURANA, H. R., PITTS, W. H. and MCCULLOCH, W. S. (1961) Two remarks on the visual system of the frog, *Sensory Communication*, ed. W. A. Rosenblith, p. 757, M.I.T. Press.

MINSKY, M. and PAPERT, S. (1969) *Perceptrons*, M.I.T. Press.
PARKS, J. R. (1969) A multi-level system of analysis for mixed font and hand-blocked printed characters recognition, *Automatic Interpretation and Classification of Images*, p. 295. Academic Press.
ROSENBLATT, F. (1962) *Principles of Neurodynamics*, Spartan Books.

The journal *Pattern Recognition* is published by Pergamon Press for the Pattern Recognition Society.

A. M. ANDREW

PLANETS, MAGNETIC FIELDS OF. The first detailed investigation into the magnetic field of the Earth is generally ascribed to William Gilbert in A.D. 1600, although the directional properties of certain iron ores (when suspended sufficiently freely) had been put to use five centuries earlier and known of possibly as early as 200 B.C. Investigation into the magnetic properties of extra-terrestrial bodies, on the other hand, is a subject which has no precedent before the last few decades and of planetary bodies in particular before about 1960.

Many of the properties of the Earth were thought to be typical of at least the higher density terrestrial planets and it has come as something of a surprise that the possession of a substantial magnetic field has only been definitely established for one other planet than the Earth, namely the giant planet Jupiter.

The two principal methods of detecting a planetary magnetic field are the direct measurement with a magnetometer in a space-probe and the indirect observation of its effect on charged particles. There would seem at first sight to be few difficulties in interpretation of the results of direct magnetometer measurement but this is not so. In the solar system three distinct situations arise, the first is of a planet with a substantial magnetic field (i.e. the Earth), the second is of a planet with a negligible field but in possession of an atmosphere and ionosphere, and the third is of a planet with neither atmosphere nor magnetic field. In attempting to predict the type of magnetic measurements which a space-probe may record near a planet of the first kind a very comprehensive determination of the morphology of the geomagnetic field is invaluable. In the absence of matter the attenuation of the geomagnetic field with distance would be as the inverse cube and the detection limit would ultimately be set by interplanetary noise. However, the effect of the solar wind, a stream of charged particles of both signs carrying with it weak remanents of the solar field, is to limit the geomagnetic field to a region known as the magnetosphere (shown in Fig. 1) the tail of which may maintain its form over a distance of 0·1 astronomical unit in the "downwind" direction. As the velocity of the solar wind relative to Earth is in excess of the velocity of propagation of disturbances in the solar wind itself, the Earth sets up a shock wave. This shockfront strongly delineates the region of irregular magnetic field outside the magnetosphere (inside which the field is strongly dipolar) from the more uniform interplanetary magnetic field. The solar wind streams away from the sun and consequently the "bow-shock" is nearest the planet at the "subsolar" point (a distance of 13 radii for the Earth) and the tail of the magnetosphere is directed away from the Sun. Space probe measurements have to be analysed to detect these divisions into the characteristic zones. It is, of course, essential that the probe approaches sufficiently close to a planet to penetrate the shockfront or no information about its field is available at all. The distance of this shockfront from the planet depends on both the magnitude of the planetary field and the strength of the solar wind, therefore the position of the shockfront relative to the planet gives an estimate of its field. If a planet has only a very weak intrinsic field but a substantial atmosphere then there is the possibility of the solar wind penetrating into the ionosphere where a shockfront will occur very much closer to the planet than in the previous case. Observation of ionospheric behaviour from spacecraft can indicate when this effect is present. When a planet has neither field nor atmosphere, measurements in the solar-wind shadow region can yield information about its electrical conductivity.

Surface probes do not necessarily give better information as the magnetization of rocks in the vicinity can grossly distort the local field, satellites in very low orbits obtain the best information. Viewed in this way the geomagnetic field is seen to be predominantly an axial dipole with surface values of a fraction of an oersted.

The Earth's moon has been most extensively examined of all the planetary sized objects. Explorer 35 has determined that the average surface field is approximately 10 gamma (1 gamma = 10^{-5} Oe) although local surface values of up to 250 gamma have been obtained by Apollo 16 instruments.

Venus has been visited by missions Mariners 2 and 5, and Veneras 4, 5 and 6. These failed to detect any shockfront at a large distance from the planet, and solar-wind interaction with the ionosphere suggests an upper limit to the surface field of a few gamma.

Mars similarly has been visited by Mariners 4, 6, 7 and 9, again as in the case of Venus there is no evidence of a large intrinsic field and similar surface values are indicated by a shockfront produced by solar wind upper atmosphere interaction.

The outer planets by virtue of their immense distance will not receive "flyby" missions until at least the mid-1980s.

The second principal tool for investigating planetary fields is the observation of their electromagnetic radiation; this consists of thermal radiation with its characteristic frequency spectrum plus a component from any charged particles in radiation belts. These originate in exactly the same way as the Earth's Van Allen belts. The radiation they emit comes from charged particles orbiting around magnetic field lines giving off cyclotron radiation with its characteristic frequency spectrum and polarization. This nonthermal radiation can tell us both the intensity and spacial distribution of the field. From the frequency spectrum the strength in certain regions of the field can be estimated, and as the plane of polarization of the electric vector in the emitted radiation is perpendicular to the magnetic field lines the field direction can be obtained. An examination of the emission from Jupiter suggests surface field strengths of 50 Oe and from the polarization an axial dipole component is the largest part of the Jovian field, although higher field harmonics are of greater magnitude than is the case for the Earth. The radio emission from Saturn has components which might be non-thermal but are not completely characteristic of cyclotron emission either. The outer planets Neptune, Pluto and Uranus are just too distant to have their radio-emission spectrum determined accurately; terrestrial planets emit only thermal radiation.

The observational data from the solar system planets is of little help in indicating the origin of their magnetic fields because almost every planet is unique in some respect. However, it is possible to rule out simpler explanations from the known physical properties of the Earth. For example, remanent magnetism in the layer of the crust and mantle above the Curie temperature cannot conceivably produce a large enough field. The electromagnetic decay time of the Earth is about 10^4 years, this would require an enormously strong field at the time of the Earth's origin if we were seeing just a remanent of a primeval field. A more plausible suggestion known as "Schusters' Hypothesis" that the magnetic moment and angular momentum were related was found not to hold for solid bodies, although it is possible that a sufficiently massive body with a liquid core may yield some relation of this sort as will be mentioned later.

The remaining alternative is that there is some

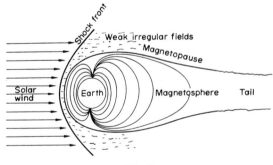

Fig. 1

mechanism which generates electric currents to produce the observed field. Thermo-electric theories enjoyed a brief popularity until palaeomagnetic evidence of frequent field reversals made them less tenable. The currently most acceptable theory of origin is in a hydromagnetic dynamo mechanism in the liquid conducting core of a planet; this theory has had a rather chequered history following the original suggestion of Larmour in 1919 that convective motions in a large conducting liquid mass might behave as a "homogeneous" dynamo and generate a magnetic field. Cowling in 1934 demonstrated that simple symmetrical types of motion could not generate a field and a further two decades elapsed until more complex motions were examined and found to be capable of acting as "self-exciting" dynamos. A major difficulty here is that there is a great deal of uncertainty about the energy source of the field (which, however, need only be comparatively modest) and even more uncertainty over the nature of fluid motions in a self-gravitating rotating sphere. Whether or not a planet has a liquid core depends to a certain extent on the planet's size. Its presence can be inferred from the value of the moment of inertia of the planet. In terrestrial planets meteoritic compositions lead to the suggestion that the cores are iron and therefore dense and highly conducting.

The usual approach to the dynamo problem is to specify the fluid velocity and describe it in terms of spherical harmonics, the magnetic field is described in a similar fashion and the interactions between field and motion examined with the important stipulation that the motion is not modified by the electromagnetic forces. It can be shown that the rate of change of magnetic field depends on two terms, one gives the rate of generation of field and the other the rate of diffusion of field lines through the conductor; the ratio of these terms is called the Magnetic Reynold's Number and it is proportional to (length) × (velocity) × (conductivity), there is a critical value of this number for dynamo action. Various theoretical dynamo mechanisms have been produced notably by Bullard and Gellman, Backus, Herzenberg, Braginski and Steenbeck, Krause and Rädler, these have all been "steady state" models. A feature of many dynamo models is differential rotation of the core (this is necessary to conserve angular momentum) and one step in the dynamo mechanism involves the generation of a toroidal field from the interaction of the dipole field with this motion. Another approach to the problem of core motions, however, considers them to be essentially turbulent but with some degree of ordering; the ordering mechanism is the coriolis force. Steenbeck and others have shown that if the motions have a preferred helicity (either left-handed or right-handed) then an axial dipolar field would be generated which was proportional to some mean angular velocity of the core. Similarly for other types of core motions the coriolis force could be responsible for the axial dipolar nature of the Earth's field, and so Schuster's hypothesis may hold if a liquid core is present.

It is not possible to deduce the time variation of the field produced by this type of mechanism without resorting to inhomogeneous analogues with each step being represented by a Faraday disc-dynamo. Computations on coupled disc-dynamos show irregular magnetic-field reversals which correlate qualitatively with the palaeomagnetic evidence but unfortunately do not scale in time correctly. Some simple laboratory homogeneous dynamos constructed by Lowes and Wilkinson both generate a magnetic field and show instability; the field reversals demonstrated are completely periodic unlike those of the Earth, the reason for this difference is not yet clear but at least it has been shown that homogeneous dynamos can be unstable and the field can reverse spontaneously.

Attempted comparison of the observational data with theory suffers from the tenuous nature of the theory and the number of variables present in the data, however the present state of knowledge proceeding from the sun outwards can be summed up as follows:

Mercury—with its very slow rotation rate and small size does not seem likely to possess a field and there is no evidence of one.

Venus—in many ways similar to the Earth, but rotating very much slower, is an obvious piece of evidence for the field/angular momentum relation. There is no evidence for any substantial field (see earlier comments).

Earth—has a high rotation rate and a liquid core and at the moment the best documented field in the solar system.

Mars—rotates at almost the same rate as the Earth but is just a little more than half its diameter; the moment of inertia factor would allow only a small core and the possibility of only a weak field is in accord with this.

Jupiter—the first of the giant planets, which differ from the terrestrial planets in not having lost all of their hydrogen shortly after formation, has a large field and a high rotation rate but the location of the dynamo process is open to speculation. One suggestion is that the major part of the core is a liquid metallic phase of hydrogen; certainly high-pressure data indicates a phase change of some sort which would occur at a reasonable depth in Jupiter. Another suggestion is that the dimensions and conductivity of the liquid upper layers are large enough to give a Magnetic Reynold's Number which would permit dynamo action.

Saturn—rather smaller than Jupiter but still rotating rapidly does not give such clear indication of possessing a field, the phase transition in hydrogen would not occur until near the centre of Saturn.

The outer planets can be grouped together in the absence of data, they are all much smaller than Jupiter and much more distant that Saturn so the problem of detecting a field is too difficult for current technology and indeed the possibility of them possessing fields is remote.

Returning to the Earth because of our extensive knowledge of the geomagnetic field, we may expect on other planets possessing magnetic fields to find similar features to those which distort the geomagnetic from a pure dipolar form. The regions of maximum and minimum residual field left when the dipole component is removed are probably produced by the interaction of small eddies in the core and near the core-mantle boundary with large internal fields; the resultant field is a dipole centred on the eddy. By comparing the geomagnetic field at different times an overall "westward drift" of these irregular features can be seen showing that the upper layer of the core is rotating less rapidly than the solid mantle. Furthermore, a net reduction in strength of the main dipole field since earliest measurements were made has been detected. This sort of detail will, of course, only be resolved in other planetary fields after longterm measurements have been carried out.

It is worth considering at this point our Moon; although it is not a planet now it is an unusually large satellite and some theories of its origin suggest a formation separate from that of the Earth. The Moon has been examined in some detail and the investigation of its surface rocks through samples returned to Earth reveal a magnetization which could not have been acquired in any plausible extralunar field. Investigation by several different groups of workers suggest a surface field at the time of formation of the rock samples some thousands of millions of years ago of 1000 to 10,000 γ. This would indicate that the Moon in the past had a molten-iron core in which a dynamo existed. This inference from the most recent lunar data lends support to the idea that the existence of a dynamo inside a planet could be a normal event at some stage of that planet's development. When oriented rock samples can be collected from a planetary surface palaeomagnetic methods will allow the determination of the magnetic-field variation over geological time, and this in turn has important consequences for any theories of planetary evolution.

Bibliography

HIDE, R. (1966) Planetary magnetic fields, *Planetary and Space Sci.*, **14**, 579–586.
HINDMARSH, LOWES, ROBERTS and RUNCORN (eds.) (1967) *Magnetism and the Cosmos*, Academic Press.
JACOBS, J. A. (1963) *The Earth's Core and Geomagnetism*, Pergamon Press.
RIKITAKE, T. (1966) *Electromagnetism and the Earth's Interior*, Elsevier.
RUNCORN, S. K. (ed.) *The Application of Modern Physics to the Earth and Planetary Interiors*, Wiley Interscience.
SMITH, A. G. and CARR, T. D. (1964) *Radio Exploration of the Planetary System*, Princeton: D. Van Nostrand Co.
STACEY, F. D. (1969) *Physics of the Earth*, Wiley Space Science text series.

I. WILKINSON

POLLUTION OF THE AIR. The atmosphere is and always has been "polluted" in the sense that various unlikely substances can be found in transient amounts due to natural causes such as volcanoes, forest fires, lightning, wind, the decay of vegetation and cosmic rays. In another sense the relative amounts of the "constant" components of the atmosphere (see Table 1) can alter with natural variations of, for instance, climate and amount and type of vegetation. Superimposed on these relatively short-term variations are the natural variations on a geological time-scale, responsible for the evolution of life, which has fairly promptly adapted to any subsequent changes in the composition of the atmosphere. With the rise of civilization, technologically produced long- and short-term disturbances in the atmosphere have become significant. All such additions and alterations are termed air pollution. When, due to meteorological circumstances, annoying or even lethal concentrations of pollutants have arisen (Los Angeles, Tokyo photochemical smogs, London smogs of 1952, 1963), legislation has arisen to limit the extent of the emission of gaseous and solid matter into the atmosphere. Concern and literature on the extent and effects of technology on the atmosphere have grown enormously in the years since 1945. Table 1 gives a comparison between "clean" and average urban air. Table 2 indicates the sources of the obvious pollutants in Table 1 where the figures refer to the United States. Legally enforceable emission control devices on all vehicles have been introduced as a result of these findings.

The Effect of the Weather

Apart from the effects of gravity on large particles, and direct absorption of gases by vegetation, meteorological processes are entirely responsible for the dispersal and indirectly resposible for the removal of most pollutants emitted into the troposphere. These processes are driven by solar energy, and as the presence of pollutants alters the vertical attenuation characteristics of the Sun's radiation, the pollutants near large industrial locations themselves tend to alter the weather unfavourably for their dispersal. The main factors determining dispersal are the atmospheric stability, and the wind/turbulence. The first of these factors is based on the behaviour of a parcel of air at ground level which has become heated to a temperature slightly above ambient by solar energy or some other factor. If the adiabatic

Table 1

(All values of concentration in parts per million (ppm) unless stated otherwise)

Clean dry air (sea level)		Average urban air		Estimated residence time
Nitrogen N_2	780,900			
Oxygen O_2	209,400			
Argon A	9300			
Carbon dioxide CO_2[a]	318	very variable		2–100 years
Neon Ne	18			
Helium He	5·2			
Krypton Kr	1·0			
Methane CH_4	1·5			
Nitric oxide NO	0·5			
Hydrogen H_2	0·5			
Carbon monoxide CO	0·1	7	100 ⎤[d]	3 years
Xenon Xe	0·08			
Ozone O_3	0·02	0·02[c]	0·1 ⎬	hours
Nitrogen dioxide NO_2	0·001	0·04	5	days
Sulphur dioxide SO_2	0·002	0·10	10 ⎦	days
Hydrocarbons	~0	3·0		4–20 years
Total dust ⎫[b]	?	100 μg m^{-3}		⎧
Dustfall ⎬	?	1% of surface covered per day		⎨ years in stratosphere, days in troposphere, depending on size
Aitken nuclei ⎭	10^3 cc^{-1}	10^5 cc^{-1}		⎩
Radioactivity[e]				

[a] The concentration of CO_2 is rising by 0·7 ppm per year.

[b] Particulate matter comprises a vast range of organic and inorganic solids and liquids both charged and uncharged. Sizes in the range 10 μ–0·1 μ act as condensation nuclei in meteorological processes, sizes below 0·1 μ are called Aitken nuclei.

[c] Ozone concentrations rise to 0·1–0·2 ppm in photochemical smogs.

[d] Values of maximum allowable concentration (MAC) which are widely accepted as industrial toxic levels and listed by Sax (1957). The maximum values currently being accepted and legalized in the United States under the Air Quality Act of 1967 and the Clean Air Amendments of 1970 are considerably lower than the MAC values, and are already exceeded by some of the pollutants.

[e] Radioactivity: the concentration of ^{14}C in the atmosphere has doubled from 1953 to 1963, the period of intensive atomic weapon testing.

The maximum increase in radioactive airborne fragments was at most 10% of the normal background during testing. Normal background approximately 250 μrad per day.

100 mC per square mile of 28 year ^{90}Sr on average has been deposited in the Northern Hemisphere so far.

Table 2

Agent	% of all pollutants
Vehicles	60
Industry	17
Power production	14
Other combustion	9
	100

cooling of this parcel as it rises (the dry adiabatic lapse rate) is less than the actual drop in temperature present with height (the environmental lapse rate) then it will continue to rise until the two temperatures become equal, around which height it becomes stable. If at a certain height the environmental temperature increases with height, then conditions are very stable and no vertical convective motion occurs through this region, thus trapping pollutants under such a so-called inversion layer. All of the more infamous smogs have occurred, and can be predicted to be likely to occur under shallow inversions in the absence of wind. The second factor, the wind/turbulence, where the turbulence is caused both by the vertical convective motion in an unstable atmosphere and by ground irregularities, determines the direction of travel as well as the vertical and lateral dispersal of pollutants from point sources such as stacks. Under the combined influence of these two factors, several identifiable and picturesquely named behaviours of stack plumes are possible (looping, coning, fanning, lofting, trapping, fumigation—see, for instance, Bach (1972)). Various empirical mathematical models are used to predict the extent of the area affected by stack plumes under these different conditions.

Finally, and sometimes after photochemical changes, rain washes away most of the pollutants with the significant exceptions of CO and CO_2, the latter being involved in a complex cycle involving vegetation, and the dissolved CO_2 in the ocean, depending on its temperature. In the stratosphere (12 km altitude and upwards), where stable conditions prevail, the weather as we know it being confined to the troposphere, the lifetimes of pollutants in particulate and gaseous form are vastly increased.

The Effects of Pollution

(a) *On health*

The main hazard is in disease of the lung, and in particular the death rates from lung cancer, asthma, and emphysema are all rising, the latter being the fastest growing cause of death in the United States. The London smog of 1952 was estimated to have caused 4000 excess deaths (and hastened the Clean Air Act of 1956) with a concentration of SO_2 of 1·2 ppm for a few days, allied with a high concentration of particulates. There is some evidence that the effects of certain pollutants together are enhanced, in particular that of SO_2 and particulate matter together may be more hazardous than the sum of their separate effects. There can be no doubt that the effect of even the average urban pollution of Table 1 hastens the death of sufferers from lung complaints, but that this is probably a smaller effect than the intermittent but comparatively vast personal pollution caused by smoking.

(b) *On vegetation*

Apart from local specialized chemical episodes, the most likely cause of damage is photochemical smog, now common in and near large cities. A complex chain of reactions involving the NO, NO_2, and hydrocarbons generated by vehicles in the presence of sunlight, generates ozone, and various other complex organic compounds, such as peroxyacetyl nitrate, which are known to cause plant damage even in minute amounts.

(c) *On the weather*

Decreases in the amount of sunlight and in visibility are directly attributable to particulate pollution, and the same cause is reported to increase by a factor of two the incidence of fog in urban areas compared to the surrounding area. On a global scale, the mean temperature, rising until 1940 has dropped by 0·25 K since. The increasing worldwide CO_2 level shown in Table 1 would suggest a rising world temperature from the greenhouse effect, (the absorption of the Sun's infra-red radiation by carbon dioxide) and some sources have suggested that the increasing dust content of the atmosphere may more than compensate for this by reflecting the incident solar energy. The picture is far from clear.

Measurement Techniques

(a) *For gases*

Many methods of gas sampling are employed both continuous and batch, as well as freezing-out techniques. A great many chemical techniques (colorimetric, conductimetric, potentiometric) are employed to give satisfactorily sensitive methods for the determination of most gases. Infra-red absorption is increasingly used, also gas chromatography and mass spectrometry. These latter at least offer the possibility of a single highly specific detector for all the common pollutants, except that sensitivities will have to be improved. Only if such detectors exist, does any legislation make sense.

(b) *For solids*

Solid material, depending on particle size, is collected by gravity in cans, filtering air and weighing the filtrate, sticky tape, electrostatic precipitation and thermal precipitation. Following collection it can be analysed physically by sedimentation, electron and optical microscopy, light scattering, and chemically by all the normal methods. More recently lasers at various frequencies and powers have been used to detect gas absorption and particle scattering using long optical paths.

Bibliography

BACH, W. (1972) *Atmospheric Pollution*. McGraw-Hill.
SAX (1957) *Dangerous Properties of Industrial Materials*. Reinhold.
STERN, A. C. (1968) *Air Pollution*, (3 vols.), 2nd ed., Academic Press. [This is the most comprehensive, multiauthor book on the subject.]
STRAUSS, W. (1972) *Air Pollution Control* (2 vols.). Wiley-Interscience.

W. S. MOORE

POWER SOURCES, ISOTOPIC. The radioactive decay of materials constitutes a source of primary power which may be used in a number of ways to produce electrical or mechanical energy. In radioisotopes the energy density may be high and the total stored energy extremely large: ^{238}Pu, for example, has a power density of 0·56 W per gram and a half-life of 87 years, therefore one gram of material will release 780 kilowatt hours during total decay; by contrast a normal car battery fully charged will give about 0·5 kWh. However, unlike the car battery the rate of release of the energy is invariant with time. The power sources are therefore most suitable for applications where power is required continuously. Isotopes may decay by the emission of alpha particles, beta particles, gamma rays and neutrons, all of which are biologically damaging and it is therefore necessary to shield the sources to prevent tissue damage. The extent of the shielding depends upon the radiation. Alpha particles are easily stopped and a thin metal container is adequate for containment of pure alpha-emitting isotopes; however, beta emitters such as ^{90}Sr may emit penetrating Bremsstrahlung which arise from the stopping of the high-energy electrons and this, like gamma radiation, requires large thicknesses of heavy materials, uranium, lead, tungsten to attenuate the radiation. Neutron sources are more difficult to shield and require large amounts of hydrogenous material such as paraffin wax for absorption of the decay products.

Ideally as a prime energy source the isotope should be long lived, cheap and easily available in a stable chemical form which is compatible with a canning material at the operating temperatures and under all reasonable accident conditions. Preferably it should be an alpha-emitter since this makes the shielding problem simple. Plutonium-238 comes closest to this ideal, but it is relatively expensive and therefore other isotopes may be preferred for economic reasons. Table 1 gives a list of materials suitable for isotopic power sources.

Radioactive materials are subject to stringent safety precautions and their use is subject to International Codes of Practice.[1] The degree of risk and the necessary proving tests depends to some extent on both the form of the power source and upon the isotope used, but in all cases it must be shown that the source will remain intact following the maximum forseeable accident. The Codes of Practice are formulated to ensure that the system is safe in this condition.

The primary power of the isotope may be used in three main ways to produce electrical or mechanical power. The most direct, and theoretically the most efficient, process is to collect the charged particles emitted by beta or alpha isotopes at one electrode whilst using the isotope source as the other. Secondly, the ionizing properties of the radiation can be used to produce ion pairs in an environment which enables charge separation; batteries operating on this radiovoltaic principle have been in operation for a number of years. This conversion mode is not theoretically as efficient as that of charge particle collection since the energy necessary for the ionization is lost. Finally, the ionizing radiation may be absorbed and the heat generated in the absorption used to drive a heat engine. This group includes thermoelectric and thermionic converters and also the Stirling, Brayton and Rankine engines. Setting aside the mechanical engines the current from the other devices is proportional to the active device area. They are all therefore difficult to use for large-scale power generation because of the excessive size involved and thus in most cases the isotope powered sources are limited to small devices giving a maximum electrical output up to 1 kW. Conceptually the engines may be used to drive generators to produce power at high levels but the difficulties of safely containing the extremely large quantities of isotope and of making the engine sufficiently reliable to justify the use of the expensive power source has limited the development of large power sources.

Table 1. Properties of isotopes suitable for power sources

Isotope	Half-life (years)	Decay mode	Power density (Wg^{-1})
^{238}Pu	89	α	0·66
^{244}Cm	18	α	2·4
^{242}Cm	0·44	α	120
^{210}Po	0·38	α	140
^{228}Th	1·9	α	170
^{90}Sr	28	β-	0·95
^{137}Cs	30	β-	0·42
^{144}Ce	0·78	β- (α)	25
^{60}Co	5·3	β- (α)	17·4
^{227}Ac	2·2	β-, α	14·5
^{232}U	74	α	4·4
^{85}Kr	10·6	β-	0·62
^{147}Pr	2·67	β-	0·27
^{250}Cf	10·1	α	5·1
^{3}H	12·4	β-	0·35

Direct-charge Collection

The direct-charge collection power source is essentially a self-charging capacitor. In the simplest embodiment the emitting electrode is coated on the inner side by a thin layer of the isotopic fuel and separated from the other electrode by an evacuated space. Some of the particles emitted in the forward direction will arrive at the collector and a charge will build up. Particles arriving later will require sufficient energy to overcome the potential energy built up on the collector by the earlier particles. The conversion is therefore one in which some of the kinetic energy of the radioactive particles is converted to potential energy. As the kinetic energy of the nuclear particles is high the equivalent cell voltage is also high; theoretically open-circuit voltages for optimum efficiency should be a few tens of kilovolts. Electrical insulators to sustain these voltages are difficult to produce and maintain for long periods of time particularly in the presence of the particle flux necessary. In practice, cell voltages range from a few hundred volts to about 10 kV. Fission fragments (^{235}U), alpha emitters (^{210}Po) and beta emitters (^{144}Ce), tritium (^{85}Kr) and ^{147}Pm, may be used as sources. Even where the source is thin some self-absorption of the nuclear particles will occur which when combined with their distribution of both energy and emission angle will limit the efficiency in a parallel plate cell to approximately 7% with practical values of about half a per cent. Theoretically, the power density in the cells can be high; thus, for example, the estimated current from a fission product powered cell is 0·8 µA/cm² which since the output voltage would be about 2×10^6 V gives a power density of $\frac{1}{2}$ W/cm³. It is the high voltage required for efficient operation that presents the major obstacle to the technological development and use of the cells and most cells built to date operate at a voltage too low for efficient operation. The bulk of the uses of these cells are, however, in instrumentation where only a small current source is required and hence efficiency being irrelevant the cells are operated in a short-circuit mode. There have, however, been a number of proposals for extremely large power sources, for example, Satanov designed a low-temperature reactor in which the fission fragments were directly collected giving 7 A at 1 MV.[2] The subject is reviewed in detail in a recent book.[3]

Ionization Cells

Rather than collect the nuclear particles directly they may be used to produce ion pairs which may then be separated; the ionization may be produced in gas, liquid or a solid. Since energy is required for the ionization process the maximum theoretical efficiency is less than that of the direct charge collection cell but in practice the efficiencies are considerably higher and values of about 1% have been obtained in ^{147}Pm-powered devices. In the solid state device a thin layer of isotope is deposited on a p–n junction, usually in silicon. The intrinsic layer must be sufficiently wide to enable the ionization to occur within its boundaries and as pure as possible so that the recombination at impurities and defects does not immediately remove the ions before they are able to drift in the junction field to the electrodes. Since nuclear particles have a higher energy than the quanta of sunlight it is possible to use with advantage semiconductors with higher energy gaps than solar cells; in this manner the leakage current, which depends both on the operating temperature and the energy gap, will be reduced. The main problem in using materials other than silicon is obtaining a long carrier lifetime to enable the ions to drift; this is principally a problem of purity. The production of low-resistance reliable contacts to the cell is also a problem, and a considerable amount of the energy loss occurring is caused by high series resistance in the circuit. The Betacell (McDonnel Douglas Company) powered by ^{147}Pm and using a p–n junction in silicon operates on this ionization principle at a voltage of 3·35 and delivers a current of 62 µA. It has a small amount of heavy alloy (tungsten alloy) surrounding the operating region in order to reduce the leakage of radiation to acceptable limits.

An electric field may also be created by using electrodes which have different thermionic work functions such as lead oxide and magnesium oxide, where the voltage difference is approximately 1 V. These are placed in a gas which is often also the radiation source (e.g. tritium or ^{85}Kr). The ion pairs created by the primary particles are then collected at the electrodes and the conversion process occurs because the kinetic energy is converted to potential energy. Despite the poor efficiency the simplicity, reliability and low cost of these devices has led to their use in thickness gauges.[4]

Thermionic Converters

In thermionic generation of power the energy of the nuclear particles is used as heat in order to drive off electrons from a metal surface and collect them on a second closely spaced surface. If the electrode space is evacuated then the build up of space charge requires the electrodes to be spaced by about 0·0001 in. if reasonable efficiency is required. Such a space is difficult to achieve and maintain under the operating temperatures, which are usually of order 1000–1500°C for the emitter and 500–700°C for the collector. To avoid this difficulty another method for reducing the space charge has been devised in which positive ions from caesium vapour are introduced into the electrode space. The electrode spacing can then be increased to about 1 mm and the emitter electrode temperature put up to about 2000°C. Under these conditions power densities of 40 W/cm² and efficiencies of up to 20% have been measured. In view of the high power density it is necessary to concentrate the heat when using isotopic power supplies. For this reason heat pipes are used to transfer the heat from a large volume of isotope to a small emitter surface. Heat pipes using silver as a transfer medium and capable of operation up to 2000°C have been specially developed for the emitter whilst for the collector more conventional materials such as sodium are used.[5] The device is, in principle, extremely simple, the encapsulated isotope, in the form of a stable compound such as strontium titanate, is surrounded by the heat pipe which is further surrounded by thermal insulation and a biological shield. The end of the heat pipe forms the emitter which is spaced about 1 mm from the collector by an insulating ring. The electrode space is filled with ions from a heated reservoir of caesium and power at about 0·5 V is taken from the two electrodes. Because of the high operating temperature of the emitter the thermal insulation surrounding the prime heat source is of considerable importance and novel methods have been developed to ensure extremely low losses. A special high-temperature insulator has been developed consisting of multiple foils of polished metals separated in a vacuum by small oxide particles. The effective thermal conductivity of these foils, which consists of forty to eighty layers of material, is about 10^{-6} W/cm/°C, several orders of magnitude lower than conventional solid insulators. Running times of 10,000 hours have been obtained at power levels of about 1 kW(e) and systems for power levels up to 10 kW are under construction.[6]

Thermoelectric Generators

Although this method of conversion has received more attention than the others the principle was established in 1823 when Seebeck found that if the junctions of two dissimilar metals are held at different temperatures then a current flows. A figure of merit (Z) is used to assess the suitability of a material for conversion purposes. The figure of merit is the square of the Seebeck coefficient times the electrical conductivity divided by the thermal conductivity which should, therefore, be low. Joule heating dissipates the power generated and the electrical conductivity should therefore be high. Unfortunately the parameters may not be varied independently and it turns out that Z is a maximum for a heavy semiconductor with carrier concentrations of around 2×10^{19} carriers/cm³. It is for this reason that materials such as bismuth telluride and lead tellurite are used to make thermopiles. However, since no thermoelectric material is equally as efficient over all temperature ranges, the applications tend to divide into three regions:

1. Low-power terrestrial and medical applications where bismuth telluride is used.
2. Space applications where, because of the high rejection temperature, lead telluride alloys are preferred.
3. High-power systems where composite converters of germanium silicon and one of the other tellurides are used.

Conceptually the generator is simple. The thermopile is compressed by a sprung heat sink against the encapsulated heat source. The remaining source area is then surrounded by thermal insulation and the whole by a biological shield.

Figure 1 shows a cross-section through one of the U.K.A.E.A. generators demonstrating these principles. In this case ^{90}Sr was used as the heat source in the form of strontium titanate, primarily for reasons of cost. As a consequence of the high-energy Bremsstrahlung a large thickness of depleted uranium is required for biological shielding. In the case of the miniature batteries shown in Fig. 2, ^{238}Pu is used as the prime heat source; this is almost a pure alpha emitter and consequently the primary encapsulation is a sufficient biological shield.

Because of the high cost of the isotope special insulators have been developed to prevent thermal losses. In these insulators the pore size is less than

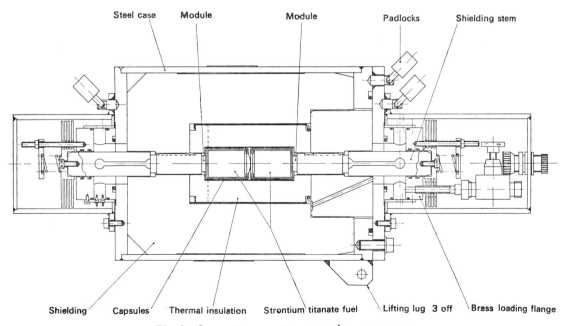

Fig. 1. One-watt generator general arrangement.

the molecular mean free path so that at normal temperatures and pressures molecule to wall collisions occur instead of molecule to molecule collisions and hence the thermal conductivity can be less than that of the still gas.[7]

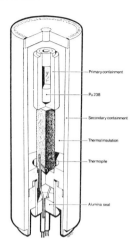

Fig. 2. Isotope battery (approximate size 4·5 × 1·5 cm).

The output from a single thermocouple is low (of order 80 mV) and therefore a generator usually has a large number of thermocouples connected in series and voltages can vary from around half a volt for the miniature generators, which therefore usually require a d.c./d.c. converter in order to allow operation of transistorized equipment, to around 30 V for large terrestrial generators. Efficiencies vary considerably from around 1% for the miniature generators to as high as 12% for high power cascaded systems. A large number of generators have been built and are currently operating various types of installations around the world. In particular, the Americans have both a space project[8] and an underwater effort.[9] Work is also being conducted by the French[10] and the Germans[11] and the U.K.[12] and a number of others.[13] Intense activity has occurred during the past few years in the medical applications and miniature batteries using bismuth telluride-type converters have been developed for implantable heart pacemaker use by both the O.E.C.D. (NEA) Study Group for Miniature Isotopic Batteries[14] and by the U.K.[15] America has developed a miniature battery based on metal thermocouples and Germany a battery based on evaporated materials, which is at present in the prototype stage.

Heat Engines

Three engine systems,
1. Rankine,
2. Brayton,
3. Stirling,

have been considered for use with radioisotopes In the Rankine system a heat source heats a condensable vapour which is then expanded through a turbine which drives the generator. The vapour is then condensed and returned by a pump to be reheated. SNAP 1 used mercury as the working fluid[16] and produced 500 W of electrical power from 5300 W of thermal power produced by 1·75 Mc of ^{144}Cs, i.e. an efficiency of $9\frac{1}{2}\%$. The problems of this type of engine are primarily concerned with corrosion produced by the working fluid and the large amount of power absorbed in the bearings. Many of these problems are overcome in the Brayton cycle, a closed-loop gas-turbine system. The gas is heated by the isotopic power source and then used to drive the turbine which powers the generator. The gas then passes through a recuperator to the heat sink from which it passes, via the compressor, and recuperator, back to the heat source. In this system inert gases are used as the working fluid.

The Stirling engine has been the subject of considerable development. Apart from the major effort in the automobile field[17] where hydrocarbon fuels are used there are two artificial heart programmes in America[18,19] which are investigating a Stirling cycle prime mover and also an artificial heart programme in Germany. The reciprocating Stirling engine has potentially the same efficiency as that of a Carnot engine working between the same temperature limits. The cycle consists of two constant-volume processes joined by two isothermal processes. The engine delivers mechanical power because the energy required to compress the gas at low temperature is less than that delivered during expansion at high temperature. The basic engine is well described in a series of papers from Phillips.[17,20] Since the engine will run equally well on any fuel so long as the working temperature is reached coupling the engine to an isotope heat source presents no problem; the difficulty as with the other engines is to make the system sufficiently reliable over a long-time scale to make the use of isotopes sensible. In the heart-assist devices sliding seals have been replaced by flextures[21] and an inert gas is used as the working medium. In the artificial heart programme the hydraulic output power is used to drive a specially developed heart pump and work is in progress to produce electric power directly by coupling the output to piezo-electric crystals. Table 2 gives comparative efficiency figures for the various engines.

In engine systems used in the body the pure alpha emitter ^{238}Pu is used as the heat source because the radiation dose to the surrounding tissues can be kept to reasonable levels without excessive biological shielding. The ^{238}Pu is made into a high-density ceramic using enriched oxygen (O_{16}) since in this fashion the gamma rays associated with the n–p reactions are removed. The oxide is extremely stable and has a high melting point which reduces the

Table 2. Comparison of dynamic converters

	Hot junction temperature °C	Cold junction temperature °C	Efficiency	Working fluid
Rankine	700	150	10%	Mercury
Brayton	600–700	40	20–30%	Rare gas
Stirling	700	50	20%	Helium

attack on container materials. However, because of the decay helium is given off which will produce very high pressures unless vented. The latest capsule designs[19] thus surround the outside of the refractory metal can with an oxidation resistant outer can and a leak is so arranged that the helium is able to escape whilst solid materials are retained. The leak takes the form of an extremely convoluted path allowing a slow diffusion of gas. Because of the body use and the possibility of cremation combined with the high upper operating temperature of the engines (order 700°C) the sources are subject to stringent tests at high temperature/pressures and also to crush, fire, bullet impact and corrosion.

Bibliography

1. *Guide to the Safe Design, Construction and Use of Radioisotopic Power Generators for Certain Land and Sea Applications*, I.A.E.A. Safety Report number 33, Vienna.
2. G. SATANOV, RM1870, The Rand Co., Jan. 1952.
3. G. H. MILEY, *Direct Energy Conversion*, published by Am. Nuc. Soc., 1970, for U.S.A.E.C., Division of Technical Information.
4. ANON: Bull. No. GC-2 (1967) The Ohmart Co., Allendorf Drive, Cincinnati.
5. See, for example, D. A. REAY, IRD 71 83, Int. Research and Development Co., 1971.
6. D. S. BEARD, *1971 Intersociety Eng. Conv. Conf. Boston*, *1971*, p. 933, Soc. of Auto. Eng. Inc., New York, 1971.
7. J. A. MCWILLIAMS and J. HUGHES (1969). *Ind. Appl. for Isotopic Power Generators*, A.E.R.E. Harwell, OECD-ENEA, p. 363.
8. See, for example, four papers in *1971 Int. Eng. Conv. Conf.*, Boston, 1971, pp. 478–512, Soc. Aut. Eng. Inc., New York, *1971*.
9. *U.S. Navy Quarterly Operating Reports for Radioisotope Devices*, 1968–72.
10. J. FREUND, paper EN/IB/21, *Power from Radioisotopes*, O.E.C.D., Madrid, May 1972.
11. V. MERGES, paper EN/IB/12, *Power from Radioisotopes*, O.E.C.D., Madrid, May 1972.
12. A. W. PENN, *Bull. Phys. Soc.* **22**, 207 (1971).
13. E. G. MAVROYANNAKIS, paper EN/IB/7, *Power from Radioisotopes*, O.E.C.D., Madrid, May 1972.
14. M. ALAIS *et al.*, paper EN/IB/18, *Power from Radioisotopes*, O.E.C.D., Madrid, May 1972.
15. A. W. PENN *et al.*, *Nuc. Tech.* **14**, 39 (1972).
16. A. D. TONELLI *et al.*, *Trans. ASME* (B) **88**, p. 129 (May 1966).
17. K. A. J. O. VAN WITTEREEN (1969) *Ind. Appl. for Isotopic-powered Generators*, ENEA-OECD, p. 525.
18. L. T. HARMISON *et al.*, Paper EN/IB/10, *Power from Radioisotopes*, O.E.C.D., Madrid, May 1972.
19. W. E. MOTT, paper EN/IB/57, *Power from Radioisotopes*, O.E.C.D., Madrid, May 1972.
20. H. B. VAN NEDERVEEN (1969) *Ind. Appl. for Isotopic Power Generators*, A.E.R.E. Harwell, ENEA–OECD 517.
21. L. T. HARMISON, *Totally Implantable Nuclear Heart Assist and Artifical Heart*. National Heart and Lung Inst., U.S. Dept. Health, Feb. 1972.

A. W. PENN

PULSARS. The pulsating sources of radio waves discovered by Professor Hewish and Miss Bell at the Mullard Radio Astronomy Observatory, Cambridge, in 1967, are now known to be rapidly rotating neutron stars located in the Milky Way. The discovery was made using a radio telescope specially constructed for the study of randomly fluctuating radio sources; the appearance of a new source with regular pulsations of intervals of 1·337 sec was a complete surprise. The shortness and extreme regularity of the pulsation period showed that the source must be some form of small condensed star such as a white dwarf or neutron star, either vibrating or rotating. Subsequently the observation that the periods of all pulsars are slowly increasing ruled out vibration, while the discovery of a pulsar with a period as short as 88 msec ruled out the white dwarf stars in favour of the neutron stars.

The pulsar with 88-msec period is located in the remains of a supernova in the constellation of Vela. A surmise that neutron stars are formed in supernova explosions led to a search for a pulsar in the young and active supernova remnant known as the Crab Nebula. A pulsar was found there by Staelin and Reifenstein, at the National Radio Astronomy Observatory, U.S.A., and measurements at the Arecibo Observatory, Puerto Rico, found the period to be 33 msec, the shortest period known so far. This led to a search for optical pulses, which were found by Cocke, Disney and Taylor at the Steward Observatory in Arizona. Pulses at the same period have also been found at infra-red and ultra-violet wavelengths, and in the X-ray and gamma-ray spectra. All other pulsars appear to radiate only at radio wavelengths; their radiation can usually be detected at wavelengths between 10 cm and 5 m.

A neutron star is believed to be typically about 20 km in diameter, with a mass between 0·2 and 1·5 solar masses. It represents a final stage of collapse of matter under its own gravitational attraction. In the earlier stage, a white dwarf star is made up of nuclear material stripped of orbiting electrons, while in a neutron star the nuclear material is broken down into closely packed neutrons. The density is of the order of 10^{15} g cm^{-3}. The outside crust is solid and crystalline, while the interior is probably a superfluid. Neutron stars are highly magnetized, as a result of their condensation from a weakly magnetized normal star: the surface field may amount to over 10^{12} G.

More than 100 are now known. Their periods range from 33 msec to 3·7 sec, with the majority lying between 100 and 500 msec. They are concentrated towards the plane of the Milky Way, at distances generally less than 2 kiloparsecs. Their ages, as determined from the rate of change of their periods, are mostly in the range of 1 to 10 million years; the shortest is that of the Crab Nebula, which corresponds to the known age of 900 years (since the supernova was observed in A.D. 1054).

The rotational energy of neutron stars, which is derived from gravitational energy during the collapse of the star, becomes available as the star slows down. Energy is transferred to charged particles around the pulsar through the rotating magnetic field, rather as the rotor of a dynamo induces currents in the surrounding stator windings. This energy supply produces the visible light and X-rays radiated from the Crab Nebula itself; only a small proportion, of the order of 0·1%, goes into the direct radiation from a pulsar.

The formation of pulses of radio waves, and for the Crab Nebula pulsar, of light and X-rays, has been accounted for by a relativistic effect. Rotating magnetized neutron stars are necessarily surrounded by a co-rotating region of dense electron and proton clouds. At a large distance from the star these clouds can only co-rotate by moving with velocities approaching the velocity of light; it is in this region that the radiation is supposed to originate. The circular motion of the source with velocity of the order of 0·8 c beams the radiation in the direction of travel, so that the observer is swept by a "searchlight" beam once per rotation. The mechanism of radiation is believed to be synchrotron radiation for optical and shorter wavelengths, and coherent gyro-radiation at radio wavelengths.

The radio pulses were found, in observations at Jodrell Bank, to be polarized practically 100%, generally elliptically but with a slowly changing form through a typical pulse. A superposition of many pulses produced an integrated pulse profile which is characteristic of individual pulsars. The polarization and width of the single pulses can be related to an individual radiating source, while the width of the integrated profile may represent a spread of sources within the co-rotating magnetosphere.

Radio waves from pulsars have proved to be extremely useful as probes of the interstellar medium. The propagation of radio waves is affected by ionized gas, which decreases the velocity of propagation of radio pulses. There is therefore a delay in arrival time of the pulses as compared with free space conditions; this delay depends on the radiofrequency of observation, so that comparison of observations at two separate frequencies can provide a measure of the delay. In this way the number of electrons along the propagation path can be measured.

Faraday rotation also occurs along the propagation path. The plane of polarization of the radio pulses is rotated by an amount proportional to the total electron number and the galactic magnetic field component along the path. Since the electrons are separately detected in the measurement of delay, the magnetic field can be derived. This is the most direct way of obtaining the value of the galactic magnetic field, which is found to be about 3 μG.

Irregularities in the density of the interstellar ionized gas cause a random diffraction of the radio pulses, so that the strength of the pulses varies randomly on a time scale typically of a few minutes. This interstellar scintillation is not seen in other types of radio source, since their diameters are larger than the linear scale of the scintillation pattern, and the fluctuations are smoothed out.

The scintillation pattern of pulsars has been observed to be moving rapidly across the surface of the Earth. This is apparently due to a rapid motion of the pulsar itself; one pulsar is found in this way to be moving at a speed of 360 km s^{-1}. Such a high speed may possibly be acquired during the formation of the neutron star at the time of the supernova explosion. The speed is so high that most of the older pulsars may be expected to be located very far from any detectable traces of the supernova in which they originated. The question of their origin and the numbers of pulsars to be found in the Milky Way therefore depends on a large statistical survey of the pulsars rather than attempts to relate them directly to visible objects. Surveys of pulsars have notably been carried out at the Molonglo radio observatory (Australia), where over thirty have been found, and at Jodrell Bank, where over forty have been found. Techniques for the search process have ranged from a simple pen recording of individual radio pulses, as in the original discovery, to some complex applications of on-line computation, in which the computer is arranged to search for any precisely periodic signal in the random signal collected by a radio telescope. Further discoveries of pulsars will depend on the use of the computer technique, since only the strongest pulsars give a recognizable pulse on a pen recorder.

Many extensions of the detailed investigations of

the radio and optical pulses are under way. Some are concerned with accurate timing of the pulses; here the main interest is in irregularities which have been seen in the monotonic slowing of the Vela and the Crab Nebula pulsars. Others are concerned with the structure of the pulses. Here the recent interest is in short-duration fluctuation within the pulses; one pulsar is known to exhibit short spikes of radiation less than 10 μsec in duration. The intensity of radiation within such a short spike is higher than that of any other known radio source.

References, and a more detailed account, may be obtained from a review article by F. G. Smith, *Review of Progress in Physics.* 35, No. 4 (1972).

See also: Neutron stars.

F. G. SMITH

Q

QUASARS. Quasars (or *quasi*-stel*lar* objects; QSO) are defined as a class of astronomical objects of dominantly star-like appearance on direct photographic plates with emission-line spectra exhibiting redshifts much larger than those of ordinary stars in the Galaxy. The property described as "star-like" means in practice that the angular diameter is smaller than one second of arc.

The nature of the quasars and their relation to other astronomical objects is still uncertain. They are not clearly associated with other objects in the sky, such as galaxies or clusters of galaxies. Some quasars emit at radio wavelengths; these radio quasars often contain components of very small angular diameter ($<0''\cdot001$ arc), although this does not appear to be a definitive property. Variations in intensity have been observed both at radio and optical wavelengths, with time scales of a week to 10 years.

The interpretation of large redshifts from astronomical bodies that are small enough to show large intensity fluctuations within a year remains a controversial matter. The cosmological hypothesis, which states that the redshifts are due to the general expansion of the Universe, requires very large amounts of energy radiated from a small volume. The corresponding large distances have not been confirmed by independent evidence. Alternative hypotheses have been advanced to explain the redshifts but these raise perhaps even greater difficulties.

Discovery

The first quasars were discovered in the early sixties when accurate radio positions determined with the Owens Valley interferometer pointed to the identification of some discrete radio sources with "stars" rather than galaxies; all of the established radio-optical associations at that time were with external galaxies or with supernovae and regions of ionized hydrogen within our Galaxy. The prototype of the *quasars*, as they are now called, is 3C273, one of the strongest radio sources in the Third Cambridge (3C) catalogue. A precise radio position for this source was measured by Hazard and his colleagues in Australia using the method of lunar occultation. This enabled an unambiguous association to be made with a thirteenth magnitude star-like object with a spectrum containing six broad emission lines. The major breakthrough came when Schmidt was able to identify the emission lines with lines in the Balmer series of hydrogen and a line of ionized magnesium at $\lambda_0 = 2798$ Å, shifted to longer wavelengths by an amount $z = (\lambda - \lambda_0)/\lambda_0 = 0\cdot158$, where λ is the measured wavelength and λ_0 is the laboratory (rest) wavelength. A redshift as large as $0\cdot158$ is normally seen only in galaxies of magnitude 17 or fainter.

Much larger quasar redshifts (up to ~ 2) were detected in 1965 as many more identifications were made. At the present time (1973), redshifts have been measured for more than 200 quasars, most of which have been found in systematic identification programs based on radio surveys. The largest redshift found to date is $3\cdot53$ for the quasar associated with the radio source $1442+10\cdot1$.

Optical Properties

At large values of $z(\sim 2)$, strong emission lines of hydrogen (Lyman α at 1216 Å) and triply ionized carbon (1550 Å) are shifted into the photographic part of the spectrum. These lines are normally inaccessible to ground-based telescopes due to the absorption of ultra-violet radiation in the Earth's atmosphere. The emission lines are usually between 20 Å and 100 Å wide; if the line widths are due to mass motions of the gas, this implies random velocities of several thousand km/s within the line-producing regions of the quasar. However, several alternative broadening mechanisms have been proposed to avoid the containment problems posed by such large relative velocities.

The optical emission from most, and perhaps all, quasars is variable over time scales of between one week and several years; intensity fluctuations by as much as a factor of 5 are not uncommon. Depending on the rapidity of the variations, this important property indicates that a large fraction of the light originates in a volume of less than a light week to a few light years across. The variations generally take the form of sporadic outbursts, in some cases accompanied by a slow secular trend. From spectra obtained for several quasars near maximum and minimum brightness, the evidence suggests that the strength of the emission lines remains constant, i.e. strong lines are visible when the quasar is faint but disappear when the continuum brightness increases.

One of the most striking results of the optical measurements is the strong correlation found between high optical polarization and variability (both optical and radio). In two quasars, 3C345 and 3C446, large changes in the total brightness have been accompanied by a rotation of the plane of polarization. This suggests that the variable components are highly polarized and has been taken as strong evidence that the optical flux is, at least in part, of synchrotron origin.

Narrow absorption lines are observed in the spectra of a number of quasars, especially in those of large redshift. Many of the lines can be identified as due to hydrogen, carbon, silicon, etc., in various stages of ionization. Their redshifts are usually a little smaller than the emission line redshift and one would tend to interpret them as caused by material expelled by the quasar. In a few quasars, however, some 50 to 100 narrow absorption lines are observed and in these cases there is evidence for many different redshift systems. A typical case is the 16·6 magnitude quasar PHL957 which has an emission-line redshift of 2·69 and in which some eighty absorption lines provide evidence for the existence of eight redshift systems, with redshifts between 2·0 and 2·7.

If the absorbing material has been ejected by the quasar then ejection velocities of some one- or two-tenths of the velocity of light, and in some cases even more, are required to explain the difference in redshifts. Furthermore, the narrow line widths impose severe dynamical constraints on the ejected material. An alternative hypothesis is that the different absorptions are due to galaxies or concentrations of matter at smaller cosmological redshift than the quasars. The observed number of absorption systems is larger than one would expect on this hypothesis.

Broadband (UBV) photometry of quasars identified with radio sources has established that they lie well above the locus of the main-sequence stars in the two-colour (U–B, B–V) diagram. This is a region characterized by a large flux of ultra-violet radiation and is also occupied by some white dwarfs, old novae and related highly evolved stars. Several authors have pointed out a strong correlation between the colours (U–B, B–V) of QSOs and their redshifts. Although this was thought earlier to be due to an intrinsic spectral energy distribution, it is now generally agreed that the emission lines themselves are largely responsible for the variations in colour with redshift, z. This is supported by the large differences found amongst individual QSOs in the form of the continuum spectra determined by narrow-band spectrophotometry.

It is the ultra-violet excess in the optical radiation of the quasars that has proved particularly valuable in radio source identification programmes and this general property has also been used to search optically for quasars. Photometric surveys of faint blue stars by Sandage and Luyten show that there are around 0·4 QSOs per square degree brighter than a blue (B) magnitude of 18. The number rises steeply towards fainter magnitudes and the total number in the sky brighter than $B = 22^m$ has been estimated to be 10^6–10^7.

In addition to the ultra-violet excess, quasars also exhibit very strong infra-red emission. In fact, most of the energy radiated by quasars appears in the infra-red. The prototype quasar 3C273 radiates 10 to 100 times more energy in the infra-red than in the radio, visual or ultra-violet regions of the spectrum. This quasar has also been observed as an X-ray source.

Radio Properties

Quasars with detectable radio emission are relatively rare. For instance, only about 1% of quasars brighter than 18th magnitude are radio sources strong enough to be included in the 3C catalogue. The reason that most of the quasars studied so far are comparatively strong radio emitters is simply that the radio emission (coupled with the ultra-violet excess) serves as effectively in singling out quasars, as it did in the discovery of these objects.

The radio properties of quasars are similar to those of radio galaxies and it is not possible to distinguish uniquely between these two classes on the basis of radio properties alone. In particular, many quasars show the double radio structure so characteristic of radio galaxies. The radio morphology of quasars has been investigated over a range of frequencies by the methods of lunar occultation, aperture synthesis and long-baseline interferometry. The structures of the well-resolved sources may be classified as follows:

(i) two similar, well-separated emitting regions, neither of which coincides with the optical quasar;

(ii) two well-separated emitting regions with the small, active component coincident with the quasar and the second larger component displaced; and

(iii) triple or more complex structure with the quasar lying close to the line joining the components.

A detailed study at 2700 MHz shows that about half of all radio quasars have structure $>5''$ arc, more than 20% are larger than $30''$ arc and some are as large as $150''$ arc. This is in marked contrast with the optical sizes which must be $\leqslant 1''$ arc to present a star-like appearance.

Below a frequency ~1400 MHz, the spectra of most radio sources can be represented by a simple power law. The flux density, S_ν, is given by $S_\nu = k\nu^{-\alpha}$, where ν is the frequency and α is the spectral index. The value of α is approximately 0·8 for most extragalactic radio sources, there being no significant difference between quasars and radio galaxies. Above 1400 MHz, there is a general tendency for α to increase for radio galaxies and decrease for quasars. This divergence at high frequency between the spectral properties of radio galaxies and quasars has several important ramifications. Firstly, a radio survey at high frequency, complete to a specified flux density limit, will contain a larger fraction of flat-spectrum ($\alpha \sim 0$) sources than the equivalent survey carried out at low frequency; the high-frequency survey will correspondingly contain a larger fraction of quasars. Secondly, the occurrence of secular intensity variations at frequencies >1000 MHz is strongly correlated with a flat or inverted radio spectrum ($\alpha \lesssim 0$). Quasars, therefore, compose the majority of variable radio sources. Finally, extensive statistics of angular diameters obtained from surveys of interplanetary scintillation show that quasars scintillate more frequently than radio galaxies. Scintillating quasars have a predominance of curved and flat spectra, confirming that the spectral enhancement at short wavelengths is a result of very small component sizes.

Observations of the linear polarization in the radio emission from quasars have been made over a frequency range of 400 MHz to greater than 15 GHz. The general trend is a decrease in percentage polarization with decreasing frequency, although some quasars show a maximum within this range. At short wavelengths there appears to be little or no correlation between the degree of polarization and other source parameters. However, at long wavelengths ($\lambda \sim 50$ cm) the flat-spectrum quasars show systematically higher polarization than steep spectrum quasars. Since this phenomenon implies a difference in the rate of depolarization (with frequency) between the two classes, it is suggested that the main cause is Faraday dispersion occurring in or near the source itself. A strong correlation has also been found between the rate of depolarization and the redshift, in the sense that quasars with $z > 1$ have an intrinsic tendency to depolarize more strongly than those of lower redshift.

One of the most interesting "correlation games" that has been played recently with the radio properties of quasars involves the relationship between redshift and largest angular size (LAS) for resolved quasars. In the general case of a two-dimensional brightness distribution the LAS refers to the greatest angular dimension of the distribution. Miley has presented rather striking evidence for a decrease of LAS with redshift. The implication of this discovery is that the sources are all of similar physical size and the correlation results from the sources being observed at different distances. This evidence therefore provides indirect support for a cosmological origin of quasar redshifts. A combined plot of LAS against redshift for both radio galaxies and quasars with steep spectra (believed to be the completely evolved or "old" sources) shows very clearly a continuity in overall angular size which is consistent with maximum physical dimensions ~400 kiloparsecs (kpc; 1 parsec = $3 \cdot 1 \times 10^{13}$ km = 3·26 light years). Since this continuity must be explained by any theory of quasar redshifts, it seems improbable that the origin of the redshifts of quasars differs appreciably from that of the redshifts of radio galaxies. This does not, however, rule out the possibility that the systematic changes in angular size may be due to a redshift-dependent property of the quasars.

The Nature of the Quasars

The outstanding property of the quasars is their large observed redshifts. Most astronomers accept the cosmological nature of the redshifts but the arguments are still indirect and warrant discussion. Alternative explanations are often referred to as the "local" hypothesis because they place the quasars at much smaller distances. Different versions of the local hypothesis assume that the redshift is due to an explosion centred on our Galaxy, that it is gravitational in origin, or leave the redshift unexplained. The three main hypotheses are now discussed:

(i) *Gravitational hypothesis.* This hypothesis assumes that the emission lines are formed close to a compact massive object. The gravitational redshift corresponds to the loss of energy suffered by a photon in leaving the gravitational field of such an object and is small in "ordinary" astronomical objects. Since it is, to first order, proportional to the ratio of mass to radius, the gravitational redshift can become large in objects of large mass or small radius. However, any gradient of gravitational potential across the region where the spectrum lines arise will produce a broadening of the lines. Taking the observed line-widths as limiting the size of this gradient, Greenstein and Schmidt have estimated the corresponding distances and masses required. For a mass of 10^5 solar masses, the quasar 3C48 ($z = 0\cdot 37$) would still be inside the solar system, while for a typical galaxy mass of 10^{11} solar masses, the distance would be ~20 kpc and 3C48 would rival the gravitational field of our own Galaxy. Larger masses will place the quasars outside our Galaxy and their contribution to the mass density of the Universe then becomes important. Several alternative models have been proposed but on close scrutiny they are all difficult to reconcile with observation. One such proposal is that small redshifts are mostly cosmological and large redshifts gravitational or "non-velocity" in origin. This entirely arbitrary procedure presents more problems than it purports to solve and is certainly less satisfying physically than either hypothesis alone.

(ii) *Local Doppler hypothesis.* In this hypothesis the redshift is of Doppler origin with the distance not specified. Since only redshifts and no blue shifts are involved, the very large velocities (up to 0·95 of the velocity of light) would have to be the consequence of an explosion from a nearby centre. Terrell, who has been the main proponent of this theory, assumed the centre of our Galaxy to be the seat of the explosion. From the upper limit to the proper motion of the quasar 3C273 he derived a minimum distance of 200 kpc, a minimum age of the explosion of 5 million years and a mass of 1000 solar masses for the average QSO. The required kinetic energy carried away by an estimated total number of 10^6–10^7 quasars is $\sim 10^{56}$ joules. At an improbably high efficiency of 10%, this would require involvement of 10^{10}–10^{11} solar masses or practically the total mass of the Galaxy!

(iii) *Cosmological hypothesis.* In the cosmological hypothesis the redshifts are explained as being due to the universal expansion. The corresponding distances are very large, requiring very high luminosities from small volumes (due to the variable emission). The physical dimensions (R) are governed by the relation $R \leq c\tau$, where c is the velocity of light and τ is a characteristic time scale for significant changes in intensity. The small-scale angular structure of many variable quasars has been measured using very long baseline interferometry (VLBI). In general, the physical sizes deduced from the VLBI angular data at the redshift distance are consistent with the light travel distances given by the equation above, although the margin is quite small in many cases. (The angular dimensions in several quasars have been shown to be $<0''·0004$ arc by this method, using an interferometer baseline practically equal to the diameter of the Earth.)

A further problem caused by the limit $R \leq c\tau$ arises through the very large radiation density which must exist within the small volume. Assuming that the radiation from quasars is of synchrotron origin (as is fairly well established for radio galaxies), Hoyle and others have suggested that the energy losses resulting from the interaction between this intense radiation field and the synchrotron electrons (the so-called "inverse" Compton effect) would quench the source very rapidly. This was taken to indicate that quasars must be "local" objects, although many other possibilities were considered. The problem was essentially resolved by the discovery that similar arguments could be applied to the variable nuclei of N-type and Seyfert galaxies. However, in the latter case there is no reasonable doubt about the cosmological nature of the redshifts. Hence the argument based on inverse Compton losses is considerably weakened and cannot be used to limit reliably the distance of non-thermal sources.

If the redshifts are cosmological there must be a strict redshift-distance relation. However, a redshift-apparent magnitude relation will only be clearly shown if the absolute magnitudes of quasars have a small spread. Such a redshift-magnitude diagram for the quasars shows only a weak relationship with much scatter, whereas a similar diagram for radio galaxies shows a much stronger correlation; in both cases the fainter objects show larger redshifts on the average. This does not argue against the cosmological distances of quasars since the scatter may simply reflect a large dispersion in absolute magnitudes.

Some suggestive statistical evidence has been adduced for a non-random distribution of quasar redshifts. Burbidge and others have claimed that there are several preferred values of z, the two most "popular" being $z = 0.061$ and $z = 1.95$. They have taken this as evidence for a non-cosmological component of quasar redshifts. However, Roeder has pointed out that this effect may be attributed directly to observational selection effects arising from the night-sky emission lines and the atmospheric "window" which determines the specific quasar emission lines used to measure z in each redshift interval.

(a) *Associations.* Several attempts have been made to establish associations of quasars with other kinds of astronomical objects. No quasars have yet been found associated with very rich clusters of galaxies. However, some associations have been reported between quasars and small groups of galaxies. One of the best examples is the quasar PKS 2251 + 11 ($z = 0.32$), which lies in the direction of a group of faint galaxies, the brightest of which was found to have essentially the same redshift as the quasar. This result constitutes strong support for the cosmological hypothesis.

On the other hand, the possibility that quasars may be associated with individual bright galaxies has been the subject of much discussion in recent years (e.g. see Arp, 1971). Most recently, Burbidge *et al.* have found that four out of a sample of forty-seven identified quasars from the 3C catalogue are much closer to catalogued bright galaxies than would be expected for a random distribution. The probability of any one of the four close pairs occurring by chance was estimated to be $\sim 10^{-3}$. Since the quasars involved in these close pairs have redshifts between 0·5 and 1·4, while the bright galaxies all have $z < 0.01$, it was suggested that some quasars may have substantial redshift components that are not due to the expansion of the Universe. However, there remains the obvious problem of deciding how to interpret formal probabilities for "unlikely" events whose existence was not predicted before they were noticed and for which probability arguments were only subsequently applied.

One indication of whether "unlikely" events are a statistical fluctuation or not is the predictive power of the relations for other observations. Two groups at Cambridge and Princeton have independently examined coincidences between quasars and bright galaxies using a much larger sample of quasars. In each case they find fewer pairings by almost an order of magnitude than would have been expected on the basis of the earlier results and they conclude that the original findings most probably represented a rare statistical fluctuation.

(b) *Similarity arguments.* Similarities have been frequently discussed between quasars and the nuclei of Seyfert and N-type galaxies in properties such as the ultra-violet excess of the continuum, the wide emission lines, the strong infra-red radiation and in some cases, the rapid variability. These arguments tend to support the cosmological hypothesis although Burbidge has disputed their value.

As far as redshifts are concerned, it is usually found that $z_{Seyfert} < z_N < z_{quasar}$. However, exceptions are known: quasars have been found with redshifts as small as 0·06, while Seyfert galaxies with $z \sim 0.1$ and N galaxies with $z \sim 0.3$ have also been identified. On the cosmological hypothesis, it has often been suggested that the morphological classifications assigned to the three types of objects may well arise naturally from a common type of object viewed at widely differing distances or alternatively be due to an evolutionary connection between quasars, N galaxies and Seyfert galaxies.

On average there is a progression of optical luminosity with the Seyfert galaxies being intrinsically the faintest objects, while the quasars are the brightest. A similar progression exists for the radio luminosities although there is a large scatter. It must also be remembered that the bulk of the quasars are not strong enough emitters even to be detected as radio sources at present.

The realization that there is such a continuity if all the redshifts are tacitly assumed to have a cosmological origin has led many astronomers to argue that this is evidence for the cosmological redshift hypothesis for quasars. Burbidge, however, has suggested that if the quasars or even the N systems have large intrinsic (i.e. non-cosmological) redshift components, so that quasars and Seyfert nuclei are at comparable distances, there is again continuity. In any case, the observed properties of Seyfert nuclei, N galaxies and quasars are so similar that the physics of all these objects must also be very similar, in the sense that the energy-generating mechanisms are likely to be the same and the differences only those of degree.

A great deal of interest has been generated recently by the results of very long-baseline radio interferometry on the compact double sources (separations $\sim 0''\cdot 001$ arc) in the quasars 3C273 and 3C279. These observations show that the components are moving apart at very high speeds. The expansion rates, computed on the basis of a cosmological redshift distance, are four and six times the velocity of light, respectively. Of course, at a smaller distance the expansion rate is reduced and one may argue that these observations speak in favour of the local hypothesis. However, subsequent observations of the Seyfert galaxy 3C120 ($z = 0.033$) showed apparent rapid changes in structure similar to those observed for the quasars 3C273 and 3C279. The apparent transverse velocity is two or three times the velocity of light. Since there is no reasonable doubt about the cosmological nature of galaxy redshifts, the argument above is greatly weakened. Nevertheless, a plausible mechanism for these "superlight" velocities has yet to be found and it seems likely that considerable effort will be devoted to this problem over the next few years.

(c) *Distribution of quasars.* The distribution of quasars over the sky appears to be isotropic. This follows from the observed isotropy of faint radio sources, 20–30% of which have been identified as quasars. Isotropy is to be expected in the cosmological hypothesis, while the various alternative hypotheses require rather contrived geometries to satisfy this condition.

The distribution of quasars in depth appears to be far from uniform. Counts of quasars detected optically, from their ultra-violet excess, show that around magnitude 18 their number increases by a factor of about 6 per magnitude. If quasars were uniformly distributed through space, then at the cosmological distances involved their number should increase by a factor of only 2 per magnitude. The observations thus suggest that at large distances there are many more quasars than nearby. Since large distances correspond to long light-travel times, this implies that at early cosmic times the number of quasars was much larger than at present, by a factor of perhaps 1000. The derivation of the space distribution of radio quasars is complicated by the fact that optical and radio selection effects operate simultaneously; the redshift of a radio quasar can only be determined at optical wavelengths, and hence requires optical identification.

As it seems likely that quasars have cosmically short life-times, of perhaps 10^6 to 10^8 years, the numbers seen at a given epoch represent the birth rate of quasars. There are indications that the increase in density may not continue beyond $z = 2.5$. This would suggest that the greatest activity in quasar births took place some thousand million years after the expansion of the Universe started. If galaxies were forming at that time, it has been suggested that quasars may have represented a stage in the formation of the galaxies or at least in the formation of their nuclei.

Bibliography

ARP, H. (1971) *Science*, **174**, 1189.
BURBIDGE, E. M. (1967) *Ann. Rev. Astron. Astrophys.* **5**, 399.
BURBIDGE, G. R. (1970) *Ann. Rev. Astron. Astrophys.* **8**, 369.
BURBIDGE, G. R. and BURBIDGE, E. M. (1967) *Quasi-stellar Objects*, San Francisco and London: Freeman.
EVANS, D. E. (ed.) (1972) *External Galaxies and Quasi-stellar Objects* (I.A.U. Symposium No. 44), Dordrecht-Holland: D. Reidel.
KELLERMANN, K. I. (1972) *Scientific American*, **226**, 72 (July).
SCHMIDT, M. (1969) *Ann. Rev. Astron. Astrophys.* **7**, 527.
SCHMIDT, M. and BELLO, F. (1971) *Scientific American*, **224**, 54 (May).

R. W. HUNSTEAD

R

RADIATION AND ENTRY OF SPACE VEHICLES INTO PLANETARY ATMOSPHERES.
As a space vehicle re-enters the Earth's atmosphere it passes through various gas-flow regimes. At very high altitudes the density is so low that the molecules do not collide with each other in the vicinity of the space vehicle. This *free molecule flow* regime occurs above approximately 150 km for a typical re-entry vehicle having a nose radius of the order of 1 m. As the vehicle proceeds to lower altitudes molecular collisions become important and a *viscous merged layer* forms about the vehicle. Then below approximately 90 km the front part of the merged layer develops into a *shock wave*. At still lower altitudes, below approximately 80 km, viscous effects become confined to a thin *boundary layer* near the vehicle surface, with the remainder of the *shock layer* being nearly inviscid.

During re-entry a space vehicle uses the atmosphere for braking, i.e., the vehicle is slowed down by the aerodynamic drag, with much of the dissipated vehicle energy being transferred as heat energy to the air which flows through the shock layer. The resulting high-temperature shock layer produces large heat-transfer rates which the vehicle surface must absorb by an appropriately designed heat shield (usually consisting of an *ablating* material). For vehicle re-entry velocities less than an Earth satellite orbital velocity (approximately 8 km sec^{-1}) the convective heat transfer from the boundary layer to the vehicle dominates over the radiative heat transfer from the shock layer. However, as the re-entry velocity increases above the orbital velocity, radiative transfer increases rapidly and soon dominates over the convective heat transfer. At the high superorbital re-entry velocities corresponding to return from planetary missions radiation is not only the dominant heat transfer mode, but *radiative cooling* can affect the temperature distribution and flow in the shock layer. Accordingly, the analysis of shock layers under high superorbital conditions must include the coupled equations of gas motion and radiative transfer, i.e. the equations of *radiative gas dynamics*. The large radiative heat load encountered at high re-entry velocities can be reduced by designing vehicles which are more pointed (smaller nose radii) and more slender than current re-entry vehicles.

Figure 1 illustrates the flow regions and radiation environment around a re-entry vehicle. The radiating shock layer can have many effects. First, even at suborbital velocities, ultra-violet radiation from the shock layer can produce sufficient photo-ionization ahead of the shock wave to result in a larger radar cross-section for vehicle detection. Also, the flow of heated shock-layer gases and ablation products in to the wake behind the vehicle results in a long ionized trail which may be observed by radar or optical instruments.

As previously mentioned, radiative heat transfer to the vehicle will dominate over the convective heat transfer for superorbital velocities. In Fig. 1 the dots in the boundary layer represent ablation products resulting from a large radiative heat transfer to the vehicle. Altough the ablation process absorbs much of the heat load, *reradiation* can also be an important means of energy accommodation. Sufficient time for appreciable heat rejection by surface reradiation can be achieved during relatively low velocity re-entry and during shallow re-entry trajectories of lifting vehicles.

Study of the effects of thermal radiation during the re-entry involves first a determination of the basic radiative properties of the constituent gases and solid materials and second, a determination of the detailed gas flow with radiative transfer and

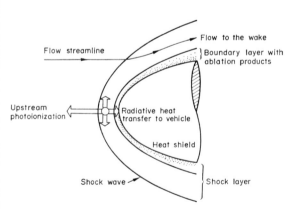

Fig. 1. Schematic drawing of the flow regions and radiation environment around a re-entry vehicle.

heat-shield ablation included. Direct measurements of high-temperature air radiation can be achieved by the use of shock tubes and ballistic ranges. However, in order to predict the radiation for the actual re-entry conditions it is usually necessary to use equilibrium statistical mechanics to calculate the population numbers of the constituent molecules, atoms and ions in the various energy states (i.e. *local thermodynamic equilibrium* is assumed). The *spectral absorption coefficient* is calculated by coupling the population numbers to the appropriate radiation constants, i.e. to the *transition probabilities* (*f-numbers* or *oscillator strengths*) and *line half-widths*, which are usually determined by experiments, but may be calculated approximately for simple species. The resulting absorption coefficient, determined as a function of temperature and pressure, is then used in the radiative transfer equation, which is coupled with the equations of motion for a computation of the flow field and heat transfer.

It should be noted that above approximately 60 km altitude the assumption of local thermodynamic equilibrium breaks down, and that appropriate rates for the non-equilibrium processes or direct experimental data must be used to determine the resulting *non-equilibrium radiation*. However, non-equilibrium effects are not very important for Earth atmospheric re-entry since large radiative heat transfer generally occurs only at altitudes below 60 km.

Figure 2 shows calculations for the continuum absorption coefficient of air at a temperature and pressure characteristic of the shock layer during re-entry. The continuum absorption coefficient arises from *bound-free* transitions corresponding to the photoionization of atoms and ions, and from *free-free* transitions corresponding to the acceleration of free electrons in the vicinity of ions and atoms. In addition to the continuum absorption coefficient, *bound-bound* transitions produce atomic lines. Also, particularly at the lower temperatures, molecular species can be important in producing band spectra. In air the atoms and ions of nitrogen and oxygen produce the greatest contributions to the absorption coefficient, with important contributions also arising from the molecular species N_2, O_2 and NO.

Radiative transfer is also important for space vehicle entry into the atmospheres of other planets. The atmospheres of Venus and Mars are composed of carbon dioxide, with relatively unknown concentrations of nitrogen and argon. Calculations and laboratory measurements on CO_2–N_2–A mixtures show that for the lower entry velocities CO and CN will be the dominant radiators, whereas for the higher velocities the atoms and ions will provide the greatest contributions to the radiation. For the low density of the Martian atmosphere, non-equilibrium radiation should be important.

The most extreme planetary environment involves the high-velocity entry into the atmosphere of Jupiter.

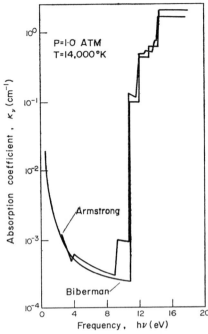

Fig. 2. Continuum absorption coefficient for air, with the detailed numerical computations of Armstrong compared with the approximate analytical expressions of Biberman; from Hoshizaki and Wilson (1967).

The high-temperature hydrogen–helium shock layer will result in such strong radiation (on the order of 100 kW cm^{-3}) from the atomic and ionic species that there will be appreciable radiative cooling in the shock layer. Even with radiative cooling the radiative heat load on the heat shield will result in massive ablation which will blow the boundary layer away from the vehicle surface. It has been suggested that the use of an appropriate dielectric heat shield may provide appreciable energy accommodation by means of *reflection* as well as by ablation.

In addition to atmospheric constituents, the ablation products are also important radiators and absorbers during atmospheric entry. Calculation of radiative transfer in the ablation layer must include the chemistry of ablation, as well as the determination of the absorption coefficients for the important species such as carbon molecules, atoms and ions.

Bibliography

ALLEN, H. J. (1964) Hypersonic aerodynamic problems of the future, pp. 1–41 in *The High Temperature Aspects of Hypersonic Flow*, W. C. Nelson (ed.), Oxford: Pergamon Press.

ANDERSON, J. D., Jr. (1969) An engineering survey of radiating shock layers, *AIAA J.* **7**, 1665-75.
HOSHIZAKI, H. and WILSON, K. H. (1967) Convective and radiative heat transfer during superorbital entry, *AIAA J.* **5**, 25-35.
LIN, S. C. (1962) Radio echoes from a manned satellite during re-entry, *J. Geophys. Res.* **67**, 3851-70.
PENNER, S. S. and OLFE, D. B. (1968) *Radiation and Reentry*, New York: Academic Press.
TAUBER, M. E. and WAKEFIELD, R. M. (1971) Heating environment and protection during Jupiter entry, *J. Spacecraft and Rockets*, **8**, 630-6.

D. B. OLFE

RADIATION AND SHIELDING IN SPACE.

Various types of nuclear radiation occur naturally in space. Three types are of major importance from biological and material damage considerations: magnetically trapped (Van Allen) belts, galactic (cosmic) radiation, and solar radiation (the solar wind and solar flare-related radiation). All three types essentially consist of charged particles and (except for solar flare radiation) are quasi-stationary in time. The 11-year solar activity cycle (last minimum in 1965) modulates the characteristics of all three types in the vicinity of the Earth. When the Sun is active the galactic particle flux is decreased (up to about 30%), the Van Allen belts are compressed by up to about 30% (especially the outer portions), and the solar wind velocity increases (up to about a factor of 3). While the Van Allen belts occupy a distorted toroid about planets with dipole magnetic fields (e.g. the Earth, Jupiter), the solar wind is found only outside a teardrop-shaped cavity (the magnetosphere) containing these Van Allen belts and the solar flare particles are unable to reach the top of the Earth's atmosphere, except near the magnetic poles. The galactic radiation, having a lower flux (2–4 particles/cm^2 sec) but higher particle energies (10^8–10^{19} eV) than the other types, penetrates not only the Earth's Van Allen belts but into the atmosphere where it produces air showers of secondary radiation.

For space travel these three types of nuclear radiation constitute hazards. The solar wind is readily stopped in surface coatings <0·025 mm thick, but for these coatings its high flux ($\sim 10^8$ protons/cm^2 sec) can combine with solar ultra-violet photons to adversely affect their transparency, reflectivity, and other surface properties. Judicious choices of materials (e.g. pure quartz, special inorganic paints, etc.) have reduced this problem to manageable proportions.

For earth-orbiting missions, the Van Allen belts constitute the major radiation hazard. This hazard is minimal for orbit altitudes of up to about 400 km, increasing to a severe hazard at 2000–4000 km, thereafter decreasing slowly as the outer surface of the Van Allen belts (7–12 earth radii) is approached. For thin shields (<1 g/cm^2) this dose rate maximum is up to about 10^3 rem/day, while for thick shields (>10 g/cm^2) the maximum is ~100 rem/day (see Table 1). Since biological doses in excess of about 50 rem may cause sickness and even lower doses are considered harmful (see Table 2) manned earth-orbit missions have been limited to low altitudes. Rapid passage through the Van Allen belts produces doses of ≲1 rem, since the region of maximum hazard is traversed in about 10 min.

For deep space missions solar flare particles con-

Table 1. Calculated skin dose rates for Earth orbital missions (rem day) for 1980

Orbit characteristics		Spacecraft shielding thickness			
Altitude	Inclination	1 g/cm^2 Van Allen	Other†	10 g/cm^2 Van Allen	Other†
300	Equatorial	<0·1	<0·1	<0·1	<0·1
	Polar	<0·1	3·0	<0·1	0·1
400	Equatorial	0·4	<0·1	0·1	<0·1
	Polar	0·1	4·0	<0·1	0·1
600	Equatorial	3·2	<0·1	1·0	<0·1
	Polar	0·6	5·0	0·2	0·1
1000	Equatorial	32·0	<0·1	8·0	<0·1
	Polar	10·0	6·0	2·5	0·1
3000	Equatorial	650·0	<0·1	165·0	<0·1
	Polar	200·0	7·5	55·0	0·1
10,000	Equatorial	18·0	<0·1	5·0	<0·1
	Polar	6·0	9·0	1·5	0·2
35,900 (synchronous)	Equatorial	1·5	25·0	0·2	0·4
	Polar	0·3	25·0	0·1	0·4

† Other includes solar flare (1% probability averaged over 6 months) and galactic radiation.

stitute the major hazard, especially at distances less than 1 astronomical unit (i.e. about 1.5×10^8 km, the distance between the Earth and the Sun) from the Sun during active periods. Solar flares can only be predicted statistically and less than 10% of the observed solar flares produce particles near the Earth. Consequently, for each space mission a solar flare particle environment is usually calculated corresponding to the expected maximum probability of occurrence (usually 1% to 0.1%). These environments are typically 10^2–10^4 rads/year, depending upon this probability, level of solar activity, and spacecraft shielding (see Table 3). Typical spacecraft have inherent shielding of 3–10 g/cm². Large single flares have produced a calculated dose of 10–100 rem inside such spacecraft (see Table 4). Fortunately, there have been no significant solar flare particle events when men were in deep space.

The galactic radiation constitutes a background dose rate of 2–5×10^{-2} rem/day, essentially independent of spacecraft shielding. This radiation environment is usually taken to be a spatially invariant when the Sun is relatively inactive while the solar flare hazard is often assumed to vary inversely with the square of the Sun–spacecraft distance.

The present state of knowledge of the natural nuclear radiation environment is such that the numbers in the tables have standard deviations up to plus or minus a factor of 3. Much of this is due to their time and spatial variations. While direct application of the numbers in the tables to specific situations is not recommended, these numbers are approximately correct and do indicate the effects of mission trajectory, duration and spacecraft shielding on the radiation doses expected. For further information, the publications listed in the bibliography should be consulted.

Table 2. Brief summary of radiation effects

<10 rem	No appreciable effect on men or materials.
10–30 rem	Possibly minor decrease in white blood cell count, some fogging of fast photographic film.
30–100 rem	Vomiting and nausea in 5–10% of exposed personnel, serious fogging of fast photographic film.
100–200 rem	Vomiting and nausea in 25–50% of exposed personnel, some fogging of slow photographic film.
200–300 rem	Vomiting and nausea in >50%, deaths in 5–10% of exposed personnel, serious fogging of slow photographic film.
300–1000 rem	Vomiting and nausea in ~100%, deaths in 10–100% of exposed personnel, some degradation of sensitive transistors.

Table 3. Calculated rem skin doses for deep space missions for 1980

Mission duration (Months)	Spacecraft shielding (g/cm²)	Van Allen†	Solar flares‡	Galactic
1	1	3.5	1000	0.9
	10	0.8	22	0.9
3	1	3.5	1400	2.7
	10	0.8	34	2.7
6	1	3.5	2000	5.4
	10	0.8	44	5.4
12	1	3.5	2600	11.0
	10	0.8	60	11.0
18	1	3.5	4000	16.0
	10	0.8	90	16.0
24	1	3.5	6600	22.0
	10	0.8	150	22.0

† Two high thrust equatorial trips assumed.
‡ 1% probability assumed.

Table 4. Calculated rem skin doses for eleven large observed solar flare radiation events at 1 astronomical unit

Event date	Spacecraft shielding thickness			
	1 g/cm²	2 g/cm²	5 g/cm²	10 g/cm²
23 Feb. 1956	975	350	130	58.0
10 May 1959	900	200	40	15.0
10 July 1959	975	260	60	24.0
14 July 1959	1270	280	50	17.0
16 July 1959	850	205	50	22.0
12 Nov. 1960	1600	470	140	55.0
15 Nov. 1960	700	205	55	20.0
18 July 1961	290	80	20	6.5
23 May 1961	180	55	18	5.0
18 Nov. 1968	100	23	6	2.5
4 Aug. 1972	4000	1200	280	110

Bibliography

BECK, A. J. (1972) *Proceedings of the Jupiter Radiation Belt Workshop*, Technical Memorandum 33–543, Pasadena, California: Jet Propulsion Laboratory.

HAFFNER, J. W. (1967) *Radiation and Shielding in Space*, New York: Academic Press.

HESS, W. R. (1968) *The Radiation Belt and Magnetosphere*, London: Blaisdell Publishing Co.

LANGLEY, R. W. (1970) *Space Radiation Protection*, SP-8054, Washington, D.C.: National Aeronautics and Space Administration.

REETZ, A. (1965) *Second Symposium on Protection Against Radiations in Space*, SP-71, Washington, D.C.: National Aeronautics and Space Administration.

SMITH. R. E. (1971) *Space and Planetary Environment Criteria Guidelines for Use in Space Vehicle Development*, TM X-64 627, Washington, D. C., National Aeronautics and Space Administration.

VETTE, J. I. et al. (1966–1971) *Models of the Trapped Radiation Environment*, Vols. I–VII, Report SP-3024 Washington, D. C.: National Aeronautics and Space Administration.

WARMAN, E. A. (1972) *Proceedings of the National Symposium on Natural and Manmade Radiation in Space*, TM X-2440, Washington, D.C.: National Aeronautics and Space Administration.

J. W. HAFFNER

RADIATION DOSIMETER, COLOUR. Large doses (kilorads or greater) of ionizing radiation are known to change the hue, shade, or colour saturation of many substances. Just as extended exposure to sunlight causes some plastics to darken or turn slightly yellow or brown, and other materials to bleach or change colour, ionizing radiations can cause even more radical changes in plastics, glasses, and certain dyed substances. As long as a critical portion of the resulting absorption spectrum or reflection spectrum is sufficiently stable and can be measured spectrophotometrically in a reproducible way, the material can be a useful dosimeter. The dosimetric response is given in terms of the change

Table 1. Some Commonly Used Colour Radiation Dosimeters
(Thin Dye Films and Plastics)

Type	Formula	Approx. density (g/cm^3)	Typical response range (rad)	Typical wavelength for read-out (nm)
Cellulose acetate butyrate	$(C_{15}H_{22}O_8)_n$	1·2	10^6–10^8	320
Cellulose acetates	$(C_6H_{10}O_5)_n$ $(C_{12}H_{16}O_8)_n$	1·2–1·4	3×10^5–10^9	280–380
Melamine	$(C_6H_{12}N_6O_3)_n$	1·5	10^5–4×10^7	380–420
Polyethylene	$(C_2H_4)_n$	0·95	10^6–10^8	236, 307, 1040, 1720, 3420, 5840
Polyamide	$(C_6H_{11}NO)_n$	1·1	10^8–10^{10}	315
Polycarbonate	$(C_{16}H_{16}O_3)_n$	1·2	10^5–10^8	290, 300, 325
Polyethylene terephthalate	$(C_{10}H_8O_4)_n$	1·4	10^6–10^9	325
Polymethyl methacrylate	$(C_5H_8O_2)_n$	1·2	10^5–10^7	305
Dyed polymethylmethacrylate	$(C_5H_8O_2)_n$	1·2	10^5–5×10^7	603, 620, 640, 651
Dyed polyvinylalcohol	$(C_2H_4O)_n$	1·3	10^4–10^6	660
Heteropolyacids in polyvinyl-alcohol	$H_3[P(W_3O_{10})_4]$ $H_4[Si(W_3O_{10})_4]$ $H_4[P(Mo_3O_{10})_4]$	2	10^4–10^7	700
Polyvinylchloride	$(C_2H_3Cl)_n$	1·4–1·5	10^5–10^7	395,480
Dyed polyvinylchloride	$(C_2H_3Cl)_n$	1·4–1·5	10^6–10^7	400 + 615
Polyvinylvinylidene chloride	$(C_4H_5Cl)_n$	1·5–1·7	10^5–10^7	260
Polystyrene	$(C_8H_8)_n$	1·05	5×10^7–10^8	330, 420
Dyed polystyrene	$(C_8H_8)_n$ halogens	1·05	10^4–5×10^7	430, 610, 460
Polytetrafluoroethylene	$(CF_2)_n$	2·2	10^3–10^{11}	ESR
Polyvinylfluoride	$(C_2H_3F)_n$	1·6	10^6–3×10^7	315
Polyvinylpyrrolidone	$(C_6H_9NO)_n$	1·3	10^6–10^8	350–380
Blue cellophane (dyed cellulosics)	$(C_3H_{10}O_5)_n$	1·4	10^6–10^8	650
Dyed cinemoids (dyed cellulose acetate)	$(C_6H_{10}O_5)_n$	1·3	10^6–10^8	437, 530
Malachite green methoxide in 4-chlorostyrene	$(C_8H_7Cl)_n$	1·1	5×10^4–2×10^7	430, 630
Triphenylmethane cyanides in various media	C, H, N, O	0·9–2·0	10^3–10^8	430, 560, 580, 600, 625
Stilbene in polystyrene	$(C_{14}H_{12}(CH))_n$	1·1	2×10^5–10^8	324
Polymers with fluorescing ingredients	C, H, N, O	1–2	1–10^7	variable

Table 2. Typical Glass Dosimeters
(colour centre production)

Type of glass	Approximate dose range (rads)
Miscellaneous industrial glasses	10^5–10^7
Activated borosilicate (Co, Cr, etc.)	10^4–10^6
Activated phosphate (Mg, Mn, Sn, Ag, etc.)	10^3–10^6
Antimonate	10^5–10^8
Bismuth silicate	10^6–10^{10}
Bismuth lead borate	10^6–10^8
Lithium borate	10^4–10^7
Manganese arsenic borate	10^5–10^8
Soda lime silicate	10^4–10^7

in optical transmission or reflection density, or change in percent transmittance or reflectance, at a given wavelength of light, as a function of the radiation absorbed dose in the material of interest. With this approach, any material susceptible to permanent radiation bleaching, colour change, or radiolytic darkening may serve as a *colour radiation dosimeter*.

Dosimetry based on changes in optical properties of dyes, plastics, and glasses is of particular importance in the use of ionizing radiations in industry, such as the radiation processing of polymers, medical supplies, building materials, dielectrics, foodstuffs, etc., and in the study of various radiation chemical and solid-state effects. For making radiation measurements, the dosimeter can often be made up of atomic constituents and a geometrical form appropriate to the material under study. Thus the response can be tailored to the radiation effects of interest for various energies and types of ionizing radiation. This helps minimize errors due to differences in degraded spectra in the irradiated media and any associated energy dependence of response.

In each of the categories, *dyes*, *plastics*, and *glasses*, there have been offered multitudes of different types of potential dosimeter, so many, in fact, that it is impossible to list them all here. Typical examples are given in Tables 1 and 2. The comprehensive articles cited in the bibliography will serve as a basis for a deeper search of the literature.

Dyes

Dye dosimeters may be coloured substances that are bleachable under irradiation, solutions containing indicators that change colour through a radiation chemical process, or radiochromic systems that are essentially colourless until radiation energy deposition releases a conjugated molecular complex capable of absorbing certain wavelengths of light. They are used in various forms, as liquid, gel, or solid solutions. Some commonly used bleachable systems are liquid methylene blue or crystal violet solutions, or cellophane films containing azo dyes. Various forms of halogenated hydrocarbons containing azo dyes (e.g. methyl yellow or red), or cresol sulfonphthaleins (e.g. bromocresol green or purple), which indicate acidity, are typical of indicator dye systems. Radiochromic dye systems may be leuco forms of thiazine or aniline dyes, or cyanides or carbinols of triphenylmethane dyes. Figure 1 shows absorption spectra of unirradiated and irradiated radiochromic dye dosimeter solutions. The typical response range of such dosimeters when measured spectrophotometrically across a 10 mm thickness is 10^3 to 10^6 rads. Here the ordinate is given in units of per cent transmittance. For this system, if one plotted optical density change at the wavelength of maximum absorption (~600 nm) as a function of absorbed dose, a linear response would result. Thus, for this system, one can simply use an optical density reading to determine the dose, by applying the G-value (number of molecules changed per 100 eV energy absorbed by the system) and the molar extinction coefficient (optical density per mole litre^{-1} across a one-centimetre pathlength) of the dye mole-

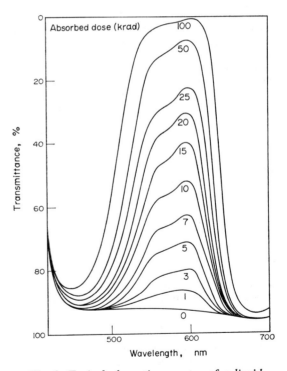

Fig. 1. Typical absorption spectra of a liquid radiochromic dye solution irradiated with various doses of cobalt-60 gamma-rays as indicated. Optical pathlength: 10 millimetres (see McLaughlin, W. L. et al. (1971) Int. J. Appl. Rad. Isotopes, 22, 135–140).

cules produced in solution. This cannot be done for all colour radiation dosimeters, because many do not have a linear response. In such cases, the dose must be determined from a calibration curve, as in the case of blue cellophane or most indicator dye systems. Radiochromic dye films of 1 to 10 mg/cm^2 area density have response ranges of about 10^5 to 10^8 rads. Even these thin systems have a broad linear response because they have such high molar extinction coefficients ($\sim 10^5$ l M^{-1} cm^{-1}) that even a relatively few dye molecules in a given area still cause a marked colouration, and saturation is attained at relatively high optical densities. Such thin dosimeter films placed at various positions in irradiated media are useful for high-resolution radiographic imaging and for measuring dose distributions and radiation field profiles.

Plastics

Some dyed plastics, such as coloured vinyls or acrylics, change or lose colour upon irradiation. These may be put into the categories discussed in the previous paragraph. Many ordinary plastics, such as polyethylene, polystyrene, vinyls, polyesters, cellulose esters, and others, undergo somewhat impermanent changes in the ultra-violet and visible portions of their absorption spectra, when irradiated in the megarad region. Although these do not always constitute visible colour changes, they should be included here, as they are becoming widely used in photometric dosimetry. Since the induced absorption spectrum is generally not stable, the period between irradiation and readings must be carefully controlled. The radiation effects are usually sensitive to oxygen diffusion, which causes large differences in response between irradiations in various atmospheres and in a vacuum. The presence of impurities, plasticizers, solvents, etc., in the plastics are other sources of error because they are apt to produce spurious optical absorption in competition with the absorption band of dosimetric interest. These impurities are difficult to trace and reproduce from batch to batch. An advantage in using polymers for dosimetry is that the optical readout can be combined with other types of analysis of radiation effects, such as measuring changes in dimensions, conductivity, viscosity, tensile strength, crystallinity, etc., to make a faithful registration of a particular radiation effect.

Glasses

Nearly all types of amorphous glass will darken to brown, green, violet or blue-grey when given large doses of ionizing radiation. In most instances, the colour centres are not very stable, which is true also of crystalline alkali halides and other ordered lattice systems. Induced photo- or thermoluminescent properties of glasses containing small amounts of metallic activators, such as silver, manganese, thallium, etc., are often used for dosimetry, because fairly low doses may be measured in this way, but for indicating doses of the order of 10^5 rads, glass colouration is a useful method as long as the instability problem is allowed for. This is often accomplished by a post-irradiation heat treatment, which bleaches some of the colour centres that are most apt to fade, leaving a less sensitive but more stable radiolytic image. Since glasses can be made to contain appreciable quantities of constituents having various atomic numbers, from lithium and boron to lead and thorium, the energy dependence of response to photons and electrons of different glasses varies over a wide range, and the neutron response can also be controlled in relation to, say, the response to gamma-rays. Table 2 gives a list of several types of glass dosimeter, and includes the approximate useful range of response.

Advantages offered by all the dosimeters listed in tables 1 and 2 are their simplicity, ruggedness, and low cost. Possible sources of error in making radiation measurements are as follows: variation of response with changing radiation spectra and dose rate; temperature dependence during irradiation, storage, or photometric reading; environmental effect such as oxygen concentration or relative humidity; instability of induced colour; batch differences; thickness variations; instrumental readout imprecision; etc.

Bibliography

BOAG, J. W., DOLPHIN, G. W. and ROTBLAT, J. (1958) Radiation dosimetry by transparent plastics. *Rad. Res.* **9**, 589.

BOLT, R. O. and CARROLL, I. G. (1963) *Radiation Effects on Organic Materials*, New York: Academic.

CHAPIRO, A. (1962) *Radiation Chemistry of Polymeric Systems*, New York: Wiley (Interscience).

CHARLESBY, A. (1960) *Atomic Radiations and Polymers*, Oxford: Pergamon.

DEAN, K. J. (1969) Photochromism, in *Encyclopaedic Dictionary of Physics* (J. Thewlis, Ed.), Suppl. 3, 295, Oxford: Pergamon Press.

FRANK, M. and STOLZ, W. (1969) *Festkörperdosimetrie ionisierender Strahlung*, Leipzig: BSB B. G. Teubner.

HART, E. J. et al. (1958) Measurement systems for high-level dosimetry, in *Proceedings of the Second United Nations International Conference on the Peaceful Uses of Atomic Energy*, Vol. XXI, p.188, Geneva: United Nations.

HOFMANN, E. G. (1963) Messung hoher Röntgenstrahlendosen durch optische Absorption in Gläsern und Kunststoffen. *Kerntechnik*, **5**, 439.

HOLM, N. W. (1968) Dosimetry in industrial processing, Chapter 33 in *Radiation Dosimetry*, Vol. III (F. H. Attix and E. Tochilin, Eds.), New York: Academic.

KIRCHNER, J. F. *et al.* (1958) Recent research in high-level gamma dosimetry, in *Proceedings of the Second United Nations International Conference on the Peaceful Uses of Atomic Energy*, Vol. XXI, p. 199. Geneva: United Nations.

KREIDL, N. J. *et al.* (1959) Measuring large radiation doses, *Nucleonics*, **17** (10), 57.

LONGSTAFF, R. M. (1969) Ionizing radiations in industry, in *Encyclopaedic Dictionary of Physics* (J. Thewlis, Ed.), Suppl. 3, 172, Oxford: Pergamon Press.

MCLAUGHLIN, W. L. (1970) Films, dyes, and photographic systems, Chapter 6 in *Manual on Radiation Dosimetry* (N. W. Holm and R. J. Berry, Eds.), New York: Marcel Dekker.

MCLAUGHLIN, W. L. (1963) Dosimetry: photographic megaroentgen range, in *Encyclopedia of X- and Gamma-rays* (G. L. Clark, Ed.), p. 307. New York: Reinhold.

MCLAUGHLIN, W. L., HJORTENBERG, P. and RADAK, B. B. (1973) Absorbed dose measurements by thin films, *Dosimetric Techniques in Agriculture, Industry, Medicine, and Biology*, Proceedings of IAEA Symposium, Vienna: International Atomic Energy Agency.

POTSAID, M. S. (1963) Dosimeters: solid phantoms, in *Encyclopedia of X- and Gamma-rays* (G. L. Clark, Ed.), p. 279. New York: Reinhold.

SCHARMANN, A. (1961) Dosimetry of large doses of radiation by coloration or decoloration of glasses and plastics, in *Selected Topics in Radiation Dosimetry*, p. 511. Vienna: International Atomic Energy Agency.

SCHULMAN, J. H., KLICK, C. C. and RABIN, H. (1955) Measuring high doses by absorption changes in glass, *Nucleonics*, **13** (2), 30.

SWALLOW, A. J. (1960) *Radiation Chemistry of Organic Compounds*, Oxford: Pergamon.

TAPLIN, G. V. (1956) Chemical and colorimetric indication, Chapter 8 in *Radiation Dosimetry* (G. J. Hine and G. L. Brownell, Eds.), New York: Academic.

TSUDA, S. (1963) Dosimetry, gamma: a comparison of all known methods, in *Encyclopedia of X- and Gamma-rays* (G. L. Clark, Ed.), p. 291. New York: Reinhold.

WILLIAM L. MCLAUGHLIN

RADIATION, NON-IONIZING, HAZARDS OF

Types of Radiation

Usually considered under this heading are electromagnetic radiations of low photon energy, that is *radiofrequencies* and *laser light* in the visible and infra-red wavelength bands, together with *ultrasonic* radiation. These are taken in contrast to X-radiation and the gamma-ray and charged particle emissions from radioactive materials or particle accelerators, where the primary interactions always involve sufficient energy to ionize (that is, detach an electron from) the atom of the medium involved.

It is in the context of the known serious radiobiological effects of these ionizing radiations and the obvious need to establish protection in this connection, that discussion of the hazards of *non-ionizing* radiations has taken place, sometimes with the assumption that all "radiation" is dangerous. Certainly the possibility of hazardous exposure must be considered, and the experience of control procedures for ionizing radiation work can be useful in planning any supervision which appears to be necessary. But the fundamental difference in the interaction mechanisms for these different radiations must be appreciated, and possible and relevant hazards identified in each case. Both the nature of any hazard and the protection procedures required will be different from those established for exposure to X-radiation.

An individual ionization is a relatively high energy local event and its occurrence is not related to total beam intensity. Thus there is no beam intensity threshold for the production of ionization, and as it is usually assumed that there is also no threshold for the biological effect, no irradiation level can be considered "safe". On the other hand, dissipation of energy from non-ionizing radiation as heat will always be proportional to intensity. Further, the body tissues will be able to survive a certain rise in temperature without suffering any permanent damage. The corresponding permissible rate of energy dissipation will depend on the heat exchange properties of the particular tissue involved. But there will be a definite intensity threshold below which heat damage will not occur, and the radiation exposure can be considered *safe*. It is conceivable that other specific (non-thermal) interactions may lead to biological damage below this threshold. Although, as discussed below, these are unlikely to be significant, such possibilities merit very careful radiobiological investigation, as a real, but low level of hazard may be difficult to appreciate under normal operating conditions. But the essential difference in interaction mechanism for ionizing and non-ionizing radiations, and the very real probability of a threshold response in the latter case, must be borne in mind in setting realistic maximum permissible exposure levels and control procedures which will not limit unnecessarily their application and usefulness of the radiations.

For many years a trefoil symbol has been used to indicate the presence of ionizing radiation. It has been felt that a similar sign is required for non-ionizing radiations, in particular to indicate controlled areas and to differentiate between these very different situations. A host of designs has been put forward and used, particularly in connection with laser beams, but the British Standards Institution (1971) has now recommended the symbol shown in Fig. 1 for *all* types of non-ionizing radiation. This choice may not be universally popular, but the general adoption of some such standard warning sign would be a useful development.

The new sign for non-ionizing radiation.

Fig. 1

Radiofrequency Radiation

In practice, because of the greater absorption in tissue at higher frequency and the intense fields which may be encountered in highly directional microwave beams, concern is mainly with radiation in the range 100–10,000 MHz. Particular attention is to be given to applications in microwave heating (about 2500 MHz radiation) and radar (3000 MHz) where high intensities may be produced respectively in areas near ovens or aerial systems accessible to operators and the general public. The electrical properties of body tissues relating to the absorption of microwave energy have been discussed in detail by Schwann (1955, 1969). The main determining factor is the water content of the tissue and Schwann concludes that in this frequency range there is always absorption of between 50 and 100% of the radiation incident on a body-shaped phantom of appropriate electrical characteristics. The penetration decreases with increasing frequency so that above 3000 MHz essentially complete absorption occurs in the superficial few millimetres thickness of tissue, but in the important range of 1000–3000 MHz the energy distribution varies considerably with frequency and with the relative thicknesses of skin and fat layers above the deeper tissues.

Combined dielectric and conductivity losses lead to production of heat in the tissues but the consequent temperature rise depends on the local conditions for superficial cooling and blood perfusion. The present U.S. regulations specify a maximum permissible intensity of 10 mW/cm^2 (100 W/m^2) for times exceeding 360 sec and an energy of 3·6 J/cm^2 (3·6 × 10^4 J/m^2) in shorter exposures. The body is normally radiating about 5 mW/cm^2 (50 W/m^2) from its surface and this emission can be increased very considerably by sweating to produce evaporative cooling. Thus, the recommended level is only of the order of the normal heat dissipation from the body and has been generally accepted, although in the Soviet Union a limit lower by an order of magnitude has been proposed (Michaelson and Dodge, 1971).

Two particular tissues have been considered in terms of possible heat damage—the lens of the eye and the testis. The lens does not have the efficient system of heat dissipation through circulating blood available to most tissues and consequently a greater temperature rise may occur. Cataract production from heat damage to the lens has been observed in glass blowers and furnace men whose eyes may be exposed to excessive infra-red radiation for long periods. Similar lens opacities have been produced experimentally in rabbits by microwave exposure at some 200 mW/cm^2 (2000 W/m^2) for periods of an hour or more. For whole body exposure, cataract production would not seem to be a problem within the limits set by general temperature effects. But the lens may certainly be the critical organ for partial body exposure and there remains some uncertainty over possible cumulative effects of chronic sub-threshold exposure. It would probably be unsafe therefore to adopt safety limits much higher than those at present recommended.

In the same way, limited blood perfusion makes the testis important from the point of view of possible temperature rise for a given rate of energy dissipation, and possible effects on fertility or even on mutation rate have been postulated. In fact, no clear experimental evidence of genetic effects of microwave exposure has been reported and the present limits would appear to be adequate to avoid any significant local temperature rise.

Schwann (1955, 1969) has considered other possible effects of radiofrequency radiation, in particular "field effect" forces, which may produce orientation or alignment of particles (so-called "pearl chain" formation) under the influence of the alternating electric field. These effects have, in fact, been demonstrated for bacteria (Heller and Teixerra–Pinto, 1959) and for erythrocytes and algae (Griffin and Ferris, 1970; Griffin, 1970). But calculations (Schwann, 1969) show that for particles the size of cells, the threshold field strength is several hundred volts per millimetre, increasing with further reduction in particle diameter. The recommended intensity limit of 10 mW/cm^2 corresponds to a field strength of the order of only 0·1 V/mm, so that such effects at cellular or subcellular level cannot occur under these conditions.

Any possible molecular resonances would be expected to be highly damped in normal tissues with high electrical loss. The only point to consider seems to be any possible effect of high field intensity in pulsed radiation with a low duty cycle to reduce the average power level below the limit set. Schwann (1969) does point out that, theoretically, field effect forces could be significant at low average power with on/off periods of optimal length in relation to the time constants involved (seconds or minutes for the wavelengths used in radar). Also, it has been suggested that orientation of molecular side chains in intense radiofrequency fields could lead to the breakage of hydrogen bonds, with consequent changes in protein solubility. Such non-thermal denaturation could possibly account for the alleged greater effectiveness of pulsed radiation in the production of cataract.

In summary, although many biological and physiological phenomena have at times been attributed to microwave exposure, there is no clear evidence for other than thermal effects under normal conditions of use. The present recommended limit of exposure of 10 mW/cm^2 appears to be a satisfactory and practicable working standard. The necessary supervision and monitoring should be provided for conditions of work near powerful radiofrequency equipment, precautions depending on the use of safety interlocks (e.g. on oven doors) and restricted access to areas of high intensity. But as a full understanding of possible damage effects is still not available, further radiobiological studies with microwave radiation should be undertaken, with particular reference to the use of high intensity pulsed radiation at low average power.

Lasers

Lasers provide intense, near parallel beams of coherent, monochromatic radiation in the visible, or near visible wavelength range. According to design, the radiation may be emitted continuously or in pulses with a duration of the order of a few milliseconds (normal mode) or nanoseconds (Q-switched mode). Very high intensities can be produced and when this radiation is absorbed there is considerable heat dissipation. Thus a 100 W infrared (10·6 μm wavelength) beam from a CO_2 laser will burn its way through a firebrick and a single ruby laser pulse (0·69 μm wavelength) of more than about 60 J/cm^2 will produce a plume of vaporized tissue on absorption. Such high intensity beams are used industrially for welding and drilling, in physics research, and in medicine for surgery or dermatological treatments.

For visible radiation, a parallel beam is focused by the eye to a very small spot on the retina, the energy concentration factor being of the order of 5×10^5. Thus the eye is very definitely the critical organ for protection purposes. The maximum permissible exposures listed in Table 1 are derived from recommendations in the Ministry of Technology Laser Systems Code of Practice (Wills, 1969), which document provides detailed consideration of exposure calculations and protection requirements. Damage is, of course, related to the local temperature rise and the energy limit must be lowest for the Q-switched pulses which are too short to allow significant heat dissipation during the exposure. The figures given are best estimates from published data from experiments on the rabbit and monkey, and for man appear to include a safety factor above the threshold of at least 100. The smallest focal spot on

Table 1. *Recommended Maximum Exposures for Laser Radiation*
(derived from Ministry of Technology Laser Systems Code of Practice, 1969)

Laser radiation	Maximum exposure at retina*		
	CW	Normal mode pulse†	Q-switched pulse†
Ruby (0·694 μm)	0·18 W/cm^2‡	0·52 J/cm^2	$1·6 \times 10^{-2}$ J/cm^2
Neodymium (1·06 μm)	1·0 W/cm^2	1·5 J/cm^2	$9·4 \times 10^{-2}$ J/cm^2
	Maximum exposure at skin		
	CW	Single pulse	
All lasers including UV and IR	0·1 W/cm^2	0·1 J/cm^2	
	Maximum exposure at cornea		
	CW	Single pulse	
U.V. (less than 0·4 μm)	$2·2 \times 10^{-4}$ W/cm^2	—	
I.R. (above 1·4 μm)	5×10^{-2} W/cm^2	0·1 J/cm^2	

* For intensity at cornea divide by 5×10^5, dilated pupil 7 mm diameter (0·5 cm^2).

† Pulse repetition rate not to exceed 10 per sec.

‡ Equivalent to 0·3 μW/cm^2 over a pupil diameter of 7 mm, 125 μW/cm^2 can be accepted for short (accidental) exposures.

the retina will be the diffraction-limited image of about 10 μm diameter (78.5×10^{-6} mm^2 area) from a parallel incident beam. The effect of sub-threshold exposures on retinal blood supply, and hence on the effect of subsequent exposure on the same area, must not be ignored.

Apart from the very evident hazard associated with high intensity beams there must still be some concern over the very widespread use of gas lasers (particularly He/Ne, 0·63 μm wavelength) providing continuous radiation at the milliwatt level. These instruments are now used extensively in laboratory (even school) optics, in range finders and for alignment systems for surveying, tunnelling, etc. It seems desirable therefore (Oliver, 1970) that for the vast majority of applications where this is, in fact, possible, the laser output should be below 125 μW/cm^2 (about 50 μW in a 7 mm diameter beam) so that the recommended limit for accidental exposure cannot be exceeded even under the worst possible conditions.

Absorption in other tissues will depend on the wavelength (colour) of the radiation, but in this range penetration is never very great, and so limits of exposure are given for skin in Table 1. For ultraviolet or infra-red radiation the hazard of exposure to the eye is heat damage to the cornea or the lens and corresponding recommended limits are given in the Table.

Non-thermal effects must also be considered. Certainly vaporization of tissue is associated with the generation of pressure waves, particularly in enclosed spaces, and these may lead to mechanical damage remote from the original site of energy absorption. With the very high intensities available (200 MW or more in a Q-switched pulse), multiphoton effects become possible, and excitation with associated fluorescence can occur. Such beams can be focused to produce electrical field strengths in excess of 10^6 V/cm, with consequent local breakdown of the air. However, there is no clear evidence of any biological damage as a result of these high radiation intensities in the absence of local heat damage.

In summary, retinal damage will be a hazard from exposure to the direct or specularly reflected beam from all but the lowest intensity visible light lasers, and for a high-power installation some protection will be necessary even from the diffusely scattered radiation. Where possible the work area should be well lit as there is a 16/1 increase in the energy accepted by the eye when the pupil is fully dilated in the dark. Dependance on protective goggles is not recommended as absorption must be precisely related to wavelength and the glass may shatter in a high-intensity beam. Except for lesions in the central (foveal) area of the retina, injury may not be appreciated as a loss of visual acuity, and regular ophthalmological examinations are usually recommended to check individuals at risk from high-energy lasers. Such damage cannot be repaired, but early detection does allow conditions of work to be checked and further exposure avoided. As the divergence of a laser beam is small, the usual inverse square law does not apply and, without appreciable atmospheric absorption, the intensity falls only very slowly with distance. In general, protection for high-intensity lasers must depend on restricted access to areas of work, enclosure of the beam, and interlocks to prevent possible exposure or accidental operation.

Ultrasonic Radiation

Ultrasonic radiation is poorly transmitted through the air so, in this connection, only situations involving direct and efficient coupling between the transducer and the body are of importance. The lower frequencies (20–50 kHz) find considerable industrial use for cleaning baths, emulsifiers, etc., and workers hands may be exposed during these operations. But the main concern is with possible tissue damage arising from the use of higher frequency (1–10 MHz) ultrasonic radiation in the diagnostic examination of patients. Two different techniques are used in medical diagnosis. With continuous radiation at a power level of 10–20 mW/cm^2, the Doppler effect is applied for the measurement of blood flow in vessels and, more importantly, for detecting and monitoring the foetal heart. Alternatively, short pulses are used in echo techniques for examination of the brain, in cardiology and in producing sectional scans of the body, particularly in obstetrics and in the localization of pelvic and kidney tumours. Here the *average* power level remains in the range 5–50 mW/cm^2, but *peak* power during the pulses of a few microseconds duration may be as high as 10–100 W/cm^2.

The mechanisms of interaction of ultrasonic radiation with tissue have been discussed by several authors (e.g. Connolly and Pond, 1967; Hill, 1968). Absorption increases with frequency and the variations in this absorption observed between different tissues appears to be related not to gross tissue structure but to resonances at the sub-cellular, macromolecular level with a wide range of relaxation time constants (Pauly and Schwann, 1971) leading to general heat dissipation.

Non-thermal effects could be expected as a result of *oscillatory forces*, of *time averaged forces* at discontinuities which may lead to streaming of fluids at the interface, or of *cavitation* which involves oscillation or collapse of small gas bubbles in the medium. Certainly cavitation can lead to chemical changes including the formation of free radicals (Hill, 1968), but it will be less effective at the high frequencies used in medical procedures. Also, it is predicted that both streaming and cavitation are unlikely to build up with the very short pulses used in echo techniques.

The *average* power level used in medical diagnostic techniques is so low that thermal effects are not observed, but it is important to know whether true threshold conditions apply. This is particularly relevant because of the increasing use of ultrasonics for the examination of the foetus in which situation a slight, but significant, hazard of X-ray examination has been established. There has been no clinical evidence of foetal damage as a result of ultrasonic obstetrical examinations (Hellman et al., 1970) and experiments with pregnant mice and rats using similar techniques have not demonstrated any significant reduction in litter size or increase in the number of abnormalities (Woodward et al., 1970; McClain et al., 1972). Results of Macintosh and Davey (1970, 1972) caused some concern as these related to the production of chromosome aberrations in human blood after a few hours exposure at power levels of only 10–20 mW/cm^2. However, these results have not been repeated and have not been confirmed in similar investigations carried out by a number of other experienced groups (e.g. Bobrow et al., 1971; Coakley et al., 1971; Buckton and Baker, 1972). On the other hand, haemorrhage and damage to the spinal cord have been produced in rats by Taylor and Pond (1972) under conditions which ensured that this was not simply a thermal effect and with total energy dissipation corresponding to several hours of diagnostic echo examination. This damage appeared to be an "all or none" effect requiring the administration at a supra-threshold intensity of a certain total energy, the contribution of exposures separated by several hours being additive in this respect. These authors considered that, in view of the frequency response, discrete cavitation was the most likely cause of this biological damage. Clarke and Hill (1969) observed killing of mouse lymphoma cells in suspension under conditions where cavitation was able to occur. But, presumably in the absence of cavitation, Bleaney et al. (1972) were unable to demonstrate such loss of reproductive integrity in hamster cells in culture except where a lethal temperature rise was produced. Recently Bleaney and Oliver (1972) have published dose response curves for inhibition of root growth in *Vicia faba* following exposure to 1·5 MHz ultrasonic radiation at 1 to 8 W/cm^2, apparently involving non-thermal damage to dividing cells of the meristem, with the effects of split exposures additive over at least 24 hr. Similar response in roots of *Zea mays* has been reported by Hering and Shepstone (1972). Mechanical effects may well be different in plant and mammalian cells but these systems do allow the investigation of the variation of a quantitative response with radiation parameters, including the pulse length.

In summary, there is no evidence for tissue damage with the low average power level, and consequent low heat dissipation, used in medical diagnostic techniques. It seems likely that any non-thermal effects will be subject to a threshold which will not be exceeded under these conditions. However, the haemorrhagic injury produced in animals and the meristematic damage demonstrated in plant roots referred to above, justify further radiobiological studies on the possible effects of ultrasonic radiation in tissue, particularly to confirm the safety of the high intensity short pulses used medically in echo techniques. In any case, care is necessary in clinical applications to avoid heat damage resulting from the use of high average power levels or accidental focusing of the beam due to reflections at, for instance, bone surfaces. There is no problem of operator protection unless direct exposure of the hands occurs, for example in the use of ultrasonic cleaning baths. Even then significant hazard is unlikely to exist without the warning of excessive tissue heating.

See also: Lasers in biology and medicine.

Bibliography

BLEANEY, B. I., BLACKBOURN, P. and KIRKLEY, J. (1972) *Brit. J. Radiol.* **45**, 354.
BLEANEY, B. I. and OLIVER, R. (1972) *Brit. J. Radiol.* **45**, 358.
BOBROW, M., BLACKWELL, N., UNRAU, A. E. and BLEANEY, B. I. (1971) *J. Obstet. Gynaec.* **78**, 730.
British Standards Institution (1971) British Standard No. 4765.
BUCKTON, K. E. and BAKER, N. V. (1972) *Brit. J. Radiol.* **45**, 340.
CLARKE, P. R. and HILL, C. R. (1969) *Exp. Cell Res.* **58**, 443.
COAKLEY, W. T., HUGHES, D. E., LAWRENCE, K. M. and SLACK, J. S. (1971) *Brit. Med. J.* **1**, 109.
CONOLLY, C. and POND, J. (1967) *Biomed. Engng.* **2**, 112.
GRIFFIN, J. L. (1970) *Exp. Cell Res.* **G1**, 113.
GRIFFIN, J. L. and FERRIS, C. D. (1970) *Nature*, **226**, 152.
HELLER, J. H. and TEIXERRA–PINTO, A. A. (1959) *Nature*, **183**, 905.
HELLMAN, L. M., DUFFUS, G. M., DONALD, I. and SANDERS, B. (1970) *Lancet*, **1**, 1133.
HERING, E. R. and SHEPSTONE, B. J. (1972) *Brit. J. Radiol.* **45**, 786.
HILL, C. R. (1968) *Brit. J. Radiol.* **41**, 561.
MACINTOSH, J. C. and DAVEY, D. A. (1970) *Brit. Med. J.* **4**, 92.
MACINTOSH, J. C. and DAVEY, D. A. (1972) *Brit. J. Radiol.* **45**, 320.
MCCLAIN, R. M., HOAR, R. M. and SALTZMANN, M. B. (1972) *Amer. J. Obstet. Gynaec.* **114**, 39.
MICHAELSON, S. M. and DODGE, C. H. (1971) *Health Physics*, **21**, 108.
OLIVER, R. (1970) *Brit. J. Radiol.* **43**, 909.
PAULY, H. and SCHWANN, H. P. (1971) *J. Acoust. Soc. Amer.* **50**, 692.
SCHWANN, H. P. (1955) *Amer. J. Phys. Med.* **34**, 25.
SCHWANN, H. P. (1969) *Non-Ionizing Radiation*, **1**, 23.
TAYLOR, K. J. W. and POND, J. B. (1972) *Brit. J. Radiol.* **45**, 343.

WILLS, M. H. A. (1969) *Ministry of Technology Laser Systems Code of Practice*. London: Ministry of Technology.

WOODWARD, B., POND, J. B. and WARWICK, R. (1970) *Brit. J. Radiol.* **43**, 719.

R. OLIVER

RECENT ADVANCES IN RADIO AND X-RAY ASTRONOMY. In recent years radio and X-ray astronomy have enjoyed rapid and vigorous growth, mainly because of the amazing variety of new objects which have been discovered by astronomers who observe at non-optical wavelengths. Pulsars, interstellar organic molecules, radio stars, and powerful X-ray galaxies are among the phenomena to burst unexpectedly on the scene since 1967. The detailed investigation of these has inspired important theoretical advances in both astronomy and basic physics. Furthermore, it has encouraged national agencies to finance the construction of several new instruments that are extremely sensitive tools for the modern astronomer.

Pulsars

The discovery of pulsars in 1967 was completely fortuitous. In August 1967 a weak, flickering, source of cosmic radio waves was discovered on a new radio telescope at Cambridge, England, by Professor Antony Hewish and Jocelyn Bell. The most outstanding property of the new object was that its radiation consisted of a series of extremely regular pulses. Within a short time several dozen similar radio sources were picked up at Cambridge and other radio observatories throughout the world.

The pulsars send out a regular pattern of emission which is repeated at a rate between 33 msec, for the fastest known pulsar, to around 4 sec for the slowest; 0·5 to 1 sec is the most commonly observed pulse period. The pulses have complex structure and are generally between 1 and 30 msec wide, depending on the observing frequency and particular object. Over periods of a year or so it is found that for many pulsars the pulse period is getting systematically longer, which suggests that an internal "clock" is running down. Two pulsars have shown abrupt changes in their periods.

Only one pulsar has been convincingly identified with a star that is visible in large optical telescopes. This pulsar is in the *Crab nebula*, the remains of a massive star which exploded in A.D. 1054. The Crab nebula pulsar is unique in many respects; it is the only one which has been detected at all energies from radio waves right through to gamma rays. At all these energies the bulk of the emission from the pulsar is in the form of sharp pulses. A major stumbling block to progress in understanding the emission mechanism is the large frequency range of the pulse phenomenon. The Crab pulsar also has the shortest period, and it is the only pulsar whose age (just over 900 years) is known with certainty.

Most of the pulsars are thought to be comparatively close to our solar system—perhaps between 100 and 3000 light years away. Apart from the Crab pulsar the lack of success in identifying pulsars with stars demonstrates that they must be exceedingly faint in their optical emission. It is thought, from theoretical considerations, that the masses of pulsars are between about 1·4 and 2 solar masses. The diameters are very small; this is required by the astonishing sharpness of the pulses at high frequencies. Estimates of the diameters are a few tens of kilometres.

Nothing remotely like pulsars had been charted in the heavens before 1967, and it was soon evident that a totally new type of object had been unveiled. It is still not known how the pulses themselves are generated. However, the "clockwork" that determines the very exact repetition of the pulses is widely accepted to be regular rotation. As the pulsar spins around, some disturbance in its outer atmosphere causes a narrow beam of radiation to sweep past us once per revolution, as if it is a cosmic lighthouse.

Theorists now accept that pulsars are the long-sought *neutron stars*—an extreme end-point in stellar evolution. When a star has exhausted all its nuclear fuel there are several ways in which further evolution proceeds; which path a particular star will take is determined by its mass. Between 1·4 and 2 solar masses it is possible for the star to reach a stable configuration in which the force of gravity is balanced by the degeneracy pressure of neutrons in the star. The stellar density is that of nuclear matter; the star consists principally of neutrons, with perhaps a core of baryons and hyperons. Its surface magnetic field is around 10^{12} G. In order to conserve angular momentum the collapsed star must rotate rapidly, and this spinning provides the pulsar with its internal clock. The interpretation of pulsars as neutron stars has had a profound effect on theoretical astrophysics. It has inspired studies on the stable configurations for large aggregates of nuclear matter; on physical processes in very large magnetic fields; on possible emission mechanisms; and also on the endpoint of more massive stars, which may collapse (indefinitely?) until the density and gravitational field are so large that no radiation can escape—the so-called "*black holes*".

As a result of advances in pulsar theory, theorists have given a fairly complete picture of the physics of the Crab nebula. Long-standing problems which have been solved include the production of X-rays in the nebula and its continuous source of energy. The period of the Crab pulsar is increasing at the rate of about 10^{-5} sec per year. Presumably this is accounted for by the star's slowing down, with consequent loss of energy and angular momentum. For many years it has been known that the Crab nebula requires a rate of energy input of about 10^{31} watts in order to keep the emission going. This value is similar to that being released by the slow-down of the pulsar, which is, therefore, in all probability the power-house of the nebula. The observed non-pulsating radiation from the nebula is caused by synchrotron radiation of relativistic electrons in the interstellar magnetic field. Electrons which produce X-rays in this way have a lifetime of only a year. Compared to the known age of the nebula this is very small, so that the pulsar probably continuously injects high-energy electrons into the nebula.

Interstellar Organic Molecules

Since the 1930's it has been known that the simple radicals CN, CH and CH^+ exist in interstellar space because their characteristic absorption spectra are superimposed on the optical spectra of bright stars. The first molecular structure revealed by radio telescopes was the hydroxyl radical, OH, discovered in 1963 at M.I.T. by Dr Sander Weinreb and his colleagues. What has caused great excitement in the last 3 years is the discovery of an astonishing variety of molecules in the gas clouds of interstellar space: H_2O (water), NH_3 (ammonia), CO (carbon monoxide), CN (cyanogen), CS (carbon monosulphide), OCS (carbonyl sulphide), SiO (silicon oxide), HNC (hydrogen cyanide), HNCO (isocyanic acid), HC_3N (cyanoacetylene), HCOH (formaldehyde), HCOOH (formic acid), CH_3OH (methyl alcohol), CH_3CN (acetonitrile), NH_2COH (acetyldehyde), CH_3C_2H (methyl acetylene). Some of the organic molecules discovered may play an essential part in the synthesis of *amino acids* in the cosmic environment. This, of course, may be an indispensable clue to our future understanding of the origin of life in the Galaxy since amino acids are necessary to the development of life as we know it. Besides this obvious importance to the study of extra-terrestrial life (*exobiology*), the molecules are of great value to astronomy also. Observed changes in the received wavelength, caused by the Doppler effect, are a great help in mapping out the structure, location and dynamics of the molecular gas clouds.

Interstellar emission from OH and H_2O molecules appears to be excited by a *maser process*. The OH line is split into four components whose relative intensities are anomalous. Furthermore, the OH and H_2O sources are often strongly polarized (both linearly and circularly). Spectacular variations in the received intensity are standard: changes by a factor of 2 in a matter of months are common. Interferometers have shown that the diameters of OH and H_2O sources may be as small as the orbit of Mars about the Sun—remarkably tiny by astronomical standards. The astonishing intensity, compactness and rapid variability imply that the emitting regions have a brightness temperature so high that it can only be accounted for by maser action.

The natural masers that occur deep in space are poorly understood at present, although a recently proposed model suggests that typical dimensions for a maser are $1 \times 1 \times 100$ astronomical units (1 astronomical unit is $1 \cdot 5 \times 10^8$ km) with an amplification of 10^{10}. Some of the masers are located at the centre of our Galaxy, and one may already have been discovered in an external galaxy, M82, which is 4 million light years away.

In 1968 a group at Berkeley announced the discovery of ammonia, NH_3, and they astonished the majority of astronomers, for the existence of complex molecules had never received much attention. The changed situation encouraged other groups to search for additional molecular species, which resulted in the present impressive list. Most of the molecules occur in a system of gas clouds that lie close to the centre of our Galaxy; possibly they make up a ring-shaped structure with a diameter of about 1000 light years. This torus of molecular gas clouds is both rotating and expanding. Its expansion velocity suggests that it was formed in a huge explosion that occurred several million years ago in the galactic centre. Molecules have also been detected in cocoons of gas that surround cool *infra-red stars*, and in the gaseous remnants of *supernova* explosions.

An interesting application of the molecules is the determination of the *isotopic abundances* of certain elements. In several cases the isotope shift of spectral lines in $H_2^{12}CO$ and $H_2^{13}CO$ has been resolved. The relative intensities of the lines from the two isotopic species of H_2CO can thus be used to infer the ratio $^{12}C/^{13}C$ in different parts of the Galaxy. Searches for deuterated species of the hydrogen-rich molecules have also been successful.

The relevance of the new discoveries to the origin of life is clear; astronomers have found that the basic components of *amino acids* are abundant in many parts of our Galaxy. Certainly we are witnessing the growth of a new science, in which chemists and biologists will co-operate with astrophysicists to elucidate further the exciting problems posed by the molecular clouds.

Galactic X-ray Sources

X-ray astronomy was effectively born in 1962, when a rocket launched by physicists in the United States, who had hoped to detect luminescence from the Moon, picked up a strong source in Scorpius. This object, known as Sco X-1, is the strongest in the sky (aside from our Sun). About 120 cosmic X-ray sources have now been detected by a research satellite, called Uhuru, which has the X-ray sky under constant surveillance. About two-thirds of the objects which Uhuru has recorded are believed to be inside our Galaxy. Observations of some of the objects are repeated regularly in order to monitor variability.

In common with radio astronomy, a key measurement in X-ray observations is that of position so that identification with an optically visible counterpart may be made. Until recently this was very difficult; fortunately the Uhuru satellite has improved the situation.

Sco X-1 is a 13th magnitude star which has similarities to old novae. It has been observed by radio, optical and X-ray instruments, and at all wavelenghts it shows an astonishing range of variability. There are rapid flickerings that are imposed on longer term variations. Its radio flux has been known to increase tenfold in a few hours. The radio structure is highly unusual, consisting of three areas spanning 4 arc seconds of sky.

Some binary star systems constitute another series of X-ray sources that exhibit marked variations. Generally they consist of one "ordinary" star and an X-ray star in mutual orbit. In some cases the X-ray star is regularly eclipsed as it passes behind its companion star. The main cause of X-ray emission may be the large-scale transfer of gas between the ordinary and X-ray star; the plasma becomes heated to a high temperature as it is pulled towards the X-ray star. A few of these objects also show variable radio emission.

Another important object is the *Crab nebula*, identified as a source of cosmic X-rays in 1964. In 1972 it was shown that the X-rays are plane-polarized, which demonstrates that they are produced by the *synchrotron process*. The pulsed X-rays from the pulsar in the Crab nebula make up almost one-fifth of the total X-ray emission from the nebula.

Despite the large amount of information on the spectra and temporal variations of the several species of X-ray stars, ideas on the underlying mechanism are still-fluid. Thermal bremsstrahlung emission, inverse Compton scattering and free–free emission have all been suggested, and in many objects are indistinguishable on the basis of present data.

Radio Stars

Since the beginning of this century astronomers have asserted that ordinary stars do not emit radio waves with detectable intensity (apart from our Sun). At the U.S. National Radio Astronomy Observatory, Campbell Wade and Robert Hjellming have dramatically turned the tables with the announcement that several bright stars can be detected with the present generation of radio telescopes. They use a sensitive *interferometer* for radio star work because it is thus possible to measure positions accurately and prove the coircidence of a radio and optical star.

Among the radio stars are *novae*—stars which have recently suffered explosions that cause sharp, but temporary, increases in their luminosity. In the gaseous envelope of a nova high-temperature plasma is ejected at around 1000 km sec^{-1}. Novae show a continuous increase in their radio intensity for many months after the optical outburst, and they should remain observable for several years.

The bright star *Antares*, which is a binary system, is a radio source. It is believed that exchange of matter between the envelopes of the two stars in the Antares system is exciting the radio emission. Two further binaries, Beta Lyrae and Beta Persei, are highly sporadic radio emitters. The presence of radio emission from some *X-ray stars*, mentioned above, is promising from the standpoint of identification work because positional measurements are more accurately made by radio techniques.

Radio Emission from Galaxies

The discovery of radio emission from distant galaxies in 1954 is widely accepted as the first breakthrough in modern astronomy. The absolute radio powers of these objects range up to 10^{37} watts, and the *minimum* energy is around 10^{53} joules for a typical radio galaxy. The origin of this large store of energy in the form of high-energy electrons and magnetic field is, perhaps, the most profound unsolved problem in cosmic physics.

During the past few years attention has centred on mapping the radio structure of these galaxies with high-resolution *interferometers*. Maps with an angular resolution of 6 arc sec have been made at Cambridge, England, and West Virginia, while structures down to 0·001 arc sec have been partially investigated by long-baseline *interferometers*. The majority of the sources exhibit a double structure with two radio "clouds" symmetrically disposed about the optically-visible galaxy. The total structure spans as much as 10^6 light years, clearly the result of a truly stupendous explosion.

Recent research suggests that rapid expansion of the plasma clouds is prevented by a shock-wave that is generated in the intergalactic gas. It is believed that the clouds have a velocity of 0·01c to 0·1c, and that ram-pressure at the front of the cloud is sufficient to restrain expansion over the lifetime of 10^6 to 10^7 years of a typical radio galaxy. Polarization studies have delineated the magnetic field configuration inside the plasma clouds of some radio galaxies.

A new type of radio galaxy has been described by astronomers at Cambridge, England, and Leiden, Netherlands, which they term an "interacting *radio galaxy*". Studies of rich clusters of galaxies have shown that some of these possess a radio source that does not exhibit the classical double structure. Morphologically the radio emission resembles a curved-sausage or tadpole shape, with the optical source being located at the bright extremity of the radio map. In three cases studied so far in detail it turns out that the optical galaxies are highly excited. For example, the radio emission from the Perseus

Cluster is dominated by 3C 84, a *Seyfert galaxy*. This has a very compact nucleus—only a few tens of light years in diameter—and it is a powerful extragalactic X-ray source. Its optical spectrum is profusely rich in highly excited emission lines. It is possible that 3C 84 has also triggered radio emission in three nearby galaxies. Observations at radio, infrared, optical and X-ray energies show that 3C 84 is a very active galaxy which is profoundly affecting astrophysical processes in its neighbourhood. Together with other examples of active galaxies in clusters it is probably a new type of object within which energy is being continuously released at the present time. How the energy is created is unknown, but the process could conceivably be related to the energy source in classical radio galaxies.

X-ray Galaxies

One of the achievements of the Uhuru project has been the discovery of around forty X-ray sources that are located outside our Galaxy. One of these is 3C 84, which is discussed above. Other important examples are: the radio galaxy Virgo A whose total X-ray luminosity is 10^{36} watts; the Large and Small Magellanic Clouds—dwarf companions to our own Galaxy; one of the nearest quasars, 3C 273; and several Seyfert galaxies. A discovery of far-reaching importance, but not yet firmly established, is that the majority of extra-galactic X-ray sources are located in rich galaxy clusters. The high-energy photons are probably generated in a plasma at a temperature of about 10^6 to 10^7 K that fills out the space between the galaxies. Thus future X-ray investigations on these clusters may confirm that hot gas exists in the intra-galactic volume. Already the observers are reasonably confident that their data show that this gaseous medium does exist. For many years theorists have been worried that the mass of visible material in clusters was only a few per cent of the total cluster mass inferred from dynamical arguments. X-ray astronomers may have located the missing 90%.

Cosmological Investigations

Recent years have been noticeably free of the bitter arguments associated with cosmology during the late 1950s and early 1960s, mainly because almost all astronomers now favour the *Big Bang* rather than Steady State model of the universe. Sir Martin Ryle and his colleagues at Cambridge have measured the number density of radio sources down to very faint flux levels. They find that there is an excess of the fainter sources over what would be expected for a uniform distribution of sources in a steady state universe. The most plausible explanation is that the faintest sources are very distant; therefore they are being observed as they were long ago, in a much earlier epoch of the universe. The implication is that radio sources were more numerous in the past, which would rule out the theory associated with Sir Fred Hoyle.

Studies of the various diffuse background emissions have also favoured the Big Bang model. Radio noise attributable to thermal emission from a universal photon gas at a temperature of 2·7 K is now convincingly established from measurements between 100-cm and 1-mm wavelength. This flux of microwave radiation is extremely isotropic, which carries the fantastic implication that the large-scale properties of our universe are identical in every direction to a precision of 0·1%. Studies of the *background* X-ray *emission* show that this diffuse flux is highly *isotropic* also. It may represent the integrated contributions of an abundance of X-ray sources at an early epoch of the evolving universe. More data are an essential prerequisite to further progress in the cosmological interpretation of the background radiations.

That cornerstone of cosmology—the *redshifted* light from distant objects—has recently come under fire from prominent astronomers such as Halton Arp and Geoffrey Burbidge. They do not accept that all redshifts, which now range up to $d\lambda/\lambda = 3.53$ (λ = wavelength) for OQ 172, are caused by the expansion of the universe. In particular they have been considering pairs of objects that are apparently close to each other on the sky, but for which the redshifts of each member of the pair differ considerably. In a handful of cases it is possible to argue that two objects with very different redshifts are related physically. This implies that high redshifts may be caused by some mechanism other than the recession of distant parts of the universe—by high gravitational fields, for example. Once again this is a problem that has a bearing on the whole of modern physics as new physical laws may be implicated, according to some authorities.

Instrumental Developments

Several radio telescopes have recently been constructed or refitted. Among the new instruments that are now operational are synthesis radio telescopes at Leiden, Netherlands (1·4-km baseline, ten antennae), and Cambridge, England (5-km baseline, eight antennae). The latter instrument has the highest resolving power available (about 1 arc sec) to map fully radio sources. In the United States a very-large-array of twenty-eight antennae, arranged in Y formation 26 miles in total extent, is planned for the late 1970s. This will be an instrument of unprecedented versatility for the investigation of radio source structures. Among the single-dish telescopes, the 1000-foot fixed antenna at Arecibo, Puerto Rico, is being refitted with a new surface which will improve its performance at high frequencies. The 250-foot fully steerable antenna at Jodrell Bank, England, has similarly been updated; a new 350-foot dish is planned for construction in North Wales. In West Germany a 300-foot telescope, completed in 1971, is currently the largest fully steerable dish in the world.

The X-ray facilities are being rapidly improved. The orbiting observatory Copernicus, launched late 1972, is functioning well; it is the first satellite to carry X-ray telescopes with high positional accuracy. The SAS-C satellite is scheduled for 1974, and this also will carry telescopes, rather than detectors. In the late 1970s NASA is proposing a series of high-energy astrophysics satellites (HEOS), that will make optical, ultra-violet and X-ray studies.

S. MITTON

REFLECTION SPECTROSCOPY

Introduction

A beam of electromagnetic radiation falling on a flat surface of a dense material divides into a reflected and a refracted beam, the amplitudes and phases of which, and consequently the powers, are determined by the specific optical properties of the material. Reflection spectroscopy is generally understood to be the measurement of the spectral variation of the absolute or relative power of the beam of electromagnetic radiation specularly reflected from the plane surface of the medium and is more correctly called reflection spectrometry. The reflected power is usually normalized to the level of the incident power and is expressed as a power reflection coefficient R. This is also called the power reflectivity or the power reflectance. The magnitude of the normalized reflected power is a function of the microscopic and macroscopic structure of the medium as described by the optical "constants" and is dependent on the angle of incidence and polarization of the incident beam. Although the reflection spectrum can be of use in its own right it is frequently necessary to evaluate the optical "constants" from the reflection spectrum. The *optical "constants"* which are, in fact, not actually constant but vary with the wave-number o of the radiation, completely specify the optical properties of the medium. They are usually combined and expressed as the complex refractive index $\hat{n}$ which we may write in terms of real and imaginary parts as

$$\hat{n} = n - i n \varkappa, \qquad (1)$$

where n is the real refractive index, $\varkappa$ is the extinction coefficient and $n\varkappa = k$ is the absorption index. A form which is frequently given, and is of special relevance and use to spectroscopists, is obtained by expressing k as

$$k = \frac{\alpha}{4\pi\sigma} \qquad (2)$$

in terms of the power absorption coefficient α; hence

$$\hat{n} = n - i \frac{\alpha}{4\pi\sigma}. \qquad (3)$$

α represents the power absorption per unit length within the medium and shows variations with wave-number or frequency that are fairly easily related to the frequencies and strengths of the oscillators of which, from a classical point of view, we imagine the medium to be composed. Resonances occur at or near the frequencies of the oscillators. They have widths which are proportional to the damping of the oscillators and integrated areas which are a measure of the oscillator strengths. Whilst the power reflectivity is also determined by these oscillators the shape of its spectrum is not usually related to their positions or strengths in an obviously recognizable way. For this reason it is one of the major aims of reflection spectrometry to calculate the absorption coefficient (and often the refractive index as well) from the measured reflectivity. However, it is shown below that the calculations are not straightforward; consequently reflection measurements are rarely made when conventional transmission measurements are possible and capable of giving acceptable accuracy. Reflection spectrometry, however, is essential when large absorption makes it necessary to use impracticably thin specimens for conventional transmission measurements. In the case of solids, this means preparing extremely thin sections of material, which cannot be extensive in area, or powdering the solid and mulling into a diluting medium, which is unsatisfactory. In the case of liquids, cells with very thin spacers are required. In both cases, the specimen thicknesses cannot be accurately measured. For reflection measurements the specimen preparation is considerably easier as it is necessary only to prepare a single flat polished surface and to ensure that the specimen is sufficiently thick to prevent radiation from reaching and being reflected from the back surface. The use of wedged specimens reduces the risk of such back reflections contributing to the detected power. The medium through which the radiation approaches the specimen is commonly air or vacuum.

Reflection spectroscopy is a relatively young subject. Apart from some early work on mid-infra-red restrahlen bands carried out around 1920 by, for example, Krebs and Boeckner little else was done in the infrared region of the spectrum until the early 1950s. At shorter wavelengths, work was done in the 1930s and 1940s and was concerned mainly with measuring values of the power reflectivity of such materials as paper, textiles and paint (see the descriptions given by Wendlandt and Hecht). Although such measurements are still made (for quality control, for example) the greatest development of reflection spectrometry has occurred at infra-red wavelengths. This is because the problems caused by surface layers and surface scatter are minimal and because the information retrieved from the spectra is of fundamental interest and importance (see Moss, 1959, and Stern, 1963).

From the experimental point of view, power reflection measurements are fairly straightforward but the analysis of the measured data can be difficult and cumbersome. This drawback provided the main limitation to the application of reflection spectrometry and has been satisfactorily overcome only recently since computers became widely available. A consideration of the principles of reflectance measurements reveals the reasons for the difficulties.

Basic Principles

When electromagnetic radiation of complex amplitude $\hat{A}_i$ strikes a plane boundary between a medium 1 and a medium 2 at an angle of incidence ψ_1, the instantaneous value of the electric field strength

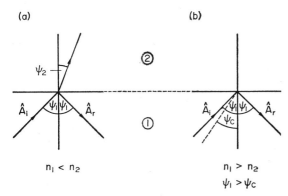

Fig. 1. Reflection at a plane boundary. (a) "External" reflection ($n_1 < n_2$). (b) "Total" reflection ($n_1 > n_2$).

amplitude undergoes attenuation r and phase shift θ on reflection to give a complex amplitude $\hat{A}_r$ at the boundary. This is related to $\hat{A}_i$ through the (complex) amplitude reflectivity

$$\hat{r} = re^{i\theta} \tag{4}$$

by

$$\hat{A}_r = \hat{r}\hat{A}_i. \tag{5}$$

The corresponding expression for the powers is

$$I_r = RI_i, \tag{6}$$

where I_i and I_r are, respectively, the incident and reflected powers and

$$R = |\hat{r}|^2 = r^2 \tag{7}$$

is the power reflectivity.

We now assume that the medium 1 is transparent and that the medium 2 is absorbing and is the substance to be studied. Thus, the complex refractive index of 1 has negligible imaginary part ($\varkappa_1 \ll 1$) and may be taken as purely real (n_1) whilst the refractive index of 2 is complex because the absorption index is not negligible compared with the real refractive index; hence $\hat{n}_2 = n_2(1 - i\varkappa_2) = n_2 - ik_2$.

When $n_1 < n_2$, as is the case when the medium 1 is air or vacuum, the medium 2 under study is optically more dense than 1 and the reflection at its boundary is partial, being accompanied by refraction in 2 (Fig. 1(a)). This is the basis of conventional reflection spectrometry. Since $\hat{n}_2$ is complex, Snell's law describing the refraction must be written in a complex form. A convenient representation of this law is

$$n_1 \sin \psi_1 = \hat{n}_2 \sin \hat{\psi}_2, \tag{8}$$

where $\hat{\psi}_2$ is the angle of refraction in the medium 2, written as a complex quantity.

When $n_1 > n_2$, as can be the case when a special (transparent) medium 1 is placed against the specimen 2, the reflection is total for all angles of incidence ψ_1 greater than a critical angle ψ_c as there is no refracted ray (Fig. 1(b)). This is the basis of a special technique called attenuated total reflection spectrometry (ATR).

Both $\hat{r}$ and R are functions of the relative refractive index $n_{12} = n_2/n_1$, of the extinction coefficient $\varkappa_2$, of the angle of incidence ψ_1 and of the polarization of the radiation. Thus

$$\hat{r} = \text{function}\,(\hat{n}_{12}, \psi_1, \text{polarization}) \tag{9}$$

and

$$R = \text{function}\,(n_{12}, \varkappa_2, \psi_1, \text{polarization}) \tag{10}$$

and both vary with wave-number because $\hat{n}_{12} = \hat{n}_2/n_1$ so varies. The general functions (9) and (10) are obtained in explicit form from Fresnel's relations used in conjunction with Snell's law. For present purposes, we write Fresnel's relations as

$$\hat{r}_\perp = -\frac{\sin(\psi_1 - \hat{\psi}_2)}{\sin(\psi_1 + \hat{\psi}_2)} \tag{11}$$

for radiation polarized with the electric field vector perpendicular to the plane of incidence (TE) and

$$\hat{r}_\parallel = \frac{\tan(\psi_1 - \hat{\psi}_2)}{\tan(\psi_1 + \hat{\psi}_2)} \tag{12}$$

for radiation polarized with the electric vector in the plane of incidence (TM). Bell has expressed each of Fresnel's relations in the convenient form

$$\hat{r} = \frac{1 - \hat{\chi}}{1 + \hat{\chi}} \tag{13}$$

in terms of a characteristic parameter

$$\hat{\chi} = \chi' - i\chi'' \tag{14}$$

which for ⊥ radiation is

$$\hat{\chi}_\perp = \frac{\hat{n}_2 \cos \hat{\psi}_2}{n_1 \cos \psi_1} \quad (15)$$

and for ∥ radiation is

$$\hat{\chi}_\parallel = \frac{\hat{n}_2 \cos \psi_1}{n_1 \cos \hat{\psi}_2}. \quad (16)$$

The real and imaginary parts χ' and χ'' of the characteristic parameter $\hat{\chi}$ are expressible in terms of r and θ as

$$\chi' = \frac{1 - r^2}{1 + 2r \cos \theta + r^2} \quad (17a)$$

and

$$\chi'' = \frac{2r \sin \theta}{1 + 2r \cos \theta + r^2}. \quad (17b)$$

As the aim is to deduce $\hat{n}_2$ from measured values of R an attempt is made to find from (10) the solutions

$$n_{12} = n_2/n_1 = \text{function}(R, \varkappa_2, \psi_1, \text{polarization}) \quad (18a)$$

and

$$\varkappa_2 = \text{function}(R, n_{12}, \psi_1, \text{polarization}). \quad (18b)$$

To evaluate these equations it is generally necessary to make at least two measurements under different conditions, for example, with a different polarization or a different angle of incidence. The need for different measurements does not arise if one of the quantities n_2 or $\varkappa_2$ is known already. The need for more than one measurement is also avoided, in principle, if the (complex) amplitude reflectivity can be found. This is because a solution of (9) gives

$$\hat{n}_{12} = \hat{n}_2/n_1 = \text{function}(\hat{r}, \psi_1, \text{polarization}) \quad (19)$$

which, being complex, is effectively two equations. Until recently, although $|\hat{r}| = r$ could be measured it was impossible to measure the phase of $\hat{r}$ directly using freewave optical techniques and square-law detectors. It is now possible to use a Fourier transform spectrometer for the purpose but instruments of sufficient precision can so far be built only for far infra-red use.

An indirect method of finding the phase of $\hat{r}$ is based on the fact that the modulus r and phase θ of $\hat{r}$ are not independent but are related on general grounds. Thus r is found from R, θ is calculated numerically, subject to some approximations, from R also and then (19) is used to give $\hat{n}_2$.

The full equations for $\hat{n}_2$ in terms of $\hat{r}_\perp$ or $\hat{r}_\parallel$ are complicated, but less so than those for n_2 and $\varkappa_2$ in terms of $R_\perp$ or $R_\parallel$. In the latter case, explicit solutions have been obtained only in a few special instances, for example by Heilmann and by Fahrenfort and Visser. It is these difficulties which caused the early workers to evaluate n_2 and $\varkappa_2$ by matching values of R calculated from (10) to their measured values, usually using pre-computed graphs or charts.

This can be a tedious and inexact process. More recently, computers have been used to ease the calculation but have been most profitably applied to "forward" calculations based on (18). This article is concerned with the measurement aspects of reflection spectroscopy; descriptions of the origins and significance of the spectra are given in accounts of the solid (and liquid) state—for example, those of Moss and Stern.

Measurement of reflection spectra

A complete system for reflection measurements will consist typically of source; *monochromator* or *spectrometer*; mount, etc., for specimen (the reflection head) and detector. The exact arrangement depends on the type of measurement to be made. Reflection measurements can be made on both solids and liquids. Conventional reflection spectra ($n_1 < n_2$) are more commonly found for solids because of the difficulties associated with taking full account of the influence of the windows used to contain a liquid. Attenuated reflection spectra ($n_1 > n_2$) are more commonly found for liquids because of the difficulty of obtaining good optical contact between the solid medium 1 and a solid specimen 2.

Power reflectivity measurements can be conveniently divided into two categories according to how the pair of independent data necessary for the evaluation of $\hat{n}_2$ from R are obtained. There are those measurements made at inclined incidence ($\psi_1 \neq 0$) and those made at normal incidence ($\psi_1 = 0$). At inclined incidence a further subdivision is possible: measurements of power reflectivity may be made for a given polarization at two distinct angles of incidence, or for both polarizations at a given angle of incidence. There is also a "hybrid" method involving changes of both polarization and angle of incidence.

Normal incidence measurement of power reflectance

At normal incidence

$$\hat{\chi}_\perp = \hat{\chi}_\parallel = \frac{\hat{n}_2}{n_1} \quad (20)$$

and

$$\hat{r}_\perp = -\frac{\hat{n}_2 - n_1}{\hat{n}_2 + n_1} = \hat{r}_\parallel \quad (21)$$

so the distinction between the powers reflected in the two planes of polarization vanishes. The equations for the power reflectivity both simplify to

$$R = |\hat{r}_\perp|^2 = |\hat{r}_\parallel|^2 \quad (22a)$$

$$= \left| \frac{\hat{n}_{12} - 1}{\hat{n}_{12} + 1} \right|^2 \quad (22b)$$

$$= \text{function}(n_{12}, \varkappa_2, 0). \quad (22c)$$

The simplicity is, however, deceptive because $\hat{n}_{12}$ cannot be found directly from this equation. In practice, normal incidence power measurements form the basis of the method of Robinson and Price who used them first to find $|\hat{r}| = R^{1/2}$ and then to find θ from the Kramers–Kronig integral relation

$$\theta(\sigma) = \pi + \frac{\sigma}{\pi} \int_0^\infty \frac{\ln R(\sigma') - \ln R(\sigma)}{\sigma^2 - \sigma'^2} d\sigma' \quad (23)$$

which is known, from considerations based on causality, to hold between the imaginary and real parts of the complex reflection attenuation defined by

$$-\ln \hat{r} = -\tfrac{1}{2} \ln R - i\theta. \quad (24)$$

The valid use of this integral strictly requires knowledge of R at all wave-numbers σ which is, of course, not possible in practice. It is found, fortunately, that θ can be calculated meaningfully provided the values of R are obtained over a suitably-chosen wide range stretching from (say) σ_1 to (say) σ_2. In the centre of this is the region of interest over which knowledge of the optical "constants" is required. In the simplest approach the power reflectivity is regarded as constant outside the range $\sigma_1 \leq \sigma \leq \sigma_2$ and equal to R_1 for $\sigma < \sigma_1$ and R_2 for $\sigma > \sigma_2$

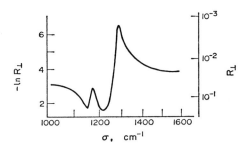

(Fig. 2) thereby enabling the integral in (23) to be split into three parts to give

$$\theta(\sigma) \simeq \pi + \frac{1}{\pi} \ln \frac{R_1}{R(\sigma)} \ln \frac{\sigma + \sigma_1}{|\sigma - \sigma_1|}$$

$$+ \frac{\sigma}{\pi} \int_{\sigma_1}^{\sigma_2} \frac{\ln R(\sigma') - \ln R(\sigma)}{\sigma^2 - \sigma'^2} d\sigma'$$

$$- \frac{1}{\pi} \ln \frac{R_2}{R(\sigma)} \ln \frac{\sigma + \sigma_2}{|\sigma - \sigma_2|}. \quad (25)$$

This can be evaluated numerically using a computer. When r and θ are known, $\hat{n}_{12}$ can be calculated graphically using, say, a Smith chart (Fig. 3). It is preferable, however, to use (17) and (20) which give

$$n_2 = n_1 \frac{1 - r^2}{1 + 2r\cos\theta + r^2} \quad (26a)$$

and

$$n_2 \varkappa_2 = n_1 \frac{2r \sin\theta}{1 + 2r \cos\theta + r^2}. \quad (26b)$$

Apart from the inaccuracies which may arise from the use of a truncated integral in the Kramers–Kronig relation there is the question of how uncertainties in $\hat{n}_2$ depend on errors in the measured values of R. Unfortunately, when R is small the absolute value of the real part $-\tfrac{1}{2} \ln R$ of the complex reflection attenuation, which is the quantity appearing in the integral (23), is large and the large fractional errors which are unavoidably associated with the small values of R lead to large errors in the computed phase. These are distributed throughout the phase spectrum.

There is frequently an additional source of error due to the use of a finite angle of incidence. This usage enables the reflected beam to be easily recovered but, unfortunately, the relations (26) are then not valid due to the omission of terms depend-

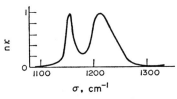

Fig. 2. Example of a power reflection spectrum (TM reflection of PTFE recorded near normal incidence). The region of interest is 1100 to 1300 cm^{-1} and is narrower than the region 1000 to 1600 cm^{-1} over which the reflectivity is recorded. Near 1000 and 1600 cm^{-1}, R has assumed nearly constant values corresponding to R_1 and R_2. Note that $-\ln R$ is large where R is small. (From Robinson, T. S. and Price, W. C. (1953) Proc. Phys. Soc. B, **66**, 969.)

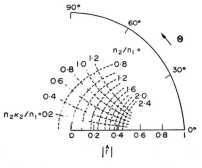

Fig. 3. Smith chart relating loci of constant $\hat{n}_{12}$ to $\hat{r}$. The point corresponding to the measured r and θ is chosen to give the required n_2/n_1 and $n_2\varkappa_2/n_1$. (From Robinson, T. S. and Price, W. C. (1953) Proc. Phys. Soc. B **66**, 969.)

ing on ψ_1. This difficulty is avoided if a beam-divider arrangement is used to ensure normal incidence (Fig. 4); however, the price paid is a considerable reduction in the radiant power reaching the detector owing to losses at the beam-divider.

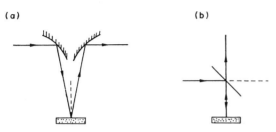

Fig. 4. Arrangements used for (a) near normal incidence ($\psi_1 \sim 10°$) and (b) exactly normal incidence.

Normal incidence measurement of amplitude reflectance

Many of the shortcomings of the Robinson–Price technique can, in principle, be avoided if $\hat{r}$ is measured directly. This can be done with a Fourier spectrometer since this includes an interferometer which permits the phase behaviour of electromagnetic waves to be studied.

Fourier transform spectrometry is the technique whereby the spectral distribution of the power incident on a detector is obtained by Fourier transformation of the interference record observed by the detector as the path-difference in a two-beam interferometer placed between it and the source is varied. The variable part of the record, called the interferogram, and the detected spectrum are a Fourier transform pair.

The interferogram observed at the output of a Michelson-type interferometer is produced by the superposition of the radiation from each of the two arms of the interferometer; when these arms are identical in all respects except length (which may be varied in one of them) the interferogram is a symmetrical function of the difference x between the lengths of the paths in the two arms. There is a grand maximum of power at zero path-difference and after some fluctuations the signal becomes roughly invariant on either side. By using a specimen to reflect the radiation from the interferometer to the detector (using, say, one of the arrangements of Fig. 4), the interferometer and, though changed, the interferogram remain symmetrical. The power reflectivity can then be deduced rather as with a conventional spectrometer. However, if the mirror terminating one of the arms is exactly replaced by a specimen (which is optically flat), the symmetry of the interferometer is disturbed since the wave in the specimen arm suffers attenuation and phase shift according to (5) on striking the specimen. The effect of this is to displace the grand maximum of

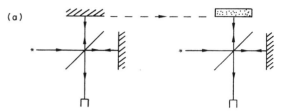

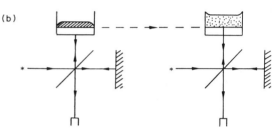

Fig. 5. Michelson interferometers for measurements of $\hat{r}(\sigma)$ (a) solid specimen, (b) liquid specimen.

the interferogram by a small amount $\bar{x}$ and to render the interferogram unsymmetrical about this maximum. Since the displacement and distortion of the interferogram are due to the substitution within the interferometer of the reflective properties of the specimen for those of the mirror the complex reflectivity of the specimen can be calculated from the Fourier transforms of the observed interferograms.

In the absence of the specimen the interferogram is the symmetrical function

$$F_0(x) = \int_{-\infty}^{\infty} B_0(\sigma) \cos 2\pi\sigma x \, d\sigma$$

$$= \int_{-\infty}^{\infty} B_0(\sigma) \exp(-2\pi i \sigma x) \, d\sigma \quad (27)$$

of the detected spectral power $B_0(\sigma)$. When the specimen is introduced the interferogram becomes

$$F(x) = \int_{-\infty}^{\infty} B_0(\sigma) \, r(\sigma) \exp[i\theta(\sigma) - i\pi]$$

$$\exp(-2\pi i \sigma x) \, d\sigma, \quad (28)$$

where $\theta(\sigma) - \pi$ is the phase shift at the specimen relative to that at the reference mirror (assumed to be a perfect reflector). The ratio

$$\frac{\mathscr{F}\{F(x)\}}{\mathscr{F}\{F_0(x)\}} = r(\sigma) \exp[i\theta(\sigma) - i\pi] = -\hat{r}(\sigma) \quad (29)$$

of the full Fourier transforms $\mathscr{F}\{F(x)\}$ and $\mathscr{F}\{F_0(x)\}$ enables the (complex) amplitude reflectivity to be found.

Bell and Chamberlain and their colleagues have pioneered this technique, Bell concentrating on solids and Chamberlain on liquids. For solid measurements (Fig. 5(a)), the interferometer mirror is removed from a specially stable support and exactly replaced by the specimen of the medium 2. For liquid measurements (Fig. 5(b)), the reflectivity at a flat interface between a transparent window of medium 1 and mercury is compared with that found when an equal weight of the specimen replaces the mercury. There is a scarcity of suitable materials for the window. The polymer poly 4-methyl pentene-1 (called TPX by ICI Ltd.) is transparent over the submillimetre waveband, but having a refractive index $n_1 \sim 1.46$ it is not suitable for use with most organic liquids which have similar refractive indices. Moreover, TPX is attacked by many of these liquids. Crystalline quartz is virtually the only suitable material, with an index $n_1 \sim 2.13$. It absorbs prohibitively above 125 cm^{-1}.

The numerical analysis yields $\hat{r}$ directly and differs from what is generally required for Fourier transform spectrometry only in detail. However, special practical difficulties concern the quality of the specimen and the stability of not only its support but of the interferometer as a whole. Moreover, it is necessary to know $\bar{x}$ which can be very small (of the order 1 μm or less). For these reasons, this form of spectrometry has so far been used only at far infrared and submillimetre wavelengths. Quite apart from yielding the phase and the modulus of the complex reflectivity, a particular advantage possessed by the Bell–Chamberlain technique is that it enhances structure in the reflectivity spectrum where the reflectivity itself is fairly small. This is because r is the important quantity and since $R < 1$ then $r > R$: therefore, in the low reflectivity regions r (measured here) shows greater variation than R (conventionally measured).

Inclined incidence measurement of power reflectance

When normal incidence is not used, two distinct and independent values of R are required. The variable chosen has generally been angle of incidence ψ_1. Care is necessary in the execution of these measurements to ensure that as the angle of incidence is increased the axis of the radiation remains centrally placed on the specimen and the projected diameter of the beam remains less than that of the specimen.

Power reflection measurements at two (or more) angles of incidence were first used recently by Simon in 1951 and subsequently modified and improved by Avery in 1952. $\psi_1 = 20°$ and $70°$ are commonly chosen values. Simon deduced $\hat{n}_2$ using pre-computed graphs of the type shown in Fig. 6 where, for the angles of incidence used, R is calculated as a function of n_2 using $\varkappa_2$ as a parameter. This method is not very satisfactory as a great change in the final value of $\hat{n}_2$ results from a small change in the measured reflectivities—particularly for $R_\perp$ where the graphs are steep. Avery devised a "hybrid" technique and measured $R_\parallel/R_\perp = \varrho$ thereby obviating the need for the replacement of the specimen by a reference detector. In this method, $R_\parallel$ and $R_\perp$ are usually recorded by rotating a polarizer. It is essential to avoid any shift of the beam during this rotation and it is necessary to know the instrumental polarization. At small ψ_1 the differences between $R_\parallel$ and $R_\perp$ are relatively small and so the method should be confined to relatively large ψ_1.

Avery, like Simon, used pre-computed graphs to estimate $\hat{n}_2$ but a more direct analysis was introduced by Lindquist and Ewald in 1963. Like Avery, they measured ϱ but they devised an ingenious way of calculating a unique value for $\hat{n}_2$ from ϱ based on geometrical construction of epicycloids in a complex plane.

In the complex plane of the complex conjugate $\hat{\varepsilon}_2^* = (\hat{n}_2^*)^2$ of the relative permittivity $\hat{\varepsilon}_2 = \hat{n}_2^2$ an epicycloid

$$\hat{\varepsilon}_2^* = (\hat{n}_2^*)^2 = A + B\,e^{i\Phi} + C\,e^{2i\Phi} \qquad (30)$$

can be plotted. In this expression Φ is the variable used to generate the curve and A, B and C are coefficients dependent only on n_1, ψ_1 and ϱ. The epicycloids corresponding to the various ψ_1 intersect at a unique point to give the required value of $\hat{\varepsilon}_2^*$ and hence $\hat{n}_2$. An example, taken from the work of Lindquist and Ewald is given in Fig. 7.

Inclined incidence measurement of power reflectance—ATR

A particular disadvantage of all "external" reflection measurements ($n_1 < n_2$) is the smallness of the power reflectivity and the insensitivity of its dependence on $\varkappa_2$ (see Table 1). Fahrenfort and, independently, Harrick showed that this disadvantage can be surmounted by making $n_1 > n_2$ and using an angle of incidence ψ_1 greater than the critical angle ψ_c for the interface.

When $n_1 > n_2$ and the medium 2 is non-absorbent, the power reflectivity is equal to unity for all angles of incidence ψ_1 greater than the critical angle $\psi_c = \arcsin n_{12}$ and the reflection is "total". There is no net penetration of the medium 2 by the radiation. If, however, the medium 2 is absorbing there is penetration and during its journey in the second medium the radiation continually changes direction and suffers attenuation before emerging into the medium 1. The greater the absorption the less the radiation that emerges. The reflection is still "total", in the sense that there is no refracted beam, but the reflectivity is less than one. This phenomenon is known as attenuated total reflection. One of its noteworthy characteristics is that R is relatively large for relatively small $\varkappa_2$ and is strongly dependent on $\varkappa_2$ as Fig. 8 shows. This makes R considerably more amenable to accurate measurement than its counterpart in conventional power reflection experiments. Fahrenfort has given some data which illu-

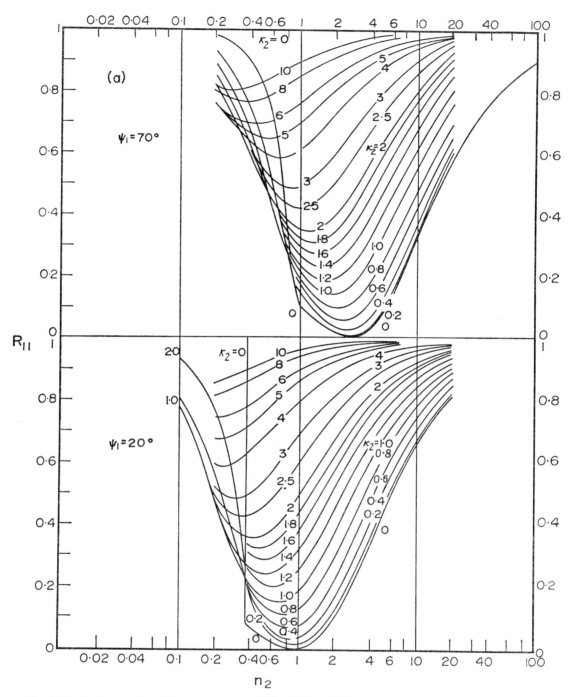

Fig. 6. Typical example of R versus n_2 graphs for (a) $\parallel$ and (b) $\perp$ radiation. (*From Simon, I. (1951) J. Opt. Soc. Am.* **41**, *336*.)

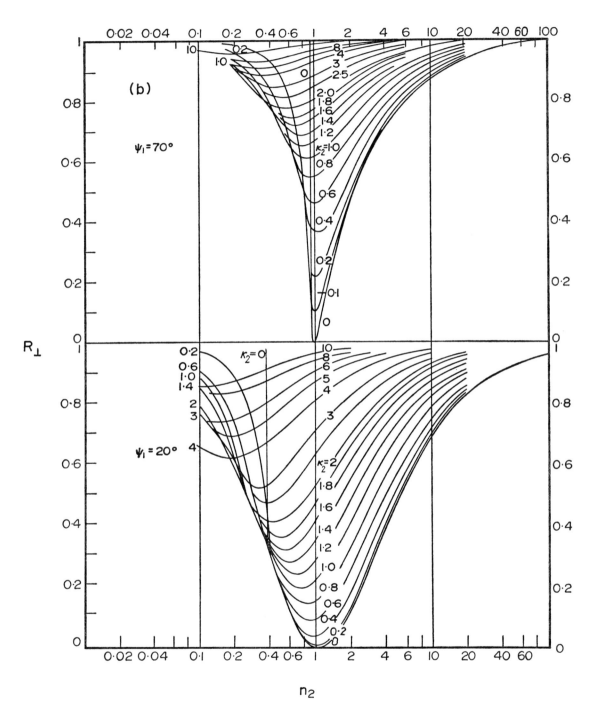

Fig. 6.

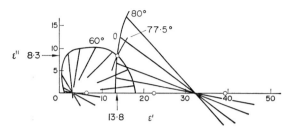

Fig. 7. Epicycloids used for "forward" calculation of n_2 from ϱ. The curves were obtained for grey tin measured at $\psi_1 = 60°$, $77.5°$ and $80°$ using 11-μm wavelength radiation. $\hat{n}_2$ is calculated from the values of the real and imaginary parts ε'_2 and ε''_2 of $\hat{\varepsilon}_2$ indicated by the point 0 at which the epicycloids intersect. (From Lindquist, R. E. and Ewald, A. W. (1963) J. Opt. Soc. Am. 53, 247.)

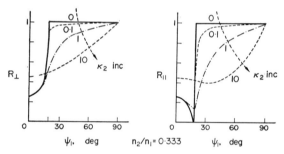

Fig. 8. Numerical example of variation of $R_\perp$ and $R_\parallel$ with ψ_1 and $\varkappa_2$ for totally reflected radiation ($n_1 > n_2$). (From Harrick, p. 20.)

strate this point well. For radiation incident from a vacuum on a medium 2 of refractive index 1·4, $R_\perp$ has the values shown in Table 1 for $\psi_1 = 20°$ and

Table 1. Comparison of power reflectivity $R_\perp$ for "partial" and "total" reflection at plane boundary between media 1 and 2 ($n_2 = 1·4$)

$\varkappa_2$	Partial (external) reflection ($n_1 = 1$)		Total reflection (ATR) ($n_1 = 2$)
	$\psi_1 = 20°$	$\psi_1 = 70°$	$\psi_1 = 46°$
0	0·03	0·26	1
0·1	0·03	0·26	0·35
0·2	0·04	0·31	0·26

$\psi_1 = 70°$. The values of $R_\perp$ differ little at a given angle of incidence for the various $\varkappa_2$. If, however, a medium of refractive index $n_1 = 2$ replaces the vacuum and observations are made at $\psi_1 = 46° (>\psi_c)$

the dependence of $R_\perp$ on $\varkappa_2$ is dramatic; in fact, as Fig. 9 shows, $R_\perp$ changes five-fold as $\varkappa_2$ varies from 10^{-4} to 0·5. This makes sensitive measurements possible on the wide class of commonly found materials having $\varkappa_2 < 0·5$ and poses a serious challenge to transmission methods commonly used to find α_2.

A variety of ways have been devised for obtaining the desired attenuated reflection coefficient and extended accounts will be found in the writings of

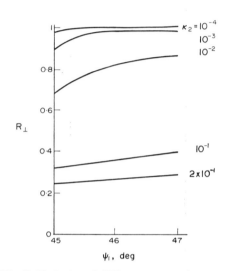

Fig. 9. Variation of ATR power near the critical angle as a function of ψ_1. (From Fahrenfort.)

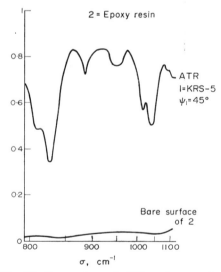

Fig. 10. Comparison of ATR and conventional reflection spectrum showing enhancement in the former case. (Example for an epoxy resin taken from Fahrenfort.)

Harrick and Fahrenfort. One of the simplest approaches is to fabricate the optically dense medium 1 in the form of a hemi-cylindrical prism (Fig. 11(a)). The various arrangements enable the angle of incidence at the interface 1–2 to be easily changed and subsequently deduced. Since the reflected power depends on the penetration of the radiation into 2, wavelengths which are only weakly absorbed can be studied through use of repeated reflections using an arrangement such as that shown in Fig. 11(b). For these measurements the radiation must be well collimated and the angle of incidence must be accurately known. One of the important limitations in ATR is the difficulty of finding materials for the ATR element that have a sufficiently high refractive index and yet remain transparent; commonly used materials are alumina ($n_1 \sim 1.7$), silver bromide ($n_1 \sim 2$), KRS-5 ($n_1 \sim 2.4$), silicon ($n_1 \sim 3.5$) and

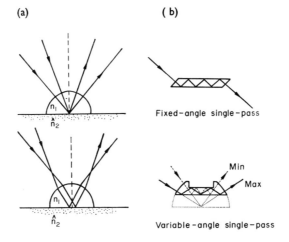

Fig. 11. *Arrangements for ATR measurements.*
(a) Single reflection (used for strong absorbers).
(b) Multiple reflections (used for weak absorbers).

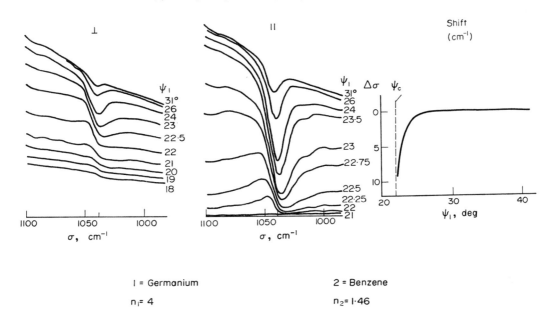

1 = Germanium
$n_1 = 4$

2 = Benzene
$n_2 = 1.46$

Fig. 12. *Variation of ATR spectra with angle of incidence for (a) TE and (b) TM propagation. The shift of the position of minimum reflection as ψ_1 approaches ψ_c is shown in (c). (Example for a germanium ($n_1 = 4$)-benzene ($n_2 \sim 1.46$) interface taken. (From Harrick, pp. 71–73.)*

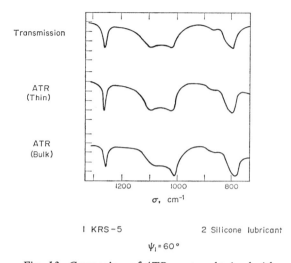

Fig. 13. Comparison of ATR spectra obtained with a thin and a bulk specimen and a transmission spectrum. (Example for a KRS-5–silicon lubricant interface taken from Harrick, p. 64.)

germanium ($n_1 \sim 4$). Full details are given by Harrick.

A significant point about ATR worth emphasizing is that since R falls as $\varkappa_2$ increases, rather as the transmitted power falls in conventional transmission measurements, ATR spectra and transmission spectra can have a striking similarity in appearance. The actual form of the reflection spectrum depends on the angle and its relation to the critical angle ψ_c, which is, in turn, governed by n_{12}. If ψ_1 is close to ψ_c the ATR spectrum is distorted by comparison with the transmission spectrum; if ψ_1 is made considerably larger than ψ_c the two spectra are virtually identical for a thin film specimen and very similar with only slight differences in the case of a bulk specimen. The differences found in the ATR spectrum are due to the dependence on wavelength of the penetration depth into 2, so bands at longer wavelengths tend to be enhanced and broad bands are noticeably strengthened on the long wavelength side with a resultant shift of peak relative to transmission measurements. Nevertheless, the ability to relate ATR spectra to conventional transmission spectra reasonably reliably makes ATR an important and useful analytical technique.

The optical "constants" can be deduced from ATR spectra recorded at two angles of incidence rather as in the techniques of Simon or Lindquist and Ewald. However, these and other cumbersome graphical procedures are rarely used now following an explicit solution by Fahrenfort and Visser of Fresnel's equations for $\hat{n}_2$ in terms of $R_\perp$ measured at two angles of incidence. The solution is unique but the algebra is complicated and a computer is virtually essential for the numerical evaluation.

ATR is, therefore, a very powerful technique which provides recognizable spectra in its own right and enables the optical "constants" to be determined reliably. Many of the potential applications are only just being realized.

Bibliography

BELL, E. E. (1967) *Handbuch der Physik* **XXV/2a**, 1.
CHAMBERLAIN, J. (1971) *Physics Bulletin* **22**, 333.
FAHRENFORT, J. (1963) *Proc. 10th Coll. Spectr. Int.*, eds. Lippincott and Margoshes. Washington: Spartan Books.
HARRICK, N. J. (1967) *Internal Reflection Spectroscopy*, New York: Interscience.
MOSS, T. S. (1959) *Optical Properties of Semiconductors*, London: Butterworths.
STERN, F. (1963) *Solid State Physics*, eds. Seitz and Turnbull, **15**, 229.
WENDLANDT, W. W. and HECHT, H. F. (1966) *Reflection Spectroscopy*, New York: Interscience.

JOHN CHAMBERLAIN

REGENERATIVE FLYWHEELS. A machine which runs at varying speed usually requires energy at a correspondingly varying rate from its prime mover. The capacity of the prime mover must be sufficient to meet the maximum demand, and its average efficiency is likely to be lower than if a steady output were required. If in addition the speed of the machine is reduced by a normal braking system a direct energy loss occurs, which is dissipated as heat to the surroundings.

One way in which these losses can be reduced is by the use of a flywheel as an energy reservoir; such a device is called a regenerative (or recuperative) flywheel. Energy is transferred between the machine and the flywheel, usually electrically or mechanically. Thus as the kinetic energy of the machine decreases there is a corresponding increase in the kinetic energy of the flywheel (apart from inevitable losses), and vice versa.

There is some difficulty in transferring energy between two components whose speeds are changing in opposite senses, and the extra cost and complication of a regenerative system is only likely to be justified in rather special circumstances. The examples quoted below are all from the field of transport, where energy conservation is more than usually important. [*Note*: The kinetic energy of a flywheel rotating at angular speed ω is $I\omega^2/2$, where I is the second moment of mass ("moment of inertia"); further, $I = Mk^2$ where M is the mass of the flywheel and k is a dimension, the "radius of gyration"; for a flywheel in the form of a uniform-thickness disc, k^2 is equal to half the square of the radius.]

The "Electrogyro"

A regenerative flywheel is the central component in a system in which the kinetic energy of a flywheel, made up periodically from external power supply points, drives the vehicle through an electrical generator and motor, and advantage is taken of regenerative braking. This system (under the name "electrogyro") has been developed by the Oerlikon Company of Switzerland for application in public service road vehicles and in railway locomotives for shunting and similar duties.

Figure 1 indicates the main features of the system. F represents the flywheel, and G and M the two electrical machines; both of these can operate as either a motor or a generator. The sequence of energy flows is as follows:

(i) From the idle condition, the flywheel F is run up to speed by machine G, to which it is directly coupled, acting as a motor, drawing electrical power from an external source; during the process the other machine M is disconnected from the rest of the system.

(ii) When the flywheel is energized, the vehicle moves off; the kinetic energy of the F–G rotor assembly is converted to electrical energy in G and supplied to M, which rotates the driving wheels through mechanical gearing.

(iii) During braking periods the energy flow is reversed; machines M and G act as generator and motor respectively, and the kinetic energy of the flywheel is increased at the expense of that of the vehicle.

(iv) When required, the flywheel is "recharged" from external electrical supply points, as in the initial activating operation (i).

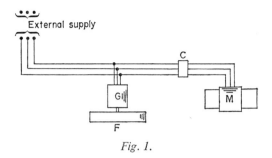

Fig. 1.

Electrical Details

The electrical machines used in the Oerlikon design are of the squirrel-cage induction type. The use of these simple machines (without slip-rings or commutators) as motors is common, but in this application they also act as generators. If a capacitor of suitable value is connected in parallel with an externally driven induction machine, the residual magnetism induces a small voltage which causes a magnetizing current to flow; the voltage then builds up as in a d.c. shunt machine.

Machine G is a 2-pole 3-phase a.c. machine, and with a 50 Hz supply the maximum speed is 3000 rev/min, the synchronous speed. The flywheel normally operates down to about 2000 rev/min. The much larger speed range required of the traction motor M is achieved by various combinations of pole-changing (i.e. altering the connections of the stator winding to vary the number of pole pairs) and mechanical gearing, depending on the requirements (see below).

In induction alternators, the phase of the output current leads the voltage. The load (the induction motor) absorbs a lagging current. It is, therefore, necessary to connect capacitors in parallel with the main circuit between the two machines M and G. Variation of the motor voltage, and therefore torque, is achieved by varying the amount of capacitance in circuit. In the simplified diagram (Fig. 1) C represents the pole-changing and capacitor control.

Road Passenger Application ("Gyrobus")

A single-deck omnibus was successfully tried out in several Swiss towns in 1950; further vehicles of similar type operated regular services in Switzerland and the Belgian Congo, (now Zaire), for several years.

In this application, the driving motor M has three electrically distinct sections separated by two mechanical reduction gears with velocity ratios of 1·24 : 1; the final drive to the road wheels is by a conventional differential gearbox on the rear axle. Two of the motors can be operated with either 2 or 4 poles, the third with either 2 or 8 poles. Six different tractive-force characteristics are thus available to cover the speed range of the vehicle, 0 to

50 km/h, as shown by Fig. 2. Selection of the appropriate motor and pole arrangement is made by means of a foot-pedal in the driver's cab.

In typical service conditions these buses covered 20 km in an hour, including twenty scheduled stops. For a total vehicle weight of 15 tonnes with a rolling friction of 1500 newtons, the calculated work required at the road wheels is 16·6 kWh/h. Making allowances for speed variations required by general traffic conditions, and the efficiency of the power transmission system, the amount of energy required from the flywheel is about 30 kWh/h. This has to be made up at the recharging points visited during the

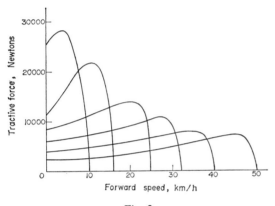

Fig. 2.

period. If recharging occurs at 10-min intervals, then the useful energy storage of the flywheel must be 5 kWh. By designing the flywheel for twice this capacity at its maximum speed, a reserve is provided for emergencies. An energy storage of 10 kWh can be achieved with a flywheel of 1500 kg mass and a diameter of 1·6 m running at about 2950 rev/min.

The capacity of the machine G is determined by the period allowed for recharging the flywheel. If, for example, the recharging period is not to exceed 2 min, a capacity of about 200 kW is required.

The accommodation of this power system makes special demands on the layout of the vehicle. The major rotating assembly, the flywheel F and generator/motor G, should be mounted vertically, so that gyroscopic moments have the least harmful effect on the stability of the vehicle. The solution adopted by Oerlikon is shown diagrammatically in Fig. 3. F and G are located in the middle of the vehicle, and the three-component driving motor M at the rear. The electrical controls are operated from the driver's cab through pneumatic circuits. The flywheel is housed in a casing filled with hydrogen at sub-atmospheric pressure. This reduces the rotor windage while allowing adequate cooling. Apart from increasing the efficiency of power transmission, this design feature decreases the rate at which the flywheel speed falls away when no power is being used;

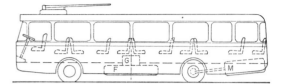

Fig. 3.

typically, it takes 2 hr for the speed to decrease from 2900 to 1800 rev/min.

Railway Applications

The intermittent nature of shunting operations suggests that this would be a suitable application for the electrogyro, which would provide a locomotive free from the restrictions of conductor rails or overhead lines, limited only by the necessity to recharge at fairly frequent intervals.

Six electrogyro locomotives were built in the 1950's for mining companies in France and South Africa. These use one flywheel-generator unit similar to that of the gyrobus. In 1953 the U.K. National Coal Board ordered an experimental locomotive for surface shunting duties. This incorporated two flywheel-generator units, mounted vertically and of about the same size and capacity as in the gyrobus. One traction motor with fixed-ratio gearing powered each of the two axles. These motors were arranged to operate with 4, 6, 8 or 12 poles to give four speed ranges up to a maximum of 19 km/h.

An earlier application of energy storage in a flywheel had been made in an electric locomotive built by Southern Railway in 1939–41. The power system of this locomotive included two motor-generator sets with heavy flywheels. The main objective was to maintain the power for the driving wheels when the electrical supply to the locomotive was interrupted by gaps in the conductor rail. Other similar locomotives were built later. The flywheels in these machines fulfilled a less obviously regenerative function than those in the electrogyro system.

The "*Gyreacta*"

In this system, designed and developed in England, the flywheel operates in conjunction with an internal combustion engine and an epicyclic gearbox. The system differs in two main respects from the electrogyro—the flywheel does not provide the sole source of energy for propelling the vehicle, and the energy flow between it and the rest of the system is effected mechanically. The main gearbox comprises two epicyclic gear trains (sun, planet and annular wheels), and further auxiliary epicyclic gearing connects the flywheel shaft with one of the annular gears in the main gearbox. By appropriate choice of the speed ratios of the main and auxiliary gears (by means of brakes and clutches operating on the annular gear wheels) the energy flow between the gearbox and the flywheel can be controlled throughout the whole speed range of the vehicle. In particular, as the vehicle slows down, the speed and kinetic energy of the flywheel increase. Conversely, the flywheel yields energy to supplement the engine power for vehicle acceleration.

A Gyreacta transmission for use in omnibuses has been built and tested; various other applications for private cars, racing cars and delivery vans have been proposed, aimed at achieving improved vehicle performance and/or economy. However, none of these has been realized. The flywheels envisaged in these designs vary in size up to 0·65 m in diameter, with maximum speeds up to 20000 rev/min; and energy storage up to about 7 kWh.

Discussion

Although the electrogyro was in regular service for several years it is clear that neither this system nor the Gyreacta will achieve widespread acceptance in present conditions. In practice the theoretical advantages are outweighed by the disadvantages. These are mainly the extra cost and complexity of these systems, and, in the case of the electrogyro, the need for fairly frequent recharging pauses, and rather poor vehicle performance on uphill sections of the route. However, increasing world fuel shortage may alter the balance again. This is particularly possible in the case of the electrogyro which has the additional advantage of pollution-free operation. It is likely to be in competition with simpler battery-powered electric vehicles, which do not need even the recharging points require by the electrogyro, but suffer from the capacity limitations of present storage batteries. Because of its regenerative operation the electrogyro should be cheaper to run, but this has to be balanced against the greater capital cost and weight of its more complicated power system.

Bibliography

Loco. Engrs. **49** (4), 396.
ANON. (1949) *Loco. Mag.* **55**, 148.
ANON. (1955) *Auto. Engr.* **45**, 559.
CLERK, R. C. (1963) Soc. Auto. Engrs, paper for Summer Meeting, Montreal, 1963.
GILLON, E. (1955) *Bulletin Société Belge des Electricians*, **71**, 9.
GREEN, T. E. and GESSLER, J. K. (1959) *J. Instn.*

S. J. PEERLESS and R. T. SMITH

REGGE MODEL OF ELEMENTARY PARTICLES. The Regge model began as a formal mathematical procedure for obtaining asymptotic bounds on scattering amplitudes. It was developed to provide a phenomenological basis for understanding high energy experiments, and in recent years it has played a large part in attempts to provide a comprehensive theory of elementary particles and their interactions.

To introduce the model we consider a scattering process in which particles 1 and 2, with 4-momenta p_1 and p_2, collide and produce a final state containing particles 3 and 4, with 4-momenta $-p_3$ and $-p_4$. Although the theory can be generalized we shall simplify our discussion by assuming all four particles have equal mass, m, and zero spin. The scattering amplitude, A, describing this process is a function of two variables for which it is convenient to take the Lorentz invariant quantities $s = (p_1 + p_2)^2$ and $t = (p_3 + p_4)^2$. The quantity s is related to the centre-of-mass 3-momentum q by $s = 4(m^2 + q^2)$. The centre-of-mass scattering angle, θ, between the spatial directions of particles 1 and 3 is given by $\cos \theta = 1 + 2t/(s - 4m^2)$. Clearly in an actual process $s > 4m^2$ and z lies between -1 and $+1$, so that t is negative.

There is good reason to believe that A is an analytic function of the variables s and t, with certain isolated singularities. Thus, although it represents a physical process only in a certain region of its variables (the "physical region"), it can be continued to the whole of the complex s- and t-planes. There are two reasons why we do this. Firstly, it permits us to use Cauchy's theorem to write integral representations for A ("dispersion relations") and secondly it allows us to continue $A(s, t)$ to the physical region of the "crossed channel", i.e. where s and t are interchanged so that $s < 0$ and $t > 4m^2$. The property of "crossing", which in some theoretical models can be derived and is always assumed to be true, then tells us that this function $A(s, t)$ describes also the crossed process $1 + \bar{3} \to \bar{2} + 4$. Here the bars over the particle labels indicate that we should change the sign of the additive quantum numbers associated with the particles, e.g. if 3 has charge $+1$ then $\bar{3}$ has charge -1. Clearly this is necessary so that $1 + \bar{3} \to \bar{2} + 4$ will be allowed by quantum numbers if and only if the original process $1 + 2 \to 3 + 4$ is allowed. The fact that the same function then describes two processes is a major constraint on the form of this function.

We now introduce the Legendre polynomials $P_l(z)$, $l = 0, 1, 2$, etc., which are complete and orthogonal in the internal $-1 \leq z \leq +1$. Thus we can expand

$$A(s, t) = \sum_{l=0}^{\infty} (2l + 1) a_l(s) P_l(z) \qquad (1)$$

where the expansion coefficients $a_l(s)$, known as partial wave amplitudes, are uniquely defined by this equation. The orthogonality property permits us to invert (1) to give

$$a_l(s) = \tfrac{1}{2} \int_{-1}^{+1} dz \, A(s, t) P_l(z), \quad l = 0, 1, 2, \text{ etc.} \qquad (2)$$

As is well known (1) gives a decomposition into contributions of given angular momentum, the latter being given by $\hbar\sqrt{\{l(l+1)\}}$.

The series in (1) converges in an ellipse in the complex z-plane given by $\text{Im } \theta < \Gamma$, for some $\Gamma > 0$. This is known as the "Lehmann-ellipse" and it clearly contains the s-channel physical region ($\text{Im } \theta = 0$). However, (2) cannot be used in the crossed-channel physical region and is not suitable for large values of $|t|$. It can be put in a form more suitable for continuation by means of the Sommerfeld-Watson transform, first used in this connection by T. Regge.† In order to make this transformation we must define an analytic continuation of $a_l(s)$ away from the "physical" values of l, i.e. 0, 1, 2, etc. In fact we define two functions $a^\pm(l, s)$ by

(i) $a^\pm(l, s) = a_l(s)$, $l = $ even/odd integer $\geq L$, for some $L \geq 0$;
(ii) $a^\pm(l, s)$ is an analytic function of l for $\text{Re } l \geq L$;
(iii) $a^\pm(l, s) < O(e^{\lambda|l|})$ in $\text{Re } l > L$ for some $\lambda < \pi$.

Although it is not immediately obvious, it can be proved, under certain reasonable assumptions, that functions defined by the above properties exist. A theorem due to Carlson tells us that, if the $a^\pm(l, s)$ exist, they are unique. Note that the same functions can be defined by taking any $L' > L$ in the definition, so for the $a^\pm$ to exist it is clearly necessary that the $a_l(s)$ in $L < l < L'$ are "determined" by the $a_l(s)$ in $l > L'$, for any L'. In cases where there are no "exchange forces", i.e. forces which in some sense exchange the two scattering particles, the two functions $a^\pm$ are equal and the distinction between the $+$ and $-$ is unnecessary. In general, however, the values of $a_l(s)$ for even and odd l can be completely unrelated so it is necessary to continue away from the even and odd integers separately as in (i) above.

Condition (iii), which is imposed in order to make the Sommerfeld-Watson transformation possible, is clearly vital in the uniqueness of the $a^\pm$ since without this condition we could, for example, add any multiple of $\sin \pi l$. The expression (2), with the usual definition of P_l for non-integral values of l, does not satisfy the condition (iii), so it cannot be used as an explicit expression for the $a^\pm$. If $A(s, t)$ satisfies a dispersion relation in t, then an explicit expression for the $a^\pm$ can be obtained (the "Froissart-Gribov" formula).

Using the functions $a^\pm(l, s)$ we can perform the Sommerfeld-Watson transformation on (1) to obtain

$$A(s, t) = \sum_{\text{Integers} < L} (2l + 1) a_l(s) P_l(z)$$

$$- \frac{1}{2i} \int_{L - i\infty}^{L + i\infty} \frac{(2l + 1) \, dl}{\sin \pi l} \{a^+(l, s) [P_l(-z)$$

$$+ P_l(z)] + a^-(l, s) [P_l(-z) - P_l(z)]\}. \quad (3)$$

---

† T. Regge, *Nuovo Cimento* **14**, 951 (1959).

The equivalence of (3) and (1) follows by closing the integration contour to the right (this can be shown not to contribute anything because of (iii)), and then using Cauchy's theorem together with the formula $P_l(z) = (-1)^l P_l(-z)$. The integral in (3) is convergent in a considerably larger t range than the sum in (1).

To proceed further we continue $a^\pm(l, s)$ into the region $\text{Re } l < L$. Apart from the unlikely possibility of a "natural boundary" this will certainly be possible. We assume for the moment that at positive integers and zero the $a^\pm(l, s)$ still satisfy (i) above. In general we expect to meet singularities of $a^\pm(l, s)$. In particular we know that there will be poles—these are called Regge poles. A Regge pole is said to have even or odd signature according to whether it occurs in a^+ or a^-. Suppose that $a^\pm(l, s)$ have simple poles at $l = \alpha_i^\pm(s)$, for $i = 1, 2, 3$, etc., with residues $\beta_i^\pm(s)$. If we move the integration contour in (3) to the left we pick up also the contribution of some of these poles, so instead of (3) we obtain

$$A(s, t) = -\pi \sum_{\substack{i \\ \text{Re}\alpha_i^\pm > M}} (2\alpha_i + 1) \beta_i^\pm$$

$$\times \frac{P_{\alpha_i^\pm}(-z) \pm P_{\alpha_i^\pm}(z)}{\sin \pi \alpha_i^\pm} + B \quad (4)$$

where B is the background integral which is the same as that in (3) but taken along $\text{Re } l = M < L$. We assume $M < 0$ so that there is no residual summation as in (3).

The chief usefulness of (4) lies in the fact that, for $\text{Re } \alpha > -\frac{1}{2}$,

$$P_\alpha(z) \underset{|z| \to \infty}{\sim} z^\alpha. \quad (5)$$

Thus, for sufficiently large z, the terms of the summation in (4) which have the largest $\text{Re } \alpha_i^\pm$ dominate, and these in turn dominate over the background integral (which behaves like z^M). Recalling that large z is equivalent to large t we see that (4) gives a high energy (i.e. high t) expansion of the t-channel process $(1 + \bar{3} \to \bar{2} + 4)$. The restriction to $\text{Re } \alpha > -\frac{1}{2}$ noted before (5) is a technical one which can be overcome, again under certain reasonable assumptions. Of course our high-energy expansion is useless unless we know something about the expected Regge poles, i.e. the $\alpha_i^\pm(s)$, and also something about the possible presence of other singularities of $a^\pm(l, s)$. We now discuss these two problems.

The functions $\alpha^\pm(s)$ (known as Regge trajectories) are real, analytic functions of s, with right-hand cuts starting at $s = 4m^2$. Unless $\alpha(s)$ is a constant then a pole in l at $l = \alpha(s)$ gives rise to a pole in s at the solution of this equation. If we take $l = 0, 1, 2$, etc., then such s-plane poles correspond to bound states, for s real and $0 < s < 4m^2$, or to resonances, when the pole must be below the s-plane cut from $4m^2$ to ∞. We refer to either bound states or resonances as "particles". The value of s is the mass squared of the particle and the value of l is its spin.

Thus it is plausible that there will be a Regge trajectory corresponding to each particle. More precisely there will be a Regge trajectory in the s-channel $(1 + 2 \to 3 + 4)$ corresponding to every particle with the quantum numbers of the s-channel, i.e. those of $1 + 2$ which must equal those of $3 + 4$. Note that a trajectory is seen as a particle, or several particles, in the region $s > 0$, whereas it contributes to the high-energy behaviour in the crossed channel for $s < 0$. Hence, to correlate these two channels we must be able to extrapolate between the two regions. This is normally done by assuming that $\alpha(s)$ is a linear function of s with a common slope (approximately 1 GeV^{-2}). The evidence for this assumption comes from the fact that, for several different sets of quantum numbers, particles have masses and spins that lie on linear trajectories of this slope (a plot of mass squared against spin is known as a Chew–Frautschi plot), and that when extrapolated such trajectories appear to dominate high energy effects in the crossed channel (see Fig. 1).

A further effect shows itself in the Chew–Frautschi plots, namely that in many cases a^+ and a^- appear to have particles lying on common trajectories. This is known as "exchange degeneracy".

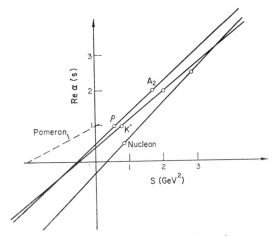

Fig. 1. *Some typical Regge trajectories.*

The approximate linearity of Regge trajectories was an unexpected feature and is hard to understand theoretically. The "quark-model" of elementary particles offers a possible explanation. The phenomena of "exchange-degeneracy" suggests that in many cases exchange forces are small. This again can be understood in terms of the quark model.

In order to explain high-energy scattering it is necessary to assume, in addition to these linear trajectories, the existence of a singularity having $\alpha(s = 0) = 1$, and occurring only in those channels with zero quantum numbers. Such a singularity gives an amplitude behaving like t^1 in the forward direction ($s = 0$). From the optical theorem this then gives the observed asymptotically constant total cross-sections. This trajectory, which does not appear to give any observed particles in the $s > 0$ region, is called the "Pomeron" after I. Pomeranchuk who presented some theorems about processes with asymptotically constant total cross-sections.

In Fig. 1 we give some typical Regge trajectories. The energy dependences of a large number of scattering processes have been observed and in all cases the energy dependence in the forward direction is compatible with $t^{\alpha(0)}$ where $\alpha(0)$ has the value corresponding to the highest lying Regge trajectory with the appropriate cross-channel quantum numbers. Thus in the process $\pi^- p \to \pi^0 n$, the cross-channel has ϱ quantum numbers so one obtains a $t^{\frac{1}{2}}$ dependence, whereas in the backward direction for this process, i.e. for the process $\pi^- p \to n \pi^0$ one has baryon quantum numbers and a $t^{-\frac{1}{2}}$ behaviour. One expects these results to hold for laboratory momenta above 5 GeV/c and data are available in most cases up to about 25 GeV/c, and in a few cases up to much higher energy.

A second prediction which follows from the dominance of a single Regge pole is known as "shrinkage". According to (4) and (5) the contribution of the leading Regge pole has the form

$$\Gamma(s) \, [(-t)^{\alpha(s)} \pm (t)^{\alpha(s)}]$$
$$= \Gamma(s) \, (e^{-i\pi\alpha} \pm 1) \, t^{\alpha(0)} \, e^{\alpha' s \log t} \qquad (6)$$

where we have taken a linear trajectory with slope $d\alpha/ds \equiv \alpha'$. The last factor in (6) gives an exponential fall-off with increasing $|s|$ (recall that in the t-channel physical region s is negative), i.e. with increasing angle. The coefficient of this exponential increases logarithmically with t; thus, as the energy is increased the forward peak becomes narrower. This phenomena, which is independent of the form of $\Gamma(s)$ (usually in fact this is also an exponential), is observed in many cases where a single trajectory is expected to dominate and the rate of shrinkage is compatible with the value of α' seen in Fig. 1. The observed shrinkage associated with the Pomeron trajectory appears to require a trajectory slope less than one-half of the slope of the other trajectories.

The residues of Regge poles are expected to "factorize", i.e. the residue of a given pole in the process $1 + 2 \to 3 + 4$ has the form

$$\beta_{12 \to 34} = \gamma_{12} \gamma_{34}. \qquad (7)$$

Clearly equations of this type enable one to relate the Regge pole contributions to different processes. In some cases these relations appear to work well but there are significant failures. In particular, when β has a zero then clearly either γ_{12} or γ_{34} must have a zero, so the residue for, say, $1 + 2 \to 5 + 6$ or $3 + 4 \to 5 + 6$, for any 5 and 6, should also have a zero. This is not always in agreement with the data.

These, and other failures, lead one to consider the possibility of other singularities in $a^{\pm}(l, s)$. It is known that there must be branch points, at positions closely associated with the poles, but it is not clear how large the appropriate discontinuities are. Various procedures for calculating them, e.g. the absorptive model, have been proposed but none is convincing when confronted with the data. Fixed poles, at special points, must also be present but it is not known whether these should have any observable effects. The possibility of a Kronecker delta (δ_{ll_0}) is interesting. Such a singularity would give rise to a s^{l_0} behaviour (with l_0 independent of t). A term $\delta_{ll_0}/(s - m^2)$ with l_0 an integer would correspond to a particle of spin l_0 not lying on a Regge trajectory. There are no obvious candidates for such particles. In fact, of the familiar strongly interacting particles, all except possibly the π (the contribution of which is not easy to isolate) must lie on trajectories.

This brief article is no more than an introduction to Regge theory, which has become a vast subject. The introduction of spin and unequal masses considerably complicates the discussion and introduces some new features. Detailed fits to the data are complicated and it is seldom that accurate *predictions* have been made—on the contrary new experiments have usually required modifications of existing ideas. Nevertheless, Regge theory has played a dominant part in the study of strongly interacting particles (not only in the high energy region) for the past 10 years. Its role shows no sign of diminishing.

Bibliography

For the early development of the subject and for non-relativistic potential scattering see

NEWTON, R. G. (1964) *The Complex j-plane*, Benjamin, New York.

SQUIRES, E. J. *Complex Angular Momenta and Particle Physics*, (1963) Benjamin, New York.

For a detailed discussion of the formalism see

COLLINS, P. D. B. and SQUIRES, E. J., (1968) Regge poles in particle physics, *Springer Tracts in Modern Physics*, Vol. 45, Springer-Verlag, Berlin.

For recent reviews and the current state of phenomenology see

BARGER, V. and CLINE, D. *Phenomenological Theories of High Energy Scattering*, (1969) Benjamin, New York.

COLLINS, P. D. B. (1971) *Physics Reports*, **1C**, 105.

JACKSON, J. D. (1970) *Rev. Mod. Phys.* **42**, 12.

E. J. SQUIRES

RESONANCE IN THE ORBITS OF THE EARTH AND AN ASTEROID

Introduction

The asteroid 1685 Toro was discovered in 1948 by C. A. Wirtanen at the Lick Observatory. It belongs to the Apollo group, which means that it has perihelion inside the Earth's orbit. Toro has recently attracted much attention because it is one of the known planetary objects which come very close to the Earth and that the Earth and Toro repeatedly appear in symmetric positions relative to each other. The consequences of the latter fact will be discussed in this article.

Present Orbit

Toro's present period is very close to 1·6 years which means that it makes five revolutions around the Sun while the Earth makes eight. The orbital elements are:

Elements 1972

Semi-major axis	1·3677 A.U.
Period	1·5995 years
Eccentricity	0·4360
Perihelion	0·7715 A.U.
Aphelion	1·9640 A.U.
Inclination	9·3719 deg
Ascending node	273·892 deg
Argument of perihelion	126·697 deg
Longitude of perihelion	40·589 deg
Latest perihelion	1972 October 5·6

The elements refer to the time after the close encounter to the Earth in August 1972. The magnitude during the close approach to the Earth was 13·5 when the distance was 0·134 A.U.

The present orbits of Toro, Venus and the Earth are oriented as shown in Fig. 1 (projection to the ecliptic plane). Due to the 8 : 5 commensurability of the periods the Earth can be in one of five different

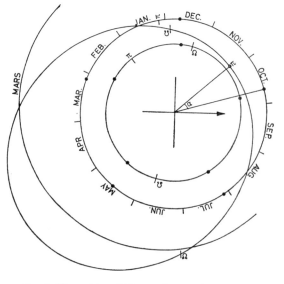

Fig. 1. The orbits of Venus, the Earth and Toro. The possible positions of Venus and Earth when Toro passes its perihelion are marked.

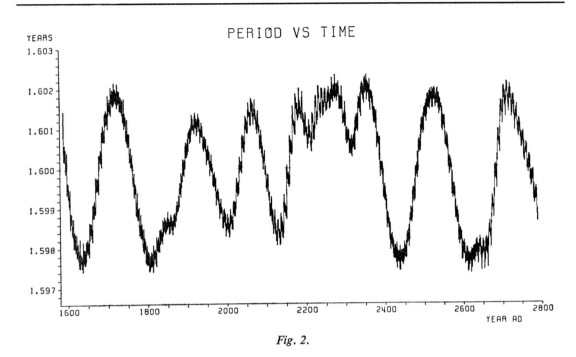

Fig. 2.

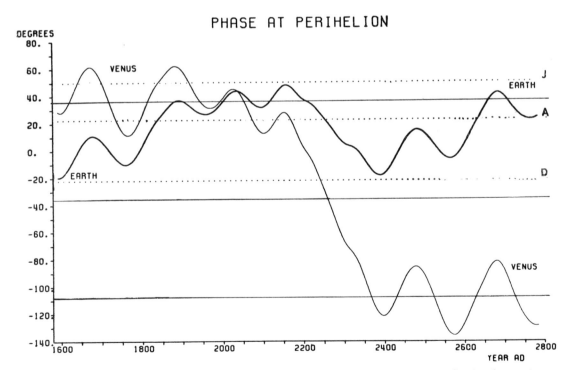

Fig. 3. The difference in longitude between Toro and the Earth and Venus at the October perihelion every 8th year.

positions when Toro is at perihelion and these five points are marked in Fig. 1. If there is a resonance the mean position of these points must be stationary. Since Toro's period in addition is rather close to a 13 : 5 commensurability with Venus' period, also this planet is expected to appear at one of five different, quasi-stable positions (also marked in Fig. 1) relative to Toro at perihelion. The longitudinal distance between Toro at perihelion and the Earth at the October position (angle α in Fig. 1) will be studied below. It can be stated that we have an orbital resonance if this angle is restricted to certain values. Since the Earth positions in Fig. 1 form a regular pentagon, the variation of the angle α must have a peak-to-peak amplitude less than 72° if there shall be a stable resonance.

The present orbital elements also imply that Toro comes close to the Earth twice in every sequence of 8 years. These close encounters will occur in the months of August and January, see Fig. 1. The balance between the counteracting forces during these close encounters could keep Toro's orbit in resonance with the Earth's but it will be seen below that the situation is more complicated.

Libration

The standard method for studying orbital resonances is based on the secular variations due to the slowly changing forces from the major planets. However, in the present case, where it is known that Toro has close approaches to both the Earth and Venus, this method is very unreliable.

To investigate the proposed resonance Toro's orbit has been integrated over 1200 years, from A.D.1580 to A.D.2800. Cowell's method of integration with automatically varied step-length was used to achieve a very high accuracy: after 100 years which requires 10^4 integration steps the relative error is about 10^{-8}, i.e. the absolute error in the position of the Earth is 10^{-5} A.U. With this method it is impossible to integrate over very long periods because the computing time gets exceedingly long and, ultimately, the error accumulates to an inacceptable level.

The 1200 year integration has produced the plots in Figs. 2 and 3. More detailed results will be published elsewhere (Danielsson and Mehra, 1973). Figure 2 shows how the period varies with time and Fig. 3 shows the relative phase between Toro at perihelion and the Earth (α in Fig. 1) and Venus (similarly defined). The oscillation of the period is caused by the repeated interactions with the Earth; the acceleration and decelerations of Toro dominate alternately. The obvious irregularities in the curve are due to perturbations from close encounters with Venus and to Earth forces which temporarily become insufficient to keep Toro's orbit in equilibrium. The latter is due to the fact that, with the present inclination and node, the ecliptical latitude of Toro's orbit is rather large for August longitudes (Fig. 1). Thus Toro can slip behind the Earth at the August encounter.

The variation of the phase angle (α) in Fig. 3 shows a quite complicated pattern. The phase of the Earth here is in a sense the integral of the curve in Fig. 2. For a stable resonance it is required that the phase is constant on the average. The horizontal lines mark possible stable equilibrium phases for both the Earth and Venus ($\alpha = 36° \pm n \times 72°$). The phase of the Earth could also have stable values between these lines ($\alpha = \pm n \times 36°$). It can be seen in the figure that the phase of the Earth is constant on the average but that it does not return to the same phase until after about 800 years. The average period of revolution taken over this 800-year period of the libration differs from 1·6 years by less than 10^{-5} years.

The most remarkable effect seen in Fig. 2 is that Toro seems to be locked in resonance with Venus for part of the time. However, Venus period is 0·11 days too short for commensurability so on the average the Venus phase is bound to drift about 18° per century. Thus it will drift from the first equilibrium phase to the third in just 800 years.

Mechanism of Libration

The Toro orbit and its very complicated libration is best illustrated in a coordinate system rotating with the Earth, see Fig. 4. It takes Toro 8 years (i.e. five revolutions around the Sun) to complete this loop (counterclockwise, see arrow). In the same figure Venus makes five revolutions (clockwise) along the inner circle in 7·993 years. (The eccentriaties of the Earth and Venus' orbits are neglected in this figure.) The Toro loop is not exactly closed; it is oscillating around the centre (Sun) as described by the variation of the angle α in Fig. 3. In Fig. 4 this is illustrated by keeping the Toro loop fixed and letting the Earth oscillate along the marked arc. At the same time Venus will move (i.e. when the Toro loop is kept fixed) so that it temporarily remains in a semi-stable phase relative to Toro in the marked regions of the Venus circle.

Attempting to visualize the libration one can say that the effect of the perturbation of Toro by the Earth and Venus is that the points representing the Earth and Venus "repels" the orbit of Toro in Fig. 4. Also Venus is then located in a (semi-)stationary point, viz. the meeting point of Venus with Toro in the rotating coordinate system. However this point moves relative to the Earth with the angular speed 18° per century. The "repelling forces" should be applied on the points of the Toro orbit which are closest to the two planets.

The repulsion can be understood in the following way. Consider the rotating coordinate system around the point A in Fig. 4 (August encounter). When the small mass of Toro passes the Earth it will be at-

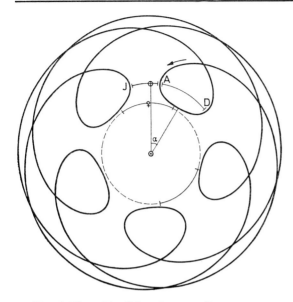

Fig. 4. The orbit of Toro in a coordinate system rotating with the Earth. The fact that this loop oscillates is accounted for by letting the Earth oscillate along the arc J–A–D. At the same time Venus moves along the inner circle, having a temporarily stable phase relative to Toro on the solid parts of the line.

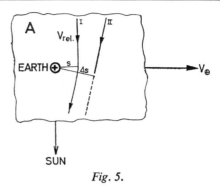

Fig. 5.

tracted and the direction of the relative velocity will be slightly changed. Since Toro here passes ahead of the Earth (see Fig. 5) the change in relative velocity will decrease the total velocity and energy. Thus the period decreases and Toro will tend to arrive to the same point earlier relative to the Earth and hence further away at the subsequent encounter. The remaining orbital parameters change so that Toro leaves and comes back to the region under consideration along parallel paths. At the January encounter (point J in Fig. 4), on the other hand, Toro passes behind the Earth so that its energy and period are increased. Thus Toro tends to come back to the same point later at the subsequent encounter and again it tends to move away from the Earth (cf. Alfvén and Arrhenius, 1970, p. 397). With this model the phase variations in Fig. 3 can be understood qualitatively.

At present (A.D.1972) the Earth and Venus are situated as marked in Fig. 4, cf. also Fig. 3. The Toro loop is oscillating so that the Earth force dominates alternately on the points A and J while the Venus force acts near the perihelion. The Toro loop performs a small amplitude oscillation with the points A and J alternately approaching the Earth. Due to the drift of the Venus phase, the point representing Venus will move towards the perihelion between A and D, and "pushing" stronger on the Toro loop with decreasing distance. However, it cannot push the J point past the Earth because the Earth pushes stronger. When Venus has passed the perihelium (A.D.2200) it pushes the Toro loop in the other direction and now it can push the point A past the Earth (A.D.2255) because the rather large ecliptic latitude at this point makes the force from the Earth smaller and less effective (directed out of the ecliptic plane). In this way the Toro loop will swing until stopped by the Earth (A.D.2390) at the point D (here the latitude is small again). When this point is reached Venus has already drifted past the next perihelium (A.D.2315). In this region Venus will lock the Toro loop to its phase and ultimately drive the point A past the Earth again (A.D.2630) and back to the starting conditions.

Conclusions

From the limited but accurate computations performed so far it appears that Toro has an average period which is almost exactly commensurable with the Earth's in the ratio 8 : 5. The main period of libration is about 800 years and the peak-to-peak amplitude of the libration is about 65°. Once every main libration period Toro is captured into a resonance with Venus instead of the Earth and this intermittent resonance lasts for about 450 years, i.e. more than half the time of the main libration period. Nevertheless Toro is on the average in resonance with the Earth and not with Venus, cf. Fig. 3.

The implication of the present results is that a completely new type of orbital resonance in the planetary system has been discovered. While in previously known resonances close encounters are avoided—the best known example is Neptune–Pluto (Cohen and Hubbard, 1964)—the resonance between the asteroid Toro and the Earth and Venus entirely relies on close encounters. This is of course a natural consequence of the very small sphere of influence of any terrestrial planet.

There is no reason to believe that Toro is a unique case; also a few other Apollo asteroids have periods very nearly commensurable with the Earth's, others show a very regular variation of the period, indicating an orbital resonance of some kind.

The stability of the Toro resonance over longer periods of time cannot yet be judged with certainty. Although the secular variation of the orbital elements can change the situation—the stability being particularly sensitive to the inclination—it is quite possible that the Jupiter interaction will be overruled by the strong periodic interactions with the Earth and Venus.

One of the objectives for studying orbits like Toro's in detail is that this is one of the possibilities for a precapture orbit for the Moon (Alfvén and Arrhenius, 1969). Another is that the existence of orbits like Toro's may be important for determining the lifetime of Apollo asteroids.

Bibliography

ALFVÉN, H. and ARRHENIUS, G. (1969) *Science* **165**, 11.
ALFVÉN, H. and ARRHENIUS, G. (1970) *Astrophys. Space Sci.* **8**, 338.
COHEN, C. H. and HUBBARD, E. C. (1964) *Astron. J.* **70**, 10.
DANIELSSON, L. and MEHRA, R. (1973) *Proceedings of the First European Astronomical Meeting, Athens, Sept. 4–9, 1972*, Springer Verlag.

LARS DANIELSSON

S

SIGNAL DETECTION, STATISTICAL THEORY OF

I. Introduction

Signal detection is the extraction of useful information or intelligence from observable data, that is data to which an observer has access. If the information is corrupted by noise which can be modelled as random, then a systematic approach to extraction of the signal can be made by applying requisite statistical tools. The synthesis of receivers to extract the information from the data in an optimum way requires such application. The application of statistical tools for both the synthesis and analysis of receivers to extract information constitutes the statistical theory of signal detection.

Signal detection is an inherent function in communication systems, an example of which is shown in Fig. 1. Here the digital communication system is modelled as consisting of a source in which information originates, a receiver at whose output the information is intended for use, and a link, the transmission medium or channel, connecting each. In this system one of M possible pieces of information is intended to be communicated. The information is encoded in a form suitable for transmission over the channel. In the channel the signal itself may undergo some distortion and, furthermore, noise may be unavoidably added to the signal causing further distortions. The resulting observable data at the receiver may then be so distorted that it is uncertain as to which signal, $s_i(t)$, and hence which piece of information, i, was sent. Communication or information theory is concerned with all aspects of the model shown in Fig. 1 including the message source and the encoding of information. Signal detection theory, however, is nominally concerned with synthesizing and analysing the receiver for a given signal encoding scheme. Therefore, for the example at hand, the receiver's function is to operate on the distorted data and decide which of the M pieces of information was sent.

Radar provides other examples. In a radar system electromagnetic energy is radiated by an antenna into space. If this energy irradiates an object some of the energy will be redirected back toward a receiving antenna. However, the receiving antenna will receive not only the signal if it is present but also noise. Therefore, from the observable data, it cannot be determined with certainty whether a signal (and consequently a target) is present or not.

In each of the above examples the form of the encoded signal, but not necessarily specific parameter values, was known to the receiver. At the receiver the uncertainty involved which of two or more pieces of information was present. In other applications the form of the encoded signal is unknown. Examples can be found in the detection of seismic events, and in sonar systems where acoustic signals are generated by marine biological sources as well as merchant vessels. In these applications it is possible that the signal is as random as the noise. As a consequence the receiver may have as an integral part an estimator of signal parameters such as amplitude, phase, frequency, or time of arrival. The random nature of both the signal and the noise do not permit an arbitrarily precise determination of the parameters.

In all of these cases a signal known precisely, or imperfectly, or so completely unknown in form that it is best modelled as random is distorted by another random process, the noise, so that the observable signal is itself a random process. It is not surprising then that a statistical theory of signal detection should be required.

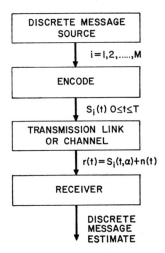

Fig. 1. Discrete message communication system.

II. Hypothesis Testing

The fundamental statistical tool of detection theory is hypothesis testing. The hypotheses are the alternative decisions which are possible. For example, in a communication system with i possible signals, each signal could be represented by a hypothesis. In the radar case there can be two hypotheses, one that there is no signal present, and the other that a signal is present. Equivalently, the hypotheses are respectively that the observable signal contains noise only, and the observable signal contains signal plus noise. For a hypothesis test, a statistical description of each hypothesis and a criterion for selecting a hypothesis are necessary. The object of the hypothesis test is to design a receiver to operate on the observable data in such a way that a specific criterion of goodness is satisfied. The resulting receiver is, by definition, optimum for that criterion and the assumptions used in its derivation, and no other non-equivalent receiver could be constructed to better satisfy the criterion. For these purposes, a receiver is a mathematical description of the operations to be performed on the observable data.

A versatile criterion is that of Bayes. Many other criteria can be structured as being a subset of the Bayes solution. To demonstrate it consider the discrete communication system depicted in Fig. 1 to be specialized to a binary system. Therefore $r(t) = s_i(t) + n(t)$ for $i = 1, 2$. Two hypotheses are possible, that

$$H_1: r(t) = s_1(t) + n(t)$$
$$H_2: r(t) = s_2(t) + n(t) \quad (1)$$

The observable data is $r(t)$ and from operations on this data a decision, D_1 or D_2, is to be made as to which hypothesis, H_1 or H_2, is true. For each hypothesis there is an assumed statistical description of the data, $p_1(\mathbf{r})$ and $p_2(\mathbf{r})$ respectively. Here $p_i(\mathbf{r})$ is the probability density function of the data, generically represented by $\mathbf{r}$, given that hypothesis H_i is true. Each hypothesis is assumed to occur with *a priori* probability $P(H_1)$ and $P(H_2)$ and, since only two hypotheses are possible, $P(H_1) + P(H_2) = 1$.

In deciding hypotheses the following circumstances are possible:

Decide D_1 when H_1 is true
Decide D_1 when H_2 is true
Decide D_2 when H_1 is true
Decide D_2 when H_2 is true

Denote the corresponding probabilities as $P(D_1|H_1)$, $P(D_1|H_2)$, $P(D_2|H_1)$ and $P(D_2|H_2)$. (The vertical bar is to denote a conditional probability.) The first and fourth terms represent correct decisions whereas the second and third represent errors.

Bayes Criterion

Each type of error may have different consequences—some errors are more serious than others. To accommodate this, cost functions if known may be assigned to the above alternatives: C_{11}, C_{12}, C_{21}, C_{22}, ($C_{ij} \geq 0$, $C_{ij} > C_{ii}$) where C_{ij} is the cost of deciding that H_i is true when, in fact, H_j is true. Then, the average cost associated with any receiver is

$$\overline{C} = P(H_1) \left[P(D_1|H_1) C_{11} + P(D_2|H_1) C_{21} \right]$$
$$+ P(H_2) \left[P(D_1|H_2) C_{12} + P(D_2|H_2) C_{22} \right] \quad (2)$$

The Bayes criterion is to design a receiver to minimize this average cost. Such a receiver will be optimum in this sense. By actually carrying out the indicated operations, including the minimization, produces the Bayes optimum receiver:

$$\lambda(\mathbf{r}) \underset{H_1}{\overset{H_2}{\lessgtr}} \lambda_0 \quad (3)$$

where

$$\lambda(\mathbf{r}) = \frac{p_2(\mathbf{r})}{p_1(\mathbf{r})} \quad (4)$$

and

$$\lambda_0 = \frac{P(H_1)(C_{21} - C_{11})}{P(H_2)(C_{12} - C_{22})} \quad (5)$$

The notation in eqn. (3) implies a decision to choose hypothesis H_2 if $\lambda(\mathbf{r}) > \lambda_0$, and to choose H_1 otherwise. The function in eq. (4) is very important in detection theory and is given the name *likelihood ratio*. Similarly the receiver is called a likelihood-ratio receiver or likelihood ratio test. The functions $p_i(\mathbf{r})$ are called likelihood functions. The term λ_0 is a threshold against which the likelihood ratio is compared and a decision made. The likelihood ratio is so important because for most criteria of interest, Bayes and those discussed below, it is the optimum receiver and only the threshold changes.

Minimum error probability criterion

In communication systems it is usual to minimize the average error probability

$$P_e = P(H_1) P(D_2|H_1) + P(H_2) P(D_1|H_2)$$

Furthermore, no cost is associated with correct decisions, and errors of each kind are equally costly. Thus $C_{11} = C_{22} = 0$; $C_{12} = C_{21} = 1$. These cost functions produce $\overline{C} = P_e$ from which it follows that this is a special case of the Bayes criterion and therefore the optimum receiver is

$$\lambda(\mathbf{r}) \underset{H_1}{\overset{H_2}{\gtrless}} \frac{P(H_1)}{P(H_2)}$$

This test is also called the *ideal observer test*. It is furthermore optimum in the maximum *a posteriori* probability sense. That is, of the *a posteriori* probabilities, $P(H_1|\mathbf{r})$ and $P(H_2|\mathbf{r})$, it selects that hypothesis corresponding to the maximum of these.

Neyman–Pearson criterion

In some applications, quite notably radar and sonar, neither cost functions nor *a priori* probabilities are known so a direct application of Bayes criterion is not apparent.

In the radar case the usual two hypotheses are that no signal is present, and a signal is present. It is conventional to denote these as H_0 and H_1, respectively. Two probabilities are of particular importance, the probability of detection $P_d = P(D_1|H_1)$, and the probability of false alarm $P_{fa} = P(D_1|H_0)$. Using this convention, it may be stated that the Neyman–Pearson criterion is to maximize the probability of detection for a given probability of false alarm. It may be shown that this criterion is yet another special case of the Bayes criterion so that the optimum receiver must be a likelihood ratio test, specifically

$$\lambda(\mathbf{r}) \underset{H_0}{\overset{H_1}{\gtrless}} \eta \qquad (6)$$

where η is chosen so that the constraint on P_{fa} is satisfied.

Multiple alternative hypothesis testing

The preceding material relating to binary hypothesis testing may be extended to M hypotheses. Using the Bayes criterion the optimum receiver selects that hypothesis H_j for which

$$\lambda_j = \sum_{i=1}^{M} C_{ji} p_i(\mathbf{r}) P(H_i)$$

is a minimum.

Composite hypothesis testing

If an hypothesis is a function of an unknown parameter which is itself a random variable for which an *a priori* probability density function is known, the hypothesis is treated much as in the preceding cases. For example, denote the unknown parameter or set of parameters by θ, then

$$p(\mathbf{r}) = \int_{\{\theta\}} p(\mathbf{r}|\theta) p(\theta) d\theta \qquad (7)$$

where $p(\mathbf{r}|\theta)$ is the conditional probability density of $\mathbf{r}$ given θ, and $p(\theta)$ is the *a priori* probability density for θ defined over the sample space $\{\theta\}$. If this sample space is single valued the hypothesis is *simple*. If it is multivalued the hypothesis is *composite*.

If the *a priori* distribution is unknown alternative procedures are sought. If an optimum test can be found independent of $p(\theta)$, that test is called *uniformly most powerful*. If it is otherwise, the maximum likelihood principle, although not necessarily optimum, may provide some guidance. For this principle, the generalized likelihood function

$$\lambda(\mathbf{r}) = \frac{\max_\theta p_2(\mathbf{r}|\theta)}{\max_\varphi p_1(\mathbf{r}|\varphi)} \qquad (8)$$

is used. In the numerator the value of θ which maximizes $p_2(\mathbf{r}|\theta)$ is used. Similarly in the denominator. Such a value is actually an estimate of the parameter.

III. Detection of Known Signals

A frequently occurring model of an observable signal is the case where $r(t) = s(t) + n(t)$, $0 \leq t \leq T$. Here, $s(t)$ is assumed completely known in form and having known parameters. The noise is additive. Further assume that the noise is white and Gaussian with an autocorrelation function $(N_0/2)\delta(\tau)$. (For white noise the power spectral density, that is, the distribution of noise power in frequency, is constant.) Then the likelihood function is

$$p(\mathbf{r}) = F \exp\left\{-(1/N_0) \int_0^T [r(t) - s(t)]^2 \, dt\right\} \qquad (9)$$

where F is a constant not generally of interest.

As an example consider the binary communication case for which, in the interval $0 \leq t \leq T$,

$$H_1: r(t) = s_1(t) + n(t)$$
$$H_2: r(t) = s_2(t) + n(t)$$

The likelihood ratio receiver is

$$\lambda(\mathbf{r}) \underset{H_1}{\overset{H_2}{\gtrless}} \lambda_0$$

For the communications case, λ_0 is usually unity. Substituting for $p_1(\mathbf{r})$ and $p_2(\mathbf{r})$ the form shown in eq. (9) the receiver may be shown to be equivalent to

$$\int_0^T r(t) s_2(t) \, dt - \int_0^T r(t) s_1(t) \, dt \underset{H_1}{\overset{H_2}{\gtrless}} V_T$$

where

$$V_T = -(\tfrac{1}{2}) \int_0^T [s_1^2(t) - s_2^2(t)] \, dt.$$

The receiver is shown in Fig. 2. Assuming that $P(H_1) = P(H_2)$, the error probability is

$$P_e = \int_\gamma^\infty \frac{1}{(2\pi)^{1/2}} e^{-z^2/2} \, dz \qquad (10)$$

where

$$\gamma = [(1 - \varrho) E/N_0]^{1/2}$$

$$E = \tfrac{1}{2} \int_0^T [s_1^2(t) + s_2^2(t)] \, dt$$

$$\varrho = (1/E) \int_0^T s_1(t) s_2(t) \, dt$$

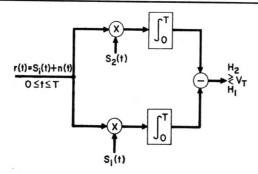

Fig. 2. Correlation receiver.

The error performance is therefore a function of only:
1. The average signal energy (E).
2. The noise spectral level (N_0).
3. The time cross-correlation between signals (ϱ).

The minimum error probability occurs for $\varrho = -1$. This is attained for $s_1(t) = -s_2(t)$ and is known as the ideal binary communication system.

The receiver shown in Fig. 2 is called a correlation receiver. An important equivalent form is shown in Fig. 3 in which the correlators are replaced by linear filters having impulse response $h_1(t) = s_1(T-t)$ and $h_2(t) = s_2(T-t)$, and whose outputs are sampled at $t = T$. These are called matched filters, being matched to time inverted versions of the signal. It may be shown that these filters maximize the

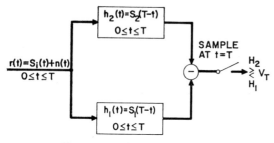

Fig. 3. Matched filter receiver.

ratio of signal power to noise power at their output when the input signal is accompanied by additive white noise. In the case where the noise is not white, the filter impulse response which maximizes the signal-to-noise ratio at its output in response to a signal $s(t) + n(t)$ at its input is given by the solution $h(\tau)$ to the integral equation

$$\int_0^T h(z) R_n(\tau - z) dz = s(T - \tau) \quad 0 \leq \tau \leq T \quad (11)$$

where $R_n(\tau)$ is the noise autocorrelation function. The solution is called the generalized matched filter.

IV. Signals with Random Parameters

It is common in practice to consider signals whose parameters are not all known. For example, in the radar case while the amplitude, frequency, phase, and timing reference of the transmitted signal is known, few of these parameters are known for the reflected signal. The most common unknown parameter is the signal phase and consideration of it is fundamental to working with the other parameters. Consider the single hypothesis for which

$$r(t) = A \sin(\omega_c t + \theta) + n(t)$$

where the phase θ is a random variable uniformly distributed over the range $0 \leq \theta \leq 2\pi$. The additive noise $n(t)$ is white, Gaussian, with autocorrelation function $(N_0/2) \delta(\tau)$. Using eq. (9) the conditional probability density function is

$$p(\mathbf{r}|\theta) = F$$
$$\times \exp\left\{-\frac{1}{N_0}\int_0^T [r(t) - A \sin(\omega_c t + \theta)]^2 dt\right\}$$

Using eq. (7) and averaging over the interval $0 \leq \theta \leq 2\pi$, produces

$$p(\mathbf{r}) = F \exp\left\{-\frac{1}{N_0}\int_0^T r^2(t) dt\right\}$$
$$\times \exp\{-A^2 T/2N_0\} I_0\left(\frac{2Aq}{N_0}\right) \quad (12)$$

where $I_0(\cdot)$ is the modified Bessel function of zero order, and

$$q^2 = \left[\int_0^T [r(t) \sin \omega_c t \, dt\right]^2 + \left[\int_0^T r(t) \cos \omega_c t \, dt\right]^2 \quad (13)$$

Now consider the radar example where, for $0 \leq t \leq T$,
$$H_1: r(t) = A \sin(\omega_c t + \theta) + n(t)$$
$$H_0: r(t) = n(t)$$

The likelihood ratio receiver is

$$\lambda(\mathbf{r}) = \exp\{-A^2 T/2N_0\} I_0\left(\frac{2Aq}{N_0}\right) \underset{H_0}{\overset{H_1}{\gtrless}} \eta$$

Since the exponential term is a constant, and $I_0(\cdot)$ is monotonic in its argument, it follows that an equivalent test is

$$q \underset{H_0}{\overset{H_1}{\gtrless}} \eta' \quad \text{or} \quad q^2 \underset{H_0}{\overset{H_1}{\gtrless}} \eta'^2$$

where η' is chosen to satisfy the Neyman–Pearson criterion of constrained P_{fa}. Thus the receiver may be implemented as shown in Fig. 4, which is called a quadrature receiver. In the same spirit as for the matched filters, an equivalent receiver (Fig. 5) called the incoherent matched filter receiver may be shown to exist. In this case the matched filter, matched to the signal with arbitrary phase, is followed by an envelope detector. The term *incoherent* refers to a lack of knowledge, or of use, of phase information.

The performance for this example is

$$P_d = \int_\beta^\infty z \exp\left[\frac{z^2 + \alpha^2}{-2}\right] I_0(\alpha z)\, dz \qquad (14)$$

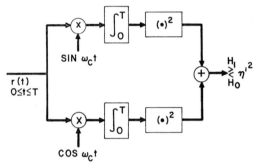

Fig. 4. Quadrature receiver.

Fig. 5. Incoherent matched filter receiver.

where
$$\beta = [-\ln P_{fa}^2]^{1/2}$$
and
$$\alpha = 2E/N_0.$$

The signal energy is E. This is one demonstration that P_{fa} cannot be arbitrarily decreased while simultaneously increasing P_d. The reason is clear if Fig. 6 is consulted. In this example the $p_i(q)$ are simple unimodal distributions. The likelihood ratio receiver has q as its output and the decision is to choose H_1 if it is larger than a threshold. It is seen that as the threshold is increased, thereby decreasing P_{fa}, it is also the case that P_d is decreased. The relationship between P_d and P_{fa} is commonly expressed in a *receiver operating characteristic* curve (ROC curve) as demonstrated in Fig. 7. The performance is a monotonically increasing function of the signal-to-noise energy ratio, E/N_0.

In many circumstances the P_d versus P_{fa} performance is inadequate to accomplish a particular detection function and, simultaneously, the signal energy is limited. In such cases it is common to

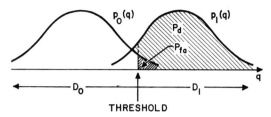

Fig. 6. Decision regions and probability density functions.

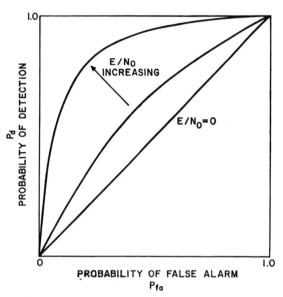

Fig. 7. Receiver operating characteristic (ROC curve).

use more than one signal, sample, or measurement for detection. In radar applications, for example, many signals are transmitted in time sequence and processed before a decision, D_1 or D_0, is made. For example, consider the hypotheses

$$H_1: r_i(t) = A \sin(\omega_c t + \theta_i) + n_i(t) \quad i = 1, \ldots, M$$
$$H_0: r_i(t) = n_i(t) \quad i = 1, \ldots, M$$

For the same set of assumptions as in the preceding problem the likelihood ratio test is

$$\sum_{i=1}^{M} \ln I_0\left(\frac{2Aq_i}{N_0}\right) \underset{H_0}{\overset{H_1}{\gtrless}} V_T$$

where V_T is a threshold chosen to satisfy P_{fa}. For each signal, the q_i are defined as in eq. (13). For low signal-to-noise ratios the likelihood ratio test statistic is approximately

$$\sum_{i=1}^{M} q_i^2$$

and for high signal-to-noise ratios it is approximately

$$\sum_{i=1}^{M} q_i.$$

The operation of summation is commonly referred to as *post-detection integration*, where "detector" in this sense refers to the nonlinear operation of extracting the envelope, q_i. The performance of such receivers improves as M increases. Consequently a reduction of per-signal energy is possible. Asymptotically, as the number of signals doubles the signal energy required to maintain the same P_d and P_{fa} is reduced by the square root of two for this example.

V. Detection of Signals in Coloured Noise

The likelihood functions and ratios expressed above were for the case where the additive noise was white and Gaussian. In most analyses this permits simplifying assumptions about the statistical independence of random variables being sampled. In many circumstances this assumption is violated and the autocorrelation function of the noise is somewhat arbitrary. Because now the distribution of noise in frequency is not uniform, additional statistical tools to handle these circumstances are necessary.

The likelihood ratio receiver requires that the probability density function of the received signal be expressed. The conventional approach is to take "samples" of the received signal, express the probability density function of these samples, and determine the limiting form as the number of samples is taken to the limit. These operations become complicated if the samples are not statistically independent. However, if the noise is Gaussian, statistically independent samples can be found by expressing it in terms of a Karhunen–Loeve expansion. The coefficients of the expansion represent the signal "samples" and may be shown to be uncorrelated which, for a Gaussian signal, also implies statistical independence. Thus, consider the signal $r(t)$, $0 \leq t \leq T$, expanded in a series

$$r(t) = \sum_{i=1}^{\infty} r_k f_k(t) \qquad (15)$$

where

$$r_k = \int_0^T r(t) f_k(t) \, dt. \qquad (16)$$

If the functions $f_k(t)$ are chosen to be the eigenfunctions of the integral equation

$$\lambda_k f_k(t) = \int_0^T f_k(\tau) R(t - \tau) \, d\tau \qquad (17)$$

then the r_k are uncorrelated. The λ_k are the eigenvalues, and $R(\tau)$ is the autocorrelation function of the process $r(t)$. Equation (15), with the r_k and $f_k(t)$ chosen as in eqs. (16) and (17), is the Karhunen–Loeve expansion.

For the case of a known signal in Gaussian noise, $r(t) = s(t) + n(t)$, the likelihood function may be shown to be

$$p(\mathbf{r}) = C \exp \left\{ -\tfrac{1}{2} \int_0^T \int_0^T [r(t) - s(t)] R_n^{-1}(t - \tau) \right.$$
$$\left. \times [r(\tau) - s(\tau)] \, dt \, d\tau \right\} \qquad (18)$$

where C is a multiplier, not normally of interest, and $R^{-1}(\tau)$ is the inverse noise kernel having the property that

$$\int_0^T R_n^{-1}(t, \tau) R_n(\tau, z) \, d\tau = \delta(t - z) \quad 0 \leq t, z \leq T.$$

It may also be expressed in terms of the eigenfunctions and eigenvalues of $R_n(\tau)$. Specifically

$$R_n^{-1}(t, \tau) = \sum_{k=1}^{\infty} \frac{1}{\lambda_k} f_k(t) f(\tau).$$

As an example, consider a receiver used for the binary detection problem of known signals in non-white, Gaussian noise. The hypotheses are, $0 \leq t \leq T$,

$$H_1 : r(t) = s_1(t) + n(t)$$
$$H_2 : r(t) = s_2(t) + n(t).$$

The likelihood ratio test may be shown to be the equivalent of

$$\int_0^T r(t) g_2(t) \, dt - \int_0^T r(t) g_1(t) \, dt \underset{H_2}{\overset{H_2}{\lessgtr}} V_T$$

where V_T is an appropriately chosen threshold. The functions $g_i(t)$ are solutions to the integral equations

$$s_i(t) = \int_0^T R_n(t - \tau) g_i(\tau) \, d\tau \quad 0 \leq t \leq T. \qquad (19)$$

The receiver is therefore analogous to that shown in Fig. 2 with the exception that the correlator inputs are $g_2(t)$ and $g_1(t)$ instead of $s_2(t)$ and $s_1(t)$. The filter functions, $g_i(t)$, are such that they take maximum advantage of where, in the frequency domain, the signal power is large relative to the noise power. In this binary communication case, the performance can be improved by careful selection of the signals, $s_2(t)$ and $s_1(t)$. The greatest performance results when $s_2(t) = -s_1(t)$, as in the white noise case, and also when $s_1(t)$ is chosen equal to that eigenfunction corresponding to the minimum eigenvalue. This has the effect of placing the signal energy where the noise energy is least.

VI. Estimation

Interest commonly extends beyond just detecting the presence of a signal or not, or detecting which of several signals may be present. Given that the presence of a signal is detected, the signal parameters may yield considerable information relative to the signal source. For example, in radar the amplitude of the signal is influenced by the size of the irradiated object, the frequency is influenced by the velocity, and the time of arrival of the signal is related to the distance of the object. Since the signal is corrupted by noise these parameters cannot be determined with arbitrary precision and statistical tools are useful for the synthesis of good estimators.

The estimation problem is not unlike the detection problem. In the event that a parameter is known to have only a discrete set of values, the parameter could be estimated by assigning a hypothesis to each value and performing a hypothesis test. When the parameter takes on a continuum of possible values, and it is not desirable to approximate by a finite set, a different approach but one with the same spirit is taken.

Estimation is also an element useful in synthesizing signal detectors. The generalized likelihood function, eq. (8), has estimation as an explicit operation. As defined below, the estimators used in this expression are called maximum likelihood estimators.

Criteria

Assume that the observable signal is of the form

$$r(t) = s(t, \alpha) + n(t)$$

where α is a set of signal parameters to be estimated. Denote the estimates as $\hat{\alpha}$. In the general case, the estimate is a random variable and will not be identically equal to the parameter so an error is introduced. A scalar cost function $C(\hat{\alpha}, \alpha)$ can be assigned to such an error. Examples include $C(\hat{\alpha}, \alpha) = (\hat{\alpha} - \alpha)^2$, the quadratic cost function, and $|\hat{\alpha} - \alpha|$, the absolute value cost function, etc. The average cost, where averaging is over all values of the parameter and all possible $\mathbf{r}$, associated with a particular estimator $\hat{\alpha}$ is

$$\overline{C}(\hat{\alpha}) = \int_{\{\mathbf{r}\}} C(\hat{\alpha}|\mathbf{r}) \, p(\mathbf{r}) \, d\mathbf{r}$$

The minimization of this cost is the Bayes criterion; the resulting estimator is the Bayes estimator. As an example, if $C(\hat{\alpha}, \alpha) = (\hat{\alpha} - \alpha)^2$ then

$$\hat{\alpha} = \int_{\{\alpha\}} \alpha p(\alpha|\mathbf{r}) \, d\alpha \qquad (20)$$

which is the conditional mean of α given the observation $\mathbf{r}$. This estimator is also optimum for a much broader class of cost functions.

Two other classes of estimators are of importance: The maximum *a posteriori* and maximum-likelihood estimators. The conditional density function for α given a set of observation $\mathbf{r}$ is

$$p(\alpha|\mathbf{r}) = \frac{p(\mathbf{r}|\alpha) \, p(\alpha)}{p(\mathbf{r})}. \qquad (21)$$

If cost functions are not available, a reasonable criterion is to maximize the *a posteriori* probability. Specifically, choose $\hat{\alpha}$ as the value which maximizes $p(\alpha|\mathbf{r})$. In contrast to eq. (20) where the mean of the function is chosen, the maximum *a posteriori* estimate corresponds to the mode of the function.

In the event that $p(\alpha)$ is a relatively flat function of α, then as can be seen from eq. (21), the maximum of $p(\alpha|\mathbf{r})$ will correspond nearly to the maximum of the likelihood function $p(\mathbf{r}|\alpha)$. Specifically, the maximum likelihood estimate is chosen as that value which maximizes $p(\mathbf{r}|\alpha)$.

Properties of estimators

Since the estimates are random variables they have a mean value and, when defined, a variance. If the mean value is equal to the true value of the parameter, then the estimator is said to be *unbiased*. If, as the number of observations used in forming the estimate increases, the estimate converges with certainty to the true value of the parameter, the estimator is called *consistent*. A *sufficient* estimator is one which contains all the information relevant to estimating a parameter; no other non-equivalent estimator can yield more information.

The variance of an unbiased estimator must satisfy he inequality

$$\sigma_{\hat{\alpha}}^2 \geq \frac{1}{E\left\{\left[\frac{\partial \ln p(\mathbf{r}|\alpha)}{\partial \alpha}\right]^2\right\}} \qquad (22)$$

or, equivalently,

$$\sigma_{\hat{\alpha}}^2 \geq \frac{-1}{E\left\{\frac{\partial^2 \ln p(\mathbf{r}|\alpha)}{\partial \alpha^2}\right\}}$$

where $E\{\cdot\}$ is the expectation operator. This bound on variance is the *Cramer–Rao* bound. Any estimator for which the equality holds is called an *efficient* estimator. If an efficient estimator exists, it will also be the maximum likelihood estimator.

For simultaneous estimation of parameters, denote the elements of Fisher's information matrix γ by

$$\gamma_{ij} = E\left\{\frac{\partial \ln p(\mathbf{r}|\alpha)}{\partial \alpha_i} \frac{\partial \ln p(\mathbf{r}|\alpha)}{\partial \alpha_j}\right\}.$$

Then the variance bound for unbiased estimators becomes $\sigma_{\hat{\alpha}_i}^2 \geq \psi_{ii}$ where the ψ_{ii} are elements of the inverse matrix $\gamma^{-1} = \psi$.

For the case $r(t) = s(t, \alpha) + n(t)$, $0 \leq t \leq T$, where $n(t)$ is white, Gaussian noise with autocorrelation function $(N_0/2)\delta(\tau)$, the elements of the information matrix can be expressed as

$$\gamma_{ij} = \frac{2}{N_0} \int_0^T \frac{\partial s(t, \alpha)}{\partial \alpha_i} \frac{\partial s(t, \alpha)}{\partial \alpha_j} dt.$$

For the case where the noise is not white the corresponding elements are

$$\gamma_{ij} = \int_0^T \frac{\partial g(t)}{\partial \alpha_i} \frac{\partial s(t, \alpha)}{\partial \alpha_j} dt$$

where $g(t)$ is the solution to the integral equation

$$s(t, \alpha) = \int_0^T R_n(t - \tau) g(\tau) d\tau \quad 0 \leq t \leq T.$$

Bibliography

DAVENPORT, W. B. and ROOT, W. L. (1958) *An Introduction to the Theory of Random Signals and Noise*, New York: McGraw-Hill.
HELSTROM, C. W. (1960) *Statistical Theory of Signal Detection*, Oxford: Pergamon Press.
KENDALL, M. G. and STUART, A. (1961) *Inference and Relationship, The Advanced Theory of Statistics*, Vol. 2, New York: Hafner.
LAWSON, J. L and UHLENBECK, G. E. (1950) *Threshold Signals*, Radiation Laboratory Series, Vol. 24, New York: McGraw-Hill.
LEHMANN, E. L. (1959) *Testing Statistical Hypotheses*, New York: Wiley.
MIDDLETON, D. (1960) *An Introduction to Statistical Communication Theory*. New York: McGraw-Hill.
MIDDLETON, D. and VAN METER, D. (1955) On optimum multiple alternative detection of signal in noise, *IRE Transactions on Information Theory* (September 1955).
NORTH, D. O. (1963) An analysis of the factors which determine signal noise discrimination in pulsed carrier systems. Reproduced in *Proc. IEEE* (July 1963).
PETERSON, W. W., BIRDSALL, T. G. and FOX, W. C. (1954) The theory of signal detectability, *IRE Transactions on Information Theory* (September 1954).
SLEPIAN, D. (1954) Estimation of signal parameters in the presence of noise, *IRE Transactions on Information Theory* (March 1954).
VAN TREES, H. L. (1968) *Detection, Estimation, and Modulation Theory* Part 1, New York: Wiley.
WOODWARD, P. M. (1964) *Probability and Information Theory with Applications to Radar*, 2nd ed., Oxford: Pergamon.
WHALEN, A. D. (1971) *Detection of Signals in Noise*, New York: Academic Press.

A. D. WHALEN

SPECTROSCOPY, BIOLOGICAL APPLICATIONS OF. The cardinal concern of current biological research is to correlate the functions fulfilled by the macromolecules and macromolecular assemblies present in all living cells—proteins, nucleic acids, and membranes—with their structures. The application of X-ray crystallography—the only direct method of structure determination—to this problem is impracticable for many reasons; also, in solutions there is no absolute method equivalent to X-ray diffraction for the solid state. Elucidation of structure and assignment of conformation is therefore carried out by collating indirect evidence gathered from several—mainly spectroscopic—techniques. Only those applications of spectroscopy which provide probes for structure determination come within the purview of this section. Analytical applications of spectroscopy for the detection and assay of compounds of biological significance will be mentioned only in passing; this aspect has been discussed in detail elsewhere in this Dictionary (Beaven, 1962).

Most of the biological giant molecules are sinuous—bent and folded into three-dimensional structures called *conformations*. In structural studies it is customary to recognize (at least) three levels of organization. The primary structure concerns the linear sequence of the molecular subunits; the term secondary structure refers to any regular and periodic folding of the backbone chain into a three-dimensional structure often held together by attractive forces between subgroups which can engage in hydrogen bonding; the tertiary structure deals with further (and irregular) folding of the secondary structure to interrelate regions of the chain not necessarily in contact with each other, and hence determines the overall molecular shape. According to currently held views, a macromolecule contains residues which are disposed so as to form an "inside", and those which constitute an "outside"; residues which are imbedded inside are not in contact with the solvent, but exposed residues on the outside are.

Spectroscopic techniques have been used mainly, but not solely, to examine the secondary and tertiary aspects of biomolecular structure; they have so far had their greatest use in following conformational changes and in monitoring interactions between the biopolymers and other, usually small, molecules. The applications considered here are based on electronic absorption and emission spectra and vibrational absorption spectra.

Applications Based on Electronic Absorption Spectra

Qualitatively, the ultra-violet absorption spectra of the biological polymers can be accounted for by the absorption characteristics of their aromatic constituents. Quantitatively, however, the spectral features are noticeably, sometimes even appreciably, modified by the secondary and tertiary structure of the giant molecule as well as by the solvent medium.

Detailed disquisition of these differences has contributed considerably to the elucidation of structure of biological macromolecules.

When a protein or a nucleic acid is hydrolysed, the additive absorption of the fragments is found to be higher than the absorption of the intact molecule. For instance, the enzymatic hydrolysis of DNA by DNase is attended by a 30–35% increase in the magnitude of the extinction coefficient at 260 nm. (This finding has been exploited for determining DNase activity.) The percentage decrease in the integrated intensity of an absorption band is termed *hypochromism*. Point hypochromism is the percentage decrease in the extinction coefficient at a given wavelength. *Hyperchromism* refers to the converse phenomenon: an increase in absorption upon degradation.

Quantitative observations on small oligonucleotides (compounds having a few nucleotide residues) show that: (a) hypochromicity attains a limiting value of 35% at chain lengths of about 15; (b) the hypochromicity depends on the nucleotide composition as well as on the base sequence; (c) hydrogen bonding is not the main source of hypochromicity. On the other hand, in some enzymatically synthesized homopolymers of high molecular weight, such as polyadenylic acid (poly-A) or polyinosinic acid (poly-I), the hypochromicity is 55% and 60%, respectively. These high values are ascribed to some change in conformation. When heated to 350 K, a solution of poly-A shows a hypochromicity of only 35%; cooling restores the original value. This effect is attributed to the existence of helical regions stabilized by hydrogen bonds. By softening these helices to a random coil, heating decreases the hypochromicity; as the temperature is lowered, the helical regions are re-formed. Single stranded natural polynucleotide chains, such as RNA from various sources, behave in a similar way. The foregoing interpretation is borne out by measurements of optical rotation (*see below*). Additional information regarding sites suspected of participating in hydrogen bonding in polynucleotides can be obtained by occluding (e.g. by methylation) one or more of the sites and preventing the formation of helical complexes.

When applied to the DNA molecule, these concepts account for more than half the observed hypochromicity in terms of interaction between base-pairs in the double stranded Watson–Crick helical structure. The hypochromicity is absent at high temperatures where the two strands sever and the DNA molecule is denatured. The narrow temperature range (*ca.* 5 K) over which a partial loss of hypochromicity occurs is characteristic for a given DNA. The mid-point of this temperature range, called the melting temperature and designated by T_m, is related to the guanine–cytosine content of the DNA. The enhanced thermal stability of the DNA helix with increasing guanine–cytosine content is attributed to the fact that guanine–cytosine base-pairs may form three hydrogen bonds as compared with two for the adenine–thymine pair. The guanine–cytosine content of an unknown DNA can be precisely determined from a knowledge of its melting temperature.

While the study of hypochromism has had a great measure of success in exploring the DNA molecule and the nucleotides, solvent perturbation techniques have been found more useful in studying proteins and peptides. The influence of various agents on the absorption spectrum of a polypeptide can be used to determine the whereabouts of the three aromatic amino acids, viz. phenylalanine, tyrosine, and tryptophan. For instance, the difference in the absorption spectrum of a polypeptide in water and that in a mixture of ethanol and water can be interpreted in terms of the absorbing side chains that are exposed—provided that the change in solvent composition causes no conformational change. (The validity of this assumption can be easily checked.) By employing a series of perturbants of increasing size but similar chemical properties, more precise information about the location and environment of the perturbed group can be obtained; e.g. if a chromophore is situated in a niche, small perturbants will reach it, but a large one will not.

Often it is more instructive to perturb the system while, or before, recording its absorption spectrum; the perturbation (not to be confused with *solvent* perturbation) may be a step, periodic, or delta function involving any physical or chemical parameter. Temperature jump, pressure jump, application of an electric field, and irradiation are some examples of the perturbations that have been employed; when the perturbation takes the form of a light flash of high intensity and short duration (or a modulated light beam) the technique is called *flash spectroscopy* (or *modulation spectroscopy*). Rates of important biological processes can be measured with the aid of these techniques.

The renaturation rates of double-stranded DNA have been measured by noting the diminishing ultraviolet absorption after the temperature is suddenly lowered below T_m. Temperature-jump methods have also been applied to study the double helices formed by oligonucleotides of adenine and uridine of various lengths and at several concentrations. These investigations provide evidence for the nucleation reaction, which requires concomitant formation of hydrogen bonds by three contiguous base-pairs, and the subsequent propagation process at about 10^7 base-pairs per second.

This type of work has been particularly helpful in unravelling the mechanism of the interaction of nucleic acids with smaller molecules such as proflavine and ethidium bromide which are thought to intercalate between adjacent base-pairs. Temperature-jump studies of the interaction between DNA and proflavine reveal two relaxation processes; one is in the millisecond range, while the other is much faster. This indicates that the smaller molecule first attaches electrostatically, in a fast bimolecular reaction, to the outside of the DNA helix, and later a slow first-order intercalation process takes place.

One can also study the absorption of plane polarized light by an ensemble of oriented biological molecules. The directions of transition moments of the different electronic transitions giving rise to various absorption bands can then be determined. If the direction of the transition moment is already known, information regarding the orientation of the chromophore can be obtained.

Even in a random sample polarized light preferentially excites those chromophores which are suitably oriented, and produces an anisotropy in the sample; the technique is accordingly called *photoselection*. By studying the transient absorption of the photoselected species, i.e. by combining photoselection with flash spectroscopy, rotational motion of the chromophore can be detected. The rhodopsin chromophore, which is a part of the visual pigment of the retina, has recently been studied by employing these techniques. Rhodopsin was illuminated by a plane polarized laser flash, and the subsequent absorption of light beams polarized parallel and perpendicular to the laser flash were recorded. Initially, the parallel absorbance was higher but, after a few microseconds, the dichroism disappeared. These results suggest that the site occupied by rhodopsin in retinal membrane must be highly fluid, and that rhodopsin may be a diffusional carrier.

More often than not photoexcitation produces triplet states or free radicals which live, at room temperature, for as long as a few milliseconds. By observing the transient spectra of the triplet states (or free radicals) of the aromatic amino acids, of the natural chromophores present in some proteins, or of suitable dyes appended to the proteins, slow rotations and other processes occurring in proteins have been investigated.

If the molecule is dissymmetric, i.e. if it is devoid of a centre of inversion, a plane of symmetry, and an alternating rotation reflection axis of symmetry, it will be optically active; the use of plane polarized light (which can also be described as an in-phase combination of left circularly polarized light and right circularly polarized light of equal amplitude) can then provide valuable data which can be correlated with structure. Optical activity is caused by a difference in refractive index for right and left circularly polarized light; this phenomenon, called circular birefringence, manifests itself by rotating the plane of polarization of the transmitted beam. A plot of the rotation of the plane of polarization against wavelength is called an optical rotatory dispersion (ORD) curve. In the spectral region in which an optically active molecule has an absorption band, the two components are also differentially absorbed, and the transmitted beam is elliptically polarized; the medium is said to exhibit circular dichroism (CD).

All important biological polymers are optically active; ORD and CD can, therefore, be applied to a wide range of biological problems. Besides studying the CD of the "intrinsic" absorption bands of the polymers, one can bind foreign molecules to the biopolymers and utilize their "extrinsic" absorption bands which may be located in a spectral region more convenient to study than the region of intrinsic absorption.

The study of optical activity provides (at least) a semiquantitative estimate of the α-helical and random coil contents of a protein in solution. Its most important use is the detection, and occasionally identification, of *small* conformational changes. Optical rotation measurements have also been combined with temperature jump and other perturbation techniques to estimate the rates of helix-coil transitions in proteins. ORD and CD are used for the investigation of conformational stability and denaturation equilibria and kinetics, for detecting stages in refolding of enzymes, and for studying the mode of interaction of heavy metal ions with enzymes.

Applications Based on Electronic Emission Spectra

The emission is called fluorescence or phosphorescence according as it arises from a spin-allowed or a spin-forbidden transition. Fluorescence can have a short or a long-lifetime, but phosphorescence is necessarily long-lived. If, during the period between absorption and emission, the excited molecule spends no time in a metastable state (which in most cases is the lowest triplet state), the resulting fluorescence decays very rapidly (lifetime $\sim 10^{-8}$ s) after the withdrawal of the exciting light, and the emission is called normal or prompt fluorescence. The term delayed fluorescence defines itself: it implies emission which is spectrally identical with prompt fluorescence but having a lifetime much longer than that of prompt fluorescence. Delayed fluorescence can be subdivided into many categories; of these only two, viz. E-type and P-type delayed fluorescence, have been observed at room temperature, and of these two only the former is of interest in this discussion. E-type delayed fluorescence is observed when a molecule in the lowest triplet state is promoted, by thermal activation, to the first excited singlet state, from where it returns to the ground state by the emission of a photon. On account of its long lifetime, the triplet state is prone to quenching; consequently, phosphorescence and delayed fluorescence of a fluid solution at room temperature are much weaker in intensity than its prompt fluorescence. Nearly all the biological work has been based on prompt fluorescence; the epithet prompt will henceforth be dropped.

As with ORD and CD, fluorescence can be studied either from the macromolecules themselves (intrinsic fluorescence) or from other molecules attached to them (extrinsic fluorescence).

Most of the intrinsic fluorescence studies have been done with proteins, chlorophylls, and the pyridine nucleotide coenzymes involved in the enzymatic oxidation-reduction reactions. The intrinsic fluorescence of proteins arises from the three aromatic amino acids phenylalanine, tyrosine, and tryptophan, and bears a close relation to the fluorescence of the parent compounds benzene, phenol, and indole. These studies have been of an empirical nature based on the location of the fluorescence maximum of the native protein and changes in the wavelength of emission maximum and fluorescence quantum yield upon changes in the environment of the fluorophore, conformational changes, and denaturation. The study of non-radiative energy transfer has also been helpful in determining some conformational details.

By the use of advanced techniques available for peptide synthesis, two series of model compounds, one with both tryptophan and tyrosine, and the other with tyrosine alone, have been synthesized, and their fluorescence has been studied to find out how pH, salt concentration, temperature, etc., affect the fluorescence of these amino acids. It is found, for example, that ionization of the phenolic group or introduction of iodo groups puts an end to fluorescence. The modified residue does not merely lose its own fluorescing ability; it quenches, through resonance energy transfer, the fluorescence of tryptophan or tyrosine residues. Some other modifications such as acetylation of the tyrosine residue of RNase with acetylimidazole also abolish fluorescence, but do not produce energy acceptors. In the absence of a conformation change, the difference in fluorescence quantum yield before and after acetylation can be attributed to the number of tyrosines residues that are exposed and, therefore, acetylated. In this way the number of exposed and delitescent tyrosines can be determined.

Solvent perturbation techniques find many applications in fluorescence spectroscopy of biological molecules. Increasing concentrations of dioxan bring about major structural changes in chymotrypsinogen, chymotrypsin, and trypsin; these changes can be easily traced by monitoring the quenching of tryptophan fluorescence by nitrate ion. These studies reveal that, above 70%, dioxan everts the protein to a hydrophilic interior and an hydrophobic exterior.

An illustration of the *range* of information that can be gathered from fluorescence studies of biological macromolecules is provided by RNase T. It is a simple protein consisting of a single polypeptide chain whose amino-acid sequence was established in 1965. It has a single tryptophan residue which is believed to play an important role for the maintenance of the enzymatically active conformation of the protein. The fluorescence spectra indicate that the tryptophan is partially exposed, and that most of the nine tyrosines are "inside". Urea exposes the tryptophan (i.e. brings the tryptophan residue from the inside of the protein into the solvent environment), while sodium lauryl sulphate exposes the tyrosines. About a third of the tyrosines must be close to the tryptophan, for their fluorescence is quenched by the latter. The others respond to changes in solvent polarity and are apparently quenched by the acidic amino acid side chains in the native enzyme. Substrate analogues which bind at the active site also quench the tyrosine and tryptophan emission. The pH dependence indicates that the tryptophan and only one or two of the tyrosines are present in the active site of the enzyme, but not directly involved in the binding of the inhibitor.

Many polycyclic aromatic hydrocarbons, though non-fluorescent in water, become highly fluorescent when placed in non-aqueous environments (e.g. when attached to a protein); 1-anilino-8-naphthalene sulfonate (ANS) and 2-p-toluidinyl-6-naphthalene sulfonate (TNS) are two examples. An understanding of their behaviour has been gained by studying the spectral dependence on the polarity of many organic solvents, and the structure of anhydrous and hydrated crystals of TNS. The quantum yield increases and the wavelength of maximum emission decreases as the solvent polarity decreases. X-ray diffraction studies show that in the anhydrous crystal TNS is planar, permitting electronic delocalization over the whole molecule, but the hydrated species is bent. The absence of fluorescence could thus be the consequence of solvent polarity and/or hydrogen bonding.

The extrinsic fluorescence of probes like ANS is used to trace structural changes, to detect hydrophobic binding sites, and to ascertain the degree of polarity of many active sites. The binding of substrates and coenzymes to proteins can be readily followed if a fluorescent probe is situated at or near the active site. The probe may be dislodged by the substrate; or the formation of a ternary complex may alter its fluorescence characteristics. Distances between, and orientations of, chromophores can be established by using the fluorescent probe as one member of the donor-acceptor pair in energy transfer studies.

Rotational motion occurring in biological molecules can be observed by studying the polarization of intrinsic or extrinsic fluorescence. The degree of polarization, p, is defined by the relation: $p = (I_\| - I_\perp)/(I_\| + I_\perp)$, where $I_\|$ and $I_\perp$ denote the steady state intensities of fluorescence with polarization directions parallel and perpendicular to that of the exciting beam. If the directions of the absorption and emission transition moments are parallel to each other, $p = 0.50$, provided that Brownian rotation during the lifetime of the excited state and concentration depolarization (occurring at high concentrations because of energy transfer to a similar but unexcited and differently oriented molecule) are negligible; if the two vectors are perpendicular, $p = -0.33$. Values markedly different from the theoretical ones betray rotation of the fluorophore before emission; the magnitude of the deviation provides a measure of the rotational freedom. There is an alternative method of studying fluorescence polarization: this is to irradiate the sample with a short-lived (a few nanoseconds long) plane-polarized light pulse, and observe the transient fluorescence intensities, $I_\parallel(t)$ and $I_\perp(t)$, in directions parallel and perpendicular to the direction of polarization of the exciting pulse. The experimentally measured emission anisotropy,

$$A(t) = [I_\parallel(t) - I_\perp(t)]/[I_\parallel(t) + 2I_\perp(t)]$$

is a decaying function of time; its rate of decay can be related to the rate at which the fluorophore rotates. Various theoretical considerations can be applied to decide whether the measured rate of decay is caused by rotation of the fluorescent moiety independently of the rest of the polymer, or by the overall rotation of the biopolymer.

Both the steady state and nanosecond polarization techniques furnish sensitive methods for studying conformational transitions occurring on the nanosecond scale; no other spectroscopic method is at present capable of studying these phenomena. Rotational relaxation times of proteins and helix-coil transitions of several polypeptides have been studied by polarization methods. In the field of immunochemistry, the antigen–antibody reaction has been investigated by utilizing these techniques.

The short-time resolution which fluorescence polarization method affords is not always advantageous. In relatively rigid structures like membranes, the fluorescent molecule may rotate but not rapidly enough to depolarize the emitted light. If the rotational relaxation time is appreciably longer than the fluorescence lifetime, fluorescence polarization experiments become inconclusive. These slow rotations may be tracked with the help of E-type delayed fluorescence and phosphorescence whose lifetimes are typically in the millisecond range. There are some organic molecules, e.g. eosin and proflavine, whose phosphorescence and/or delayed fluorescence can be easily detected even in fluid solutions; other things being equal, the emission will be still more intense in rigid membranous structures. The degree of flexibility of chlorophyll in chloroplast membranes can also be investigated by means of the intrinsic E-type delayed fluorescence of chlorophyll.

Applications Based on Infra-red Absorption Spectra

The examination of biological polymers by infra-red spectroscopy is made difficult by their complex spectra. However, at least in peptides, there is a redeeming feature: each amino acid residue has the amide bond

$$-C\begin{matrix}\diagup O\\ \diagdown NH-\end{matrix}.$$

The most intense absorption transition of this bond are exhibited by the amide I band at 1600 to 1700 cm^{-1} due to C=O stretching and the amide II band at 1500 to 1550 cm^{-1} due to N—H in-plane bending and C—N stretching. Unfortunately, water absorbs strongly at 1650 cm^{-1}, and studies must be carried out in D_2O. Another difficulty is the requirement that for precise analysis the molecule is best observed in the solid phase as a cast film; some molecules cannot be oriented easily.

A good deal of structural information has, nevertheless, resulted from studies of oriented specimens of elongated molecules prepared by mechanical rolling, stretching, or by other shearing processes; many of these samples are found to be dichroic. Measurements of dichroism have also been made on non-aqueous solutions of polypeptides oriented by streaming; the helical form of these macromolecules is dichroic, but the random coil configuration is not.

Formation of helical base stacked conformations of polynucleotides is usually accompanied by a blue shift of the purine and pyrimidine carbonyl vibrations and a diminution in the extinction coefficients of the bands associated with the ring vibrations of the purines. These data can be calibrated against the spectral characteristics of homopolymers of known structure, and utilized for identifying unknown conformations; they have been employed for studying the interaction between polynucleotides and their complementary nucleosides and nucleotides which form base stacked structures of known stoichiometry.

Hydrogen bonding associations in low molecular weight purine and pyrimidine bases in solution can also be spectroscopically studied by following the dependence on concentration of the disappearance of the N—H stretching vibrations of the bases.

The eclectic survey presented above should serve to illustrate the multifarious applications of spectroscopy in biophysical research. The most appealing quality of spectroscopic measurements is that they can be carried out at sufficiently low concentrations at which even biological polymers exhibit ideal thermodynamic behaviour. The ambit of utility is wide and impressive; yet the chances of inferring macromolecular structure directly from spectroscopic observations seem slender. The biophysicist can only strive to correlate the spectroscopic properties with the other parameters of the system for deducing their unique interrelationship, and construct models that depend only on the investigated parameters. These models can then be compared with their opposite numbers found in the functionally intact subcellular units such as the chloroplast, the mitochondrion, the ribosome, etc. The fabrication of bilayer lipid membranes is a fecund step in this direction and illustrates the type of work that is likely to be done in the future.

Bibliography

ALBRECHT, A. C. (1970) in *Progress in Reaction Kinetics*, Vol. 5 (G. Porter Ed.), Oxford: Pergamon Press.

BEAVEN, G. H. (1962) Spectroscopy, biological and medical applications of, in *Encyclopaedic Dictionary of Physics* (J. Thewlis, Ed.), **6**, 738, Oxford: Pergamon Press.

CONE, R. A. (1972) *Nature* **236**, 39.

CRABBE, P. (1965) *Optical Rotatory Dispersion and Circular Dichroism in Organic Chemistry*, San Francisco: Holden-Day.

EDELMAN, G. M. and MCCLURE, W. O. (1968) *Acta Chem. Research* **1**, 44.

GOEDHEER, J. C. (1966) in *The Chlorophylls* (L. P. Vernon and G. R. Seely Eds.), New York: Academic Press.

GRATZER, W. B. and COWBURN, D. A. (1969) *Nature* **222**, 426.

KONEV, S. V. (1967) *Fluorescence and Phosphorescence of Proteins and Nuclein Acids*, New York: Plenum Press.

MCLAREN, A. D. and SHUGAR, D. (1964) *Photochemistry of Proteins and Nucleic Acids*, Oxford: Pergamon Press.

RADDA, G. K. (1971) in *Current Topics in Bioenergetics*, Vol. 4 (D. R. Sanadi Ed.), New York: Academic Press.

SCHECHTER, A. N. (1970) *Science* **170**, 273.

STRYER, L. (1968) *Science* **162**, 526.

URRY, D. W. (Ed.) (1969) *Spectroscopic Approaches to Biomolecular Conformation*, Washington: American Medical Association.

WEBER, G. and TEALE, F. W. J. (1965) in *The Proteins*, 2nd ed., Vol. III (H. Neurath Ed.) New York: Academic Press.

K. R. NAQVI

SPECTROSCOPY, PHOTOELECTRON. Photoelectron spectroscopy is concerned with the ejection of electrons from atomic systems according to the Einstein photoelectric equation

$$h\nu = E + \tfrac{1}{2} mv^2. \qquad (1)$$

E is the initial binding energy (ionization energy) of the electron and $\tfrac{1}{2} mv^2$ the kinetic energy it acquires when ejected by a photon of frequency ν. Ordinary (photon) spectroscopy is governed by the equation

$$h\nu = E' - E \qquad (2)$$

where both E' and E are bound states and E can only be obtained by extrapolating the excited states E' to zero energy, that is when the electron becomes free. To use eq. (2) it is only necessary to measure the photon frequency which of course is done with high accuracy by conventional spectroscopy. However, to use eq. (1) it is necessary to measure the kinetic energy of the photoelectron (ejected by a photon of known frequency), with a precision comparable with that of optical spectroscopy. This is has been possible only in recent years as a result of the development of sensitive electron detectors and improved electron energy analysers (both electrostatic and electromagnetic) which are capable of giving high accuracy and good resolution.

The reward for these developments has been that by using the methods of photoelectron spectroscopy it has become possible to pick out each electron from a molecule or a solid whether it is an inner (core) or outer (valence shell) electron and to determine accurately the energy with which it is bound into the system. Many other features which reveal the function of the electron in the electronic structure of the system also appear in the photoelectron spectrum. It was previously not possible to obtain this information for inner electrons by optical spectroscopy except in a few simple cases because of the interaction of their excited states E' with the ionization continua of the outer electrons. By using a high-energy photon it was found that the inner energy could be ejected so rapidly past the outer electrons that this interaction was largely eliminated. For example, by conventional spectroscopy it had been possible to study only the outer two non-bonding electrons in the water molecule and it was not until the advent of photoelectron spectroscopy that any detailed information became available on the other eight electrons in this relatively simple and vitally important molecule.

Photoelectron spectroscopy is conveniently divided into the two major fields in which it has been developed over the last decade. The first uses X-ray irradiation and is largely concerned with core electrons. It was pioneered by K. Siegbahn and his co-workers at Uppsala who used their experience on β-ray spectrometry to make the necessary technical development. Extensive details of the work of this school are given in two major publications (Siegbahn et al., 1967, 1970). The second field employs ultra-violet radiation, mostly utilizing the resonance line of helium at 58·4 nm (21·22 eV). It was developed in London by D. W. Turner and his associates who used it to study the outer valence shells of molecules. Their results are collected in Turner et al. (1970). The range was later extended by W. C. Price and collaborators who employed the shorter wavelength line of ionized helium (30·4 nm, 40·8 eV) as a photon source which gave deeper penetration into the valence shell (Price, 1970, Potts et al., 1970). The X-ray and ultra-violet irradiation techniques are conveniently abbreviated as XPS and UPS, respectively. While the former gives information on all electrons the latter gives much more highly resolved spectra for the more loosely bound outer electrons since the helium linewidth (about 0·001 eV) is less than that of the X-ray lines (about 0·5 eV) use by Siegbahn. Further, because the photoelectron velocity is much lower in UPS than XPS the precision of its determination is correspondingly greater and more highly resolved spectra showing detailed vibrational structure are obtained when ultra-violet photons are used.

Instrumentation

A schematic diagram of the equipment used in photoelectron spectroscopy is shown in Fig. 1. The

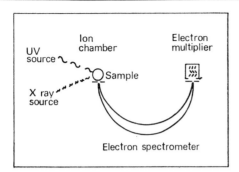

Fig. 1. Schematic diagram of ultra-violet X-ray photoelectron spectrometer

energy spectrum of the photoelectrons emitted by the sample is obtained by plotting the count rate of electrons arriving at the detector against a continuously increasing energy setting of the electron spectrometer. Thus photoelectron energy increases from left to right along the abscissae and binding energy from right to left. This presentation is used in all published XPS spectra and in some UPS spectra. There is an increasing tendency with UPS spectra to plot binding energy (ionization potential) increasing from left to right since the energy scale is always calibrated against gases of spectroscopically known ionization potential. This is because the photoelectron energy is affected by contact potentials and other factors and its absolute value cannot be determined accurately from the potentials applied to the deflecting plates.

X-ray Photoelectron Spectroscopy

Most X-ray photoelectron spectroscopy has used the soft Kα X-ray lines of Mg or Al (1254 and 1487 eV, respectively) because these are narrower than, for example, the lines of Cu Kα and the give photoelectrons whose energies are lower and can therefore be measured more accurately. These lines permit examination of the electronic structure down to 1000 eV which covers the K-shells of the light atoms and the L- or M-shells of heavier atoms. The XPS lines from the K-shells of the atoms in the second period of the periodic table are shown in Fig. 2. Their binding energies increase from lithium (55 eV) to neon (867 eV), the differences between successive elements gradually increasing from 56 eV (Li–Be) to 181 eV (F–Ne). The photoelectron lines are thus well separated from one another and can be used as a method of identifying the elements present in the sample. For the heavier atoms the binding energies of electrons in L-, M- and higher inner shells can be used for this purpose and thus all the elements of the periodic table can be identified by X-ray photoelectron spectroscopy. It was for this reason that Siegbahn called his method ESCA, the letters standing for "electron spectroscopy for chemical analysis". However, this rather restrictive title is now being dropped.

Although the binding energy of an electron in an inner shell is mainly controlled by the charge Z within the shell it is also affected to a lesser extent by the cloud of valence electrons lying above it. The XPS line of an element is found to vary over a few electron volts according to the state of chemical combination of the atom and the nature of its immediate neighbours. This is illustrated in the spectra of acetone $(CH_3)_2CO$, ethyl trifluoracetate $CF_3COC_2H_5$ and sodium azide shown in Fig. 3. The C (1s) line of the two methyl carbons in acetone is 2–3 eV lower in binding energy than that of the more positive carbonyl carbon and is, as might be expected from the number of C atoms, twice as strong. Similar remarks apply to the two outer and the inner N (1s) electrons of the azide ion. The binding energies of the

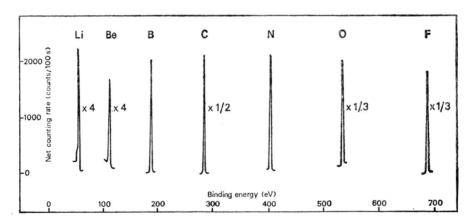

Fig. 2. Photoelectron lines from the K shells of elements of the second period when irradiated with Al Kα plotted against binding energy (i.e. 1487 eV minus the photoelectron energy).

four different carbons in ethyl trifluoroacetate similarly reflect the different environments of the four carbon atoms in this molecule (note that the formula in the diagram is placed so that the lines lie vertically below the carbon atoms with which they are associated). These "chemical shifts" have been related to the electronegativity difference of the atom from its neighbours and provide a means of estimating the effective charge on the core of each atom. The method has been extended to complicated molecules such as vitamin B_{12} where, using Al ($K\alpha$), the photoelectron line of one cobalt atom could be studied in the presence of the lines of 180 other atoms. Similarly both the sulphur and the iron electron line in the enzyme cytochrome c can be recorded. It is clear that the technique is a powerful direct method of studying chemical electronic structure.

In addition to the work on mainly covalent materials, ionic and metallic solids have been studied. For metals the conduction band structure can be found and the photoelectron energy distribution curves compared with calculated densities of states. Because of the small depth of penetration of the radiation and the necessity for the photoelectrons to emerge from the material the method is clearly a surface effect. Depths down to at most 3 nm (30 Å) are involved in the process and ultra high vacuum techniques must be employed to eliminate the effects of surface contamination, Surface properties such as are important in catalytic and adsorption processes have received considerable attention by workers in the field. It should be mentioned that although XPS can produce valuable information about the electrons in the valence shell the cross-sections for this radiation of valence shell orbitals are almost an order of magnitude less than those of core orbitals. For example, in hydrocarbons the cross-sections of C ($2s$) orbitals are about a tenth those of C ($1s$) orbitals and the cross-sections of orbitals built from C ($2p$) and H ($1s$) are still lower by almost another order of magnitude. This is a limitation not suffered by UPS where the cross-sections particularly of $2p$ orbitals are generally appreciably higher than those of $2s$ orbitals. This fact is likely to seriously affect the prospects of XPS for detailed outer electron studies.

Ultra-violet Photoelectron Spectroscopy

Ultra-violet photoelectron spectroscopy using line sources obtained from discharges in helium or the other inert gases has produced an enormous amount of new detailed information on virtually all types of molecule and chemical species. This has been such an exciting time for chemists that nearly all the effort has so far been absorbed in studying these substances as gases. Its application to the solid phase which involves greater experimental and interpretative difficulties has scarcely begun though there have been some interesting studies by UPS on the conduction bands of metals such as gold in the solid and liquid phases (Eastman, 1971). Apart from the fact that the inner valence electron orbitals of diatomic and polyatomic molecules which cannot be studied by conventional spectroscopy appear directly as features in the photoelectron spectrum an additional attraction of the technique is that the UPS record itself reveals in a very convincing and readily understandable manner the orbital shell structure of the electrons in a molecule.

For monoatomic gases only electronic energy is involved in photoionization and the photoelectron spectrum of atomic hydrogen (obtained by passing the products of a discharge in H_2 into the ion chamber) gives a single sharp peak corresponding to a group of photoelectrons of energy 7·62 eV. This corresponds to the difference in energy of the helium resonance line (21·22 eV) and the ionization energy

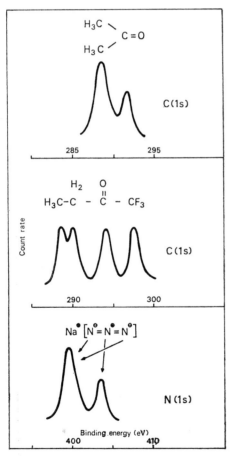

Fig. 3. XPS spectra of C (1s) shells of acetone and ethyl trifluoroacetate and the N (1s) shell of sodium azide. Note that the C atoms in the formula are arranged to be above the lines to which they correspond.

of the H (1s) electron (13·60 eV). For argon the outer electrons are in p-type orbitals and the $(3p)^5$ configuration of the ion has the two states $^2P_{3/2}$ and $^2P_{1/2}$ with statistical weights in the ratio 2 to 1 and energies differing by spin–orbit coupling. The photoelectron spectrum (see Fig. 4) corresponding to this $(3p)^{-1}$ ionized state is thus a doublet, one component of which is twice as strong as the other. Similar doublets are found for the other inert gases, the spin–orbit coupling increasing with their atomic weight.

For the ejection of an electron from an orbital in a molecule we have the equation

$$I_0 + \Delta E_{vib} + \Delta E_{rot} = hv - \tfrac{1}{2} mv^2$$

where I_0 is the "adiabatic" ionization energy (i.e. the pure electronic energy change) and ΔE_{vib} and ΔE_{rot} are the changes in vibrational and rotational energy which accompany the photoionization. To explain the nature of the information which can be obtained we shall discuss the photoelectron spectra of some simple molecules.

Figure 5b gives the potential energy curves of H_2 and H_2^+. When an electron is removed from the neutral molecule the nuclei find themselves suddenly in the potential field appropriate to the H_2^+ ion but still separated by the distance characteristic of the neutral molecule. The most probable change is thus a transition on the potential energy diagram from the internuclear separation of the ground state to a point of the potential energy curve of the ion vertically above this. This is the Franck–Condon principle and it determines to which vibrational level of the ion the most probable transition (strongest band) occurs. The energy corresponding to this change is called the vertical ionization potential I_{vert}. Transitions of I_{vert} are weaker, the one of lowest energy corresponding to the vibrationless state of the ion corresponding to I_{adiab}. The photoelectron spectra of H_2 and D_2 are given in Fig. 5a and that of H_2 is also plotted along the ordinate of Fig. 5b.

From the intensity distribution of the bands in the photoelectron spectrum it is possible to calculate the change in internuclear distance on ionization. Clearly when a bonding electron is removed, the part of the photoelectron spectrum corresponding to

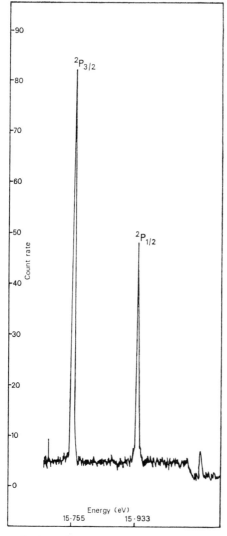

Fig. 4. Photoelectron spectrum of argon ionized by Ne (736 Å).

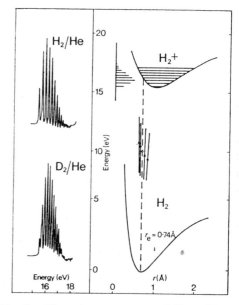

Fig. 5a. Photoelectron spectra of H_2 and D_2. 5b. Potential energy curves showing spectra plotted along ordinate.

this will show wide vibrational structure with a frequency separation which is reduced from that of the ground state vibration. The removal of relatively non-bonding electrons on the other hand will give rise to photoelectron spectra rather similar to those of the monoatomic gases and little if any vibrational structure will accompany the main electronic band. The type of vibration associated with the pattern obtained when a bonding electron is removed can usually be identified as either a bending or a stretching mode and this can throw light on the function of the electron in the structure of the molecule, that is either as angle forming or distance determining. From the band pattern it is frequently possible to calculate values of the changes in angles as well as changes in internuclear distance on ionization and so the geometry of the ionic states can be found if that of the neutral molecule is known.

The Intensities of Photoelectron Spectra

Apart from the quantitative data concerning orbital energies which are to be obtained by photoelectron spectroscopy, the intensities of the bands obviously contain much information about the nature and extent of the orbitals themselves. In this connection we must consider the transition moment of an electron from the initial orbital φ to the state ε in the ionization continuum into which it is ejected by the incident photon. This is illustrated in Fig. 6 where it can be seen that the integral representing the transition moment depends upon the spread and nodal character of the wavefunction of the initial state and the wavelength and phase of the photoelectron (i.e. the final state). For the purposes of this illustration the wave functions are taken as those of a one dimensional oscillation of an electron in a $V(1/r)$ potential field, that is pseudo Rydberg bound states. Plane wave functions are used for the continuum states ($\lambda[1 \text{ eV}] \cong 12 \text{ Å}$). The simplest fact that this illustrates is the fall-off of the photoionization cross-section with higher energy (shorter wavelength) photoelectrons. This explains the relative transparency of electrons in valence shells to X-irradiation where the area of the graph for the transition moment is divided into small compensating positive and negative regions. It can be appreciated that the cross-section appropriate to any orbital will vary with the dimensions and nodal character of this orbital and the wavelength of the photoelectron. As a rough generalization it might be expected that the photoionization cross-section would maximize when $\lambda/4$ of the photoelectron is not less than the orbital dimensions. This would explain experimental observations that near the threshold of ionization the cross-sections of molecular orbitals built from p atomic orbitals are greater than those arising from s orbitals. This situation is reversed for high photoelectron energies where, because of the smaller radial spread of s-orbitals, the

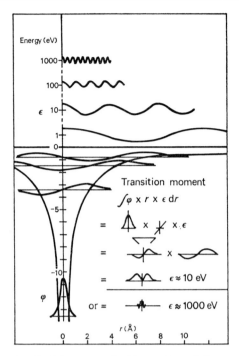

Fig. 6. Diagrammatic illustration of features of importance in the photoionization of an electron from a bound state λ to a continuum state ε.

transition moment integral maximizes with shorter wavelength photoelectrons. The study of the variation of the relative intensity of the photoelectron spectra of different orbitals with varying irradiating wavelengths is clearly going to yield much information on the nature of orbital wave functions. A summary of the information available for gases on this topic has been given by Price et al. (1972). Photon sources of different energies have already been used by Eastman and Grobman (1972) to study the band structure of gold and the densities of intrinsic surface states in Si, Ge and GeAs.

Many interesting new developments are reported in the ASILOMAR Conference Proceedings (1972) and a new journal entitled *Electron Spectroscopy* has been founded in which photoelectron spectroscopy is a major topic. Studies have been made by photoelectron spectroscopy of atoms and transient molecular species produced by electrical discharges as well as the photodetachment of negative ions of atoms and molecules. The angular distribution of the photoelectron emission is another subject of considerable theoretical interest on which experimental information is being accumulated. It is certain that exciting revelations on the electron structure of matter will be forthcoming as this field is further developed.

Bibliography

ASILOMAR Conference Proceedings–*Electron Spectroscopy* (D. A. SHIRLEY, ed.), North Holland Pub. Co., 1972.
EASTMAN, D. E. (1971) *Phys. Rev. Lett.* **26**, 1108.
EASTMAN, D. E. and GROBMAN, W. D. (1972) *Phys. Rev. Lett.* **28**, 1327 and 1379.
POTTS, A. W., LEMPKA, H. J., STREETS, D. G. and PRICE, W. C. (1970) *Phil. Trans. R. Soc.* A **268**, 59.
POTTS, A. W. and PRICE, W. C. (1972a) *Proc. R. Soc.* A **326**, 165.
POTTS, A. W. and PRICE, W. C. (1972b) *Proc. R. Soc.* A **326**, 181.
PRICE, W. C. (1970) *Molecular Spectroscopy IV* (P. W. HEPPLE, ed.), London: Inst. Petroleum, distrib. Elsevier, p. 221.
PRICE, W. C., POTTS, A. W. and STREETS, D. G. (1972) *Electron Spectroscopy* (D. A. Shirley, Ed.), p.187, Amsterdam: North Holland.
SIEGBAHN, K. *et al.* (1967) Electron spectroscopy for chemical analysis, *Nova acta Reg. Soc. Sci. Uppsala*, Ser. IV, **20**.
SIEGBAHN, K. *et al.* (1970) *ESCA Applied to Free Molecules*, Amsterdam: North Holland.
SIEGBAHN, K. (1973) *Endeavour* **32**, 51.
TURNER, D. W. *et al.* (1970) *Molecular Photoelectron Spectroscopy* (London: Wiley–Interscience).

W. C. PRICE

SPEECH SPECTROMETRY

Introduction

The perception of speech sounds is almost independent of the phase characteristics of the transmission path between speaker's lips and listener's ears unless these are such that some frequency components of the speech are severely delayed compared with others. Hence a description of the distribution of the power density of speech as a function of frequency, i.e. the power *spectrum*, contains most of the information relevant for perception, and in fact may be used for resynthesizing realistic speechlike sounds (see speech synthesis). A speech *spectrogram* is a visual record indicating the power spectrum (shown as density of shading) as a function of time and frequency (Fig. 1). Such spectrograms are very useful in studies of the qualities which distinguish speech sounds from each other, both from the points of view of speech production and perception, as a diagnostic aid in development of speech communication systems, and as an aid in teaching deaf people to talk (Potter, Kopp and Green, 1947). They offer advantages over other methods of presenting speech information to the experimenter in that (usually) irrelevant details of waveforms are averaged out, and the various parts of an utterance can be compared simultaneously visually, whereas audible methods of presentation are transitory.

The following paragraphs from an introduction by J. R. Pierce and E. E. David, Jr., to the 1966 edition of *Visible Speech* Potter, Kopp and Kopp) are still valid and give a good idea of the contribution made by speech spectrometry:

"The work leading to *Visible Speech*, and its general availability through the publication of that book in 1947, worked a revolution in acoustic phonetics, a revolution which is still in progress. Literally for centuries, men had puzzled over the riddle of the relation between the vowels and consonants of spoken language and the acoustic signal emerging from a talker's mouth. Early workers, including Hermann L. F. Helmholtz, Alexander Graham Bell, and Sir Richard Paget, were aware of the existence in vowel sounds of concentrations of acoustic energy, known as formants, but the formants could be detected only through acute listening or by cumbersome analysis with resonators. The very number of formants in a vowel was in question. Early applications of electronics were scarcely more revealing.

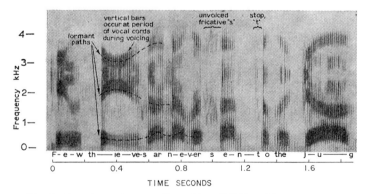

Fig. 1. Speech spectrogram (broad band). "*Few thieves are never sent to the jug.*"

"Then suddenly, through the work of Potter, Kopp, Green and their many associates, the frequency and time structure of speech was made strikingly apparent to the eye, and subject to easy measurement. The formants were clearly visible as snaky bars. These moved smoothly during diphthongs, and bent up or down toward fixed or variable hubs (later called loci) characteristic of preceding or succeeding consonants. The bursts of energy characteristic of plosives or stop consonants were clearly apparent as vertical stripes, and the extended blurs of fricatives were equally clear. These visual features, made so clear by the visible speech spectrograph instrument, have been assiduously measured for the sounds of many languages, English, Swedish, Russian, Bantu, and many others, so that we now know a great deal about the acoustic counterpart of perceived sounds.

"Today no phonetician is without his sound spectrograph or sonograph, and the use of these devices has extended into many other fields, from anthropology and medical diagnosis to electromagnetic radio propagation and underwater sound. In the field of speech itself, the work disclosed in *Visible Speech* has formed a foundation for work on speaker identification, automatic recognition of speech (alas, not yet usefully accomplished), the improvement of electrical speech communication facilities and voice coding devices, the interpretation of important, garbled, or noisy speech messages, and the construction of speaking machines. The use of visible speech in training the deaf and in helping them to communicate is being revived. It is here that we fondly hope much more can be accomplished."

The properties of speech evidenced in a spectrogram are best understood through a simple description of speech production (Fletcher, 1953). There are three main contributors to the characteristics in the spectrograms as shown in Figs. 1 and 6:

(a) The nature of the excitation, i.e. voiced (Fig. 2b, c) or fricative (by air friction as in s, f, th) (Fig. 3). The latter excitation has a random or noiselike waveform and a spectrum without the line structure.
(b) The pitch in voiced sounds, i.e. the repetition frequency of the vibration of the vocal cords (Fig. 2b, c), usually in the range 50–300 Hz.
(c) The vocal tract (Fig. 2a) transmission whose resonances or *formants* cause selective enhancement of some frequency ranges (Fig. 2d, e).

Example of a Spectrograph

A speech *spectrograph* (Flanagan, 1972; Koenig, Dunn and Lacey, 1946) is an instrument for producing speech spectrograms. Briefly it consists of three elements (Fig. 4): Storage for the speech signal

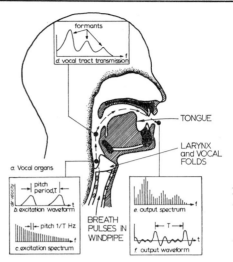

Fig. 2. Principles of speech production (voiced sounds) (a) Cross-section of head, showing vocal organs. (b) Pulsed air flow through vocal cords. (c) Spectrum of pulsed air flow. (d) Frequency selective enhancement by vocal tract resonances. (e) Speech spectrum [result of (d) on (c)]. (f) Speech waveform.

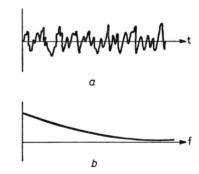

Fig. 3. Fricative or frictional excitation. (a) Waveform. (b) Spectrum.

during analysis, frequency-selective filtering and detection, and display.

A typical instrument (e.g., Kay "Sonograph" or Voiceprint "Sound Spectrograph") operates in principle as follows (Fig. 5). A record of speech of 2·5 sec duration is stored on a magnetic disc. Coupled to the disc is a drum around which is wrapped sensitized paper (usually Teledeltos) on which the spectrogram will be formed. The disc and drum rotate, and the speech record is played back through a bandpass filter of bandwidth 300 Hz. The filter output is applied (sometimes after rectification and smoothing) to vary the intensity of a line formed on the moving sensitized paper by a stationary stylus. After each revolution the stylus is moved 0·4 mm along the drum and the centre frequency

Fig. 4. Elements of a speech spectrograph.

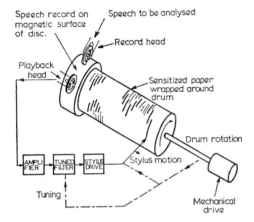

Fig. 5. A typical speech spectrograph.

of the filter is changed by 20 Hz and the process is repeated. Thus after 250 revolutions the centre frequency of the filter has been stepped through 5 kHz, the whole frequency range of interest in speech, and the stylus has moved 10 cm.

The spectrogram thus obtained (Fig. 1) consists of about 250 parallel lines each modulated in intensity according to the power contributed by the speech in the corresponding 300 Hz band.

Usually the playback and formation of the spectrogram are carried out at a speed up to 10 times faster than the original recording, with corresponding scaling of filter frequency and bandwith, to save time. Thus a typical spectrogram of 2·5 sec of speech is formed in about a minute.

Frequency versus Time Resolution

It is common practice in speech spectrometry to analyse with two different filter bandwidths, 300 Hz as above, and also 45 Hz. The latter filter permits resolution of the harmonic components of the voiced excitation (Fig. 2c, e and Fig. 6). However, the higher frequency-resolution filters tend to display detail which is confusing to the eye, and fundamentally must have poorer time resolution (see below). Hence

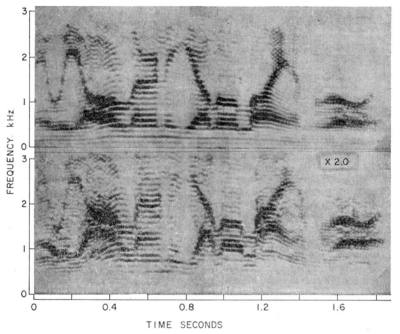

Fig. 6. Narrow band spectrogram (45 Hz resolution). The nearly horizontal bars correspond to harmonics of the pitch frequency. The lower half of the picture shows certain speech features have been modified by a particular form of processing (broadband frequency multiplication by two). The strongest formant (the lowest frequency one) has been doubled in frequency while the frequency spacings between it and other components have been maintained.

they tend to blur stop sounds (t, p, etc.) which are very brief, and do not show the time-periodic pattern corresponding to the pitch period. The relation linking frequency resolution Δf and time resolution Δt is fundamental in all spectrum analysis devices, and it is impossible to make a device in which the product $\Delta t \cdot \Delta f$ is usefully less than unity (Gabor, 1946).

There are practical difficulties in displays in which the spectral density is represented as intensity of marking—the process is not quantitatively reliable, and visual judging of intensity is imprecise. *Sectioning* is an analysis with better amplitude resolution, relevant to a particular time. This may be carried out with a spectrograph of the type described above, but with modification as follows. The normal density marking by the stylus is suppressed. The rotating drum is equipped with an adjustable trigger which initiates at a chosen time on each revolution a sampling of the output of the filter-rectifier, and simultaneously the stylus is caused to mark. The stylus marking is stopped by an electronic circuit after a time proportional to the sampled rectifier output. The aggregate of the 250 or so marks of this nature forms the section which corresponds to a profile of the spectral density plotted against frequency at the chosen time (Fig. 7).

the speech is stored as numerical samples, usually taken at 10,000 or 20,000 per second, converted to digital form as 12- to 16-bit binary numbers, and stored in the core store of a special or general purpose computer.

The filtering system is one in which there is great variety in characteristics and in implementation. The fundamental limit that the $\Delta t \cdot \Delta f$ product has a minimum of the order one is always true, so that real-time spectrometry can never provide frequency resolution of Δf Hz in less than $1/\Delta f$ sec. Changes of time and frequency scales are available through storage and speed-changed playback. Another related fundamental relationship is that the effective complexity of a spectrograph to yield analysis of a band F (about 5 kHz for speech) with resolution Δf in T sec is never less than $F/(\Delta f^2 T)$ degrees of freedom, e.g. resonators or inductor-capacitor sets, or words of memory in a digital system. Often, more are used to provide desirable responses.

Within the above restrictions there are many implementations. The most popular analogue one is the constant bandwidth heterodyne analyser (Fig. 8), used in the spectrograph described above. The speech signal

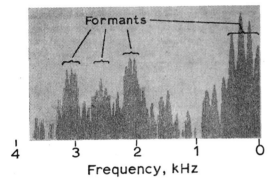

Fig. 7. *"Section" of a spectrogram, taken during "ie", just after "th" for the sentence of Fig. 1.*

Techniques

Each of the components of spectrographs may be implemented in various ways, dictated by economics, desired performance, or type of output display required. Digital techniques are becoming economic for storage and filtering, but conventional analogue systems are by far the commonest as yet.

Storage of the speech signal in analogue form is on a loop of conventional magnetic recording tape (polyester, for durability during the many playings) or on a nickel–cobalt plated disc. The recording process is the same as for conventional tape recording, as it is not necessary to maintain waveforms or frequency characteristics precisely. In digital systems,

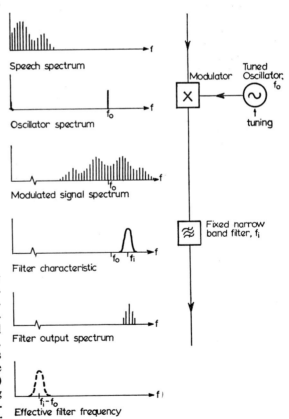

Fig. 8. *Heterodyne analyser.*

is modulated by the frequency f_0 of the variable frequency oscillator, producing a shift of its spectrum by f_0. The shifted-spectrum signal is filtered by a fixed frequency narrow-band filter at frequency f_i (typically about 30 kHz). The part of the translated speech spectrum which is in the passband near f_i depends on the value of f_0, and corresponds to the frequency $f_i - f_0$ kHz in the input. Thus, f_0 can be used as an adjustment of the effective frequency of the bandpass filter, but does not affect its shape (see below). This type of spectrograph has the simplest filtering system—one filter only, used many times, but it is very slow for many purposes because only one filtering frequency is used at a time. Because only one filter is used, it may be made fairly complex without unduly increasing the cost of the system.

Real-time spectrographs (Potter, Kopp and Green, 1947) usually do not need storage of the input speech, but use many filters (15 to 40) simultaneously. These are usually expensive, but may be necessary to provide an almost immediate display as for the training of deaf speakers and in other learning situations in which vocal gestures must be related to spectra. Each filter operates a separate output indicator—stylus, light beam or cathode-ray tube track. Earlier real-time analysers used analogue (inductor-capacitor) filters, but digital techniques, including the fast Fourier transform, are likely to make future real-time spectrographs quite different (see below). One basically simple modification is in the practice of digital storage of an input signal, and subsequent very rapid playback, allowing apparent real-time analyses with single, swept filters. This is a well-established practice in spectrometry for lower-frequency physical signals.

The spectrographs already discussed use constant-resolution filtering, i.e. each part of the frequency range is explored with a filter of the same bandwidth. This is necessary, for example, if the pitch harmonics are to be resolved over the whole frequency range, but may be inefficient if the purpose is to study formant patterns, or coarse features of the spectrum. Then it is appropriate to use filters of larger bandwidth at higher frequencies—constant-proportional-bandwidth filters are often used with bandwidths of one-third octave (i.e. bandwidth 0·23 of the centre frequency) above about 1 kHz. Then the whole frequency range of speech (up to 8 kHz) can be covered with about 12 filters (250 Hz bandwidth up to 1 kHz and $1/3$ octave beyond that). Such characteristics are consistent with the reduced resolution of the ear at high frequencies, so that the resolution of the spectrogram is matched to human perceptual ability.

Filters—Characteristics and Realizations

On average, speech spectra decrease in intensity with increasing frequency above about 600 Hz, at a rate of about 6 dB per octave. Thus it is common practice in spectrometry to include a simple filter (equalizer) to compensate for this effect, before the actual analysis. This practice makes visible the details in the spectrum at the higher frequencies which might be lost otherwise because of the limited dynamic range of display systems.

The characteristics of the filters are not critical in detail, but there are differences between the performances of filters of different classes (Harris and Waite, 1963) (Fig. 9). Most of the available (ana-

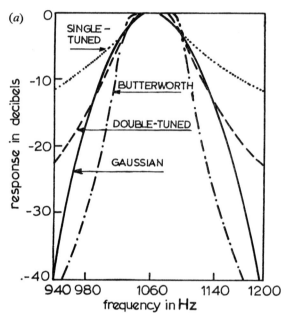

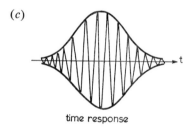

Fig. 9. *Typical speech spectrogram filters.* (a) On decibel scale, after Harris and Waite, by courtesy of the Journal of the Acoustical Society of America. (b) *Gaussian, frequency response.* (c) *Gaussian, time response.*

logue) instruments have used filters consisting of about 2 to 4 resonators (i.e. pole-pairs) adjusted to give fairly flat in-band characteristics with sharp cutoffs.

For the most critical applications it has been found that gaussian filters, i.e. with frequency responses of the form $\exp[-(f-f_0)^2/b^2]$ (f being frequency, f_0 the centre frequency and b a bandwidth measure), with substantially linear phase, give the best apparent resolution in time and frequency simultaneously. This is consistent with the fundamental relations discovered by Gabor (Gabor, 1946) and also corresponds to a gaussian envelope in time (Fig. 9b, c). Such filters are realizable by cascading a number of identical tuned circuits (5 to 8). More critical attention to phase characteristics may be necessary if it is desired to be able to modify the filter characteristics by parallel connection (Shafer and Rabiner, 1971) and this is especially true if the filters are not of equal bandwidths.

Digital processing in speech spectrometry could be carried out with digital filters emulating the analogue system. However, the fast Fourier transform (q.v.) is an algorithm for carrying out a numerical Fourier analysis and has proven to be so efficient that other approaches are unlikely to compete. The operations performed on the speech signal may be shown to be equivalent to filtering.

In a numerical or "discrete" Fourier transform, Fourier coefficients $X_k, k = 0 \ldots N-1$ are calculated for a given record consisting of N samples of the signal x_n, $n = 0 \ldots N-1$ according to the formula

$$X_k = \sum_{n=0}^{N-1} x_n \exp(-j2\pi nk/N), \quad k = 0, 1 \ldots N-1$$

If the samples x_n are taken T sec apart, then the value X_k corresponds to the Fourier component at k/NT Hz. The actual shape of the frequency response is of the form (Fig. 10a)

$$\frac{\sin[\pi NT(f - k/NT)]}{\sin[\pi(f - k/NT)]}$$

which is seen to have undesirably large "sidelobes". More suitable responses are available by windowing the signal x_n before transformation, i.e. the values x_n are weighted by a function w_n which increases smoothly from 0 at $n = 0$ until $n = N/2$, and then decreases smoothly again to zero (Fig. 10b). It is convenient to use a gaussian window, achieving approximately the desirable gaussian performance noted above. Oppenheim (1970) describes a digital computer realization of a speech spectrograph operating on these principles. Performance is similar to that of conventional analogue systems, including analysis time, but the decreasing costs of digital processing, the flexibility, and the elimination of the need for specialized equipment are making the case for the digital approach more attractive except

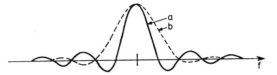

Fig. 10. Frequency responses obtained with discrete Fourier transform filters. (a) Without weighting. (b) With weighting by a raised-cosine.

where large numbers of spectrograms have to be produced.

Display Systems

The display systems used in speech spectrometry may take a variety of forms. By far the most usual is the form shown in Fig. 1, with rectangular time and frequency axes, spectral density appearing as darkness of shading. Special processing can give contour maps of the same form. For real-time displays, somewhat similar presentations are used, but are produced as moving displays on special cathode-ray tubes or phosphor-coated belts (Potter, Kopp and Green, 1947). For digital-computer realizations, cathode ray tube displays with modulated intensity are often used, but other displays on cathode ray tubes or graph plotters are available, such as three-dimensional graphical displays and contour maps which may be produced in stereoscopic form to aid visualization. These graphical methods are preferable when magnitude information is critical, but this is rarely the case and the sectioning technique noted above is usually adequate. Quite useful spectrograms may be made by suitably programming line printers with combinations of symbols and overprinting.

Other Related Analysis Methods

The spectral analysis described above is essentially in terms of sine waves—the various filtering systems respond to appropriate sinusoidal "components" in the signal. Many attempts have been made recently to analyse, describe and transmit speech and other signals in terms of Walsh functions (q.v.). These are orthogonal rectangular waves which take the values +1 and −1 and are thus very suitable for digital calculations. However, since they are not physically related to the responses of real systems, it is not surprising that the resulting spectra are not capable of simple interpretation, and it appears unlikely that this aspect of speech spectrometry will develop.

A function closely related to the power spectrum $S(f)$ is the *autocorrelation function* $R(\tau) = \text{ave}[x(t) \times x(t+\tau)]$ where, for practical speech purposes the average may be taken over intervals of about 10 ms. $R(\tau)$ and $S(f)$ are related to each other by a Fourier transform, and this property is referred to as the

Wiener–Khintchine theorem. Thus $R(\tau)$ contains the same information as the spectrum. Some studies have been made of the use of *correlograms*, displaying the variations of $R(\tau)$ with time in a manner similar to the display of $S(f)$ in spectrograms, but interpretation appears to be not as convenient as that of spectrograms for most purposes.

However, from the autocorrelation it is possible to derive certain coefficients which describe precisely the resonances of the vocal tract (Atal and Hanauer, 1971). These coefficients can represent speech so accurately that they may be used to predict the speech waveform in detail a few milliseconds ahead, and are thus often called *predictor coefficients*. It is probable that signal processing along these lines will significantly influence speech spectrometry in the next few years.

Bibliography

ATAL, B. S. and HANAUER, S. L. (1971) Speech analysis and synthesis by linear prediction of the speech wave. *J. Ac. Soc. Am.* **50**, No. 2 (2), 637–655.

FLANAGAN, J. L. (1972) *Speech Analysis, Synthesis and Perception*. Springer-Verlag.

FLETCHER, H. (1953) *Speech and Hearing in Communication*. Van Nostrand.

GABOR, D. (1946) Theory of communication. *J.I.E.E.* **93** (3), No. 26, 429–457.

HARRIS, C. M. and WAITE, W. M. (1963) Response of spectrum analysers of the bank-of-filters type to signals generated by vowel sounds. *J. Ac. Soc. Am.* **35**, No. 12, 1972–1977.

KOENIG, W., DUNN, H. K. and LACEY, L. Y. (1946) The sound spectrograph. *J. Ac. Soc. Am.* **18**, No. 1, 19–49.

OPPENHEIM, A. V. (1970) Speech spectrograms using the fast Fourier transform. *IEEE Spectrum*, August 1970, pp. 57–62.

POTTER, R. K., KOPP, G. A. and GREEN, H. C. (1947) *Visible Speech*, Van Nostrand, Also as POTTER, R. K. KOPP, G. A. and KOPP, H. C. (1966) *Visible Speech*. Dover.

SHAFER, R. W. and RABINER, L. R. (1971) Design of digital filter banks for speech analysis. *Bell System Tech. J.* **50**, No. 10, 3097–3115.

R. E. BOGNER

SUPERCONDUCTING MACHINES.

1. Introduction

The availability of superconductors enables new forms of electrical machines to be produced because of their capability of carrying very high current

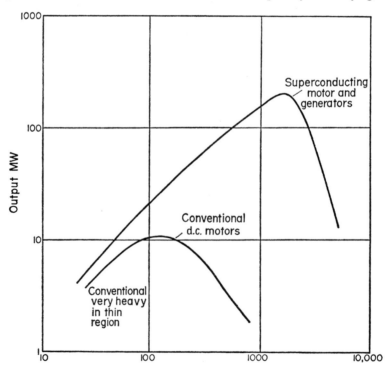

Fig. 1

densities. All rotating electrical machines require a magnetic field which links with an electrical conductor; usually the magnetic field is provided by means of an iron magnetic circuit which is excited by a copper (or some other good conductor) winding. The power consumption of the excitation winding is normally a small fraction of the output of the machine and the magnetic field is limited by saturation of the iron. However, if the iron is removed, the ampere-turns required in the excitation winding to produce even the same magnetic field is enormously increased, and the power consumption of the copper excitation winding for a machine with an air magnetic circuit is completely prohibitive. Thus in a typically conventional d.c. machine the copper field windings may have a few times 10^4 A-turns and dissipate a few tens of kilowatts; if the iron is removed the number of ampere-turns required to produce the same machine flux will increase by two orders of magnitude and the power consumption will be several megawatts. This situation is, of course, quite impractical and the only method which dispenses with the iron magnetic circuit is by the use of superconductors. The latter can not only produce the required ampere-turns without the use of iron and with zero power dissipation, but make possible magnetic flux densities several times those possible with iron magnetic circuits.

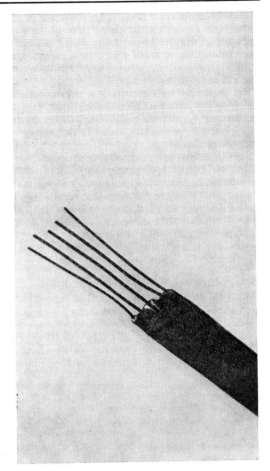

Fig. 3

Fig. 2

A feature, therefore, of all superconducting machines is that they may be produced with ratings far beyond those which are possible with conventional techniques (Fig. 1). The change which we may expect to see as this new field of electrical engineering expands is a competitive situation between conventional and smaller, lightweight, superconducting machines for present ratings and the ability to produce ratings hitherto impossible.

There exists the remarkable situation that within a few years of the commercial availability of superconductors we are able to put these new machines into production. Of course, since few machines have been produced, we cannot answer, categorically, questions of reliability, maintenance and general design "goodness"; however, we can make positive responses on these topics. The overall objective is to demonstrate that the problems which exist for the exploitation of superconducting machines involve straightforward engineering development, and are no more difficult than the development of other systems of comparable magnitude; for example, new gas turbines for air or marine propulsion. Potential users need therefore be no more afraid of applied superconductivity than of say, a new aircraft, even though it involves technologies and techniques with which they might not be familiar.

The potential applications for superconducting machines are numerous. The more important for d.c. motors are marine propulsion, drives to steel rolling mills, mine winders, boiler feed pump drives, compressor drives and, perhaps eventually, drives for airships; appropriate applications for d.c. generators are power for marine propulsion motors, chlorine production and aluminium smelting. As the relative costs of these machines are reduced by improved designs and production methods, the range of applications will be extended. The major interest for superconducting a.c. generators is for central power stations, and rapid developments are expected over the next few years.

2. *Superconductors for Electrical Machines*

Since the discovery of the high field superconductors in the late 1950s there has been a steady improvement in the performance of the superconductors that are available commercially. The earliest material was niobium–zirconium (Nb–Zr) in the form of wires of about 0·010 in. diameter; a large number of small magnets were built using this material, but the results were seldom wholly satisfactory. The reasons for this were due to instabilities in the superconductor (due to flux movement), and—generally—an unsatisfactory design for the cooling of the superconductor. The first electrical machine (Fig. 2) produced at International Research and Development Co. Ltd. employed a seven-strand Nb–Zr cable which was partially stabilized by the addition of copper wire and was completed in 1966. Soon after this, the fully stabilized niobium–titanium (Nb–Ti) superconductor became available commercially. Figure 3 shows an example of the latter which was employed in the large superconducting motor (3250 h.p., 200 rev/min) produced at I.R.D. in 1969; the superconducting winding of this machine (known as the Fawley motor) is shown in Fig. 4 and the machine itself in Fig. 5.

The next development was the very fine filament Nb–Ti subconductor which is intrinsically stable; a micrograph of this is shown in Fig. 6. The stability is created by using small-diameter strands of superconductor embedded in a copper matrix; the diameter of strands may be as low as a few microns. The advantages of this material are very significant and include the ability to change the current in a winding very rapidly; this is very important for d.c.

Fig. 4.

Fig. 5

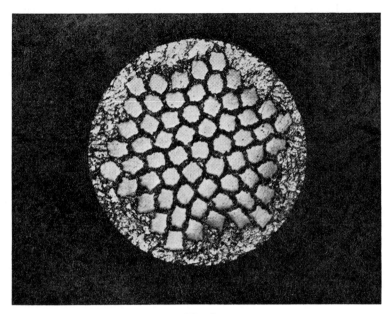

Fig. 6

Fig. 7

generators because it enables the output to be varied with the short response times that are required for many applications. Another important benefit is that the superconductor no longer has to be in intimate contact with the helium coolant, and the winding may be encapsulated in an epoxy resin as shown in Fig. 7.

3. *Superconducting Homopolar Machines*

Over the past few years a considerable amount of effort has been devoted to the superconducting homopolar machine, starting with the first machine produced at I.R.D. on behalf of the Royal Navy in 1966 (Fig. 2). This was followed by the Fawley Motor (Fig. 5) in 1969 and a complete superconducting marine propulsion system in 1973. The latter consists of a motor and generator, both superconducting, together with all the necessary auxiliary and control equipment.

The concept of these homopolar machines is elegant and simple (Fig. 8) and has been adequately covered in the literature (Appleton, 1969; Tinlin and Ross, 1969). The important topics are those which affect the nature of the machine and indeed the very nature of the development programme required to produce a practical machine. There are three major aspects which come in this category: current collection, machine voltage and magnetic screening. According to the decisions which are made in respect of these aspects, very different machines may be evolved, albeit to the same fundamental concept. Other important aspects, though of a less critical nature are the superconducting winding, the refrigeration system and the method of speed control.

Current collection has received a great deal of attention, and at I.R.D. a new method (McNab and Wilkin, 1972a, b) has been developed based upon the use of metal plated carbon fibres; using these techniques, current densities in excess of 800 A/in^2 have been achieved with low voltage drop (typically 0·3 V) and good wearing properties. Homopolar machines using liquid-metal current collection systems have been manufactured, and this method is important particularly where very high currents are required to be collected from a slipring.

The voltage of a homopolar machine is $\emptyset N/60$, where $\emptyset$ is the total magnetic flux linking the armature conductors between a pair of sliprings and N is the speed in rev/min. The terminal voltage may be increased by arranging a large number of Faraday discs in series or alternatively the segmented slipring concept developed by I.R.D. may be employed.

The screening of the flux outside the machine may be achieved by using iron, but this may not be desirable if a high power to weight ratio is required. An alternative method is to use a separate superconducting winding as indicated schematically in Fig. 9.

The superconducting windings in the machines must be cooled to liquid-helium temperature and must be protected from heat in-leak by a vacuum vessel called a cryostat. With the encapsulated windings which are possible with multifilament superconductor, the coolant may be either liquid helium at nominally atmospheric pressure, or supercritical helium which may be at a pressure of about 10 atmospheres.

During 1973 the major activity in respect of superconducting homopolar machines was for marine pro-

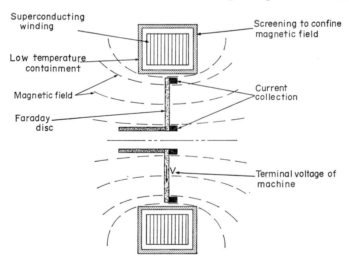

Major features which affect entire nature of development / design programme

Fig. 8

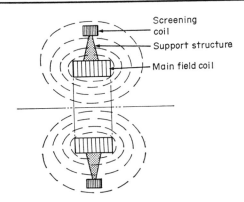

Fig. 9 Screening coil on homopolar machine

pulsion systems, and substantial developments in this field are anticipated; the other applications mentioned above will gradually begin to employ superconducting d.c. motors and generators.

4. Superconducting a.c. Generators

The ratings of conventional a.c. generators have been increasing continuously. Since World War II they have increased from 60 MW to 660 MW and 1300 MW machines are under manufacture. 1300 MW represents the largest machine that can be made at present with a single piece rotor forging; above this it becomes difficult to transport the heaviest portion of the machine, namely the stator core. Extrapolation of existing designs may well require that some of the manufacturing procedures be carried out at site, and a consensus of different manufacturers would probably yield different limiting ratings based upon conventional designs, but this is likely to be in the region of 2000 MW.

Superconductivity provides the means for an extension of manufacturing techniques, and studies have indicated that the limit of the ratings of superconducting a.c. generators is far in excess of 2000 MW. Thus, if it can be shown that the superconducting machines may be produced and operated reliably, and that their capital and operating costs are at least no higher than conventional generators, then a case exists for their development.

Studies have indicated that the breakeven point on capital cost between conventional and superconducting generators is around 500 MW and the advantage in favour of the superconducting machines increases as the size of the generators increases. It is also expected that the superconducting machines will show a slight advantage in respect of efficiency.

Consider a 500 MW superconducting generator which is required to meet all the conditions laid down by the British Central Electricity Generating Board for conventional machines. Most of the problems associated with the design of even bigger generators are likely to appear at this rating, but this will not be true for relatively small machines because of scaling effects.

The machine output is governed by

$$S = kBAND^2L \text{ volt amperes}$$

where k = constant, B = average flux density at the armature winding (tesla), A = specific electrical loading of the stator winding (A/m), L = active length of the machine (m), D = armature winding diameter (m), N = speed (rev/min).

In a conventional machine D is closely tied to the rotor diameter which in turn is controlled by the maximum permissible centrifugal loading. A is limited by the cross-section of copper that can be placed in the slots. Hence, increased output has to be obtained by increasing L, and, if necessary, running the rotor above its first critical speed. In a superconducting machine the restriction upon the rotor diameter remains. However, the restriction upon the diameter of the armature winding is removed and D may be selected from other considerations. Thus the machine may have a large effective air-gap and this is made possible by the large number of ampere-turns available on the rotor. One design possibility is a machine in which the flux density at the stator winding and the electric loading are similar to a conventional machine. Thus, in spite of similar rotor diameters on both conventional and superconducting machines, the stator winding diameter is greater on the latter, allowing the active length L, to be greatly reduced for the same output. Hence, the superconducting machine will be shorter than its conventional counterpart.

Figure 10 shows an outline drawing of one design which has been considered at I.R.D. It has two poles, with a rotating superconducting field winding and a stationary armature winding at ambient temperature. The machine has a double rotor structure consisting of an inner low-temperature member which supports the field winding and an outer rotor at ambient temperature which is designed to withstand the short-circuit torque and to screen the field winding from negative sequence and rapidly changing fluxes. Associated with the inner rotor structure is an intermediate temperature radiation screen and liquid and gaseous helium cooling ducts. Helium is led into the rotor through a rotating seal and is taken to the winding through a radial duct and heat shunt. After cooling the winding, the helium is used to cool the cone end support structure before being led back into the refrigerator.

Alternative design concepts in which all the windings are superconducting or which have a stationary superconducting field have been considered and rejected because (i) there is no superconductor capable of operating with the required performance at power frequencies, and, even if there were, it is doubtful whether the additional cost would be worth the improvement in efficiency obtained; and (ii) a sta-

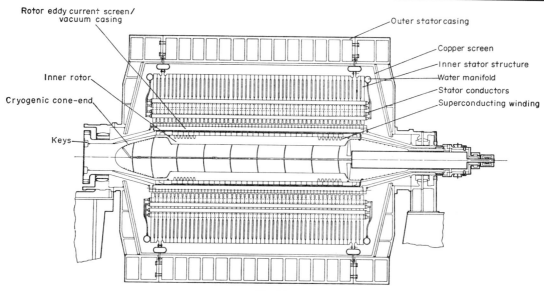

Fig.10

tionary field winding would mean that all of the c.a. power would have to be removed from the rotor through sliprings, and there would therefore be a very great risk of short circuits occurring at the machine terminals. This is not the only objection to this approach, and other workers have reached the same conclusion.

It is virtually certain that the superconducting a.c. generator will be developed and go into production; the time required to reach the production status is at least 15 years because a number of experimental and pre-production machines must first be produced.

Bibliography

APPLETON, A. D. (1969) Motors, generators and flux pumps, *Proc. Int. Conf. on Low Temperature and Electric Power*; Royal Society, London, March 1969 (under auspices of Int. Inst. of Refrigeration in conjunction with Brit. Cryogenic Council).

HARROWELL, R. V. (1969) Reciprocating superconducting generator-motor, *Nature*, **222**, 598.

HAYDEN, J. T., Superconducting generators for aircraft, *Proceedings of 4th Int. Cryo. Eng. Conf., Eindhoven, May 1972*, Paper M4 (Pergamon Press).

LORCH, H. O. (1973) Feasibility of turbogenerator with superconducting rotor and conventional stator, *Proc. IEEE*, **120**, 221.

MCNAB, I. R. and WILKIN, G. A. (1972a) Carbon fibre brushes for superconducting machines, *J. Inst. Elect. Eng. Electronics and Power*, pp. 8–10, January 1972.

MCNAB, I. R. and WILKIN, G. A. (1972b) Life tests with carbon fibre brushes, *6th Int. Conf. on Elect. Contact Phenomena*, Chicago, 5–9 June 1972.

TINLIN, F. and ROSS, J. S. H. (1969) Cryostat and refrigerator for a 3250 hp superconducting motor, *Proc. Int. Conf. on Low Temperature and Electric Power*; Royal Society, London, March 1969 (under auspices of Int. Inst. of Refrigeration in conjunction with Brit. Cryogenic Council).

WOODSON, H. H., SMITH, J. L., THULLEN, P. and KIRTLEY, J. L., The application of superconductors in the field windings of large synchronous machines, *Trans. IEEE, Power Apparatus and Systems*, **PAS-90**, March/April 1971.

A. D. APPLETON

SUPERCONDUCTIVITY, APPLICATIONS OF.

There are two quite different kinds of superconductor, known as Type I or soft superconductors and Type II or high field superconductors. Each type exhibits characteristic properties which make it useful in certain applications.

Type I superconductors will allow a magnetic field to penetrate only a very small distance below the surface (London penetration depth $\sim 1/10$ μm), the bulk of the superconductor is completely screened. Electric currents are also carried only in the surface layer and alternating currents may be carried with very little power loss. At relatively low values of magnetic field, however (0.01 to 0.1 tesla), the superconductor will revert to the resistive state.

Type II superconductors can allow bulk penetration of magnetic field and are thus able to remain superconducting up to much higher fields. The field penetration is a dissipative process, however, and Type II superconductors therefore exhibit a marked hysteresis loss when exposed to changing magnetic fields, so they are not very useful for alternating currents. Nevertheless their high field capability has proved very attractive to magnet constructors and magnets have so far accounted for almost all the large-scale applications of superconductivity.

Applications of Type I Superconductors

Probably the most useful single attribute of Type I superconductors is their ability to carry alternating currents with very little power dissipation. And probably the most natural application of this attribute is in the distribution of electric power—*the superconducting cable*. Many detailed design studies for superconducting cables have been published, measurements have been made on losses in the superconductor, dielectric strengths at low temperature, etc.; a few small prototypes have been built. The general conclusion seems to be that, although the superconducting cable will probably never be able to compete economically with conventional overhead lines, it should show a cost saving in situations where the cable must go underground and where very high power levels are involved. Even in the most favourable situations, however, the construction of such a cable presents severe technical problems. The very shape of the cable, with its large surface area and great length, makes it difficult to provide adequate thermal insulation and refrigeration. If the cable is manufactured in sections, joints will have to be made "in the field" and this means joining not only the conductor but also the outer vacuum jacket, heat shields and cold helium supply lines. Provision must be made for differential thermal contraction. And finally the cable must be as reliable as its conventional counterpart and able to withstand considerable overload currents. It seems likely that in time all these problems can be solved but it is not at present clear whether they can ever be solved economically.

A *superconducting transformer* would, in principle, be free from many of the problems listed above. It would be a compact unit which could be assembled in the factory and would be relatively easy to refrigerate. A major difficulty arises, however, from the low critical fields of Type I superconductors. This means that the fields throughout the transformer must be kept down to a low level ~ 0.05 T and so makes it very hard to arrive at an economic design. Type II superconductors would allow much higher fields but the a.c. dissipation would be excessive. For this reason, superconducting transformers are not presently exciting much interest.

At very much higher frequencies, in the microwave range, *superconducting cavities* are now being evaluated at several laboratories for use in linear accelerators. The most popular superconductor for this application is pure niobium which, although it is strictly speaking a Type II superconductor, should nevertheless be counted Type I when, as in this case, it is used below its lower critical field. The improvement factor of a cavity is defined as the reduction in r.f. power dissipation compared with a room-temperature copper cavity of similar size and shape. Improvement factors of $2-5 \times 10^4$ have been observed in prototype cavities and this reduction in power requirement could enable the accelerator klystrons to be run "continuous working" rather than pulsed. The yield of accelerated particles could thus be enormously increased.

On a much smaller scale, a whole range of electronic applications for superconductors is developing. The *cryotron* is a superconducting switch which could be useful as a computer memory element. The *superconducting bolometer* has been used as a radiation detector in the far infra-red; it depends for its operation on the very rapid variation of resistance with temperature in a superconductor which is just below its critical temperature. The *Josephson effect* is now being used to produce sensitive detectors for use at infra-red and microwave frequencies and is also the basis of some remarkably precise instruments for measuring current, voltage and magnetic field. The measurement of voltage depends only on fundamental constants and may be used as a standard. The variation of penetration depth with magnetic field may be used to vary the self-inductance of a thin film device and thus form the basis of a *parametric amplifier* for use at microwave frequencies. The non-linear characteristics of superconductors and superconducting junctions can also lend themselves to the detection, mixing or amplification of microwave signals. For a more detailed review of "cryoelectronics" see the paper by Kamper.

Technology of Type II Superconductors

High field superconducting materials were first discovered and developed at the Bell Telephone Laboratories in the early 1960s. Figure 1 shows how the current-carrying capacities of these new superconductors can persist up to very high fields. Conventional magnets, using copper conductors, are usually restricted to rather low fields by power dissipation in the copper. In order to ease this restriction, an iron yoke is generally used to reduce magnetic reluctance and allow a given field to be produced by a smaller current in the windings. This reduction in current is typically a factor of 10 and so the power requirement is reduced by 100. Iron yokes are therefore used in almost all conventional electromagnets, dynamos, motors, transformers, etc. At fields of more than 2 T, however, the beneficial effect of the yoke falls off rapidly as the iron saturates and most conventional electromagnetic devices are therefore limited to fields of about 2 T. This is a purely economic limit but is nevertheless a very real one and the only conventional magnets producing fields significantly higher than 2 T are a few research solenoids which consume many megawatts of power to produce fields of up to 20 T in volumes of about a litre. Superconductors, by removing the problem of power dissipation, allow high fields to be produced more economically, in

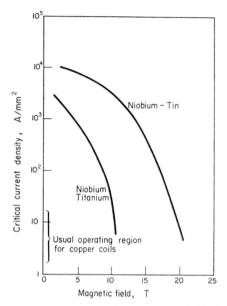

Fig. 1. Critical current density versus fields for two high field superconductors, niobium titanium and niobium tin.

large volumes, without the need for an iron yoke. The upper limit is now the critical field of the superconductor, although practical experience has shown that a more realistic limit is about 75% of the critical field. Figure 1 shows that these fields are very much higher than the 2 T iron limit.

The properties shown in Fig. 1 are for the two superconductors of major technological importance: niobium tin and niobium titanium. Current densities of this magnitude are without precedent in electrical engineering—two or three orders of magnitude higher than normal copper conductors. Although niobium tin has the best electrical properties, it has very poor mechanical properties; it is a weak, brittle intermetallic compound. For this reason it has not been so widely used as niobium titanium which is a strong ductile alloy.

The early attempts to built superconducting magnets soon showed that the level of performance indicated in Fig. 1 (measured on short samples) was not normally reached. Magnets generally became resistive or "quenched" at currents which were typically an order of magnitude below those shown in Fig. 1; the effect was called "degradation". It is now known that degradation is caused by "flux jumping"—a sudden lossy movement of magnetic flux into the superconductor which releases energy inside the magnet.

Cryostatic stabilization

The first reliable cure for degradation was devised in 1965 and has since become known as cryostatic stabilization. It makes use of a composite conductor in which a good normal conductor such as copper is connected in parallel with the superconductor. When flux jumps occur in the superconductor, the magnet current is able to switch temporarily to the copper and a quench is avoided. Cooling channels are provided so that the joule heating in the copper can be carried away by the liquid helium. The temperature rise resulting from the flux jump is thus restricted to a fraction of a degree and the magnet current can soon switch back to the superconductor.

In recent years the principle of cryostatic stabilization has been used in the construction of several very large superconducting magnets; a good example is the bubble chamber magnet at the Argonne National Laboratory in the U.S.A. This magnet has been in regular operational use since 1970 and produces a field of 1·8 tesla in a bore of 3·6 m diameter. Figure 2, which shows part of the winding during assembly, gives some impression of the enormous scale of this magnet. A conventional magnet of the same size would consume about 10 MW of electric power. The only power needed by the superconducting magnet is for the refrigerator which, if it were quite separate from the other cryogenic equipment of the bubble chamber, would use about 200 kW to extract about 400 W from the low-temperature enclosure. The decision to make the magnet superconducting was made on purely economic grounds; the power savings over the expected 10 years of operation should easily outweigh the additional capital costs of superconducting materials and refrigeration.

Cryostatic stabilization has in fact opened up the whole range of large magnet applications and magnets can now be designed which are reliable and economic. It has, however, one great disadvantage: the large amount of copper which must be used in parallel with the superconductor. This means that the high current densities shown in Fig. 1 are diluted, sometimes by as much as 100. To a first approximation, a fixed number of ampere turns is needed to produce a given field, so that

Fig. 2. The lower coils of the Argonne Bubble Chamber Magnet during assembly. (Courtesy Argonne National Laboratory.)

a magnet winding working at low current density will be much more bulky than a winding working at high current density. A bulky winding is not a serious problem in very large magnets where the winding is not a large proportion of the total volume. But in small magnets or magnets which must produce high field gradients or where weight is important, a high current density is essential. The recent development of filamentary composite superconductors has now made this possible.

Filamentary superconductors

Degradation of magnet performance is usually caused by flux jumping and this phenomenon will first be described briefly. Figure 3 illustrates the pattern of shielding or magnetization currents which are set up in a high field superconductor when a magnetic field is applied to it. Magnetization currents are set up in the windings of a superconducting magnet when the field is increased. These currents are unstable and when they collapse flux moves suddenly into the superconductor—a flux jump.

The reason for this sudden collapse is perhaps most easily seen in terms of a feedback loop. Let a small temperature perturbation be applied to the conductor of Fig. 3. This causes the current-carrying capacity of the superconductor to be reduced; more flux is thus able to penetrate the wire and the pattern of field across the diameter changes from the solid line to the dotted line. The flux change is roughly represented by the shaded area. As has already been mentioned, the penetration of magnetic flux into a Type II superconductor is a dissipative process which releases heat. This causes a rise in temperature and thus completes the (positive) feedback loop:

where ΔT, ΔJ, $\Delta \Phi$ and ΔQ are the increments of temperature, current density, magnetic flux and heat. If the ΔT resulting from the loop is greater than the original perturbation, the disturbance will grow, feeding on the energy of the magnetic field, until the temperature rise is sufficient to quench the superconductor. This is in general the case and flux jumps will thus occur throughout the windings of a superconducting magnet and cause degradation.

To overcome this problem it is necessary to weaken one or more of the links in the feedback loop to such an extent that the resulting ΔT becomes less than the initial perturbation. If this is done the disturbance will simply die away and the conductor will be stable against temperature disturbances. Similar arguments can also be used to show that it will also be stable against disturbances in the field or current level.

The most successful approach so far has been

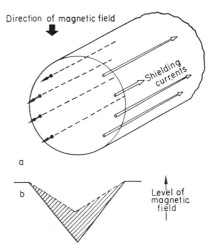

Fig. 3. (a) Cross-section of single superconducting wire, showing how shielding currents depress the field level within the wire. (b) Magnetic field plotted across diameter of wire, showing that the effect of a temperature perturbation is to reduce the shielding currents and cause a change in magnetic flux.

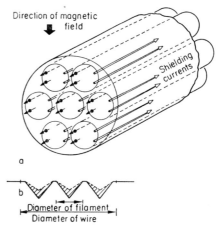

Fig. 4. (a) Cross-section of a wire consisting of a number of superconducting filaments, showing the shielding currents associated with each filament. (b) Magnetic field plotted across the diameter of the wire, showing the change in magnetic flux for each filament. The flux change for a given temperature perturbation is less than Fig. 3b. (The unbroken line represents the initial field; broken line the perturbed field; and the shaded area the change in magnetic flux, in each case.)

to subdivide the conductor into fine filaments. This modifies the pattern of shielding currents as shown in Fig. 4 and greatly reduces the flux change (represented by the shaded area) caused by a given reduction in current density. The link $-\Delta J \to \Delta \phi$ is thus weakened and it may be shown that, below a certain filament diameter (which for niobium titanium is about 50 μ) the loop becomes stable, flux jumps do not occur.

Figure 5 shows the cross-section of a typical filamentary composite superconductor; it contains 361 36-μ diameter filaments of niobium titanium superconductor embedded in a matrix of copper. Magnets made from these conductors have been found to behave in a stable fashion and to reach

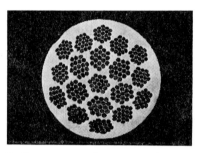

Fig. 5. Cross section of a superconducting composite containing 361 filaments of niobium titanium in copper (mag. × 27). (Photo Imperial Metal Industries Ltd.)

full current and field without degradation. The copper matrix makes it possible to handle such fine filaments both during manufacture and magnet construction. It does, however, have the disadvantage of causing the filaments to be coupled together in a changing magnetic field. The effect of this coupling is to make the pattern of shielding currents tend towards that shown in Fig. 3 rather than Fig. 4. Unstable behaviour is again exhibited by this pattern and the advantages of fine subdivision are lost. It has been found, however, that the coupling can be greatly reduced and stable behaviour restored if the composite is simply twisted about its axis so that the filaments follow helical paths, like the strands of a rope.

Another useful function which the copper matrix can perform is to protect the magnet from burn-out in the event of a quench. Type II superconductors have a very high normal state resistivity and operate at very high current density. If a section of the winding should for any reason revert to the resistive state, very high levels of voltage and joule heating can result, and will be sustained by the magnet self-inductance. Burn-out or voltage breakdown can thus occur. The copper matrix, by greatly reducing the normal state resistance, can reduce this voltage and heating and so protect the magnet. The quantity of copper required is much less than needed for cryostatic stabilization so that the advantages of high current density are still retained.

A further advantage of fine subdivision of the superconductor is that it reduces the hysteresis loss caused by changing currents and fields. Unfortunately the reduction is not sufficient to make the operation of Type II superconductors at 50 Hz a practical proposition, but is does allow magnets to be pulsed at 0·1–1 Hz.

A good example of a potential application which would fully utilize the attributes of filamentary superconductors is the superconducting synchrotron accelerator. This requires magnets which can be pulsed from zero to maximum field in a few seconds. Present machines use conventional magnets with iron yokes producing transverse fields of about 2 tesla over an aperture of about 10 cm diameter. The magnets are several metres long and up to a thousand units are needed in a large machine. The use of superconducting magnets could increase the field and hence the energy of accelerated particles by a factor of 3 or more. Filamentary conductors offer two advantages here: firstly, their reduced hysteresis loss enables the magnets to be pulsed without causing an undue refrigeration load; secondly, their high overall current density allows the magnets to be very compact. Prototype magnets have now been constructed in many laboratories and it is quite possible that the superconducting synchrotron will eventually prove to be both feasible and economic.

Applications of Type II Superconductors

The following list, which is by no means complete, includes some activities where the use of superconductivity is well established, some which are still at the prototype stage and some which are only design studies. This is, however, a technology which is still at a very early stage of development and it is hoped that the range of application will broaden considerably over the next few years.

Plasma physics and nuclear fusion

Superconducting magnets are already being used to produce a variety of special field shapes needed for plasma physics research. An interesting example is the "Levitron" in which a superconducting ring carrying a persistent circulating current is levitated inside a vacuum tank, containing a plasma, by an external system of superconducting magnets. No mechanical supports or current leads to the ring can be tolerated because they would interfere with the plasma. This requirement can only be met by superconductors because of their unique persistent current property. Two levitrons are presently in operation at the Livermore and Princeton laboratories in the U.S.A. and a third is being commissioned at the Culham laboratory, U.K. If controlled nuclear fusion eventually becomes feasible, extremely large superconducting magnets are going to be needed for plasma containment. Recent studies indicate that fields of at least 10 T over volumes of 1000 m^3 will be needed for an economically viable fusion reactor.

Rotating electrical machinery

Conventional machinery makes extensive use of iron to reduce magnetic reluctance and this accounts for most of the weight of the machine. With a superconducting winding the iron would not be needed and the size and weight could be reduced by an order of magnitude. The option of working at higher fields also allows a further reduction in size. Both d.c. and a.c. machines are now being studied. The most popular d.c. machine is the homopolar type which has been extensively studied by International Research and Development Ltd., U.K. A large 3250-horse-power prototype motor has recently been tested on full load in a power station. The motor uses a copper armature disc at room temperature, carrying a radial current, which is made to rotate in a 3·5-T axial magnetic field produced by a cryostatically stabilized superconducting solenoid magnet of 2·5 m inner diameter. A generator working on the same principle is now under construction.

The only a.c. machine to have received serious attention is the alternator with a rotating, d.c., superconducting field winding and a stationary copper armature. An 87-kVA prototype has been successfully operated at the Massachusetts Institute of Technology and a 2-MVA machine is currently under construction, also at M.I.T.

Magnetohydrodynamic power generation

If a hot ionized gas is forced through a transverse magnetic field, a potential difference is generated across the gas stream. This is the basis of MHD power generation and prototype units have been built in Europe, the U.S.A. and Russia. It is not yet clear whether MHD power generation is going to be able to compete effectively with more conventional techniques but if it can, superconducting magnets will be needed to provide the large volumes of 5–10-T field which are required.

High-energy physics research

Two examples have already been quoted from this field of research: the large bubble chamber and the synchrotron. Two more large bubble chamber magnets have recently been completed at CERN (European Organization for Nuclear Research) in Geneva and at the National Accelerator Laboratory, Illinois, U.S.A. Both magnets have bores of 3·7 m diameter and use cryostatically stabilized conductors to produce fields of about 3·5 T. The "Omega" magnet, another large solenoid also at CERN, uses a slightly different form of cryostatic stabilization in which a hollow conductor is force-cooled by cold supercritical helium pumped through the conductor.

Many smaller magnets are also used in a large accelerator laboratory to guide and focus the beams of high-energy particles. Conventional magnets are used at present but superconducting replacements, producing higher fields, are starting to be built.

High-energy physics can also be studied using cosmic rays and balloon-borne experiments incorporating superconducting magnets are now being attempted. Lightness and compactness are the main reasons for using superconductors in this application.

Solid state physics research

On a much smaller scale, superconducting magnets have had a considerable impact on experimental solid state physics by making fields of up to 10 T readily available in the laboratory. Small superconducting solenoids have been sold commercially for several years and many hundreds are now being used for research into nuclear magnetic resonance, Mossbauer effect, magneto optical effects and many other phenomena. Figure 6 shows a typical commercial solenoid intended for research use. Electron microscopes also are gaining in resolving power by the use of high strength superconducting lenses.

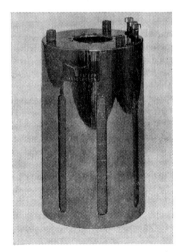

Fig. 6. Superconducting solenoid for research use, producing a field of 8 T in a 50-mm diameter bore. (Courtesy Oxford Instrument Company Limited.)

Aerospace applications

Lightness and compactness are of course at a premium here. Alternators for aircraft have been proposed using superconducting field windings. An airborne MHD plant operating in the exhaust of a jet engine is planned by Ferranti-Packard in Canada.

Large magnets could provide space craft on long missions with a lightweight shield against charged particles. It may also be possible to use the MHD idea in reverse as a propulsion system. An even more spectacular idea is the "interstellar ramjet" which would use an extremely large superconducting magnet to sweep up ionized interstellar matter. This material would then be burnt in a fusion reactor and thus used to propel the spacecraft.

Energy storage

At 5 T, the energy density in a magnetic field is 10^7 J/m^3. This is considerably higher than a capacitor bank, for example, and superconducting energy storage magnets are undoubtedly going to be used to provide pulsed power for applications such as high-power lasers, plasma physics experiments and electromagnetic metal forming. An energy-storage system for powering a superconducting synchrotron has been proposed at the Rutherford Laboratory, U.K., and a small prototype is now under construction.

High-speed ground transport

At speeds above 200 m.p.h., all wheeled vehicles start to encounter severe difficulty and yet it seems certain that the industrialized nations will eventually require some kind of high-speed ground-transport system. The magnetically levitated train has been proposed as a way of getting to much higher speeds by using a magnetic field to support and guide a vehicle so that it makes absolutely no contact with its track. The track is a sheet of normal conductor such as copper, at room temperature. Lightweight superconducting magnets are mounted on board the train so that, as the train moves along, eddy currents are induced in the track and a repulsive force is set up. Wheels are still needed at low speeds but at speeds above about 50 m.p.h. the train floats above the track. A linear motor could be used to drive the vehicle, giving a very elegant support and propulsion system based solely on electromagnetic fields and having no moving parts. An experimental four-passenger vehicle has already been tested in Japan and it could be that superconductors will eventually provide a fast, clean and quiet way of travelling between our cities.

Bibliography

FISHLOCK, D. (ed.) (1969) *A Guide to Superconductivity*, Elsevier.
HADLOW, M. E. G. *et al.* (1972) *Proc. IEE, IEE reviews*, **119**, No. 8 R.
KAMPER, R. A. (1969) *Cryogenics*, **9**, No. 1, 20.

M. W. WILSON

SUPER-HEAVY ELEMENTS—ELEMENTS BEYOND THE KNOWN PERIODIC TABLE.

Approximately 300 isotopes of nearly all the elements containing from one to ninety-four protons have been found to exist in nature. Furthermore, during the past 35 years some additional 1200 isotopes have been produced artificially, the heaviest consisting of 105 protons and 157 neutrons. Synthesis of increasingly heavier elements is quite difficult. With increasingly larger numbers of protons in a nucleus, the disruptive Coulomb repulsive forces between the protons begin to overcome the cohesive nuclear forces and nuclei become unstable toward disintegration into several fragments, i.e. alpha decay and fission. As late as 1965, it appeared that knowledge of the periodic system could not be extended to appreciably higher atomic numbers due to this instability (short half-lives) and limitations in the means to produce heavier nuclei. However, since 1965, many theoretical calculations have predicted that small regions of relatively stable nuclei, super-heavy nuclei, may exist far beyond the present limits of the periodic table. These predictions, coupled with recent advances in accelerator technology, have provided the impetus for nuclear scientists to search for these new nuclei with the expectation of being able to study an entirely new class of elements possessing nuclear and chemical properties quite different from the known elements.

Estimated Half-lives for Super-Heavy Nuclides

Nuclei possess neutron and proton shell structures analogous to the shell structures of electrons which determine atomic properties. When a shell of neutrons or protons is filled, the nucleus has a relatively lower energy and hence an increased stability. Such closed nucleon shells and enhanced stability have already been well established for proton and neutron numbers of 2, 8, 20, 28, 50 and 82 as well as 126 neutrons. It was previously assumed that the next closed-shell configuration for protons would occur at atomic number 126, analogous to the 126 neutron closed shell. This prediction was quite disappointing, since nuclei with such large numbers of protons are generally considered to be too far beyond known elements to be produced by experimenters at this time. However, more recent theoretical calculations almost uniformly conclude that the next closed proton shell will occur for a lower, and perhaps attainable, atomic number of 114 protons. These same calculations also predict that the next closed neutron shell should occur at 184 neutrons. Thus the nuclide $^{298}114$, containing both closed-shell configurations of 114 protons and 184 neutrons, is expected to have substantially increased nuclear stability. Furthermore, the stabilizing influence of closed-shell nuclear configurations extends over a range of proton and neutron numbers. Consequently, it is expected that a number of the isotopes of ele-

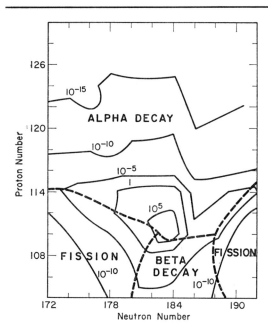

Fig. 1. Theoretically calculated total half lives (in years) for super-heavy nuclides containing even numbers of protons and neutrons.

ments in the vicinity of $^{298}114$ will also be relatively stable.

Several theoretical calculations of the total half-lives of super-heavy nuclides have been performed considering all the possible decay modes, i.e. alpha decay, beta decay and spontaneous fission. The results of one such calculation by Fiset and Nix are shown in Fig. 1. Total half-lives for isotopes with even numbers of protons and neutrons are shown as contours as a function of proton and neutron number. The most probable modes of radioactive decay are also indicated in Fig. 1 as labelled areas enclosed by dashed lines. Half-lives of nuclides containing odd numbers of particles are much more difficult to calculate since they depend to some extent on the exact quantum numbers of the odd particle and, therefore, are not shown here. However, in general the half lives of nuclides containing odd numbers of particles are longer than neighboring nuclides with even numbers of particles. As can be seen in Fig. 1, isotopes in the range of ~110–114 protons and ~180–184 neutrons are predicted by this calculation to have half-lives greater than one year, the longest calculated half-life being $10^{9.4}$ year for $^{294}110$. As can also be seen in Fig. 1, the half-lives of isotopes slightly removed by a few nucleons from this small island of enhanced stability decrease drastically.

Closed-shell nucleon configurations have also been predicted to occur for 164 protons and 318 neutrons. However, these latter predictions have not generated a great deal of interest since they are far more tenuous than those already discussed and these isotopes are experimentally unattainable at this time.

Chemistry and Electronic Configurations

The chemistry of the transactinide elements (atomic numbers greater than 103) should prove to be quite interesting. Many predictions of their chemical properties have already been made, primarily in conjunction with the design of experiments to search for, isolate and identify these elements. Chemical properties of elements are strongly influenced by the order of filling of the electronic orbitals. The electronic configurations of the transactinide elements have been predicted theoretically by many investigators using self-consistent field calculations (Hartree–Fock calculations). The predicted positions of the transactinide elements in the periodic table resulting from these calculations are shown in Fig. 2. Predictions of specific chemical properties based upon these calculations are difficult to make since the correlations between electronic ground-state configurations and the chemistry of elements are not straightforward. However, extrapolations of chemical properties can be made for elements with similar electronic configurations. The transactinide elements with atomic numbers 104–112 are expected to fill the $6d$ electron shell, analogous to the filling of the $5d$ shell for the elements hafnium (72) through mercury (80). Therefore, these elements are expected to have chemical properties similar to their next lighter congeners shown in Fig. 2. For example, element 112 is predicted to have chemical properties similar to mercury with an atomic number of 80, element 114 similar to lead (82), etc. Elements 113 to 118 are expected to fill the $7p$ electron shell similar to the elements thallium (81) through radon (86) which fill the $6p$ shell; elements 119 and 120 should fill the $8s$ shell analogous to the filling of the $7s$ shell for francium (87) and radium (88). The electron configurations and chemical properties of elements with atomic numbers from 121 to approximately 153, designated by Seaborg as the "super-actinide" series, are more difficult to predict because of the close proximity in energy of the $5g$, $6f$, $7d$ and $8p$ subshells. It is expected that after addition of some electrons to the $7d$ and $6f$ subshells, the $5g$ subshell will be filled with possibly some involvement of the $8p$ subshell. Since the binding energies of these electrons are quite similar, the elements in this series will probably exhibit a multitude of oxidation states leading to very complicated chemistry.

Nuclear Properties

The nuclear properties of the more stable super-heavy elements in the range of atomic numbers 110–114 are predicted to be quite different from the known actinide elements. Actinide nuclei are known

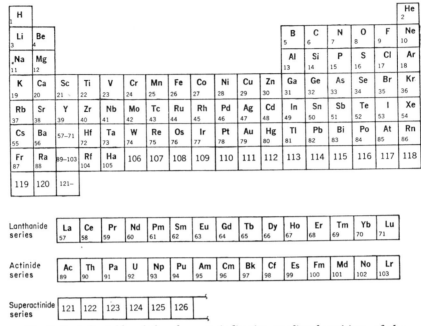

Fig. 2. Periodic table of the elements indicating predicted positions of the transactinide elements.

to have an approximately ellipsoidol shape with the major axis being approximately 20% greater than the minor axis whereas these super-heavy nuclei are expected to be nearly spherical. The super-heavy elements are also expected to be much more unstable than the actinides to small shape changes or shape oscillations at excited energies which may cause considerable difficulties in the production of these nuclei using nuclear reactions. Because of the increased Coulomb energy of the super-heavy nuclei, considerably more energy will be liberated during fission of these nuclei as compared to known nuclei. For example, in the thermal-neutron induced fission of a uranium nucleus approximately 200 MeV is released whereas in the spontaneous fission of element 114, calculations indicate that over 300 MeV will be released. Most of this released energy appears as kinetic energy of the fragments. However, an appreciable fraction of the energy release will also appear as internal excitation energy leading ultimately to emission of neutrons from the fragments. In the fission of actinide elements, ~ 3 neutrons are emitted per fission, whereas 10.5 neutrons are calculated to be emitted in the fission of a super-heavy element such as element 114. This large neutron yield per fission has many practical implications and has provided additional incentives to search for these elements.

Search for Super-Heavy Elements

Many searches for transactinide, super-heavy elements, have already been performed based upon the predicted chemical and nuclear properties of these elements. Comprehensive review articles describing these studies have been recently published by S. G. Thompson and C. F. Tsang (1972) as well as I. Zvara (1970). Since several theoretical calculations indicate that the half-lives of some isotopes of these elements could be as long as 10^9 years, there is a small probability that these elements could have been formed at the same time ($\sim 4.5 \times 10^9$ years ago) as the other elements of the Earth and still exist on the Earth today. Furthermore, small numbers of particles with atomic numbers greater than 100 have been observed as cosmic ray events at high altitudes by Fowler et al. (1967). If super-heavy elements are produced in cosmic events, such heavy nuclides could conceivably have been deposited and exist in small amounts on the surface of the Earth with much shorter half-lives ($\sim 10^4$ to 10^5 years). Based upon these assumptions, many minerals and ores containing elements such as platinum, gold, lead, etc., have been examined in detail for the presence of super-heavy elements. Presumably, the super-heavy elements would concentrate in those ores and minerals which contain high concentrations of elements with similar chemical properties to the super-heavy elements. Using a great variety of experimental methods such as X-ray fluorescence, mass spectrometric analysis, neutron activation analysis, spontaneous fission detection, fission-neutron counting, etc., upper limits have been established for the presence of super-heavy elements in these ores and minerals as well as a few lunar samples of generally less than 10^{-11} grams per gram of material and in a few cases as low as 10^{-17} grams per gram. The lowest limits have, in general, been established by Price *et al.* (1970) who searched for fission tracks from the spontaneous fission of possible super-heavy elements accumulated over periods of $\sim 10^8$ years in ancient minerals.

Due to the lack of success in finding super-heavy elements in nature, many attempts are now in progress to produce these elements artificially by use of nuclear reactions. The elements plutonium through fermium have in the past been most effectively produced by multiple neutron capture reactions in high-flux reactors. However, it now appears that elements beyond fermium (atomic number 100) cannot be produced by multiple neutron capture since the build-up of elements by this method terminates at ^{258}Fm, an isotope which decays by spontaneous fission with a short half-life of $3 \cdot 8 \times 10^{-4}$ sec. Elements 101 through 105 have been produced by bombardment of small quantities of heavy elements such as uranium through californium with accelerated ions of helium through neon. The production of heavier elements using such heavy-ion nuclear reactions has been hampered during recent years since existing charged-particle accelerators are not able to accelerate large intensities of heavier projectiles to the energies required to initiate nuclear reactions with heavy-element target materials. However, the interest in super-heavy elements as well as the scientific potentials of future heavy-ion research has led to the funding and construction of several new accelerators, the Super-HILAC at the Lawrence Berkeley Laboratory in the United States and the UNILAC at the Gesellschaft für Schwerionen, Darmstadt, West Germany, which will be capable of accelerating heavy-ion projectiles over the entire range of the periodic table to energies sufficiently high to interact with all available target materials. Table 1 shows examples of types of nuclear reactions that might be used to synthesize isotopes of elements near the region of stability. Reactions 1 and 2 represent conventional types of heavy-ion nuclear reactions employing neutron-rich targets and projectiles, but yielding isotopes of element 114 that are neutron deficient with respect to the region of enhanced stability and thus having short half-lives. However, the product in reaction 2 may well have a half-life long enough to be detectable. Reactions 3 and 4 are illustrative of nuclear reactions in which the product nuclei are formed with low excitation energies, and perhaps with large cross-sections, since the competitive fission reaction would be suppressed. In general, the use of conventional heavy-ion reactions with existing combinations of projectiles and targets will always lead to production of trans-actinide elements on the neutron-deficient side of the island of stability and consequently produce isotopes with very short half-lives. Unusual, and as yet unstudied, multinucleon transfer reactions such as reactions 5 and 6 might yield neutron-rich isotopes of element 114 in or near the desired region, but they may prove to have formation cross-sections that are too small. Reactions such as 7 involving the bombardment of heavy targets with very heavy projectiles have also been suggested. These reactions are in essence fission reactions, in which super-heavy elements could conceivably be produced as fission products. Considering in detail reaction mechanisms as well as the specific half-lives of super-heavy elements, Fiset and Nix (1972) have suggested reactions of the type shown as reaction 8 in Table 1. In this type of reaction, the projectile-target combination is chosen to over-shoot the desired region of stability. The complex decay path of an initial reaction product such as $^{302}120$ by multiple alpha particle and electron capture decays is calculated to lead to the formation of the more neutron excessive and consequently longer-lived isotope $^{290}112$.

Table 1. Possible heavy-ion reactions leading to production of super-heavy elements

1. $_{94}Pu^{244} + {}_{20}Ca^{48} \rightarrow 114^{288} + 4n$
2. $_{96}Cm^{248} + {}_{20}Ca^{48} \rightarrow 114^{290} + 2n + {}_2He^4$
3. $_{50}Sn^{124} + {}_{64}Gd^{160} \rightarrow 114^{284}$
4. $_{54}Xe^{136} + {}_{60}Nd^{150} \rightarrow 114^{286}$
5. $_{64}Gd^{160} + {}_{70}Yb^{176} \rightarrow 114^{298} + 20p + 18n$
6. $_{48}Cd^{116} + {}_{92}U^{238} \rightarrow 114^{298} + {}_{26}Fe^{56}$
7. $_{92}U^{238} + {}_{92}U^{238} \rightarrow 114^{298} + {}_{70}Yb^{174} + 4n$
8. $_{90}Th^{232} + {}_{32}Ge^{76} \rightarrow {}^{302}120 + {}_2He^4 + 2n$

$$\begin{array}{c} {}_2He^4 \\ EC \\ {}^{302}120 \rightarrow {}^{290}112 \end{array}$$

Bibliography

FISET, E. O. and NIX, J. R. (1972) *Nucl. Phys.* A **193**, 647.
FOWLER, P. H., ADAMS, R. A., COWAN, J. M. and KIDD, J. M. (1967) *Proc. Roy. Soc.* A **301**, 39.
KELLER, C. (1971) *The Chemistry of the Transuranium Elements*, Nürnberg: Grossdruckerei Erick Spandel.
MILLIGAN, W. O. (Ed.) (1970) *XIII The Transuranium Elements: The Mendeleev Centennial*. Houston: Robert A. Welch Foundation.
NIX, J. R. (1972) *Physics Today*. April 1972.
PRICE, P. B., FLEISCHER, R. L. and WOODS, R. T. (1970) *Phys. Rev.* C1, 1819.
SEABORG, G. T. (1968) *Ann. Rev. Nucl. Sci.* **18**, 53.
THOMPSON, S. G. and TSANG, C. F. (1972) *Science* **178**, 1047.
ZVARA, I. (1970) *Nuclear Reactions Induced by Heavy Ions*, Amsterdam: North Holland Publishing Co., p. 784.

J. P. UNIK and P. R. FIELDS

SWEEP CIRCUITS. Sweep circuits, also called timebase circuits, produce either a voltage or a current waveform which increases linearly with time to a maximum (the sweep), followed by a rapid (ideally instantaneous) decrease to the initial value (the flyback). They may be repetitive, in which case the cycle repeats automatically, or single-shot, requiring a triggering mechanism to initiate the sweep.

Voltage Sweep Generators

Voltage sweep generators are based on the charging of a capacitor (C) by a constant current (I), the voltage $v(t)$ on the capacitor being given by

$$v(t) = It/C + v(0),$$

where $v(0)$ is the initial voltage on the capacitor. Practical circuits endeavour in various ways to keep the charging current as constant as possible during the sweep period.

The simplest constant current generator is a resistor in series with a constant voltage. Using this to charge the capacitor yields the sweep circuit of Fig. 1a. The sweep is initiated and terminated by the switch (S) which will in general have closed resistance R_s and will usually be an electronic switching device such as a transistor, field effect transistor or silicon-controlled rectifier.

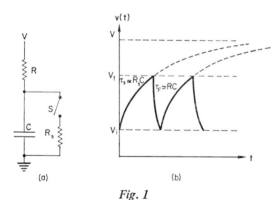

Fig. 1

The capacitor voltage $v(t)$ at time t after opening the switch is given by

$$v(t) = V_i + (V - V_i)[1 - \exp(-t/(RC))]$$

where

$$V_i = (R_s/R_s + R)]V.$$

The exponential can be expanded in series, and, for $t \ll RC$, only the linear and quadratic terms need be retained, therefore

$$v(t) \cong V_i + (V - V_i)t/RC - (V - V_i)t^2/2R^2C^2.$$

For a perfect switch $R_s = 0$ in which case

$$v(t) \cong Vt/RC - Vt^2/2R^2C^2.$$

Provided only a very small portion of the exponential is used, i.e. if the capacitor voltage is always small in comparison with the supply voltage, the quadratic term is small and a nearly linear sweep results. Compromise is inevitable between linearity, required sweep amplitude, available supply voltage, etc.

The sweep is terminated by closing the switch and discharging the capacitor rapidly to V_i when it has charged to a predetermined voltage V_f. The waveforms existing in the circuit are shown in Fig. 1b.

Feedback Voltage Sweep Generators

The sweep obtained in the simple RC circuit above can be linearized by the use of feedback. The principle can be explained with the aid of Fig. 2, which shows the basic Bootstrap sweep circuit.

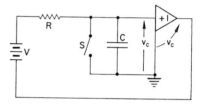

Fig. 2

The voltage across the charging capacitor (v_c) is applied to the input of a unity-gain non-inverting amplifier, the output of which, equal to v_c, is fed back in series with the supply V. The net voltage applied to the resistor R is then V at all times and the charging current is maintained strictly constant. The capacitor voltage will then be

$$v_c = Vt/RC.$$

A second method of applying feedback to linearize the sweep is the Miller sweep circuit of Fig. 3. An infinite gain inverting amplifier (or operational amplifier) is employed. The charging capacitor supplies feedback round the amplifier and maintains the input at virtual earth. The current into the input node is therefore V/R and the capacitor voltage becomes Vt/RC once more. The amplifier output will be the negative of this.

The two schemes (Bootstrap and Miller) are formally identical if the amplifier is considered to be a three terminal device, any pairs of terminals being available for use as input or output (see Millman and Taub, 1965).

Amplifiers with stable exact unity gain or negative infinite gain do not exist. Analyses of the sweep circuits with amplifiers of gain A yield the following expressions for the output voltage $v(t)$ from the amplifiers:

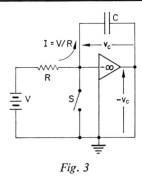

Fig. 3

Bootstrap:

$$v(t) = AVt/RC - A(1-A)t^2V/2R^2C^2 + \ldots$$
$$\cong Vt/RC - (1-A)Vt^2/2R^2C^2 \text{ for } A \cong 1.$$

Miller:

$$v(t) = AVt/RC(1-A) - AVt^2/2R^2C^2$$
$$\times (1-A)^2 + \ldots$$
$$\cong -Vt/RC + Vt^2/2|A|R^2C^2 \text{ for } |A| \gg 1.$$

Comparison of these results with the output of the simple circuit shows that while the linear terms (ignoring sign) are identical, the quadratic terms are reduced to an extent which depends on the nearness with which the amplifier gain A approaches the ideal value.

Finite input and output resistances of the amplifiers modify the results given above. In the case of the Bootstrap circuit, the output impedance R_o has little practical effect, merely adding to the current determining resistance R so that the sweep becomes $v(t) = Vt/(R + R_o)C$. The input impedance R_i is more significant since it causes non-linearity even in the case of exact unity gain. R_i appears in shunt across the capacitance C and a constant charging current gives an exponential capacitor voltage $v(t) = IR_i(1 - \exp(-t/R_iC))$, where I is the charging current. In the case of the Miller circuit, the situation is reversed. R_i is of lesser importance, merely modifying the effective supply voltage V and charging resistor R to $VR_i/(R + R_i)$ and $RR_i/(R + R_i)$, respectively. Finite output resistance R_o, however, results in a step in the output voltage at the moment of sweep initiation. This is because at the start of the sweep, current $I = V/R$ flows through the capacitor and into the output of the amplifier and so sets up a voltage $\Delta V = IR_o = VR_o/R$ at the output before the negative sweep commences. If necessary this step can be eliminated by including a resistance $v = R_o/A$ in series with the capacitor in the feedback loop.

Two further points of practical importance should be mentioned here. The first is that if, as is usual, the common terminal of the amplifier is earthed, the Bootstrap circuit requires a floating potential supply V. The second relates to the position of the switch in the Miller circuit. The position shown in Fig. 3 is not ideal, for the closed resistance (R_s) of the switch is effectively multiplied by the gain of the amplifier as a result of the feedback, so the minimum discharge time constant is $R_sC|A|$. If a floating switch is available—this fortunately is relatively easy with transistors—it is preferable to put it directly across C in which case the discharge time constant is simply R_sC.

Practical Circuits

Bootstrap sweep. Figure 4 shows a typical transistor bootstrap sweep generator. The unity gain amplifier is approximated by the emitter follower T_2. The sweep components are the resistor R and capacitor C. Capacitor C_F is used as the floating constant voltage supply. Transistor T_1 is the electronic switch to initiate and terminate the sweep, while diode D_3 and transistor T_3 form a fast recharge path to replace the charge lost by C_F during the sweep. In the waiting condition a positive potential is applied to diodes D_1 and D_2 which are therefore non-conducting, and transistors T_1 and T_3 are saturated. Capacitor C is discharged through T_1, and C_F is charged to approximately V_{cc} through diode D_3 and transitor T_3. To start the sweep the input potential is made negative, which cuts off T_1, T_3. C starts to charge, and the emitter of T_2 follows, raising the potential of the bottom end of C_F. Since C_F can discharge only slowly, the junction of C_F

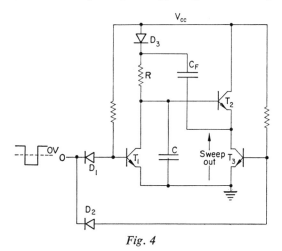

Fig. 4

and D_3 rises and D_3 cuts off. The normal bootstrap sweep then begins, C charging from C_F. The sweep is terminated by cutting off diodes D_1, D_2 with a positive input. Transistors T_1, T_3 turn on, C discharges and C_F recharges. For good linearity C_F must be much greater than C so that the effective supply voltage during the sweep is as constant as possible.

11 D.P.

Miller sweep. There is little point in giving detailed circuits for Miller sweep generators, because the amplifier will almost always be one of the many commercially available operational amplifiers, and details of connections for these to ensure stability depend on the particular device characteristics. Manufacturers usually supply information on the use of these amplifiers as integrators (essentially the Miller generator is just that). In addition sophisticated sweep generators exist using several operational amplifiers (see, for example, Burr–Brown (1971), p. 377). Elementary circuits using discrete components are given by Millman and Taub (1965) and Strauss (1970).

Current Sweep Generators

If virtually pure inductors existed (as do virtually pure capacitors) current sweeps could be generated by circuits analogous to those of voltage sweep generators, with capacitance replaced by inductance and charging current by constant voltage. For if a constant voltage V is maintained across inductance L the resulting current is

$$i(t) = Vt/L.$$

The presence of inevitable series resistance and shunt capacitance makes this impracticable. The solution to the problem can be obtained in the following way. Figure 5 shows the approximate equivalent circuit L, C, R of a real inductor connected to a voltage waveform generator of output impedance r.

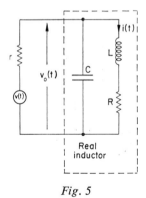

Fig. 5

It is required to determine the waveform $v(t)$ generated if the current $i(t)$ through the inductance L is of the required form, which is

$$i(t) = kt$$

where k is some constant. If this current flows through the L,R branch of the network it follows that the waveform $v_o(t)$ at the output of the generator must be

$$v_o(t) = kL + kRt,$$

in other words, a trapezoidal waveform. If C were zero, $v(t)$ would also be trapezoidal equal to $v_o(t) + krt$. The presence of C makes the production of the initial voltage step kL difficult. The capacitor has to be charged to this voltage instantaneously, requiring $v(t)$ to include an initial voltage impulse of amplitude $E(\to \infty)$ and duration $\delta t(\to 0)$. The charge supplied by the impulse is $E\delta t/r$ and the charge required to raise the potential of C to kL is kCL. Equating these gives $E\delta t = krCL$.

The ideal waveform for $v(t)$ consists then of an impulse, a step and a linear rise (Fig. 6a). The impulse is usually approximated by a fast decaying exponential (Fig. 6b). In principle the presence of C requires modification to be made to the form of the sweep but this is not necessary in practice since the charging time constant of C is usually much less than the sweep duration.

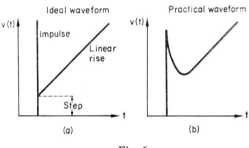

Fig. 6

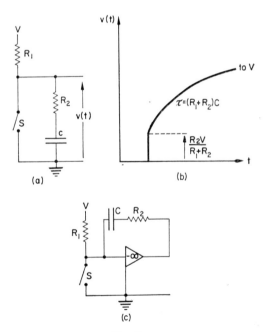

Fig. 7

The trapezoidal waveform can be generated approximately by the circuit of Fig. 7a, which gives an output (Fig. 7b)

$$v(t) = R_2 V/(R_1 + R_2) + [R_1 V/(R_1 + R_2)] [1 - \exp(-t/(R_1 + R_2)C)].$$

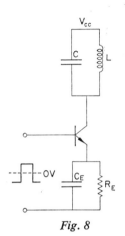

Fig. 8

For $t \ll (R_1 + R_2)C$ and $R_1 \gg R_2$ this becomes

$$v(t) \cong R_2 V/R_1 + Vt/R_1 C$$

from which it follows that

$$R_2 V/R_1 = kL, \quad V/R_1 C = kR$$

enabling R_1, R_2, V to be chosen once the coil parameters are given. For precision circuits the exponential voltage rise can be linearized by feedback (Fig. 7c).

The impulse required to charge the shunt capacitor of the coil may be obtained by passing the trapezoid through the circuit of Fig. 8. The time constant $R_E C_E$ is short compared with the sweep time, so that during the linear voltage rise the resistor R_E is unbypassed and the transistor acts as a good emitter follower, supplying a linearly rising current to L. At the initial step, however, C_E is uncharged, a large current impulse flows to charge it, and at the same time charges stray capacitance C of the coil. C_E is chosen experimentally for optimum linearity.

Recovery Problems

When the current sweep is terminated, the current changes abruptly and large oscillatory voltages are developed across the inductance which can cause damage to circuit components, especially transistors. Figure 9b shows the current in the inductance and

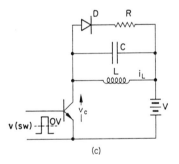

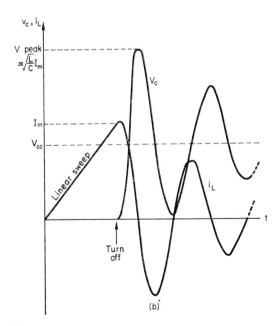

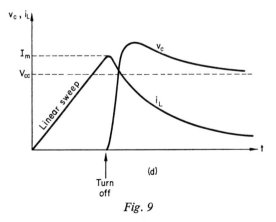

Fig. 9

the collector potential for the simple switched inductance circuit of Fig. 9a. The magnitude of the oscillations can be controlled, at the expense of recovery time, by shunting the inductance with a series diode-resistance combination (Figs. 9c, 9d).

Repetitive Current Sweep

If a repetitive current sweep is required, as for example in television line timebase circuits, the oscillation at turn-off can be utilized to give rapid fly-back. The principle is as follows. In Fig. 9a is an inductance connected to a constant voltage supply via a transistor switch driven by a repetitive waveform, $v(sw)$. When $v(sw)$ is positive the transistor is on and a linear current is generated in the inductance. The sweep is terminated when the current reaches I_m by $v(sw)$ going negative (see Fig. 10). Current oscillations are set up in the LC circuit after half a cycle of which the current in L is $-I_m$. At this time $v(sw)$ goes positive again, the oscillation ceases and the linear sweep recommences from $-I_m$ to $+I_m$.

Other Sweep Sources

Linear sweeps of voltage and current also occur during the operation of circuits not specifically designed to generate them. Notable examples are the Multivibrator and the Blocking Oscillator.

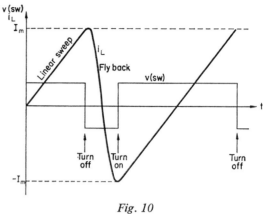

Fig. 10

Bibliography

BURR–BROWN RESEARCH CORP. (1971) *Operational Amplifiers*, McGraw-Hill.

CHANCE, B. *et al.* (1949) *Waveforms*; M.I.T. Radiation Lab. Series, vol. 19, McGraw-Hill.

MILLMAN, J. and TAUB, H. (1965) *Pulse, Digital and Switching Waveforms*, McGraw-Hill.

STRAUSS, L. (1970) *Wave Generation and Shaping*, McGraw-Hill.

J. D. WEAVER

T

TACHYONS

Introduction

At the beginning of the twentieth century, shortly before the appearance of Einstein's theory of relativity, Arnold Sommerfeld discussed the problem of accelerating bodies to velocities greater than that of light. Sommerfeld found that bodies with such superluminal velocities would behave in a very strange fashion—accelerating, rather than decelerating, as they lost energy. In the following year, 1905, Einstein's immensely significant and influential work on the special theory of relativity was published; among the predictions of this highly successful theory was the implication that no body travelling at subluminal velocities could be accelerated to the speed of light itself. The latter thus assumed the role of a limiting velocity in subsequent developments of physical theory and further speculation on the possibility of faster-than-light motion seems to have remained inhibited until 1962. In that year, three physicists at the University of Rochester, New York—O. M. Bilaniuk V. K. Deshpande, and E. C. G. Sudarshan—published a paper entitled "Metarelativity", discussing the possible existence of an exciting new class of particle, travelling always with velocities greater than that of light. It should perhaps immediately be explained that the discussion in "Metarelativity", and subsequent developments in the past decade, in no way invalidate the conclusion from special relativity stated above; it remains true that no particle moving with subluminal velocity can be accelerated up to, or through, the "light barrier". However, this conclusion does not necessarily exhaust the possibilities of the relativistic equations. Contemporary physics contains a familiar class of zero mass particle, the photons and neutrinos, which travel *always* at the speed of light and are never subject to accelerations from lower velocities: in "Metarelativity", Bilaniuk *et al.* argued that, in addition to the long-accepted major groups of particle, namely those with velocities always less than that of light, and those with velocities always identically equal to it, there might exist an altogether new class of body, which would travel with velocities *always greater than that of light*. The velocity restrictions on the three classes of particle are summarized in Table 1.

Table 1

Class	Velocity limits	Examples
1	$0 \leq v < c$	proton, electron
2	$v = c$	photon, neutrino
3	$c < v < \infty$	as yet unobserved

In "Meterelativity" Bilaniuk *et al.* discussed at some length the implications of superluminal velocities, and the place of such velocities within the accepted relativistic framework. One major objection to superluminal velocities for an "ordinary" body is that one must introduce imaginary values of the energy E and momentum P associated with the body, since these two quantities are related to the rest mass m_0 of the body by the familiar relativistic equations:

$$E = m_0 c^2 / [1 - (v/c)^2]^{1/2}$$
$$P = m_0 v / [1 - (v/c)^2]^{1/2}$$

where v is the velocity of the body.

Bilaniuk *et al.* pointed out that imaginary values of the energy and momentum might be avoided by allowing the proper mass of a superlumnary body to be imaginary instead.

The initial response to the exciting ideas of "Metarelativity", although limited, was quite favourable. The paper stimulated the first systematic experimental search for faster-than-light particles, conducted by a group of physicists at Stockholm's Nobel Institute. However, the results of this experiment were negative, and little other work, either experimental or theoretical, seems to have appeared until Feinberg's 1967 paper in *The Physical Review*. Feinberg re-examined the equations of relativistic quantum theory, and arrived at basically the same conclusions as the Rochester physicists, the essential result being that the existence of particles with velocities always in excess of that of light is consistent with the long-accepted equations of special relativity.

Feinberg also introduced the now widely adopted name of "tachyons" for the new class of hypothetical faster-than-light bodies, (from the Greek *tachys*, meaning swift). Some two years after, in a further paper on the subject, Bilaniuk and Sudarshan proposed calling particles such as the photon and neutrino *luxons*, and those which travel always slower than the speed of light, *tardyons*. (Those who are concerned by the mixed Latin–Greek derivation of these labels insist on calling the latter class "bradyons".)

Properties of Tachyons, and Further Searches

Feinberg's useful work of 1967 spelt out some of the more obvious properties which tachyons might possess, in addition to imaginary mass and superluminal velocity. It was proposed that tachyons might be created by a process analogous to the pair production of positrons and electrons, thus circumventing the difficulties associated with acceleration through the "light barrier". Tachyons would also possess rather odd energy–velocity relationships. For the well-understood subluminal tardyons, as the velocity of a particle increases, its energy and momentum also increase, tending to infinity as the velocity approaches that of light. For tachyons the reverse should be true, the energy and momentum decreasing as the tachyon accelerates until at infinite velocities the energy would fall to zero. The relativistic equations for tachyons also indicate that under certain relativistic transformations the sign of a tachyon's energy may switch from positive to negative or vice versa: that is, two observers, viewing the tachyon from different reference frames, might arrive at apparently contradictory signs for the tachyon's energy. Further, the number of tachyons detected by different observers is not necessarily invariant—another departure from the familiar world of subluminal velocities.

Two of the major objections to the tachyon hypothesis, or to searches for these exotic states, are firstly the conflict that the hypothesis forces with our established concepts of causal behaviour, and secondly the absence of a theory of the interaction properties of tachyons. The latter undermines confidence in any experimental attempt to detect tachyons, and remains an unresolved problem. The objections to the tachyon hypothesis based on *causality* arguments, while serious, have elicited a large number of possible solutions in the scientific literature. For example, in his 1970 paper entitled "The Theory of Particles Travelling Faster than Light" E.C.G. Sudarshan states that, if by causality we mean the invariance of the temporal sequence of cause-effect assignments, then tachyon processes can indeed violate causality: however, Sudarshan also points out that no single observer would witness violations of causal sequences, although different observers might well disagree about the cause–effect assignments for specific tachyon processes. This fact is considered by a number of physicists, including Sudarshan, to be perfectly acceptable within a wider conception of causality than is usually applied. Nevertheless, the relativity of the cause–effect assignments for tachyon processes does yield a number of "paradoxical" situations, and Sudarshan discusses three of these, namely Bohm's paradox, Ehler's paradox, and Newton's paradox. The question of acausality in tachyon processes remains a vexing one, and it is not clear at the present time whether the tachyon hypothesis can be reconciled with causality in a fully consistent fashion.

Despite the bizarre properties expected of tachyons, and continuing problems with the theoretical analysis of associated acausal phenomena and interaction behaviour, several tachyon searches have followed the 1963 Stockholm efforts. The first of these searches was carried out in 1968 by Alvager and Kreisler, working in the United States at the Princeton-Pennsylvania accelerator laboratory. Their experiment attempted to detect the Cerenkov radiation which, at that time, it was expected would be emitted by charged tachyons passing through a vacuum. Alvager and Kreisler hoped that tachyon pairs might be produced by the interactions in lead of gamma radiation from a caesium-134 source, and that subsequent observation of Cerenkov radiation emerging from a vacuum chamber would reveal the existence of the exotic particles. However, despite a detailed search, no sign of such radiation was found.

Soon after, a team of physicists from Columbia looked for neutral (uncharged) tachyons that might be produced in bubble chamber experiments involving the interactions of anti-protons and K-mesons with protons. The experiment used the Columbia-Brookhaven 30″ hydrogen chamber, and the team looked specifically for the interactions:

$$K^- + P \rightarrow \Lambda^\circ + T^\circ$$
$$\mathbf{P} + P \rightarrow \pi^+ + \pi^- + T^\circ$$

where T° represents a neutral tachyon. Again, no evidence for the existence of tachyons was obtained.

A second bubble chamber tachyon search has been reported by Danburg and Kalbfleisch, of Brookhaven National Laboratory. Since the proton should be unstable against the decay

$$P \rightarrow P + T^\circ$$

in which the tachyon has negative energy, Danburg and Kalbfleisch investigated the limits that could be placed on such processes, which apparently violate energy conservation. After a careful examination of bubble chamber film from a special exposure, together with a reinterpretation of earlier data related to baryon and electron non-conservation studies, these authors found that the above decays, should they occur at all, must possess very long lifetimes, and that the interaction mediating such processes must be much weaker than the gravitational interaction. (It is interesting to note that, in arriving at these conclusions, the two authors also considered the influence that tachyon-involving decays might have on the magnitude of the heat flow emanating from the earth. Such considerations lead to a very high value for the lifetime of a bound nucleon with respect to elastic tachyon decays.)

Present Situation and Conclusions

At the time of writing, no experimental evidence has been advanced in support of the tachyon hypothesis. In the past decade, and particularly in the last five years, a large number of theoretical papers have examined the tachyon question and its consequences, notably causality and radiation problems, and work in these areas is continuing. Several authors have questioned the assumption that charged tachyons should emit Cerenkov radiation, and the existence of even more exotic entities—the so-called *dybbuks*—has been postulated. It may well be that conclusive theoretical arguments denying the existence of tachyons will appear and become accepted within the physics community: if so, the tachyon episode will nevertheless have been extremely useful in forcing a re-examination of our ideas of the physical universe, and, in particular, of the limits of special relativity. However, until such conclusive evidence is available, the fascinating concept of the tachyon will doubtless continue to attract the attention of physicists, both experimental and theoretical.

Bibliography

ALVAGER, T. and ERMAN, P. (1965) *Annual Report of the Nobel Research Institute*, Stockholm.
ALVAGER, T. and KREISLER, M. N. (1968) *Phys. Rev.* **171**, 1357.
BALTAY, C., Feinberg, G., YEH, N. and LINSKER, R. (1970) *Phys. Rev. D.* **1**, 759.
BILANIUK, O. M. P., DESHPANDE, V. K. and SUDARSHAN, E. C. G. (1962) *Amer. J. Phys.* **30**, 718.
BILANIUK, O. M. and SUDARSHAN, E. C. G. (1969) *Phys. Today*, May issue.
BILANIUK, O. M., BROWN, S. L., DEWITT, B., NEWCOMB, W. A., SACHS, M., SUDARSHAN, E. C. G. and YOSHIKAWA, S. (1969) *Phys. Today*, December issue.
DANBURG, J. S. and KALBFLEISCH, G. R. (1971), *BNL Report* No. 16,394.
FEINBERG, G. (1967) *Phys. Rev.* **159**, 1089.
Fox, R. (1971) *Nature (Phys. Sci.)* **232**, 129.
NEWTON, R. G. (1970) *Science* **167**, 1569.
SUDARSHAN, E. C. G. (1970) *Symposia on Theoretical Physics* Vol. 10, p. 129. Plenum Press.
SWETMAN, T. P. (1971) *Phys. Education* **6**, 1.
WIMMEL, H. K. (1972) *Nature (Phys. Sci)* **236**, 79.

T. P. SWETMAN

TECHNOLOGICAL FORECASTING

Forecasting: an Aid to Decision-taking

Forecasting can be used to describe any exercise aimed at predicting the likely situation in the future; the term often applies to studies of growth of sales, growth of population, etc., and can be concerned with either the immediate future (as is likely to be the case with sales) or the more distant future (as would be the case with population). Technological forecasting is the recent extension of the concept of forecasting to include an appraisal of the interaction of social, economic and technological progress in long-range forecasting exercises.

A technological forecast is now becoming a necessary adjoint to the planning process; its value stems from the attitude of mind becoming prevalent in the technological world that *planning* and *control* are the key management functions and that *forecasting* is the first step in the sequence:

Forecasting—to provide a basis for a plan,
Planning—to pave the way for the implementation of the plan,
and *Implementation*—(through control of the plan) to achieve the goal.

A technological forecast is of value in assisting management decisions regarding allocation of resources *now* to meet the needs of "*tomorrow*"; "tomorrow" might be 10 or 20 years hence and the "management" might be involved at national level or at company level.

Growth of Confidence in Controlling Science and Technology

It is of interest that as of now (early 1970s) technological forecasting is more widely accepted in the U.S. than outside; this is due, no doubt in part, to the fact that the subject originated there, but there is an ever-increasing awareness of its value in Western European countries and, indeed, in most countries at an advanced stage of development; perhaps the most valuable use of these forecasting techniques has yet to come in assisting the developing countries, but to-date there is little publicized of any application in such areas.

Prior to World War II the prevailing attitude to Science and Technology was a fatalistic one: Science was like a giant entity which could not be stopped; it kept on coming and affected everything in its path; sometimes it provided benefit like the motor car which it created but inevitably the benefits were accompanied by disadvantages like the increased ability to cause death and pollution; it was accepted by the majority of people that no one could alter the direction of the progress of Science and Technology. However, the demonstrations of the control of Science and Technology during World War II—particularly in the U.S. by the Manhattan (Atomic Bomb) Project—have formed a base from which has grown the other attitude in which there is a desire to control the progress of technology to meet preconceived needs. There has been another brilliant demonstration of the control of Science and Technology in the U.S. during the 1960s with the space projects which involved control of a sizeable proportion of U.S. national resources and in the process made use of technological forecasting techniques.

Methods of Analysis

General approach to technological forecasting

Technological forecasting can be defined as an attempt to assess the present rate of progress in a technology to help identify the general direction of the progress in order to highlight the areas where encouragement (or discouragement) should be given. As a subject Technological forecasting is in its infancy but it is possible to identify some methods which are beginning to be accepted as the "tools" of the technological forecaster.

In general the forecaster searches for data which may be of a social, economic or technological nature with the aim of identifying the rate at which his technology is developing; at the same time he tries to assess the strength or vulnerability of his technology vis-à-vis other technologies; his chief aim is to present the data in a manner which will be of value in the planning process; he will also be expected to include comment or analysis of implications of any political decisions—such as agreement among European nations to expand the European Economic Community or the decision of the U.K. to use metric standards—or other decisions, like the provision of a national colour television service, which may be apolitical but none the less can have considerable social and economic consequences. Each forecasting exercise will tend to have its own particular combination of the various methods which are open to the forecaster; at the outset of an exercise, however, the forecaster will usually approach his problem by using as many of the methods as seem appropriate. The following summarizes five main methods which lie dominantly in the field of technological forecasting.

Exploratory and normative forecasting

The exploratory method looks at present trends and attempts to extrapolate these into the future—this is the natural approach to forecasting in that it ensures continuity between past, present and future; the normative method takes a less natural approach by carrying out the forecast regardless of the boundary between the present situation and the future—this method calls for the forecaster to specify the future, in the same way that an engineer or scientist might specify his requirements—the value in this less natural approach lies in creating a "blue print" from which one can identify the decisions required regarding allocation of resources *now* in order to make the links between the present and the desired future compatible with each other. A technological forecast will judiciously balance the uses of these two methods.

Examples of extrapolative forecasts are shown in Figs. 1 and 2. Figure 1 was an attempt by Lenz (1962) to indicate trends in commercial aircraft speed and was based on the observation that military aircraft speeds tend to lead commercial aircraft speed by about 10 years. Figure 2 illustrates how R. S. Isenson (1966) has summarized trends in computer speeds. These exercises indicate the rates at which developments have been taking place and serve as a guide to the future. In all extrapolative forecasting, however, the forecaster must keep asking himself to decide where the top limits lie since he must guard against making unattainable forecasts; in Lenz's case he has put the limit of aircraft speed at orbital speed; Isenson did not insert a limit but it is generally thought that the capabilities of the human brain must be the ultimate limit.

The U.S. man on the moon project can be regarded as an example of the use of normative forecasting—John F. Kennedy set the goal for the American people, the goal was reasonable since it appeared that it should be possible to divert resources in the

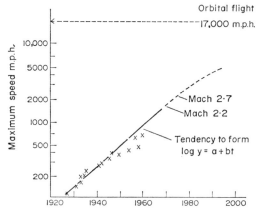

Fig. 1. Lenz's observations of trends in aircraft speed.

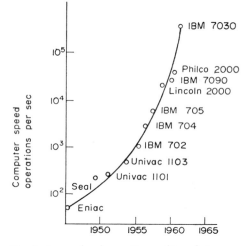

Fig. 2. Isenson's observations of trends in computer technology.

early 1960s to enable it to be achieved; from the simple definition of the goal detailed specifications were worked out and used to decide how to create the required future condition by taking steps to encourage Science and Technology in a clearly defined direction.

Technological mapping

The forecaster will attempt to use a "map" to illustrate "the state of the art" in the technology concerned. The map may be of a highly technical nature as in the illustration of aircraft skin temperature as a function of aircraft speed given by Boorer (1969) in the form shown in Fig. 3; there are occasions where the "map" will be used to summarize some simple relationships which may not be obvious until a juxtapositioning of various pieces of data is brought about; an illustration of this latter type of map is seen in the way in which Gunston (1967) illustrated the competition between modes of transport, see Fig. 4. In using this method the forecaster is striving for a way of summarizing the situation in a form suitable for stimulating thoughts about future possibilities and preferably in a way which will assist the planning process at the same time.

Morphological analysis: idea generation

Part of a technological forecast must be devoted to the generation of ideas concerning likely developments in the future. Morphological analysis is being introduced as the formal method of generating ideas.

The method was first suggested by Zwicky (1948) as a method of unfettering the thought processes of the forecaster to enable him to think as broadly as possible about solutions to problems. The basis of the idea is the construction of a matrix in which a large number of possible solutions is generated.

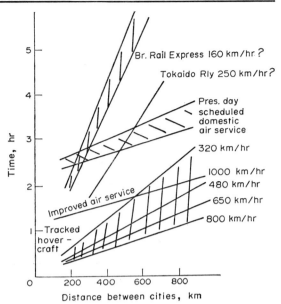

Fig. 4. Example of "map" used to portray simple relationships—Gunston's work.

The method can be understood by imaging the way in which Hawthorne (1969) developed the matrix shown in Fig. 5. The matrix shows a large number of different ways of metal forming; it was built up step by step by selecting each of the topics in the left hand column in turn and using each to generate ideas; the forecaster has gone through thought processes as follows (regardless of the feasibility of the results produced):

For example:

Question 1—what *processes* exist for making shapes of metal?

Answer 1 —removal of metal,
deformation of metal,
formation of metal.

Question 2—(sparked off by the thought that one needs energy to do the job under consideration) what sources of *energy exist*?

Answer 2 chemical energy,
force (mechanical) energy,
heat energy,
biological energy—this last possibility is added in the true spirit of morphological analysis, the person generating the matrix has allowed himself a "flight of fantasy" simply because at that time he remembered that microorganisms can cause corrosion in materials; thus here is a method of removing metal. (Of course in this instance nothing practical arose from including this "box" in the matrix.)

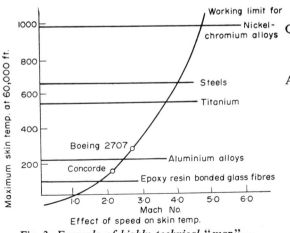

Fig. 3. Example of highly technical "map"—Boorer's work.

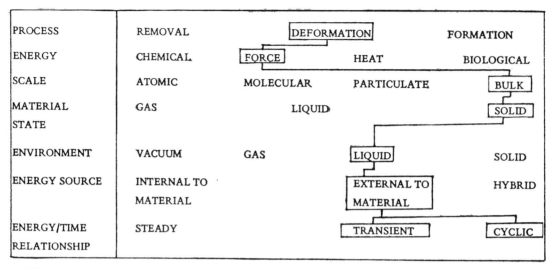

(ref Hawthorne, Aston University T.F. Conference September 1969)

Fig. 5. Morphological matrix for metalworking showing "discovery" of explosive forming.

The continuation of this procedure enabled the forecaster to produce Fig. 5. It should be stressed that initially the thought processes should not be rigidly bounded to enable many ideas to be developed; it is likely that some literature searching will be beneficial in the early stages and it is recommended that blank spaces be left in the matrix to encourage thinking about other possibilities at a later date.

Following the idea generation stage the analysis has to become more critical to help select the projects which are to be given more serious consideration. The advantage of the method arises through being uncritical during the creative stages since by this means some of the less obvious possibilities can be more easily identified.

The method is illustrated diagrammatically in Fig. 6 where the idea generation phase is followed by the evaluation phase in which various criteria are used to serve as a filter—in the example the criteria indicated are compatibility, economics and design considerations. Sometimes a "random walk/ judicious choice" approach to selecting attractive possibilities can be applied as is illustrated in Fig. 5.

Delphi method: consensus of experts

Some techniques of making use of the opinions of experts in different fields of technology is desirable and Olaf Helmer (1963) and others have been advocating the Delphi method which attempts to give a balanced appraisal of the future as seen through the eyes of the experts. The method is usually applied to long-range forecasting and, as originally conceived, takes the form of several rounds—perhaps five or six—of questions and answers using a panel of experts. The size of the panel will vary depending on the availability of experts in the field being studied. The method is illustrated diagrammatically in Fig. 7.

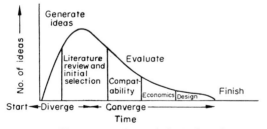

Fig. 6. Illustration of morphological analysis.

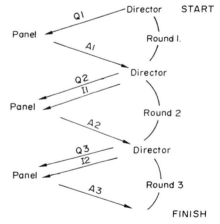

Q = question, I = subsidiary information, A = answer

Fig. 7. Illustration of Delphi method.

Basically the method is designed to avoid face to face contact between the experts participating; this is necessary to eliminate the psychological interaction which would take place were the participants to communicate in person—he who states his case most forcefully might sway those who have equally valid alternative views. The task of the director of the study is to formulate questions which force the experts to probe the future constructively; in doing this he has to be prepared, when asked by panel members, to disseminate subsidiary information—this is illustrated in Fig. 7; this is sometimes necessary in order to create a uniform base of general background information onto which the specialized knowledge is grafted.

One of the earlier applications of the Delphi method was in looking at future developments in automation and was carried out by Ozebekan. The results have been included in Wills' book (1972).

Trend analysis

When the forecaster is looking 10 or more years ahead he must make himself aware of relevant progress in overlapping technologies or in adjacent technologies which might be of advantage to his technology. Above all, however, he must make himself aware of the limits likely to be imposed by the rate at which either industry in general or the public at large are capable of adopting changes or innovations; he must then find ways of interpreting this into commercial terms bearing in mind especially that normally there is only a finite number of potential customers capable of consuming a finite quantity of material.

Methods are continually being developed by which the significance can be assessed of the interaction of relevant social factors such as

(i) rate of increase in numbers of households, etc.,

and (ii) rate at which households (or other appropriate category of consumer) adopt the use of a commercial product

with commercial factors such as

(iii) lifetime of product

and (iv) annual demand.

In some cases the allocation of Research and Development effort today will depend on estimates of the commercial viability of the likely end product of the research, a viability which can only be meaningfully assessed by inclusion of all the above factors plus knowledge of the strength of competition from others operating in the same field.

The adoption process usually takes the shape shown in Fig. 8(a) but varies widely from product to product as in Fig. 8(b). The forecaster has somehow to estimate the appropriate rate of adoption. A simple example of assessing the situation concerning future demand for tubes for colour television

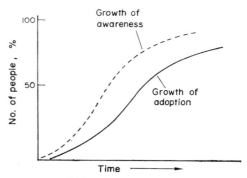

Fig. 8(a). The adaption process.

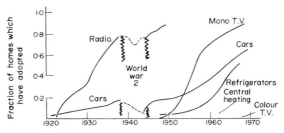

Fig. 8(b). Adoption of various items by U.K. households.

Fig. 8. *Features of the adoption process.*

might be as illustrated in Fig. 9 where number of households in the U.K. has been extrapolated into the future (Fig. 9(a)) and operates on the curve of estimated rate at which the population will adopt colour television (Fig. 9(b)) to give annual demand for new sets during the initial periods (Fig. 9(c)); television-set lifetime can then be used to estimate demand for replacement units and hence the ultimate total annual demand (Fig. 9(c)).

The long-range forecaster may want to assess the significance of a move to introduce a revolutionary new method of picture presentation which eliminates use of the high-voltage tube as we know it today; he could then feed in his estimates of the *rate* at which the population will adopt the new device and its likely *date* of introduction and use this to decide the implications of a break-through in the next few years. It becomes obvious that break-throughs do not change the course of events by causing commercial discontinuities—the processes of innovation and adoption take time and the task of the forecaster is to point out the innate momentum in these processes and weigh them alongside the more technical knowledge which is relevant.

Other methods

Technological forecasting is still developing and other methods of tackling the problems are being evolved. Among these is the use of Relevance Num-

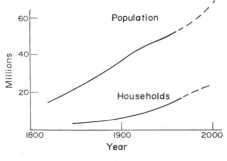

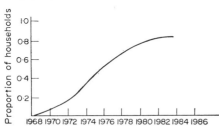

Fig. 9(a). *U.K. population and households.*
Fig. 9(b). *Likely adoption of colour TV in the U.K.*

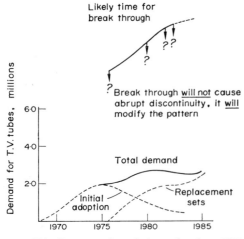

Fig. 9(c) *Pattern of evolution of colour TV demand in the U.K.*

Fig. 9. *Relating social and economic change to technical change.*

bers to help decide the relative importances of different major development projects within an organization. Hubert (1970) has indicated how this can be done. The method involves identifying the major fields of interest of the organization and their importance relative to each other; the major projects are then identified and their importance relative to each other within each field has to be ascertained. The scale of importance is normally such that at each decision level the sum of importances is unity. The resulting matrix is shown in Table 1.

Table 1

Relative Importance	Field A W1	Field B W2	Field C W3
	Relative Importance	Relative Importance	Relative Importance
Project (a)	w_{1a}	w_{2a}	w_{3a}
Project (b)	w_{1b}	w_{2b}	w_{3b}
Project (c)	w_{1c}	w_{2c}	w_{3c}

The total importance of project (a) is

$$w_{1a}W1 + w_{2a}W2 + w_{3a}W3$$

and of project (b) is

$$w_{1b}W1 + w_{2b}W2 + w_{3b}W3$$

and of project (c) is

$$w_{1c}W1 + w_{2c}W2 + w_{3c}W3.$$

It is obvious that this process of weighting can be carried out to lower levels to include the minor projects within the major projects and has considerable value in providing a systematic method of deciding priorities of the different facets of development work.

Bibliography

BRIGHT, J. R. (1969) *Technological Forecasting for Industry and Government*, N. J.: Prentice Hall Inc.
BOORER, N. W. (1969) *Futures* **1**, March (Iliffe and Technology Pub. Ltd.).
GORDON, J. T. and HELMER, O. (1963) *Mgt. Sci.* **9**, 3.
GUNSTON, W. T. (1967) *Sci. J.*, August.
HAWTHORNE, J. (1969) *Aston Univ. T. F. Conf.*
HUBERT, J. M. (1970) *R and D Mgt.* **1**, 1.
ISENSON, R. S. (1966) *Mgt. Sci.* **13**, 2.
JANTSCH, E. (1967) *Technological Forecasting in Perspective*, Paris: O.E.C.D.
LENZ, R. C. (1962) ASD-TDR-62-414 (Aeronautical Systems Division, USAF Systems Command).
WILLS, G. (1972) *Technological Forecasting*, Penguin.
ZWICKY, F. (1948) *Courant Anniversary Volume*, N.Y.: Intersci. Pub. Inc.

Useful collections of papers:
Scientific American, Sept. 1958.
Science Journal, Oct. 1967.

I. C. HENDRY

THERMAL DESORPTION SPECTROSCOPY.

A thermal desorption spectrometer (first described by Hagstrum (1953)) is essentially an ultra-high vacuum (uhv) device. A surface (usually that of a filament of a refractory metal such as tungsten) is rendered atomically clean by heating it in a uhv to about 2500 K. This surface is then maintained at, say, 300 K for a measured period of time, termed the cold interval, t_c. Molecules of the residual gas which are incident upon and retained by the surface during the time t_c are subsequently desorbed by raising the surface temperature rapidly to 2500 K again. The pulse of gas pressure due to this desorption is recorded with a total pressure vacuum gauge, usually a Bayard–Alpert type of hot-cathode ionization gauge (B.A.G.). As the number of molecules retained over the significant cold interval are effective, this forms an integrating device which enables the detection of gases present at a partial pressure of about 10^{-12} Pa within a system at a total pressure of about 10^{-8} Pa. In this respect, the performance is inferior to that of a modern mass spectrometer but advantages are afforded of high sensitivity and simplicity of structure.

Consider a surface of area A m^2 in the gas at a pressure p Pa maintained for a cold interval of t_c sec. The number of molecules retained by the surface is

$$n = 2 \cdot 9 \times 10^{22} At_c sp \quad (1)$$

where s is the sticking probability, i.e. the number of molecules retained at the surface divided by the number incident per unit time.

As there are N_A molecules in $22 \cdot 4 \times 10^{-3}$ m^3 of gas at s.t.p., where N_A is the Avogadro constant ($6 \cdot 025 \times 10^{23}$), the number of molecules in V m^3 of gas at a pressure p Pa is

$$\frac{6 \cdot 025 \times 10^{23} \, Vp}{22 \cdot 4 \times 10^{-3} \times 10^5} = 2 \cdot 7 \times 10^{20} \, Vp.$$

Therefore, if these n molecules are desorbed into a volume of V m^3, the pressure increase Δp is $n/(2 \cdot 7 \times 10^{20} V) p$. This will cause an increase Δi^+ A of the positive ion current of the ionization gauge given by $\Delta p \cdot K$ where K is the gauge sensitivity in A Pa^{-1}.

Therefore

$$n = 2 \cdot 7 \times 10^{20} \, V(\Delta i^+/K). \quad (2)$$

From eqs. (1) and (2)

$$2 \cdot 9 \times 10^{22} At_c sp = 2 \cdot 7 \times 10^{20} V(\Delta i^+/K).$$

Therefore

$$p = \frac{9 \cdot 2 \times 10^{-3} \, V\Delta i^+}{KAt_c s}.$$

For example, if V is 2×10^{-3} m^3, $A = 5 \times 10^{-5}$ m^2, $s = 0 \cdot 5$ and $K = 1 \cdot 5 \times 10^{-3}$ A Pa^{-1},

$$p = \frac{9 \cdot 2 \times 10^{-3} \times 2 \times 10^{-3}}{1 \cdot 5 \times 10^{-3} \times 5 \times 10^{-5} \times 0 \cdot 5} \left(\frac{\Delta i^+}{t_c}\right).$$

$$= 4 \cdot 9 \times 10^2 \, \Delta i^+/t_c.$$

If $t_c = 300$ sec and $\Delta i^+ = 10^{-13}$ A is the minimum detectable change of i^+, the minimum pressure detectable, $p_{\min}$, is given by

$$p_{\min} = \frac{4 \cdot 9 \times 10^2 \times 10^{-13}}{300} = 1 \cdot 6 \times 10^{-13} \, \text{Pa}.$$

If the surface after the cold interval has its temperature increased with time in a known manner, it is found that the *rate* of desorption has specific maximum values at certain temperatures. This suggests that either more than one type of gas is present and each constituent is desorbed at a different temperature or molecules of the same gas are retained with different specific energies within particular sites in the surface. Both these phenomena occur and both are identifiable.

A graph of desorption rate against the temperature of the desorbing surface—the thermal desorption spectrum—can therefore be used to investigate the gas-to-metal bond provided that some prior knowledge of the possible gas–surface interactions is available.

Thermal desorption spectra are displayed either by providing optimum pumping of the desorbed gas species and recording the pressure as the filament surface temperature is varied in some controlled manner or by ensuring negligible pumping and differentiating with respect to time by electrical means the output signal from the B.A.G. located in the system. In the latter case, the desorption rate of any gas constituent is recorded directly. In either case, satisfactory shape and separation of the desorption peaks is the aim with consequent identification and assessment of the partial pressure contribution.

The position of a peak on the temperature scale in the desorption spectrum is a function of the activation energy of desorption, E_d. Ehrlich (1961), Redhead (1962a) and Carter (1962) show that, for a first-order desorption (defined as the direct release of a given gas) the temperature T_p at which the peak occurs is given approximately by

$$E_d/RT_p = \log_e (v_1 T_p/\beta) - 3 \cdot 64 \quad (3)$$

the filament surface temperature T being increased linearly with time t above a temperature T_0 in accordance with $T = T_0 + \beta t$ where R is the universal gas constant and v_1 is the first order rate constant ($v_1 = 10^{13}$ s^{-1}).

Figure 1 (Redhead, 1962b) shows this relationship for β between 1 and 10^3 K/s.

Two examples of desorption spectra are shown in Fig. 2(a) from Redhead et al. (1968) which enables the residual gas components to be identified in a system in which the total pressure is about 10^{-8} Pa with hydrogen and nitrogen the only chemically active gases present. The cold interval t_c was 218 min, the linear temperature rise rate was 40 K/s, and the estimated partial pressures were $p_{H_2} = 3 \cdot 4 \times 10^{-11}$ Pa

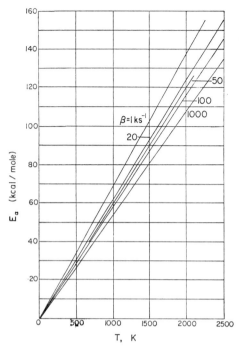

Fig. 1. Activation energy of desorption (E_d) plotted against the filament surface temperature T.

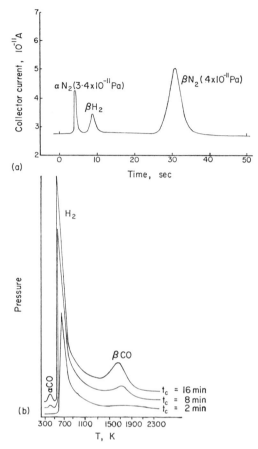

Fig. 2. (a) Thermal desorption spectrum where hydrogen and nitrogen are the only chemically active gases present in a system at a total pressure of 10^{-8} Pa. (b) Thermal desorption spectrum of residual gas consisting of hydrogen and carbon monoxide in a system at a total pressure of 10^{-8} Pa.

and $p_{N_2}^m = 4.1 \times 10^{-11}$ Pa, but to an accuracy claimed to be no better than within a factor of 2. Figure 2(b) from Redhead *et al.* (1962) shows desorption spectra of residual gases consisting of hydrogen and carbon monoxide where complete separation of the peaks due to these two gases is relatively simple. The various spectra are for increasing cold intervals t_c. The increase in peak amplitude with t_c is usually linear for small coverage up to about 0·3 monolayer but non-linear in the approach to saturation near monolayer coverage.

Provided that the temperature of a surface is sufficiently low, any gas atom incident upon the surface will be retained even though no chemical interaction occurs. As this physisorption (adsorption) occurs at a specific temperature for a given atom/surface combination, measurement of the pressure as the surface temperature is increased allows the gas components to be identified and their partial pressure contributions calculated. Surfaces operated at these low temperatures (i.e. below ambient) are often termed "physical desorption spectrometers" to distinguish them from those in which a tungsten or other metal surface is maintained at ambient or higher temperature and on which active gases are chemisorbed. Although, in theory, a physical desorption spectrometer provides a partial pressure analyser for all gases, the necessary apparatus conditions are not easy to realize experimentally, because the surface temperature then needs to be varied between that of liquid helium and that of the room (4 K to about 300 K).

Several problems remain to be solved in relation to thermal desorption spectroscopy. A number of these problems are considered.

1. The desorbing surface of which the temperature is varied is frequently of polycrystalline wire (e.g. tungsten or rhodium). Evidence so far suggests that different samples of the same material give different results so that each desorption spectrometer must therefore be calibrated with the gases for which it is to be used. Are these differences due to lack of purity of the sample or is the crystalline form the important variable? The work of Van Oostrom (1966)

suggests that metals other than tungsten and rhodium could possess properties of special interest leading to a desorption spectrometer with perhaps six or more filaments where a comparison of the desorption spectra could prove very informative.

2. The temperature of the desorbing surface needs to be known accurately, controlled and maintained uniform over the surface. Frequently long thin filaments of tungsten are employed, and the temperature found from resistance measurement. Redhead et al. (1962) used a constant current at radio-frequency to monitor the filament resistance. This r.f. component produced negligible heating of the filament and the feedback signal derived from this resistance probe controlled the d.c. heating which imposed a linear temperature–time schedule on the tungsten filament. Heating by electron bombardment has been used only in the desorption of gases implanted as energetic positive ions (Kornelsen and Sinha, 1966).

Temperature monitoring and control by means of fine thermocouples (of 0·005 inch tungsten/tungsten– 26% rhenium wires) capacitor welded to the desorbing filament surface operate satisfactorily up to 2300 K and provide signals to control thyristors. Power control of the desorbing filament heating current is readily achieved by a.c. phase control using silicon controlled rectifiers.

Other circuits of interest which depend on the resistance-temperature relationship of the desorbing filament include the resistance-follower network (Close et al., 1968) and the self-balancing bridge, of which an example is shown in Fig. 3. Using a high-gain differential amplifier with a dual FET input, voltage controlled resistive elements and modern power transistors, a variety of temperature–time schedules can be imposed on the desorbing filament. Desorption spectra obtained for tempering schedules other than linear are rare even though their form has been investigated theoretically.

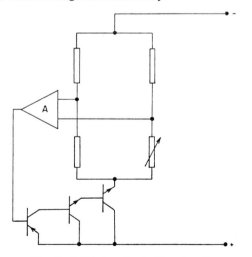

Fig. 3. Self-balancing bridge circuit.

3. Pressure and desorption rate measurement. The ionization gauges (usually B.A.G.) used have been investigated from the points of view of (a) the use of low-temperature emitters to avoid anomalous results due to hydrogen dissociation (Hickmott and Ehrlich, 1956), (b) tube versus bare structures, (c) the optimum electrode potentials for high sensitivity with minimum pumping (Redhead et al., 1962). However, the gauge is being used in thermal desorption spectroscopy under non-equilibrium conditions. Is then the position of the gauge head relative to the desorbing surface of importance?

On the desorption spectrum, well-defined peaks occur if desorption rate rather than pressure is recorded. Although it is possible to control the pumping speed and hence obtain good resolution, the pumping speed is not constant for each constituent and quantitative work is difficult. It is simpler to differentiate with respect to time the output signal from the ionization gauge. If the necessary electrical differentiating circuit utilizes passive components, adequate sensitivity demands long time constants resulting in poor resolution. By using an operational amplifier of open loop voltage gain A in a differentiating configuration, the final output voltage is reached in a time approximately $1/A$ of that of the conventional RC network which provides the same output.

Bibliography

CARTER, G. (1962) *Vacuum*, **12**, 245.
CLOSE, K. J., HODGES, E. B. and YARWOOD, J. (1968) *Brit. J. appl. Phys.* **1**, 1509.
EHRLICH, G. (1961) *J. appl. Phys.* **32**, 4.
HAGSTRUM, H. D. (1953) *Rev. sci. Instrum.* **24**, 1122.
HICKMOTT, T. W. and EHRLICH, G. (1956) *J. Chem. Phys.* **24**, 1263.
KORNELSEN, E. V. and SINHA, M. K. (1966) *Appl. Phys. Letters*, **9**, 112.
REDHEAD, P. A. (1962a) *Vacuum*, **12**, 267.
REDHEAD, P. A. (1962b) *Vacuum*, **12**, 203.
REDHEAD, P. A., HOBSON, J. P. and KORNELSEN, E. V. (1968) *The Physical Basis of Ultrahigh Vacuum*, Chapman & Hall; London.
REDHEAD, P. A., KORNELSEN, E. V. and HOBSON, J. P. (1962) *Can. J. Phys.* **40**, 1814.
VAN OOSTROM, A. G. J. (1966) *Philips Res. Rep. Suppl. No. 1.*

J. YARWOOD and K. J. CLOSE

THERMOLUMINESCENT DATING. In principle this technique can be applied to a wide range of ceramic materials in order to find the time that has elapsed since last heating (to around 300°C or more). At the time of writing (1972) it has become well established for dating archaeological pottery and for testing the authenticity of terracotta works of art, but the validity of other applications is still under investigation. These include baked soil, burnt flint, burnt stone, glass, enamel, volcanic lava and obsidian; it is also possible that bone and shell can be dated in this way even though unburnt. The following outline is based on the method as applied to pottery dating and authenticity testing. It follows the approach developed at the Oxford Universtity Research Laboratory for Archaeology, where from 1961 onwards the main research into the technique has taken place.

1. Principle

Thermoluminescence (hereafter referred to as TL) is a phenomenon exhibited to varying degrees by many minerals. It is the emission of light when a substance is heated; this light is additional to ordinary red-hot glow and usually occurs at a less elevated temperature. It represents the release of energy which has been stored as trapped electrons in the crystal lattice of the mineral. This stored energy is acquired by absorption from any ionizing radiation to which the mineral may have been exposed; consequently the amount of TL observed is proportional to the overall dose of radiation which has been received.

In terracotta and most types of pottery there are mineral constituents (e.g. quartz and feldspar) that have this property of accumulating TL and they receive a small but significant dosage of ionizing radiation which over thousands of years adds up to an appreciable total. This dosage comes from radioactive impurities (a few parts per million of uranium and thorium, a few per cent of potassium) in the clay of the object itself, and to a lesser extent from the radioactive impurities in the surroundings; cosmic rays also make a small contribution.

Heating to above 300°C removes the geologically accumulated TL and consequently the firing of the clay sets the "thermoluminescent clock" to zero. Thereafter the TL grows with time. The growth is dependent on the particular TL constituents in a given sample as well as on the radiation dose-rate. By laboratory measurements the TL carried by a sample can be expressed as an "equivalent radiation dose". This is determined by exposing another equal portion of the sample to radiation from an artificial radioisotope and finding the amount of radiation required to induce a level of TL equal to twice the "natural" TL. By measuring the amounts of uranium, thorium and potassium present in the sample and in the burial soil, the radiation dose received by the fragment each year can be evaluated. The age is then directly obtained as:

$$\text{Age} = \frac{\text{Equivalent Radiation Dose}}{\text{Dose per year}}$$

Although simple in principle, various complicating factors make accurate dating tedious and time-consuming, both in respect of evaluating the Equivalent Radiation Dose and the Dose per year. These two aspects are now discussed separately.

2. Evaluation of the Equivalent Radiation Dose

2.1. Measurement of TL

Three considerations dominate the design of the measuring apparatus. First, the intensity of the TL is low—typically of the order of 10^5 photons $s^{-1}Sr^{-1}$, roughly 10^{-11} lumens Sr^{-1}. Secondly, when the sample temperature exceeds 300°C there is a significant amount of incandescent light (i.e. thermal radiation), which masks the TL signal unless colour filters are used to discriminate against it. Thirdly, in addition to the TL that is induced by exposure to radiation, there is also a "non-radiation-induced" or "spurious" TL. One type of this is the tribo-TL that results from crushing the pottery into a form suitable for measurement; there are other more subtle types too, and the phenomenon is not well understood. If the TL measurement is made with the sample in air, the spurious TL is liable to swamp the true radiation-induced TL. Fortunately, however, by making the measurement in an atmosphere of nitrogen (or argon) the spurious TL is reduced by several orders of magnitude.

A suitable form of apparatus is outlined in Fig. 1. As will be discussed in section 3.1, the sample for measurement is usually either in the form of a thin layer of fine grains (1 to 8 microns) deposited on a 10 mm diameter aluminium disc, or it may consist of between 1 and 100 mg of quartz or other mineral grains which have been separated out from the pottery. In either case the sample is placed on a 0·3 to 0·5 mm thick strip of nichrome that is heated electrically. Rapid heating (e.g. 20°C per second) is advantageous since the TL intensity is proportional

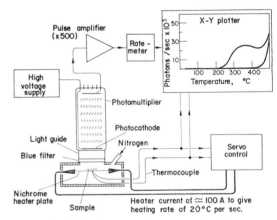

Fig. 1. Block diagram of apparatus for thermoluminescence measurements. The glow curve drawn by the X–Y recorder is typical of the natural TL from ancient pottery. Also shown is the light observed from a second heating of the same sample; only thermal radiation is emitted, all the TL having been released in the first heating.

to the heating rate, whereas the thermal radiation is independent of it.

The TL is detected by a low-noise photomultiplier with a high blue-to-red sensitivity. To improve discrimination against thermal radiation further (and also against spurious TL), blue/violet colour filters are used. The pulses at the anode of the photomultiplier, each of which corresponds to the arrival of a photon at the photocathode, are fed through a fast amplifier to a ratemeter, the output of which activates the Y-axis of a coordinate plotter. The X-axis of this is driven by a thermocouple attached to the heater plate.

2.2. Stability

For a sample which is composed of one mineral having only a single trapping level, the glow-curve consists of a single peak. The type of quartz inclusion often found in pottery exhibits a peak at 375°C having a width at half-height of about 100°C; kinetic studies indicate a trap depth of about 1·7 eV (see Fig. 2) and a lifetime for retention of trapped electrons of the order of 10 million years. However, it is more usual for the observed glow-curve to be composed of a number of overlapping peaks resulting from the presence of several types of trap within a single mineral, or from the presence in the sample of several different minerals. It is not then possible to determine lifetimes, and for evidence that the stability of the trapped electrons is sufficient, reliance is placed on the *plateau test*.

The glow-curve temperature at which lattice vibration causes the release of electrons from traps depends on the depth of the traps concerned (and also on the associated *frequency factor*). The deeper the traps the higher the temperature of release; the absence of TL below 200°C in the glow-curve of Fig. 1 is because the traps that empty in that temperature region are not deep enough to hold their electrons stably over the centuries of burial. A freshly-irradiated sample shows strong TL in this region and by comparison of the shape of the glow-curve for the natural TL with the shape of the glow-curve for this "artificial" TL it is possible to determine the glow-curve temperature at which the associated trap-depth is sufficient for stable accumulation of electrons (with respect to the age of the sample). This is illustrated in Fig. 3 in which the ratio of TL intensity for a given glow-curve temperature rises sharply to a plateau at 350°C. The sharp rise is consistent with theoretical prediction for typical minerals that the traps emptying at about 250°C should have an electron lifetime at 17°C of the order of 10 years whereas the actual lifetime is 30,000 years. In all

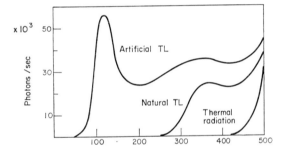

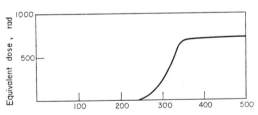

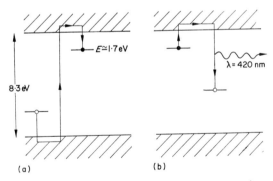

Fig. 2. The thermoluminescence process in quartz: (a) Ionization by natural radioactivity gives rise, during antiquity, to the accumulation of trapped electrons. For dating purposes only traps deep enough to retain electrons over thousands of years are of interest. The lifetime of an electron in the trap shown is of the order of 10^7 years. (b) On heating in the laboratory the trapped electrons are released at about 375°C (for the trap shown) and some of these are captured at luminescent centres with the emission of light.

Fig. 3. The "plateau test". The upper part shows the glow curves corresponding to the natural TL, the artificial TL after 1000 rads of beta irradiation, and thermal radiation. The lower part shows the ordinate ratio of the natural and artificial TL as a function of temperature. The onset of the plateau is indicative that a sufficiently high glow curve temperature has been reached for the TL to be associated with traps that are deep enough to retain their electrons with negligible leakage during archaeological times. The onset of the plateau is usually the range 350–400°C.

the discussion following it is to be taken for granted that the term "natural TL" refers to the intensity in this stable plateau region. It is the *existence* of a plateau that is critical rather than the actual temperature of onset. The lifetime of 30,000 years just quoted does not necessarily imply a limiting age for the method—if the age of the sample is several orders of magnitude greater then the onset of the plateau will be at a noticeably higher temperature. In practice the limiting age is usually determined by the approach of saturation (the filling of all available traps) and this varies from mineral to mineral.

2.3. *Linearity of accumulation*

The effect either of leakage from traps or of saturation would be to cause the growth of TL to be a saturating exponential of the form $[1 - \exp(-\lambda t)]$. In practice for pottery dating it is non-linearity in

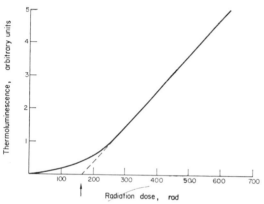

Fig. 4. Typical "*supralinear*" *dependence of TL on radiation dose for pottery samples irradiated with beta or gamma radiation. The arrow indicates the intercept of the linear portion on the dose axis.*

the early part of the growth curve that is the more commonly troublesome. This takes the form of "supralinearity", as illustrated in Fig. 4. Various mechanisms have been proposed to explain this non-linearity—creation of new traps, enhancement of the probability of released electrons reaching luminescent centres, activation of luminescent centres, quenching of "killer" centres competing for electrons either with the traps or the luminescent centres. An allied phenomenon is the "pre-dose" effect—the enhancement of TL sensitivity by a previous radiation dosage despite the fact that the TL induced by it has been erased by heating.

The occurrence of supralinearity makes accurate dating more difficult and is particularly serious for young samples. The obvious way to deal with it would seem to be to plot out the growth curve of each sample after the natural TL has been thermally drained from it in the course of the natural glow-curve, and to determine the dosage necessary to induce an amount of artificial TL equal to the natural TL. Unfortunately because of the pre-dose effect (and because of mineralogical changes occurring during drainage) the sensitivity after drainage is often different from that before drainage, and it is more reliable to determine the natural TL sensitivity by the "additive" procedure. In this the additional TL induced by a known artificial dose is found by subtracting the natural TL measured with one portion of the sample from the natural plus artificial TL measured with another portion. Correction for supralinearity is made by assuming that the dose *intercept* (see Fig. 4) is unchanged by drainage despite the likelihood of change in sensitivity. This empirical approach has been justified by tests with known-age samples.

Supralinearity (and pre-dose effect) is exhibited only when the tracks of the ionizing radiation are sufficiently numerous for their effective volumes to overlap. For the range of dosages relevant to archaeology this occurs for beta- and gamma-radiation but not for alpha-radiation.

2.4. *Quality of radiation*

For a given amount of absorbed energy, irradiation by alpha-particles is less efficient in inducing TL than irradiation by beta-particles or gamma-radiation. The ratio is usually called the "k value" and it ranges from 0·05 to 0·5 for different materials, averaging 0·15 for pottery. The poor effectiveness is probably due to saturation of traps (with consequent wastage of ionization) by the high local ionization density in the core of the track. Substances with a low k value also show early saturation in the growth induced by beta-irradiation.

The differing effectiveness means that the TL sensitivity of a sample must be determined for alpha-particles as well as for beta-particles (or gamma-radiation). Because the k-value depends on the alpha-particle energy it is also necessary to take into account the alpha-particle spectrum received by the sample.

3. *Evaluation of Annual Radiation Dose*

The typical dose rates received by a fragment of pottery buried in soil are shown in Fig. 5. It will be seen that when account is taken of the reduced effectiveness of alpha-particles in inducing TL, the beta and gamma dose-rates become important, and because of their greater range this means that the dose received by the pottery is determined not only by its own radioactive content but also by environmental radiation. The major environmental contribution is the gamma dose from the burial soil; for the beta dose there is a transition layer of between 1 and 2 mm at the surface of the pot in which the dose is partially determined by the radioactive content and partially by that of the soil. To avoid uncertainty from the latter effect the material for TL measurement must be taken from the inner part of the fragment. The gamma dose-rate is estimated by laboratory determinations on a sample of the relevant soil, or more directly, by on-site TL dosimetry (see section 3.2).

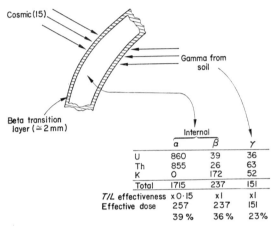

Fig. 5. *Typical annual radiation dose for fragment of pottery buried in soil, both having U 3 ppm; Th 12 ppm; K 2%.*

The laboratory evaluation of radioactive content is made on a sample after it has been dried in the course of preparation. When buried, the pottery fragment carries a significant amount of water and this has the effect of diluting the radioactivity. The saturation water content of most types of pottery is between 10 and 20% by weight; this can be measured for each sample and for the wet climates of northwestern Europe it is fairly safe to assume that the fragment has been permanently saturated. Where the climate is intermediate, uncertainty about the past climate significantly worsens the limits of accuracy of the date obtained.

It is usual to evaluate the dose rate from uranium and thorium by alpha-counting and to determine the potassium content by chemical analysis (e.g. flame photometry). Alpha counting is preferable to chemical determination of uranium and thorium because of the gross errors which would be introduced by the latter procedure if the decay chains were not in equilibrium. Because of the 267 000 year half-life of ^{234}U this may be the case for some glacial and postglacial clays. Another cause for the disequilibrium in the uranium series is the escape of the gas radon (^{222}Rn). A more direct method of evaluating the beta dose-rate is by means of a sensitive TL dosimetry phosphor: the phosphor is exposed to the beta-particle flux from the sample (but shielded from alpha-particles by a thin absorber) and knowing the sensitivity of the phosphor the dose-rate is determined by measurement of TL accumulated in a known time (typically a few weeks).

3.1. *Grain size*

The mineral inclusions in pottery are often up to several hundred microns in diameter. Since the uranium and thorium is carried predominantly in the fired clay matrix in which these inclusions are embedded rather than in the inclusions themselves, the alpha-particles, having ranges between 20 and 50 μm, reach only a thin surface layer of the inclusions. The result is that the average dose received by an inclusion is less than that received by fine grains, to a degree dependent on the size of the inclusion. The TL sensitivity of the inclusions is usually several orders of magnitude higher than that of the fired clay and consequently the effect is important, even though their percentage by weight may be small. Two techniques for overcoming this difficulty have been developed—the "inclusion" technique and the "fine grain" technique. In the former the TL measurements are made on grains for which the alpha-particle contribution is negligible whereas in the latter the measurements are made on grains small enough for the alpha-particle contribution to have its full value. As explained earlier, these TL measurements yield the "equivalent radiation dose". Then in using the age relation (section 1) only the beta and gamma contributions to the dose per year are included in the former technique, whereas for the latter the alpha contribution (multiplied by the k value—see section 2.4) is included in addition. It is evident straight away that the inclusion technique is much more affected by environmental radiation, but on the other hand it does avoid the complication of the k value.

For the inclusion technique the crystalline components of the sample are separated from the clay matrix using a magnetic separator, and grains in the size range of 90 to 105 μm are selected by sieving; larger grains are avoided because of beta attenuation effects. Grains of the selected size still carry a skin of alpha-irradiated material and the major part of this is removed by etching with hydrofluoric acid.

In the fine grain technique it is essential, in crushing the fragment, to avoid breaking up any inclusions. This is to ensure that the fine grains obtained are genuine rather than fragments of larger ones. A satisfactory crushing technique is to squeeze the fragment in a vice. After this, grains in the size range 1 to 8 μm are selected by their settling rate in acetone and deposited on a number of aluminium discs (10 mm in diameter and 0·45 mm thick). Each disc carries a 1 mg layer of fine grains, about 4 μm thick and these are used for the TL measurements, the reproducibility between discs being about 5%.

3.2. The environmental dose rate

It will be seen from Fig. 5 that for fine grain dating the gamma-ray dose is typically 23% of the total effective dose, and for inclusion dating about 39%. If the radioactivity of the soil is stronger than that of the pottery (particularly in respect of the thorium series, on account of its high gamma activity) then the contribution becomes more important still. The effective attenuation coefficients for the gamma-rays from uranium, thorium and potassium are not dissimilar and calculations indicate that 95% of the dose originates from the soil within a 0·3 m radius. It is therefore important, first that the pottery fragment should have been buried to a depth of at least 0·3 m for the major portion of the burial time, and secondly that the material lying within 0·3 m of the fragment should be homogeneous. The ideal context is the middle of a pit or ditch. Heterogeneous contexts, such as destruction layers and tombs, are less satisfactory.

Even in the homogeneous situations there are a number of difficulties associated with the evaluation of the gamma dosage obtained by laboratory measurements on an extracted soil sample. First, as with the pottery fragment itself the radioactive concentration is diluted by the presence of ground water (which is low in radioactivity) and this will vary seasonally and possible on a longer term basis also. Secondly, the gamma dosage from the uranium series is strongly affected by the emanation of the gas radon and it is well known that this is an appreciable effect, being enhanced in dry weather and attenuated in wet. The loss of radon in this way means not only the loss of the dose rate due to radon but also that due to the daughter products that follow radon in the uranium decay chain. In fact 98% of the total gamma-ray energy associated with the uranium chain lies beyond radon. The uncertainties due to this emanation would be far more important were it not for the fact that the uranium series supplies quite a modest part of the gamma dose-rate.

A superior technique for the evaluation of the environmental gamma dosage is direct on-site measurement using a TL dosimetry phosphor. This involves placing a small capsule of the phosphor in as similar a burial situation as possible to that from which the pottery fragment has been removed. Phosphors are available with sufficient sensitivity for accurate measurement to be made within a few weeks, but to average out seasonal variations in wetness and in radon emanation the capsule is left in position for a year when possible. Doubt about the gamma dosage is then centred on the extent to which short-term measurements correctly reflect the situation throughout the burial period of the fragment.

The importance of environmental radiation imposes a serious restriction on TL dating, both in respect of accuracy and in respect of the circumstances from which samples are acceptable. For samples in which the effective alpha dose-rate constitutes a high proportion of the total dose-rate it is possible to remove this limitation by using the fine grain technique and the inclusion technique in conjunction; by subtracting the equivalent radiation dose measured by the latter from that measured by the former, one is left with the effective dose due to alpha-radiation thereby avoiding the need to evaluate the environmental dosage.

4. Pre-dose Dating

The pre-dose phenomenon mentioned earlier has found important application as a dating method covering the last thousand years or so when the conventional TL method is usually too insensitive. The quartz inclusions found in pottery usually have a highly-sensitive peak at 110°C corresponding to a trap depth close to 1 eV. This peak is only seen immediately following artificial irradiation since the lifetime of electrons in these traps is only a few hours. However the *sensitivity* of the peak to artificial irradiation is strongly dependent on the total radiation dosage received since the pottery was fired—as long as activation by heating to 400° or 500°C precedes measurement of sensitivity. Measurement before activation gives the value of the sensitivity corresponding to zero pre-dose and the extra sensitivity observed after activation is exactly proportional to the pre-dose. An increase by a factor of 2 for a dose of 100 rads is not uncommon and consequently this technique can be used for quite recently-fired pottery. On the other hand saturation of the effect usually sets in before 1000 rads thereby precluding application to samples much older than 1000 years (though the method is extremely valuable in authenticity testing in establishing a *minimum* age for such samples). The same age relation (see section 1) applies as in conventional TL. As with other types of pre-dose phenomenon alpha-particles are negligibly effective so that in the evaluation of "dose per year" only the beta and gamma contributions are included—as for the inclusion technique in conventional TL. An important application in authenticity testing is that the technique distinguishes between clay that has been recently fired for the first time and clay that was first fired in antiquity and recently refired.

The effect is due to a subtle enhancement of the number of activated luminescent centres. This enhancement does not occur until holes stored in non-luminescent centres are transferred by heating to the luminescent centres (Zimmerman, 1971).

5. Application

For accurate results it is desirable to have half-a-dozen or more fragments from each context being dated, each fragment being at least 5 mm thick and

20 or 30 mm across. In addition, unless the "subtraction technique" is being used, the environmental dose-rate must be evaluated as indicated in section 3.2. In good circumstances, and as long as the TL characteristics of the pottery are satisfactory, an absolute accuracy of between 10% and 5% of the age is obtainable.

Evaluation of environmental dose-rate usually necessitates deliberate collection of samples while excavation is in progress. However in testing ceramic works of art and detecting which are modern forgeries accuracy is not usually important and the error introduced by not knowing the environmental dose-rate does not matter. For such application a prime consideration is the minimization of damage to the object; a sample of between 20 and 50 mg obtained by drilling a small hole in an unobtrusive part of the object is usually adequate. In reputable antiquities salerooms, TL testing is routine for any pieces for which there is doubt about authenticity.

See also: Archaelogical dating by physical methods.

Bibliography

AITKEN, M. J. (1974) *Physics and Archaeology*, Oxford University Press.
AITKEN, M. J. and ALLDRED, J. C. (1972) The assessment of error limits in thermoluminescent dating. *Archaeometry*, 14, 257–68.
AITKEN, M. J. and FLEMING, S. J. (1972) Thermoluminescence dosimetry in archaeological dating. *Progress in Radiation Dosimetry* (F. H. Attix, Ed.), Vol. 1, pp. 1–78.
AITKEN, M. J., MOOREY, P. R. S. and UCKO, P. J. (1971) The authenticity of vessels and figurines in the Hacilar style, *Archaeometry*, 13, 89–141.
AITKEN, M. J., ZIMMERMAN, D. W. and FLEMING, S. J. (1968) Thermoluminescent dating of ancient pottery, *Nature*, 219, 442–4.
ATTIX, F. H. (Ed.) (1967) *Luminescence Dosimetry*. U.S. Atomic Energy Commission (available as CONF-650 637 from National Bureau of Standards, Springfield, Virginia).
AUXIER, J. A., BECKER, K. and Robinson, E. M. (Eds.) (1968) *Proc. Second Int. Conf. Luminescence Dosimetry*. U.S. Atomic Energy Commission (available as CONF-680920 from National Bureau of Standards, Springfield, Virginia).
FLEMING, S. J. (1970) Thermoluminescence dating: refinement of the quartz inclusion method. *Archaeometry*, 12, 135–46.
FLEMING, S. J., JUCKER, H. and RIEDERER, J. (1971) Etruscan wall-paintings on Terracotta: a study in authenticity, *Archaeometry*, 13, 143–67.
McDOUGALL, D. J. (Ed.) (1968) *Thermoluminescence of Geological Materials*, Academic Press: London and New York.

MEJDAHL, V. (1970) Measurement of environmental radiation at archaeological excavation sites, *Archaeometry*, 12, 149–61.
MEJDAHL, V. (Ed.) (1971) *Proc. Third Int. Conf. Luminescence Dosimetry*, Danish Atomic Eneregy Commission (available as Risö Report no. 249 from Jul. Gjellerup, 87, Selvgade, DK-1307 Copenhagen K).
TITE, M. S. (1966) Thermoluminescent dating of ancient ceramics: a reassessment, *Archaeometry*, 9, 155–69.
ZIMMERMAN, D. W. (1971) Thermoluminescent dating using fine grains from pottery, *Archaeometry*, 13, 29–52.
ZIMMERMAN, D. W. (1972) Relative thermoluminescence effects of alpha- and beta-radiation. *Radiation Effects*, 14, 81–92.
ZIMMERMAN, J. (1971) The radiation-induced increase of 100°C thermoluminescence sensitivity of fired quartz, *J. Phys. C.: Solid St. Phys.* 4, 3265–76.

M. J. AITKEN

THERMONUCLEAR POWER—THE PRESENT POSITION. In the last few years, after a period of uncertain results, a much more optimistic view has emerged of the prospects of achieving thermonuclear power reactors. Substantial progress has been made in solving the containment physics problems, although the plasma conditions achieved are still generally at least an order of magnitude below the requirements of a power reactor.

Alongside encouraging experimental results has come an increasing awareness of the environmental and economic advantages of fusion reactors. These are that:

(i) they would make available a very large and hitherto untapped energy source (lithium and deuterium) with the long term prospect of limitless resources from deuterium alone;

(ii) the fuel cost is negligible;

(iii) the reaction products are harmless and therefore present no major radioactive waste disposal problems;

(iv) as fusion reactors contain only small quantities of fuel in the reaction zone there is no prospect of a nuclear explosion.

The basic requirements for the exploitation of fusion reactions are discussed by Robson (1962). For the most accessible case, the reaction $(D + T \rightarrow n + \alpha)$, a plasma must be confined at ion temperature of about 10^8 K for a time τ, such that $N\tau > 10^{20}$ sec m^{-3} (the Lawson criterion), where N is the number density of ions. Most work has been devoted to systems where the plasma is confined by magnetic fields to give $\tau \sim 1$ sec with $N \sim 10^{20}$ to 10^{21} m^{-3}. There has recently been renewed interest in "inertially confined" pulsed systems where the density is very high and the confinement time is just the expansion time of the plasma.

1. Magnetic Confinement

Plasma can escape across and along magnetic fields in a number of ways. "Macroscopic" instabilities of a fluid type, which arise from unfavourable curvature of the lines of force were originally the most serious cause of plasma loss, but it has been found that the most dangerous of these can be suppressed by ingenious choices of magnetic field geometry. (For example, in "magnetic wells" the magnitude of the magnetic field increases in all directions away from the plasma; with "magnetic shear" the lines of force are twisted into helices of varying pitch angle.) "Classical" diffusion, arising from collisions between ions and electrons is, of course, always present: its importance also depending upon the geometrical characteristics of the system. "Micro-instabilities", which stem from the detailed shape of the particle velocity distribution functions, also give rise to transport effects (termed "anomalous" in the literature, as they cannot be accounted for directly in terms of interparticle collisions). To some extent micro-instabilities can also be suppressed by an appropriate choice of field geometry and of the magnitude of density or temperature gradients, etc., but it seems improbable that any containment device could ever be rendered wholly stable by any means.

The complex problem of minimizing the total effect of these loss mechanisms has attracted a variety of solutions. We shall discuss only three systems: tokamaks, stellarators and magnetic mirrors. Accounts of other systems may be found in the works listed in the bibliography.

1.1. Tokamaks

A Tokamak consists of an axisymmetric toroidal vacuum chamber (typically, 70 cm in major radius and 25 cm in minor radius) linked to an iron-core transformer. A varying current in the primary circuit of this transformer induces an electric field which drives a current in the plasma, flowing parallel to the major circumference of the torus. Confinement is due to the poloidal magnetic field generated by the current. The current also heats the plasma. However to render the plasma macroscopically stable a much larger toroidal magnetic field (typically, 15 kG) is also required and this is generated by currents flowing in a set of coils surrounding the vacuum chamber.

Tokamak research originated in Russia in the 1950s and at first attracted little interest elsewhere. Since 1970 there has been a landslide towards Tokamaks in the United States and Western Europe and many such devices are now operating or under construction.

The closest approach to overall reactor conditions in the past few years has been in tokamak systems. Ion and electron temperatures of 6×10^6 K and 3×10^7 K respectively have been obtained with densities around 5×10^{19} m^{-3}. Moreover, it is found that electron temperature scales as the square of the current and as the ions are heated by collisions with the electrons, higher ion temperatures are expected in the next generation of experiments (with higher discharge currents) already in construction. However, the ion temperature cannot, by this method alone, be raised to the value needed in a reactor and methods of auxiliary heating, such as the injection of energetic neutral particles or the adiabatic compression of the plasma, are being devised.

Particle and energy containment times have also been very encouraging, although still an order of magnitude less than expected on the basis of binary collision theory. The highest value of the energy containment time yet achieved in tokamaks is 20 msec. This time has been found empirically to scale as $a^2 n^{1/3} I$, where a is the plasma radius, n is the density and I the current. If this scaling extends to reactor conditions it would give containment times which are an order of magnitude better than required.

From a reactor point of view a disadvantage of the tokamak is that the current will only flow while the magnetic flux in the iron yoke of the transformer is changing. Thus it is basically a pulsed device. However, theoretical studies of transport processes in toroidal geometry indicate that in the presence of radial diffusion a toroidal current can flow with no electric field present. This "bootstrap" current offers the possibility of steady-state operation, but at the time of writing a search for this effect in several present-day experimental devices has proved negative. Other proposals for steady-state tokamaks have been put forward. If any of these could be successfully developed the tokamak, because of its simple magnetic field configuration, could become the quickest and cheapest route to a fusion reactor.

1.2. Stellarators

Stellarators are also toroidal systems but the confining magnetic fields are generated solely by currents flowing in external conductors, without the need for currents flowing in the plasma. They can thus be operated as steady state systems. To ensure equilibrium of the plasma the magnetic field includes a contribution from windings which spiral helically around the torus.

Research on stellarators has had some setbacks, but in recent years experiments have yielded improvements in confinement times and a much better understanding of the nature of the losses. The densities and temperatures achieved so far are several orders of magnitude below the values required in a reactor.

Stellarators have suffered a great decline in popularity compared to tokamaks. To some extent the greater attraction of tokamaks is a technological one. For a given expenditure of money and effort the tokamak one obtains is more attractive as a reactor than the stellarator.

1.3. *Magnetic mirrors*

Magnetic mirrors are linear, open-ended steady-state systems in which particles are prevented from flowing freely along the lines of force by reflection in end regions of strong magnetic field. The earlier, simple mirror systems were macroscopically unstable but it has been found possible to suppress the most serious instability through the use of a particular magnetic field geometry ("minimum-B" geometry) in which the strength of the magnetic field increases in all directions away from the region in which the plasma is contained.

Unfortunately even in a stable system there is a serious loss mechanism. Particles whose momentum is mainly parallel to the magnetic field escape through the magnetic reflecting fields and since such particles are continually being replenished by particle–particle scattering, the resulting loss rate may be higher than could be tolerated in a fusion reactor. In fact the situation is worse than this: observed loss rates are at least an order of magnitude higher than could be accounted for by binary collisions. The discrepancy is thought to be due to the effects of micro-instabilities.

It has been suggested that the problem created by these particle losses in magnetic mirrors could be overcome by a process termed "direct conversion". In this the energy of the escaping charged particles would be directly converted to electricity, most of which would be used to reinject fast particles into the plasma. Because of the high level of circulating power, compared to useful energy output, the efficiency of direct recovery and injection would have to be very high (probably unrealizably high, unless the loss due to micro-instabilities can be overcome).

Mirror machines do, however, have some advantages from a reactor point of view. They may be more economic in smaller unit sizes than toroidal reactors, the geometry is simpler and the open ends allow easier access. Further progress depends on improving energy containment by devising a means of reducing the end losses and developing very efficient direct conversion and particle injection schemes.

2. *Inertial Confinement*

In addition to the magnetic confinement schemes described above there has very recently been increased interest in alternative systems equivalent to a succession of controlled micro-explosions. High-power lasers (or possibly pulsed relativistic electron beams) would be used to ignite pellets of DT fuel. The duration of the reaction will be limited by its free expansion at the ion thermal velocity.

It is not difficult to show that, if the radius of the DT target (and hence the inertial confinement time) is large enough to satisfy the Lawson criterion, the energy needed to heat the plasma to the required temperature scales inversely as the square of the density. At the density of solid DT, the laser energy needed is estimated to be at least of order 10^8 to 10^9 J. Lasers can at present deliver 10^3 J (and electron beams 10^5 J) in pulses of the correct duration, so clearly it will be necessary to compress the target pellet before heating it. Compression might be achieved by heating the surface of the pellet as symmetrically as possible with laser beams, causing ablation of the surface material. The reaction to the ablation causes an implosion. Computer calculations indicate that compressions to densities of ten thousand times solid density can be achieved by this method. However, these calculations are based on a number of over-simplified and unproven physical assumptions and it is difficult to say how much reliance should be placed upon them.

The attraction of the laser fusion concept is that it bypasses all the problems of magnetic confinement and stability as well as several technological problems such as the construction of large superconducting magnets. The physical problems involved have hardly begun to be explored either experimentally or theoretically, so a realistic assessment of the prospects of this idea is unlikely to be made for several years.

3. *Technology*

It is important to bear in mind that containment physics is only one part of the problem of building a thermonuclear reactor. Success in the confinement of high-temperature plasma will uncover a great many technological, engineering and economic problems. Consideration of these problems is necessarily still at a very early stage, but an important feature of such studies as have been done is that a major component of the overall generating cost will be the capital cost of the superconducting magnets. This emphasizes the great potential importance of research on systems in which the magnetic field is provided largely by currents flowing in the plasma itself, or is dispensed with all together.

4. *Conclusions*

There have been substantial improvements in the temperatures, densities and confinement times achieved in experimental assemblies and, equally important, our understanding of basic plasma physics has made striking advances. There is good reason to believe that the escape mechanisms—macroscopic instabilities, diffusion due to collisions and diffusion due to micro-instabilities—can be controlled so that the parameters required will be achieved simultaneously in experiments in the future.

Our present state of knowledge does not enable us to say that one system is pre-eminently suitable for development as the basis of a reactor, although tokamaks and laser driven fusion currently appear to be the most popular contenders. It will be necessary to carry out some large-scale confinement experiments before the construction of an electricity-generating prototype reactor could be justified. On present indications such prototypes will be operating in the last decade of the century.

Bibliography

COPPI, B. and REM, J. (1972) *Scientific American*, July 1972, 65.
I.A.E.A. (1971) *Proceedings of the Fourth International Conference on Plasma Physics and Controlled Thermonuclear Fusion Research*, Madison 1971.
I.A.E.A. (1972) *Proceedings of the Fourth International Conference on the Peaceful Uses of Atomic Energy*, Geneva 1971.
NUCKOLLS, J. *et al.* (1972) *Nature*, **239**, 139.
ROBSON, A. E. (1962) Thermonuclear reaction in *Encyclopedic Dictionary of Physics* (J. Thewlis, Ed.), **7**, 326. Oxford: Pergamon Press.
SCHMIDT, G. (1966) *Physics of High Temperature Plasmas*. London: Academic Press.
THOMPSON, W. B. (1967) Plasma physics in *Encyclopedic Dictionary of Physics* (J. Thewlis, Ed.), Suppl. **2**, 252. Oxford: Pergamon Press.
TUCK, J. L. (1971) *Nature*, **233**, 593.

I. COOK

THICK FILM TECHNOLOGY. The technology is concerned with the formation of electronic circuits by a modified printing process. In its simplest terms this is carried out using woven-cloth-patterning "screens" and viscous pastes containing powdered metals and glasses. Some of the mesh openings in the screen are blocked using a photo-sensitive filler and when the paste is forced through the remaining open holes on to a flat substrate a replica of the pattern on the screen is obtained. Virtually any material that can be formulated into a viscous paste may be deposited using the technology, including most liquids and solids, the latter being carried as powders in suspension. Substrate materials may be plastic, glass or ceramic and many types of these are in common usage.

The term "thick film", as applied to microcircuit manufacture, is generally accepted, somewhat illogically, as describing a particular method of circuit deposition. Originally the term was used to distinguish the method from the equally unsatisfactorily termed "thin-film" process which generally refers to deposition by evaporative means, since there was about an order of magnitude difference in the thickness of the deposits. The terms are still used but have little of their original significance, having become accepted in this research field as describing the processes themselves. Thin film, then, is concerned with deposition by some vacuum process and thick film by the direct printing of circuit elements. "Screen printing" and "print and fire" are also used to describe this latter process.

Initial work on the technique dates back 2000–3000 years when it was used in China for graphic display and decoration—a field, incidentally, which accounts for the largest proportion of the activity today. Although a wide range of materials and techniques can be used in the process of screen printing this discussion will be concerned only with those aspects relevant to electronic circuitry, in other words, to that area commonly known as thick-film technology. This particular application dates from the mid-1940s, with widespread commercial usage taking place in the mid-1960s.

Basis and Outline of the Technology

Basically the technique we are concerned with relies on the sintering of powders together to form conductive or insulating patterns. This feature makes it an elevated temperature process and particle sizes and size distributions are chosen such that a common temperature range of 750–1100°C can be used for all the materials. A number of compromises are necessary in arriving at this range: on the one hand, mainly between the advantages of forming stable, high-density, metal layers and continuous glass layers of good electrical performance, and on the other hand, the increasing risk of deleterious materials interaction.

Deposition

The main component of the printing equipment is the screen which usually consists of a piece of woven mesh, of stainless steel or man-made fibres mounted under tension on a metal frame. The circuit pattern is formed photographically on the mesh using a ultra-violet-sensitive emulsion which may be applied to the mesh before or after the pattern is developed. In either case the resulting finished screen is such that it has open-mesh holes only over the circuit pattern area. The appropriate paste is then placed on one side of the printing screen, which is suspended about 0·5 mm above the substrate. A scraper blade which is known as a squeegee traverses the screen under pressure bringing it into contact with the substrate and at the same time driving paste through the open meshes of the screen. The natural elasticity of the screen material peels it off the substrate: the paste, having wet the substrate, is drawn out of the screen holes and left in position. The scraping step can by carried out by hand, but a range of commercial machines is available, enabling the print to be achieved in a reproducible manner. Such a machine consists essentially of a substrate holder and squeegee traverse mechanism, these components being arranged with some precision within a rigid structure such that their position and movement with respect of each other is accurately defined and controllable. The substrate holder has three-dimensional position adjustments and a central vacuum chuck which is often fitted with internal illumination for pattern registration purposes. The holder is normally carried on a pair of linear bearings allowing movement from the loading position to the working position beneath the screen. The squeegee traverse mechanism can be quite complex, as control of factors such as traverse speed, blade pressure and angle, are necessary to ensure uniformity of paste deposition. Pneumatic-hydraulic combinations or mechanically spring-loaded systems are generally used, the former giving a greater range of controllable pressure and speed values. The squeegee mechanism is also mounted on a pair of linear bearings: the three positional planes for squeegee, screen and substrate must be maintained parallel and in a fixed position throughout the printing sequence, so the whole machine structure has to be rigid. There are many variables in the deposition sequence, but the main factors associated with the machine are rigidity, mechanical precision and uniformity of speed and pressure at the squeegee screen interface. To achieve a dimensional precision in the printed pattern of better than 250 μm nothing is critical and a complete equipment need not cost more than £ 100. At the same time, for the most precise work, where tolerances of less than 25 μm may be required, a machine may cost £ 3000 and a screen £ 50.

It is clear from the foregoing that the printing sequence is very crude and there are many sources of dimensional error in the pattern formed. Available ceramic substrates are rarely very flat and if costs are to be minimized the print will necessarily be made onto a curved surface. This can affect the amount of paste deposited unless care is taken to ensure that the screen follows all the undulations in the substrate surface. This is an easy condition to satisfy with the mesh material usually used, but this normally brings with it a number of disadvantages. It is a woven cloth and hence has many crossover points and no real flat surface. The stresses which arise due to friction at the crossovers will gradually relax in use, with consequent loss of pattern size and mesh tension, so steps must be taken to relieve these stresses before developing the pattern. This is not very difficult either and it is possible to hold pattern dimensions on a screen to a few micrometers over its life.

Typically the thickness of the dried film is 20 to 30 microns, the actual value depending on mesh choice, emulsion thickness and the solids content of the pastes. Deposits are generally large in width compared to this figure and hence the main errors arising are in edge definition and deposit thickness.

The complex shape of the mesh, as seen by the paste printed through it, results in poor line-edge definition. A good-quality screen filler emulsion will give straight edges, which tend to minimize this effect, but it is difficult to obtain edge errors of less than 10 μm by attention to the emulsion. A high build-up of emulsion, which also decreases the edge effect, introduces a more significant problem of deposit thickness control as the suspended mesh will bend down under the scraper pressure. The result of this is a variation of deposit thickness dependent on the size of the unsupported area; combined with substrate bow this can be a serious restriction on the control of resistor values. A partial answer to this predicament is to use one of a number of different types of solid metal screen which have been developed over the last few years. The most common of these has the pattern chemically milled in a 25–75-μm-thick sheet of metal. The pattern is taken halfway through the sheet and an array of feeder holes is formed on the reverse side. This results in a screen which is altogether cleaner than the mesh type and gives a better pattern but the price of this improvement is a loss of mechanical flexibility of the screen. Such a screen demands flatter substrates, which have to be ground or selected, and the loss of the natural peeling action of the screen off the substrate gives some paste plugging of the screen holes leading to a certain amount of porosity in the print. At the present time mesh screens are used for the majority of work where definition is not crucial, and for those particularly demanding applications where fine lines and spacings are needed, for example, in printing integrated circuit cell patterns, the more rigid solid screens are used. More detailed discussion on the properties of the printing screen have been presented by Savage *et al.* (1969) and Caronis (1967).

By and large the printing process has developed to the point where it is under control and dimensionally it is capable of meeting most of the requirements. Circuit patterns with line widths down to 125 μm over areas up to 15 cm^2, with short feeder lines, for active device connection, down to 25 μm widths, all with a width tolerance of ± 10 μm and a positional tolerance of ± 25 μm have been achieved. At the limit there is a fall-off in yield: around 25 μm more effort is necessary, and hence more cost, than can be justified in any but the most sophisticated applications.

Materials Utilized

The high firing temperature essentially limits the substrate choices to glasses and ceramics. The former are usually used only in specialized applications such as optoelectronics or in low wattage uses; by far the majority of circuits are formed on ceramics, usually alumina, because of their higher thermal conductivity, strength, stability and good electrical properties. Purities of 85–98% are commonly used; higher purity requires higher sintering temperatures which can lead to loss of dimensional control, and the usual practice is to incorporate some glassy phase in the ceramic body to facilitate fabrication. This may be formed by additions of, for example, magnesia and calcia. These components also have a role in the bonding process of the circuit elements to the ceramic substrate, the large majority of commercial metallization pastes being designed to react with the glass phases in this type of material. Compositions outside this range of purity are hence rarely used, except in microwave applications which demand a higher purity level for electrical reasons. Special provision has then to be made to improve the adhesion of the fired circuit.

Circuit Elements

The metals used for conductor patterns are mainly noble, the list including gold, palladium, platinum, silver. These are used in either pure or alloy form, as they have good electrical conductivity and can be fired in an air atmosphere. The cost of using these more expensive materials is more than outweighed by their ease of firing, subsequent stability in air, and ease of attachment by soldering and welding.

The metals that are to form the metallization patterns are made up in the form of viscous pastes, somewhat like syrup in consistency, for application to the substrate. The pastes contain four principal components: the powdered metal; a glass, also in particulate form, which is present as the bonding agent for joining the metal to the substrate; an organic vehicle in which these elements are suspended and whose purpose is to carry the usable components into position on the substrate; and a diluent to control the viscosity of the whole. These last two components should have a low temperature coefficient of viscosity and be constituted such that they burn off in air without residue. These are minor constraints and hence there is a very wide choice of materials for the vehicle application. The mean particulate size of the powders used is about 5 μm, as small as possible in the case of the higher melting point metals to lower the sintering temperature.

Insulant materials are similarly constituted using a variety of glasses but have no metallic component. The constraints on insulant material properties include restricted and controlled flow when laid down as films, such that through hole sizes and other dimensions may be held through several firing cycles, and a very high degree of freedom from porosity in the fired film to fulfil severe isolation requirements. They must also have controlled, reproducible, electrical properties particularly when they are being used as capacitor dielectrics. These requirements make insulant materials considerably more difficult to formulate than conductor metallizations and they can often be as complex as resistor mixes.

Resistor pastes generally have very complex compositions which are necessary to meet the requirements of a wide range of resistivities combined with low temperature coefficients of resistance. They consist of non-noble metals and compounds, often in combinations which have considerable thermodynamic instability at the firing temperature, making such pastes critically sensitive to firing conditions. The number of parameters which require to be optimized for resistors is large and the trend is for resistor tolerances to tighten. The preparation and use of some of these pastes is hence quite difficult.

If we add to these four main elements, substrates, conductors, insulants and resistors, special formulations for solder pads, resistor overcoats and hermetic sealing, we can easily reach hundreds of possible combinations. It is not surprising therefore that some of the problems which arise in using the process are associated with materials compatibility.

The firing of the deposited paste produces, in sequence, the burn-off of the organic vehicle at 200° to 500°C, and then the sintering of the metal particles and melting of the glass particles at temperatures in the range 700–900°C. The glass forms a mechanical key on the rough underside of the sintered metal and wets the glass phase in the alumina. A continuous bond is thus established between the metal and the substrate. Although the details of the firing step are most critical for resistors, care has still to be taken with conductors as the alumina can readily and rapidly take up the bonding glass: this can eventually disappear entirely from the interface, resulting in a degradation of strength if the arrays are overfired.

Generally speaking, the components in the "as fired" condition do not have sufficiently precise electrical property values for most applications and a trimming step to adjust the values is necessary. This is particularly true for resistors which can be 5–15% out of value as fired and less so for capacitance values. Trimming is achieved by removal of surplus material, using abrasion (airborne powder jet) or evaporation (using lasers or direct-heating).

The materials used in thick film have developed, and are developing, very quickly. Generally, improvements have lagged behind development of the other aspect of the process as the materials are more complex and require much more research, but at the same time they have a high value as circuit elements and hence can be relatively expensive without limiting their use. In the past 2 years conductor pastes have improved to within a factor of 2 of the bulk conductivity, the hardness has become controllable for bonding purposes and the edge flow is now within the error due to the printing process. A factor of 2 in these parameter values is often unimportant for conductors and insulators but resistor materials present a constant problem as there are more critical demands made on their electrical properties. One per cent tolerance as fired, and 10 ppm°C^{-1} TCR, is not an unusual requirement and this cannot be achieved at the present time. Less than 5% and 50 ppm is currently available, and both these figures are a factor of 10 better than those available 3 years ago. It is reasonable to expect that better figures will be achieved in 2 to 3 years time.

The other main material used, the insulant, has developed greatly in the last 3 years, much to the advantage of thick-film technology as a whole. Mention was made earlier of the presence of glass phases in the substrate, the metallization pastes and, of course, the insulants, and also of the general trend towards smaller circuitry involving multiple layers of conductors. Now, using the normal glasses available 4 to 5 years ago it was not possible to utilize more than a single layer of insulant as all the glass phases interacted, usually to form lower melting-point mixtures, resulting in a loss of dimensional control, electrical continuity and layer to layer insulation as the component materials moved out of position on firing. Two kinds of glass mixture have been developed to overcome this problem, both designed to limit interaction to the bonding surfaces, that is, to allow just sufficient interaction to achieve a bond and no more. There are devitrifying glasses, which precipitate a high melting-point crystalline phase on firing, making the layer more rigid, and loaded glasses which contain ceramic powder and rely on a high viscosity to limit interaction and flow. The difference between the performance of the two types is only important at the smallest dimension, the loaded glass moving less on an edge due to the absence of the flow which occurs prior to the re-crystallizing process in the devitrifying type. With both these materials the compatibility limitations are largely removed and it becomes possible to refire a number of times and produce multi-layered structures.

Summarizing the materials situation at present, there are resistor materials available that are capable of covering the range $1\,\Omega$ to $10\,M\Omega$ per square with as-fired tolerances in the 5% to 20% range and temperature coefficients of between 50 and 500 ppm$°C^{-1}$, insulant materials capable of 99.99% crossover yields and with dielectric constant values in the range 10 to 1000 and conductor materials with resistance values from 2 to 100 mΩ per square. These component data have been summarized by Holmes and Corkhill (1969).

Applications

Having discussed some aspects of the process it is of interest to move on to the philosophy of circuit production to demonstrate the current state of the art in the application of the technology. As there is a very wide range of circuit types it is useful to take examples from opposite ends of the spectrum, one a low-cost circuit that is currently being manufactured in large quantities and one a high-value circuit for which the volume at the moment is in the hundreds per year range. Neither represents the limit of the art at development level and some indication will be given as to where production circuits will move in the next year or so in each case.

It is perhaps worth noting, as both production technique and final forms are different, that two main groups of circuit type exist, the multi-component hybrid, containing conventional resistors, capacitors, inductors and integrated circuits, and the digital application type which consists of a high-density conductor network carrying integrated circuits only.

The first example is a motor-car voltage regulator which is a good example of the application of the technology in a number of senses. Using thick-film techniques here firstly has the advantage of replacing discrete resistors by printed ones. The environment provided by a motor vehicle is normally one of high vibration and here the reliability is greatly increased by a decrease in the number of flying leads. The small size of the unit allows the manufacturer to incorporate it within the generator body which then constitutes the outer package; this feature also enables the user to avoid the expense of connecting the generator and the regulator together in the vehicle. The main benefits of such an application, to which can be added other examples such as transistor radios, television sets or washing machines, all of which are incorporating similar arrays at present, derive from the replacement of manual steps by automated ones, but it is clear from this application that advantages can be gained from apparently incidental environmental considerations which can be dominant in some cases. The best example of this is in the domestic floor polisher. There would appear to be little value in the use of a control circuit here until it is appreciated that the polisher has to be constructed to a specification well above the normal operating rating to withstand prolonged stall conditions. Incorporation of a current limiting device, which again is possible within the normal body, enables a much lighter, lower-rated motor to be used.

Small size of itself, then, opens up new applications, and in the next example does so through the factor associated with small size, the operating speed of a circuit. At present one of the principal limiting factors in high-speed circuits is the transit time of signals between the active devices which themselves can operate in times of less than 1 nsec. Speed, which is cost, is the dominant factor in computer applications and there has been much effort in this field on the miniaturization of circuitry through the application of thick-film technology. The achievement of the shortest possible interconnection lengths necessitates about four levels of circuitry and up to double this number if intermediate earth planes are required.

It is a matter of no great difficulty to form two-layered circuits; the difficulties which arise in moving beyond two levels are centred on the materials interaction problems which have already been discussed, and special types of paste and some modification to the printing process itself, are necessary for complex multilayer arrays. Circuits of this type with five metallization layers and four insulant layers and containing 1000 crossovers are now in production whilst advanced development types have been made with several metres of interconnect and up to 10,000 crossovers on substrates of 25 cm^2 area. The development of these multilayer techniques has been discussed in rather more detail by Loasby (1969).

In summary, the last 5 years have brought the thick film, screen printing, process to the point where it is capable of meeting all existing circuit requirements in terms of density and dimensions. The current limitations arise from the materials used, particularly those for producing precision resistors and for forming reliable, high-strength bonds. These latter aspects have been accentuated by the success of the developments in the basic process and are the ones receiving the bulk of attention in the current research phase.

Definitions

Thick film. A term relating to the technique of forming electrical circuitry by screen printing the elements which are to form electrical components. The thickness of these components is typically 10–30 microns. The term originated from the need to distinguish the process from the then more common technique of thin-film vacuum deposition.

Screen printing. A process common to the graphic arts field utilized in the thick-film process for depositing circuit forming materials into distinct patterns or designs.

Pattern(s). In the first instance the electrical circuit is drawn out using a number of different line diagrams corresponding to the conductor routes, resistor, inductor and capacitor elements and dielectric layers, etc. This is carried out such that the combination of these patterns would specify the electrical circuit required. Each pattern is referred to as a layer of the circuit, several of which are required to form a circuit of average complexity. Hence the term *multiple-layer* array.

Screen. The pattern for each layer is reproduced on a photographic positive which is used to form the printing screen. *Woven mesh* which is like cloth but has a carefully controlled hole size and fibre diameter such that the open area is fixed and uniform, is attached rigidly onto a frame as for a picture and coated, or filled, with a photosensitive fluid, or *emulsion*, which sets in position. Exposure to light through the pattern positive enables the desired pattern to be reproduced in the form of open holes in the mesh, the remainder being blocked forming a screen. Alternatively the screen may be formed using a sheet of metal coated with photosensitive emulsion. After exposure and washing, metal is removed by dissolution, a process often called *chemical milling.* The resulting screen may be used to form the pattern it bears using any material that can pass through the open mesh holes. Materials may be solid, for example, in powder form, but are more generally basically liquid when they are referred to as inks or pastes.

Pastes. The material that is ultimately going to form the electrical elements themselves. They consist of powders of metal, for the conductors, or compounds, for resistors and dielectrics, dispersed in a thick medium which is usually organic such that it may subsequently be burnt off. Because of this temporary function the medium is often referred to as a *vehicle.* These pastes, a different one for each ayler of the circuit, are formed into the appropriate pattern by placing some onto the screen and forcing it through the open holes onto a suitable substrate with a hard rubber-faced scraper blade, more commonly called a squeegee.

The squeegee or metering blade functions by applying a uniform pressure onto the paste and screen over the pattern area such that a constant thickness deposit is obtained on the substrate.

The substrate is the circuit board or base on which the circuit is formed. In this technology the material principally used is a ceramic.

Ceramic is a term derived from pottery and is essentially a crystalline inorganic insulating material. The most common ceramic used for substrates is alumina which is characterized by its chemical and physical stability, good mechanical and electrical strength and reasonable thermal conductivity.

Vacuum chuck. To hold the substrate precisely in position while the pattern is printed a locating platen is normally provided. The substrate is held in this chuck or recess by drawing a vacuum behind it. Such *vacuum chucks* are common in various areas of the technology for achieving positional accuracy of components.

Sintering. After the paste has been positioned on the substrate in the pattern required the electrical properties are produced by *firing* the circuit. This consists of heating until the vehicle has been burnt off, at about 500°C, and the conducting powder particles have fused together by solid-state diffusion, a process known as *sintering*. By control of conditions this is usually arranged to take place at around 850°C and results in the pattern having the desired conducting properties.

Overfiring. The conducting components or the *metalization,* as it is referred to, are formed by this controlled sintering process. If carried on for too long a period the process goes too far and some of the advantageous properties are lost. This condition is known as *overfired*.

Devitrifying. In the case of glass-bearing pastes, used for insulating layers, overfiring can result in the loss of some components of the glass by reaction with other materials present, for example the substrate, and crystallization or *devitrification* of the glass may occur. Some reaction with the substrate is a necessary feature as adhesion of the conductors and resistors, etc., to the ceramic base is essential, both to hold these two items together and to give support to any attached components such as leads or semiconductors. The joining or *bonding* of these elements together is concerned with the mutual compatibility of a wide range of materials and is a matter of some complexity.

Print and fire. This is an alternative descriptive term for the technology of making circuitry by printing patterns with a metallizing fluid and forming the electrical properties by heating the circuit to a high temperature, or firing it. Other common equivalent terms are *thick film* or *screen printing*.

Thin film. Refers to an unrelated technology for forming circuits, normally utilizing vacuum evaporation but also encompassing other techniques such as plating. The layers deposited are typically below 1·0 micron in thickness. A similar range of components, conductors, resistors, insulants etc., is achievable within this technology.

R. G. LOASBY

Bibliography

CARONIS, H. L. (1967) Screen tension and other aspects of emulsion and metal screens, *Proc. 2nd Symp. on Hybrid Microelectronics*, Solid State Technology Book Division, 14 Vanderventer Ave., Port Washington, N.Y. 11,050.

CORKHILL, J. R. and MULLINS, D. R. (1969) Properties of thick film conductors, *Electronic Components* (May).

FISHER, H. D. and CORKHILL, J. R. (1969) Advanced packaging for the Black Arrow X3 satellite, *Proc. Int. Conf. on Electronic Packaging and Production*.

HOLMES, P. J. and CORKHILL, J. R. (1969) Thick film integrated circuits, *Electronic Components* (Oct. and Nov.).

LOASBY, R. G. (1969) The technology of multilayered thick film circuit arrays, *Microelectronics* (April).

LOASBY, R. G. (1969) Aspects of multilayering thick film circuits, *Proc. ISHM (UK) Conf. on Multilayer Interconnection Technology*, Paper No. 4.

SAVAGE, J. (1969) Factors affecting the quality of screen printed conductors, *Thin Solid Films* **4**, 137.

SAVAGE, J., HOUGHTON, D. G. and WILLIS, D. (1969) Precision screens for microcircuit printing, *Proc. ISHM (UK) Conf. on Multilayer Interconnection Technology*, Paper No. 6.

Further papers on aspects of the screen-printing process, material and the construction of circuits can be found in the various Conference proceedings of the U.S. and U.K. branches of the International Society for Hybrid Microelectronics or recent review papers such as "Hybrid Thick Film Circuits", Brook and Baldwin, *Wireless World*, **79** (1973). Early work on the process has been described by Hughes, *Screen Printing of Microcircuits* (DAN MAR Publishing Co.). More recent volumes include *Thick Film Hybrid Microcircuit Technology* by D. W. Hamer and J. V. Biggers, Wiley Interscience, and *Thick Film Technology and Chip Joining* by L. B. Miller published by Gordon & Breach.

TIME, ATOMIC. In the Système International d'Unités (S.I. units) the second is defined as the duration of 9,192,631,770 periods of the radiation corresponding to the transition between the two hyperfine levels of the ground state of the caesium-133 atom. The adoption of this definition resulted from the increasing application and precision of caesium frequency standards approaching parts in a million million, since the first such standard was operated at the National Physical Laboratory in 1955. The use of these standards, which are now manufactured commercially in the U.S.A., soon confirmed that the rate of rotation of the Earth varied, see Fig. 1, and it became increasingly difficult to sustain a definition of a unit of time that was based on the Earth's rotation, but for some astronomical and navigational purposes a time scale based on the rotation of the earth was still required. For the past 10 years, following the introduction of the system of Co-ordinated Universal Time (UTC), the majority of time-signal emissions throughout the world have operated on an adjusted atomic basis. Thus the stations did not emit atomic time (AT) but an approximation to Universal Time (UT2) which was achieved by applying an offset to the time-signal generator in order to keep the time signals in step with astronomical time.

These adjustments in the rate of standard frequency transmissions proved inconvenient both to

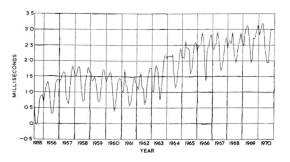

scientist and navigator, for example the Loran-C navigation system has been developed, which requires the synchronization of atomic clocks at a number of widely dispersed stations. The modified universal time system, which was introduced on January 1, 1972, is therefore operated at a constant rate and instead of frequency adjustments, changes are being made in 1-second steps, as the astronomical and atomic time scales diverge, so that the difference between the two time scales does not exceed ± 0.7 sec. The maximum difference of ± 0.7 sec corresponds to a navigational error of ± 10 sec of arc in longitude (~ 0.3 km at the equator). The error is too great for many purposes and so in addition to the seconds markers the broadcast time signals contain simply coded information which enables the difference between atomic and astronomical times to be reduced to ± 0.1 sec with a corresponding reduction in position error.

Leap Seconds

The operation of atomic clocks over the past decade has shown that the rate of divergence of the two time scales is of the order of a second per year at present and so, to encompass likely future changes, provision has been made for the introduction of up to two 1-sec step-adjustments per year. These adjustments, of plus or minus 1 sec as required, are made if necessary at the end of the last minute of the last hour of December 31 and June 30. The adjustments are to be known as "leap seconds" in analogue with the leap or intercalary day which is used to maintain the Gregorian calender in agreement with the tropical year. The leap seconds, however, may be positive or negative, a positive leap second begins at $23^h 59^m 60^s$, delaying the start of the next day by 1 sec, and a negative leap second at $23^h 59^m 58^s$, thereby advancing the start of the next day by 1 sec. Conventions have been agreed for designating events in these intervals as shown in the example, Fig. 2(i). The first such leap second was positive and was introduced on June 30, 1972. At present not enough is known about the causes of the fluctuations in the rate of rotation of the Earth to enable predictions of the occurrence of a leap second to be made more than about 3 months in advance.

Corrections to Broadcast Marker Pulses

There are therefore in a sense two "atomic" time scales, the international atomic time scale (IAT) maintained by the International Time Bureau (Bureau International de l'Heure in Paris—BIH) which is the atomic time scale, and the UTC scale. There is always an integral number of seconds difference between the two scales and they differ according to the addition or subtraction of leap seconds to UTC. On January 1, 1972, when UTC was introduced, IAT was exactly 10 sec in advance of UTC.

In order to achieve the 0·1-sec accuracy in the difference between the astronomical reference time system (UTI) and the universal coordinated time system based on atomic time (UTC) a simple coding signal is included along with the time signals which requires no additional instrumentation. This code gives the difference DUTI = UTI − UTC (capital "D" rather than the more usual Greek Δ being chosen for ease of transmission by telegram or Telex). The magnitude of the difference is obtained quite simply by counting the number of emphasized seconds markers which follow the minute marker, DUTI being positive for emphasized second markers 1 to 7 and negative for emphasized markers 9 to 15, see Fig. 2(ii). In the case of the U.K. emissions of MSF and GBR the emphasized markers take the form of a double seconds pulse, with the appropriate seconds pulses being followed by a further pulse 0·1 secs later. The International Time Bureau will announce the leap seconds at least 8 weeks in advance and make retrospective announcements of the correct values of DUTI.

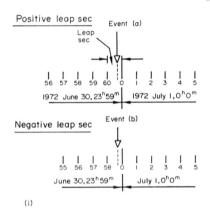

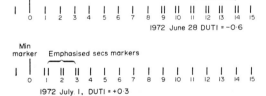

Fig. 2. (i) Showing how leap seconds are introduced and the convention for designating events during the critical positive or negative leap second. (ii) The coding for DUT1 for positive and negative values. The examples show what actually happened when the first leap second was introduced (at the same time DUT1 was changed by -0.1 sec).

The Loran-C Navigation System

The main method of dissemination of atomic time is by standard frequency transmissions but the Loran-C navigation system provides a convenient method of obtaining atomic time in the areas in which it is available. This system functions primarily as a precision hyperbolic navigation system which provides determination of position to 100 metres at distances approaching 2000 km. It is organized in a total of eight chains, with each chain comprising a master and associated slave stations. Each chain shares a common carrier frequency of 100 kHz but mutual interference is avoided by the use of different group repetition periods for the pulses transmitted by each chain. The slave pulse group comprises a series of eight pulses which are spaced 1 msec apart and the master stations are characterized by a ninth pulse. Each station has its own code and so may be distinguished by the navigator. Since electromagnetic radiation travels about 300 metres per microsecond a timing precision of 0·5 μsec suffices for most navigation purposes. The pulses have a sharp leading edge, of the form $t^2 \exp(-\alpha t)$ where α is chosen to maximize the amplitude at 65 to 70 μsec. If measurements are confined to the start of the received signal the full stability of the ground wave reception is realized, for this portion of the pulse is not contaminated by the more variable sky-wave component. Each station maintains its own time scale by means of a commercial caesium-beam-frequency standard and synchronizes its time with respect to other stations by means of travelling clocks. These calibrations may be used to eliminate errors due to propagation and receiver delay. In this way uniform time scales are established throughout the world and enable the Bureau International de l'Heure to produce and disseminate an International Time scale of extremely high accuracy.

The Einstein Clock Paradox

The very high accuracy of these travelling clocks has enabled a terrestrial version of the famous Einstein clock paradox to be demonstrated experimentally by comparing the time indicated by a clock after circumnavigating the globe on a commercial flight with that indicated by a clock which has remained "stationary". The interpretation of the observed differences is rather difficult and, with subsonic aircraft, the time difference is not very great, the clock losing ~40 nsec on an eastward trip and gaining ~275 nsec on a westward trip. Such changes are barely observable at present but it is hoped to repeat the experiment with a supersonic aircraft.

Atomic Frequency Standard

The "pendulum" in an atomic clock is the caesium atom. These atoms, as with other alkali atoms, have a single valency electron whose spin may be aligned parallel or antiparallel to the nuclear spin. The energy difference between these two states corresponds to a well-defined energy change E, with an associated frequency change f_0 which is governed by the Planck equation $E = hf_0$, h the Planck constant. The magnetic moments of the atoms in the two states are nearly equal and opposite so that in the atomic clock, Fig. 3, the beam of atoms emerging from the heated oven, ~370 K, is deflected in one or other of the two directions indicated in the figure by the magnetic field gradient of the first magnet. Those atoms, which pass through the collimating slit, are then deflected further by the second deflecting magnet. If, however, the spin directions can be reversed in the region between the two magnets then the beam deflections are also reversed and the current in the detector increased. This change of spin directions requires the induced emission or absorption of energy and this energy is supplied by the microwaves at the two cavity resonators. The detected signal as a function of the applied microwave frequency in the NPL caesium frequency standard is as indicated in Fig. 4. The sharpness of the resonance signal increases inversely as the separation of the microwave cavities and in the NPL standard the separation is 2·6 m. Portable commercial standards use a much smaller separation and hence have a wider resonance.

The caesium atom has a nuclear spin of 7/2 and in the presence of a magnetic field there is additional splitting into the various Zeeman levels. This splitting, if unresolved, would distort the resonance signal and so a small uniform field of around 5 μT is applied in the region between the two cavities. This bias field moves the fine structure-splitting frequencies to well outside the desired resonances curve and, although the field also shifts the ground state transition by about 1 Hz, this shift is well understood and so a correction may be measured and applied.

The microwave signal is synthesized from a 5-MHz crystal oscillator. The crystal frequency is itself very stable and comparatively slow rates of correction are required to its frequency. The necessary servo-signal is derived by applying an alternating voltage to a voltage-dependent capacitor which is situated

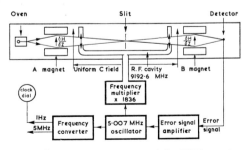

Fig. 3. Schematic diagram of the NPL caesium frequency standard.

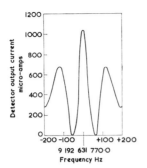

Fig. 4. The resonance curve observed with the NPL caesium frequency standard.

at a convenient point in the frequency multiplication chain. The variations in capacitance produce a variation in the phase of the signal thereby producing a small modulation of the frequency. If the output of the beam detector is detected synchronously with the modulation frequency then the resulting signal is essentially the first derivative of the resonance curve and passage of this signal through zero corresponds to the exact centre of the resonance. In the NPL standard this frequency modulation has been progressively reduced and is now only 1 Hz. In this standard the resonance frequency width at half height is only 50 Hz so that setting the resonance centre to 0·01 Hz corresponds to a frequency discrimination of about 1 in 10^{12}.

Although atomic time is at present realized through the caesium atom it does not follow that this will necessarily always be so. Recently frequency multiplication techniques have been extended right through the infra-red region to the visible region of the electromagnetic spectrum which opens out a much larger range of future possibilities. On the other hand, we already have the hydrogen maser which relies on a similar magnetic transition to that in the caesium atom. Although the frequency is about 6·6 times smaller than the caesium frequency a precision of parts in 10^{13} is potentially possible. However, as with Lasers, the properties of the surrounding cavity are important and this cavity must be tuned very carefully to the exact resonance. In addition the hydrogen atoms bounce around the cavity for about 0·5 sec before radiating and the wall collisions result in a frequency shift, $\sim 0\cdot 04$ Hz, for which a correction must be applied. These factors have so far prevented the realization of the full potential precision of the hydrogen maser, but the frequency has, for example, been measured at NPL as 1,420,405,751·766 Hz with an estimated uncertainty of 0·001 Hz, or less than 1 part in 10^{12}.

It is not, of course, necessary for everyone to possess an absolute frequency standard for other frequency standards, such as commercial caesium and rubidium standards, may be calibrated against absolute caesium standards which have known frequency shifts. For many purposes the precision of a few parts in 10^{11} achievable from broadcast signals suffices and this precision may be obtained from the low-frequency carriers such as Rugby 16 kHz(GBR) and the U.S.A. 24 kHz (NBA) which are not greatly affected by movements in the ionosphere. The latter cause Doppler shifts of signals in the megahertz range except in those areas where the ground wave predominates.

Bibliography

DERRINGTON, M. G. (1961) Clock, atomic, in: *Encyclopaedic Dictionary of Physics* (J. Thewlis, Ed.), **1**, 701, Oxford: Pergamon Press.
ESSEN, L. (Oct. 1971) The measurement of time and frequency, *Contemp. Phys.* **21**, 186.
ESSEN, L. (1958) Atomic clocks, *J.I.E.E.* **4**, 647.
ESSEN, L. (1968) Time scales, *Metrologia*, **4**, 165.
PETLEY, B. W. (1971) Quantum metrology, in *Encyclopaedic Dictionary of Physics* (J. Thewlis, Ed.), Suppl. **4**, 354, Oxford: Pergamon Press.
RICHARDSON, J. M. Time standards, in *Encyclopaedic Dictionary of Physics* (J. Thewlis, Ed.), Suppl. **1**, 351, Oxford: Pergamon Press.
SADLER, D. H. (Jan. 1972) The new system of co-ordinated universal time, *J. Navigation* **25**, 31.
SMITH, H. M. Time interval measurement, *Encyclopaedic Dictionary of Physics* (J. Thewlis, Ed.), **7**, 365, Oxford: Pergamon Press.
STEELE, J. McA. (Oct. 1971) Radio time signals, *Survey Review* **21**, 186.

B. W. PETLEY

U

ULTRASONICS IN MEDICAL DIAGNOSIS. Ultrasound has found a number of fields of application in medicine. These applications may be classified as surgical, therapeutic and diagnostic. In surgery, high intensity ultrasound is used to destroy tissue, e.g. in Mênière's disease which affects the balance mechanism of the ear, 3 MHz ultrasound of intensities up to 25 W/cm² has been found suitable. The therapeutic value of ultrasound is also being assessed in physiotherapy. In this case 1 MHz ultrasound of intensity 1 W/cm² might be applied to the painful area. However, both the surgical and therapeutic techniques are not widely used and it is in diagnostic medicine that ultrasonic techniques make their main contribution.

The object of diagnostic ultrasonic techniques is to obtain information concerning internal tissue structures of the human body. This is usually done by producing an image of the structures and in this respect the techniques are analogous to those of X-rays. There is great complementarity between the two techniques since ultrasonic methods are best at imaging soft tissues where X-rays are at their weakest. The case for using ultrasonics in a particular situation is a combination of the following aspects of the method:

(a) the ability to image soft tissue structures;
(b) the detectability of moving surfaces and blood flow within the body;
(c) the yield of information about tissues from their effect on the propagation of ultrasound;
(d) the non-invasive nature of the ultrasonic examinations;
(e) the lack of hazard.

These five properties result in medical ultrasonics being a powerful tool which although still at an early stage in its development is spreading rapidly.

Internal tissue structures are examined with ultrasound by detecting echoes produced at discontinuities in their acoustic properties. These discontinuities are usually largest at boundaries between organs and thus the techniques find their greatest success in imaging major tissue masses. The technology employed to produce and detect these ultrasonic echoes is similar to that found in the more familiar fields of radar and sonar. Indeed the history of medical ultrasonics can be traced back to these subjects via industrial flaw detection. The actual first application of ultrasound in medicine did not involve reflection but attempted to differentiate between absorption of ultrasound in tissues as is done with X-rays. This approach was not successful and is not actively pursued at present.

Ultrasonic Scanning Beams

Both pulsed and continuous wave ultrasound beams are utilized in diagnostic ultrasonics. It is desirable that these beams are highly directional and of uniform intensity distribution. Pulsed beams are used in scanning applications which localize tissue interfaces. Continuous wave beams are generated by instruments in which the Doppler effect indicates motion within the body. At frequencies in the range 1–15 MHz, i.e. those of medical ultrasonics, it is possible to construct simple small transducers which propagate directional beams. Figure 1 is a diagram of a typical transducer and the beam shape which it would produce for an ideal piston action of the crystal face at one frequency. The active element of the transducer is a disc-shaped piezoelectric crystal which is set into mechanical vibration by applying an electric signal. Detection of echoes is by the converse effect in which pressure fluctuations in the returning echo produce corresponding electrical voltages across the crystal. For the ideal beam shape illustrated, the beam is cylin-

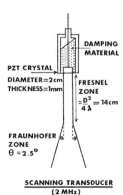

Fig. 1. Basic structure of pulse-scanning transducer and idealized beam shape.

drical in the Fresnel zone and slowly diverges in the Fraunhofer zone. For abdominal scanning, a range of 25 cm is required and it can be seen that suitable beams can be produced. In practice for pulsed scanning a band of frequencies is present, resulting in a less simple beam profile. This frequency bandwidth is beneficial in that it reduces intensity fluctuations in the beam due to wave interference effects. Doppler instrument transducers emit continuous wave ultrasound and their beams can exhibit intensity fluctuations.

Special transducers are often constructed for particular purposes, e.g. for insertion within the body or ophthalmic studies. The most common modification from a standard type is to shape the crystal in a concave manner so as to produce focusing of the beam. Weak focusing provides a narrow beam and correspondingly better azimuthal resolution. The axial resolution, i.e. the resolution along the beam, is determined by the pulse length which depends on the amount of damping the crystal experiences. For abdominal scanning at 2 MHz, the azimuthal resolution is typically 0·5 cm to 1 cm and the axial resolution about 3 mm. It is difficult therefore to image structures of dimensions less than 1 cm at this frequency. In ophthalmic examinations high frequencies, e.g. 10 mHz, are used giving azimuthal resolution of the order of 2 mm and axial about 0·5 mm.

Ultrasonic A-Scanning

The simplest way in which ultrasound is utilized to examine tissue structure is known as A-scanning. Fixed so as to point in one direction the transducer transmits a pulse through the tissue structures and echoes are generated when it interacts with tissue boundaries. The echoes are detected by the transducer crystal which recovers quickly from the transmitting action. For display purposes, amplified echo signals are presented as rectified pulses on a cathode-ray-tube. The distance of each echo along a horizontal baseline on the display screen is a measure of the time taken for that echo to return to the transducer. By assuming an average for the velocity of sound in tissue, medical instruments are calibrated to read tissue depth directly. A block diagram of a basic A-scan system is shown in Fig. 2. A-scanning is of value when the tissue structures examined are geometrically simple since it is then possible to identify the tissue interfaces. Examples of A-scan applications are in locating the mid-line of the brain or measuring the dimensions of parts of the eye.

Ultrasonic B-Scanning

When the anatomy is complex it is necessary to produce a two-dimensional image in order that tissue boundaries may be identified. This is achieved by scanning the ultrasound beam in a plane and detecting echoes for each direction of the beam. Normally scanning is performed by moving the transducer in contact with the patient's skin. Where this is not possible a water bath is used to convey the ultrasound from the transducer to the patient. During the scanning procedure the time when each echo is received and the corresponding direction of the ultrasound beam are measured. The tissue interfaces detected can then be registered in the correct positions on a storage cathode-ray tube screen. The process is similar to that commonly observed in radar, however the scanning action is more complex in diagnostic ultrasonics as weakly reflecting interfaces are often difficult to record. Figure 3 is a block

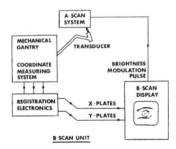

Fig. 3. Schematic diagram of B-scan machine. Considerable variation is found in the electronics and mechanics of commercial machines.

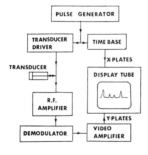

Fig. 2. Block diagram of electronics of A-scan unit. Commercial devices include additional pulse-processing circuits.

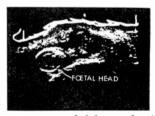

Fig. 4. B-scan image of abdomen showing foetal head, the size of which can be used to measure maturity.

diagram of a B-scan instrument. It is essentially an A-scan device with additional coordinate measuring and computing facilities. A typical B-scan image of abdominal structures is shown in Fig. 4. Commercial instruments are more complex than that illustrated, incorporating a number of additional electronic features, e.g. for the accurate measurement of organ dimensions or compensating the effect of absorption of ultrasound. B-scanning is the most important technique in medical ultrasonics at present.

Time-position Scanning

Pulsed ultrasonic techniques can examine the motion of surfaces such as those of the heart or blood vessels. The technique is similar to that of A-scanning in that the ultrasound beam is held fixed in one direction while the echoes are recorded. To record motion, the echoes are first displayed as bright spots on a line along one side of the display screen. When the desired structures have been detected the line of spots is made to sweep at a uniform speed across the screen. Echoes from static boundaries produce straight lines on a photograph or storage oscilloscope screen whereas those from moving structures give a fluctuating trace. Heart-valve action is readily studied using this method. A time-position trace, or a time-motion trace as it is often called, of a stenosed malfunctioning mitral valve is presented in Fig. 5. Ultrasonic techniques can therefore provide directly information about cardiac function.

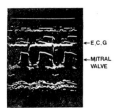

Fig. 5. *Time-position scan of a stenosed mitral heart valve. Normal valve shows greater range and speed of movement.*

Doppler Techniques

The particular configuration giving rise to Doppler shifted ultrasound in medical ultrasonics is one in which both the source and receiver are stationary and the ultrasound is reflected from a moving object. The formula for such a Doppler shift f is

$$f = 2f_0 \frac{u}{V} \cos \theta.$$

Where f_0 is the frequency of the incident beam, u is the speed of the moving object and V is the speed of sound in tissue. The angle θ is the angle between the direction of the sound beam and the velocity of the object. For a typical situation in which blood

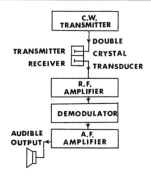

Fig. 6. *Block diagram of ultrasonic Doppler unit.*

flow is detected, e.g. $f_0 = 10$ MHz, $V = 1540$ m/sec, $u = 20$ cm/sec and $\theta = 60°$ ($\pi/3$ rad), the Doppler shift is 1·3 kHz. The fact that such shifts turn out to be in the kHz range means that a Doppler instrument can be easily made with an audible output, the pitch of which gives information about the velocity of blood flow.

A block diagram of an ultrasonic Doppler unit is shown in Fig. 6. One crystal of the transducer transmits continuous wave ultrasound while the other receives ultrasound of the transmitted frequency and of the Doppler shifted frequencies. The instrument extracts the Doppler frequency shifts from this composite signal and feeds them to an output device. Doppler instruments are exceedingly sensitive and find application in detecting foetal heart beats in early pregnancy. Another area of increasing importance is in the study of venous and arterial blood flow. Although at present they are used in a qualitative way it is hoped that in future they will provide quantitative values for blood flow. Research is also being undertaken to utilize instruments which record the direction of blood flow and with pulsed Doppler techniques localize the site of blood flow.

Reflection of Ultrasound in Tissue

The acoustic impedance discontinuities encountered in diagnostic ultrasonics vary considerably in magnitude and result in a corresponding variation in the echo amplitudes. Some appreciation of the percentage wave amplitude reflected may be obtained from the following examples:

Blood—Kidney	0·3%
Muscle—Blood	3%
Fat—Kidney	8%
Fat—Bone	70%
Muscle—Air	99·9%

The first three examples show that at soft-tissue interfaces only a small fraction of the incident beam is reflected. It is therefore possible with an ultrasound beam to examine a succession of tissue layers.

Conversely a high reflection occurs at bone which presents a number of problems for the application of ultrasound in medicine. There is also high absorption and refraction so great difficulty is experienced in attempting to visualize soft tissues lying beyond bone, e.g. in brain scanning. Similarly air or gas within the body presents a barrier to ultrasound. These examples emphasize the complementarity of X-rays and ultrasound.

Velocity of Ultrasound in Tissue

Knowledge of the velocity of sound in tissue is chiefly of value in calculating the dimensions of organs within the body. Like many physical quantities, the velocity of sound is not well documented for biological material. Values for the velocity of sound in commonly encountered materials are presented below:

Brain	1540 m/sec	Fat	1450 m/sec
Water	1480 m/sec	Blood	1570 m/sec
Muscle	1580 m/sec	Bone	4080 m/sec

It is reasonable therefore to assume a value of 1540 m/sec for the velocity in soft tissue. Refraction is not a problem when scanning soft tissues since they have neighbouring values for the velocity of sound. The same cannot be said for bone–soft tissue boundaries where beam deviations of 20° can be expected.

The variation of the speed of sound with temperature or frequency in the ranges of these parameters used in diagnostic ultrasonics is either negligible or can be readily compensated.

Absorption of Ultrasound in Tissue

The high absorption of ultrasound in tissue has great influence on the electronic techniques employed and on the final results obtained. Perhaps the greatest influence is the limitation on the resolution. In an X-ray image of the abdomen it is possible to depict objects of dimensions less than 1 mm, to do the same with ultrasound would require using a frequency of 50 MHz. Absorption at this frequency is prohibitive and frequencies in the range 1 to 5 MHz are used for abdominal scanning. A second difficulty is that echoes from superficial structures are detected much more strongly than those from greater depths. The electronic compensation of this effect, is necessarily approximate since the tissues traversed by the beam are unknown. Below are examples for some common tissues of the thicknesses required to reduce the intensity of 1 MHz ultrasound by half:

Blood	17 cm	Bone	0·15 cm
Brain	3·5 cm	Muscle	2·3 cm
Fat	5·0 cm	Water	1400 cm

Absorption is highly frequency-dependent in these materials. For soft tissue the average absorption coefficient of intensity can be taken to be $0 \cdot 9 f$ decibel/cm, where f is the frequency in MHz.

The mechanisms of absorption of ultrasound in biological material have not been extensively investigated. At present gross variations such as dependence of absorption on frequency or temperature are explained in terms of a few general mechanisms. Since absorption provides potentially useful information it is expected that in future more fundamental studies will be undertaken.

Scattering of Ultrasound in Tissue

The contribution of scattering to the attenuation of an ultrasound beam on passing through tissue has not been measured. This is another phenomenon which might provide useful data in future techniques. Scattering is exploited, however, in Doppler ultrasonic blood-flow instruments. These instruments detect ultrasound scattered by the red blood cells. The dimensions of red cells are on average 5 μm and the wavelengths of ultrasound greater than 150 μm. Conditions are therefore correct for the fulfilment of the Rayleigh criterion, i.e. the wavelength is much larger than the dimensions of the scattering object.

Intensity of Ultrasound

The intensity of ultrasound is important both from the point of view of instrument design and radiation dose to the patient. Pulsed scanning machines commonly emit an average acoustic intensity of 1 or 2 mW/cm^2 and a peak pulse intensity of 1 or 2 W/cm^2. On the other hand, Doppler instruments produce continuous wave ultrasound of around 15 mW/cm^2. The low values of ultrasonic intensity radiated by diagnostic equipment create problems in monitoring their output. Calibration is normally performed using a water bath in which it is essential to avoid convection currents and standing waves. Types of thermal, electrical, optical and mechanical methods for measurement of intensity are described in the references. Most often employed is a mechanical method in which a highly sensitive chemical balance measures the radiation pressure of the ultrasonic beam.

Hazards in Diagnostic Ultrasonics

Ultrasonic diagnostic techniques are considered to be safe and this is one of the reasons that they have found widespread application particularly in fields such as the study of foetal development. However, the question of safety has not been completely resolved. Surveys of infants at birth for harmful effects due to ultrasonic examinations have shown no increased incidence in abnormalities. To date it has not been possible to survey large samples since the techniques are relatively new. Another approach has been to tackle the problem at a more fundamental level by examining directly the effect of ultrasound on tissue cells. The possibility of chromosome damage has been explored and the general conclusion has been that at the power levels currently employed diagnostic techniques are safe. Nevertheless, since it is known that 3 MHz ultrasound at an intensity of a few W/cm² can destroy tissue, the subject of hazards in diagnostic techniques will remain the subject of research for many more years.

Future Developments

Medical diagnosis by ultrasonics is still at an early stage in its development and there are a number of projects being pursued which will greatly increase the power of the technique. The use of computers to store and analyse echo information will reduce the wastage of data as happens in the present generation of machines. Techniques such as holography are also being considered as a means of producing three-dimensional images. Improvements will also be made in the resolution achievable and in the methods of displaying images. Finally more detailed studies of the interaction of ultrasound with tissue should lead to increased diagnostic capabilities for medical ultrasonics.

See also: Radiation, non-ionizing, hazards of Medical ultrasonics.

Bibliography

BROWN, B. and GORDON, D. (1967) *Ultrasonic Techniques in Biology and Medicine*, London: Iliffe Books.

WELLS, P. N. T. (1969) *Physical Principles of Ultrasonic Diagnosis*, London and New York: Academic Press.

W. M. MCDICKEN

V

VOID FORMATION IN FAST-REACTOR MATERIALS

Introduction

During the 1960s and early 1970s, we saw rapid development of Fast Reactor technology. Until about 1966, fast reactors presented no materials problems new to general reactor technology. However, in that year C. Cawthorne and I. Fulton at Dounreay made a discovery which had important consequences. On very careful electron-microscopical examination of the irradiated stainless-steel cladding from experimental fuel pins in the Dounreay fast reactor (DFR) Cawthorne and Fulton found that the steel contained a large number of small cavities, about 100 Å diameter. Such cavities were shown to be essentially empty and were therefore called "voids", see Fig. 1. The creation of these cavities, within the solid, implies that the external dimensions of the solid must increase in order that the total quantity of material stays constant. It was soon found that certain fast-reactor components had indeed increased in size. Such a phenomenon has since come to be called "*void swelling*".

Because of its influence on the dimensional stability and mechanical properties of cladding and structural components of the fast reactor system, the phenomenon of void swelling has tended to overshadow temporarily the previously recognized effects of exposure to fast-neutron fluxes, such as *low-temperature hardening*, *high-temperature embrittlement* and the acceleration of diffusion-controlled processes which have tended to give rise to less concern as more experimental evidence has been accumulated. In the case of void formation the technological importance of the phenomenon in reactor design depends largely on the effectiveness of radiation creep in relaxing the stresses built up by non-uniform swelling of restrained core components.

The acquisition of data relevant to fast-reactor design problems by fast-neutron irradiation is, by its nature, a tedious and time-consuming business: the swelling data for satisfactory design of components, expected to dwell for 1, 5 or 25 years in a fast-neutron environment, can only be obtained directly by exposure for the same periods in that fast-neutron environment. An attempt to obtain void-swelling

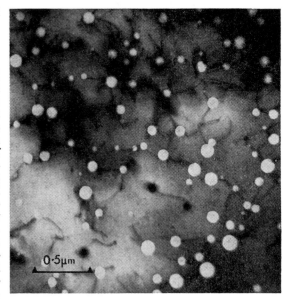

Fig. 1. Typical voids in DFR irradiated steel at 510°C (40 DPA).

data over a shorter time scale has been made, using charged particle accelerators. Due to their larger mass and correspondingly higher collision cross-section with the atoms of a solid, a beam of energetic charged particles can produce irradiation damage at a rate many thousand times faster than neutrons in a fast reactor. So in principle, the damage density produced during many years' irradiation, within a reactor, can be simulated in a few hours using ion beams.

Theoretical Basis

It is well known that fast neutrons created during nuclear fission interact with the atoms of those materials which form an integral part of the nuclear reactor—in particular, the fuel cladding and the main structural material within the reactor core, where the neutron flux is greatest. Such interaction results in the displacement of atoms from their equilibrium sites, creating what are known as "*interstitials*" and "*vacancies*". Interstitials are displaced atoms which occupy sites not normally occupied

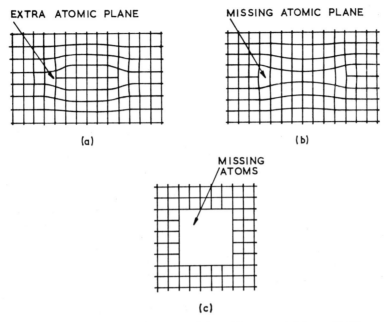

Fig. 2. Typical clusters of point defect. (a) Interstitial loop. (b) Vacancy loop. (c) Void.

within the solid, and vacancies are empty sites which are normally occupied by atoms. At any one instant large numbers of both interstitials and vacancies are freely migrating at random throughout the irradiated components in the reactor. Such migrating defects can suffer a variety of fates: an interstitial and a vacancy can annihilate one another by recombining, i.e. an interstitial can drop into a vacant site; both interstitials and vacancies can be captured and lost at extended defects within the solid such as dislocations or grain boundaries; interstitials can cluster together to form plates of extra atoms within the solid (Fig. 2[a]); and finally, vacancies can cluster together to form either plates of missing atoms or large holes such as voids (Fig. 2[b], [c]). Plates of vacancies are not stable and mechanical collapse of atoms above and below the plate closes the gap leaving only a ring of edge dislocation and a stacking fault. Plates of extra or missing atoms give rise to circumferential strains within the solid which, when observed from above, such as by transmission electron microscopy, are manifest as rings or loops. Plates of extra atoms are therefore called interstitial loops and collapsed plates of missing atoms are called vacancy loops.

Under normal circumstances when steels or nickel-based alloys are irradiated at around room temperature, the expansion of interstitial loops and the contraction of vacancy loops to all intents and purposes counterbalance one another. However, during irradiation at slightly elevated temperatures, say 500°C, it appears that a substantial fraction of vacancies agglomerate to form three-dimensional cavities rather than loops. Such cavities produce vanishingly small strains within the solid which can no longer balance the expansion produced by interstitial loops. As agglomerating interstitials can only form loops, we are left with a net expansion of the solid.

A major parameter affecting void formation is the irradiation temperature. Voids have now been observed in a large number of pure metals and their alloys, and a general feature which has been observed is that voids appear within a band of temperatures, ranging from approximately one-third to one-half the melting temperature in degrees Kelvin. There are two reasons why voids are not thought to occur at temperature below this. Primarily, at temperatures of less than one-third the melting temperature, vacancies move only rather slowly and are therefore rapidly annihilated by migrating interstitials before they are able to reach a sink such as a void. Secondly, it is a common occurrence that during neutron or heavy ion irradiation, where radiation damage is created from energetic recoil atoms in dense collision cascades, vacancy loops are created in the wake of the cascades. Such loops remain stable against thermal dissociation at these temperatures, and therefore reduce the number of vacancies available for void growth. It is implicit in the above discussion that the dynamic vacancy concentration which exists as a consequence of irradiation exceeds the vacancy concentration dictated by thermodynamic equilibrium. If the temperature was sufficiently high that the thermal vacancy concentration exceeds the irradiation value, there would be no driving force for void growth and all vacancies would be used in satisfying thermodynamic equilibrium. The temperature at which the irradiation and thermal concentrations are equal therefore approximately defines the maximum temperature at which voids will be expected to grow. It should be noted that the void cut-off temperature will depend on such parameters as the damage rate and the defect sink density, as both contribute significantly to determining the irradiation-induced vacancy concentration. For this reason we would possibly expect a shift in the void temperature range when comparing reactor results with those of simulation experiments.

Current theories of void formation suggest that trace quantities of gaseous elements, either occurring naturally within the solid or created during irradiation by transmutation, are necessary to nucleate three-dimensional void-like agglomerates. One such gaseous source is helium atoms which are continually created within the reactor components as alpha particles via nuclear reactions. Helium atoms are highly insoluble within solids and rapidly collect together to form minute pockets of gas. It is these pockets of gas which are thought to act as perfect nuclei around which the migrating vacancies grow into voids.

In practice, in order to display significant swelling, which is detectable either as a density change or by direct observation of voids within the solid by electron microscopy, a reactor component must be exposed to a dose of at least 10^{22} neutrons/cm². Typical neutron flux levels in fast reactors are between 10^{15}–10^{16} neutrons-cm²/sec and so void swelling is only significant after about a year's irradiation. Such a dose is equivalent to every atom within the solid having been displaced at least 10 times. At these high damage levels, it is convenient to express the degree of damage in units which relate to such displacements, and it has now become common practice to describe a specific irradiation dose in terms of displacements/atom (DPA). Such a description must rely on calculation and this in turn depends on the validity of atomic collision theory. This in itself is another story, but in this instance, it will suffice to say that we are now working towards international agreements on radiation-damage models.

A void is thought to be an equally perfect sink for both vacancies and interstitials and in order for voids to grow, by vacancy condensation, the interstititals must have a preferential sink somewhere within the system. Theoretical considerations have shown that because there is a larger stress field around an interstitital it has a slightly larger drift interaction with a dislocation line than does a vacancy, but only by a few per cent. However, it is now generally accepted that it is this preference of dislocations for interstitials which is the driving force for void growth. Detailed mathematical calculations based on these ideas can be found in the literature (see Bibliography).

Reactor Data

The majority of data generated within the U.K. has been obtained from fuel-pin cladding from DFR. The reactor core geometry is such that each fuel pin experiences a temperature profile rising from about 230°C to, in a typical core, 650°C along its length. The fast-neutron core, on the other hand, has a peak at the reactor core centre level.

Within the limit of the experimental data and up to the highest dose so far experienced, it has been found that in any temperature range the relationship between volume increase and dose can be adequately represented by assuming that a threshold, below which essentially no volume change is observed (though small numbers of very small voids are seen by electron microscopy), is followed by swelling at a linear rate with increasing dose. There is too little evidence as yet to indicate whether the threshold dose is a function of temperature or if, at a given temperature, it varies from material to material, so it is convenient to use a threshold of 6 DPA, corresponding to a total DFR dose of about $1·2 \times 10^{22}$ neutrons cm⁻².

Swelling rates for a group of twelve pins clad in M316 and M316L steels are compared in Fig. 3. The most obvious difference between the various groups of swelling rates vs temperature profiles lies in the higher swelling rates of solution-treated material as compared with cold-worked material. There are also significant differences in the shapes of the profiles, those for M316 peaking at a little over 600°C, while those for M316L cladding peak at about 500°C. The attribution of the different temperature dependencies to differences in carbon level is suggested by the observation that this is the most significant compositional difference; however, the reason for this is still not known.

A similar analysis of volume change data from samples of FV548 and Nimonic P.E.16 is shown in Fig. 4. The profiles show that FV548 in both pre-irradiation conditions is less susceptible to void formation than corresponding samples of M316 cladding. Nimonic P.E.16 clad (Fig. 4) suggests that the presence of finely dispersed γ' Ni(Ti, Al)$_3$ particles may be in this case an effective void suppressant.

Simulation Experiments

Due to the long times necessary to acquire data on void swelling from reactor irradiations, it would be advantageous if the damage process could be accelerated, such that the dependence on dose, temperature and materials variables could be assessed in times of the order of a day. For some time now, it has been possible to study radiation damage in metals using ion beams from versatile particle accelerators, such as the variable energy cyclotron at

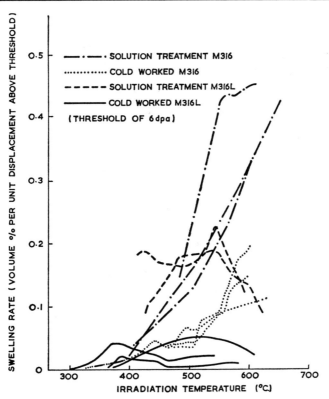

Fig. 3. Swelling rates for various M 316 and M 316 L fuel pins plotted as a function of irradiation temperature. Swelling rates are approximated to a linear rise with dose from an effective swelling threshold of 6 DPA.

Harwell (VEC), to simulate collision events which occur during neutron irradiation. Due to their larger size and stronger interaction with the atoms of a solid, a beam of energetic charged particles can produce atomic displacements at a rate many thousands of times faster than can neutrons. So, in principle, the damage density created during many years irradiation within a reactor can be simulated in a few hours using ion beams. One basic drawback is that due to their larger collision cross-section with atoms of the solid, heavy ions produce damage rather close to the surface of the bombarded specimen. However, by a suitable choice of ion species and ion energy, the damage zone can be made to penetrate sufficiently deep for quantitative measurement. Furthermore, helium can be previously implanted uniformly throughout the sample to reproduce the transmutation gases produced during fast neutron irradiation.

Figure 5 shows typical electron micrographs of voids in two nickel specimens irradiated in DFR with fast neutrons and in the VEC with charged particles respectively. These samples have been irradiated to the same calculated damage dose and at corresponding temperatures. The similarity is readily apparent and accurate determinations of the percentage void swelling in both samples yields essentially the same value to within the expected experimental uncertainty. It was then possible to proceed with confidence in attempting to answer some of the vital questions posed below.

Figure 6 shows the temperature dependence of the void swelling in 316 steel plotted for a fixed dose of 40 DPA. Whereas, on the other hand, Fig. 7 represents the dose dependence of swelling at a fixed temperature of 525°C. In such experiments, it has been possible to attain doses equivalent to 8 years irradiation in DFR in times of up to 1 day. In Fig. 8 we show the dose dependence of swelling in P.E.16 plotted at 525°C for comparison with Fig. 7. It is readily apparent that P.E.16 exhibits markedly reduced swelling compared with 316 steel under these conditions.

Recently the high-voltage electron microscope (HVEM) has been used for the study of void formation. A 1-MeV electron can transfer sufficient energy in a near head-on collision with, say, an Fe atom in steel to result in its permanent displacement. Then using the highly focused electron beams which

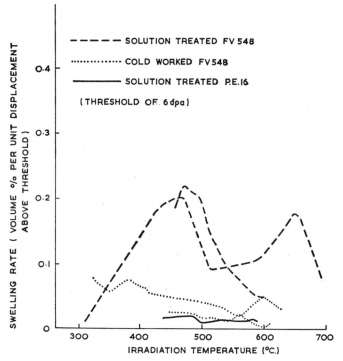

Fig. 4. Swelling rates for FV 548 and P.E. 16 fuel pins plotted as a function of irradiation temperatures.

are readily obtainable in the HVEM, it is possible to create a displacement rate within a specimen comparable with those obtained using heavy ion irradiation. The foil thickness is typically 1 μm and the irradiated area is of the order of 1 μm in diameter.

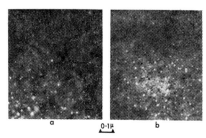

Fig. 5. Comparison of swelling in a Ni specimen containing 10^{-5} He irradiated to the same calculated dose of 10 DPA. (a) DFR. (b) VEC.

Pioneering work using this technique has been carried out by D. I. R. Norris and by M. J. Makin. Some advantages are, the ability to observe damage during irradiation, the histories of individual voids and the role of moving dislocations can be followed and the dose dependence can be studied in just one run on a single sample. As pointed out by Norris these advantages are bought at the price of having to use a thin foil where surface effects can play a major role in determining the actual behaviour of voids For instance, due to its role as a sink for migrating vacancies and interstitials, the presence of a surface results in a reduction in defect concentration which in turn produces a region denuded of voids. The thickness of this region depends on the dislocation density and the temperature. For instance, in low dislocation density material at high temperatures the surface denuded zones on either side of the foil become sufficiently large that they touch, so preventing the formation of voids. The free surfaces can also perturb the situation in respect to the mechanical or irradiation produced dislocation structures; for due to the elastic image force interaction with a free surface, dislocations preferential slip and climb towards the surface, so leaving a region of reduced dislocation density. Numerous authors have attempted to evaluate the importance of this effect for void studies in the HVEM, and estimates—depending on material—vary from 0·1 to over 1 μm. However, by judicial choice of the experimental conditions it has been possible to produce an environment which approaches that typical of the bulk material. As the published results from the HVEM are similar to those from accelerator irradiation, in this instance it will suffice to refer the

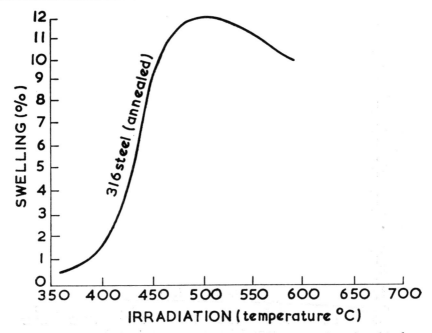

Fig. 6. The temperature dependence of swelling in 316 steel irradiated in the VEC, the curve has been shifted by 100°C to lower temperatures to account for the differences in dose rate between the VEC and DFR.

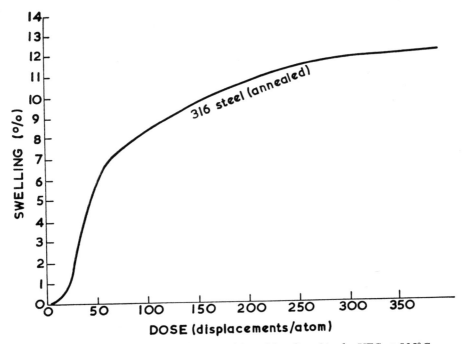

Fig. 7. Dose dependence of swelling in 316 steel irradiated in the VEC at 525°C.

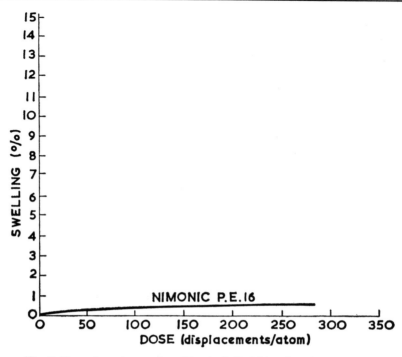

Fig. 8. Dose dependence of swelling in P.E. 16 irradiated in the VEC at 525°C.

interested reader to the conference proceedings listed in the Bibliography.

Conclusions

The reactor irradiations in DFR together with the simulation experiments using particle accelerators, and more recently electron irradiations in the high-voltage electron microscope, have provided valuable design data necessary for the economic operation of the commercial fast reactors. The likely endurance of conventional steels such as M316 has been assessed and new materials such as P.E.16 have been identified as possible alterations in the construction of fast reactors.

Bibliography

CAWTHORNE, C. and FULTON, E. J. (1967) *Nature*, **216** (5115) 575.

See the collection of papers in

(a) Conference on "Voids formed by irradiation of reactor materials" (BNES, Reading, March 1971).

(b) Conference on "Radiation induced voids in metals" (USAEC Conf—710,601, Albany, U.S.A., 1971).

R. S. NELSON

W

WANKEL ENGINE. Felix Heinrich Wankel was born at Lahr, now in West Germany, on August 13th, 1902. The engine that bears his name is a rotary internal combustion engine, operating on the four-stroke cycle. The differences between the Wankel engine and conventional reciprocating engines are derived purely from the planetary motion of the rotor, or rotary piston.

Description and Principles of Operation

Essentially the engine consists of a rotor, a housing (with flat side plates) and a shaft (Fig. 1). A gearwheel (1) is positioned on one side of the rotor, centrally mounted in the housing and fixed rigidly to it; this is the reaction gear. The rotor carries an internally toothed gear wheel (2), of larger diameter than the reaction gear, with which it can mesh. If the rotor and housing gears are constrained to remain in mesh, the rotor may revolve around the stationary reaction gear. This movement of the rotor is analogous to the rolling of a large hollow cylinder, without slip, around a fixed, smaller internal cylinder. If the gear ratio is 3:2 then, as the rotor turns, the locus of each apex (3) forms a closed curve, called a two-lobed epitrochoid. This is arranged to coincide with the inner wall of the housing so that the three apices of the rotor are always in contact with the housing, forming three distinct chambers (4, 5, and 6).

As the rotor revolves around the reaction gear, the locus of the centre of the rotor (7) is a circle, whose centre is also the centre of the housing (8), i.e. the rotor revolves both around its own centre and around the centre of the housing. Inside the rotor, alongside the annular gear (2), lies a journal bearing (9) mounted eccentrically on a shaft (10) whose axis coincides with the centre of both the reaction gears and housing. The eccentricity of the bearing is equal to the radius of rotation of the rotor centre (7). The rotor is mounted on this bearing, whose function it is to maintain the reaction gear and annular gear in mesh, and to transmit torque from the rotor to the shaft.

The chambers formed between the rotor flanks and the housing successively increase and decrease in volume as the rotor turns. It is possible to adapt these changing volumes to the four-stroke engine

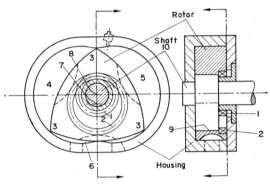

Fig. 1

cycle by inserting ports in the housing (Fig. 2). As the rotor turns, one of its apices (1) moves past the intake port (2) and connects it to a chamber (3) (Fig. 2a). Rotation draws fresh air/fuel mixture into the increasing volume of the chamber (Fig. 2b, intake phase), until apex (4) closes the intake port (Fig. 2c). Further rotation reduces the volume of the chamber to a minimum, compressing the charge (Fig. 2d, compression phase). Just previously, a sparking plug in the chamber wall has ignited the mixture. The energy released by combustion results in a high gas pressure acting on the rotor flank (5) (Fig. 2e, combustion and expansion phase). Because the rotor is eccentrically mounted relative to the shaft (6), the high pressure on this single flank exerts a torque on the rotor. The rotor turns, forcing the output shaft (6) to rotate, at a greater rate, in the same direction. The exhaust phase commences when apex (1) opens the chamber to the exhaust port (7). Continued rotation reduces the volume of the chamber forcing combustion products out of the exhaust port (Fig. 2f). A flywheel on the output shaft conserves momentum from the expansion phase to drive the rotor through the following exhaust, intake and compression phases.

Geometry

There are numerous different geometrical configurations of rotor and housing that can provide chambers of successively increasing and decreasing

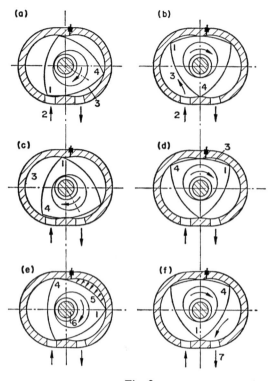

Fig. 2

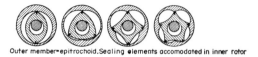
Outer member = epitrochoid. Sealing elements accomodated in inner rotor

Inner rotor = epitrochoid. Sealing elements accomodated in outer member

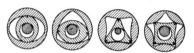
Outer member = hypotrochoid. Sealing elements accomodated in inner member

Ratio 1·2   2·3   3·4   4·5

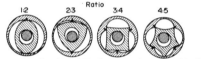
Inner rotor = hypotrochoid. Sealing elements accomodated in outer member

Fig. 3

volume. Wankel studied many possibilities before developing the three-sided rotor in a two-lobed epitrochoid housing. Some of the examples that he considered are shown in Fig. 3 (based on epitrochoids and hypotrochoids; a hypotrochoid is the locus of a point on a circle rolling around the inside of a stationary circle). The geometrical configuration determines the maximum compression ratio that can be used without preventing the rotor from turning. Epitrochoids permit higher compression ratios than hypotrochoids. The simplest epitrochoid configuration that is compatible with the four-stroke cycle (i.e. increasing and decreasing chamber volume twice—or some multiple of 2—per rotor revolution) is the two-lobed epitrochoid with a three-sided rotor of the Wankel engine. The flanks of the rotor are of cycloidal form.

The ratio of rotor and reaction gears (3 : 2) ensures that the rotor revolves around its centre once per three revolutions of the output shaft. It has been mentioned that the maximum compression ratio is governed by the geometrical configuration. Within this limitation, the maximum compression ratio is further restricted by the ratio of the rotor radius R (centre to apex) to the rotor eccentricity e (actual compression ratios are lower than those permitted by the above, due to a recess formed in the rotor flank to improve the combustion chamber shape).

Rotor tip seals are located radially and only lie perpendicular to the housing when passing its major or minor axes. At all other times the seals lie at an angle to the housing. The maximum leading angle of the apex tip seals is a function of R/e and must not become excessive or sealing will be impaired. The basic equations of the two-lobed epitrochoid housing also depend on R and e, namely:

$$x = e \cos \alpha + R \cos (\alpha/3) \quad y = e \sin \alpha + R \sin (\alpha/3)$$

(where α is the angle of rotation of the shaft).

The volume V swept by one chamber of the rotor is given by:

$$V = 3\sqrt{3}\, eRb \quad \text{(where } b \text{ is the rotor width).}$$

Classification of the capacity (swept volume) of the original Wankel engine caused heated discussion. A single-rotor engine can be considered equivalent to a two-cylinder, four-stroke engine. Consider one rotor revolution: during this revolution each of the three chambers passes through all four phases of the four-stroke cycle (i.e. each completes its operating cycle). Meanwhile the output shaft has completed three revolutions. This is equivalent to two chambers completing their cycles in two revolutions, analogous to two cylinders of a four-stroke engine completing their cycles in two crankshaft revolutions. Thus the capacity classification should be based on twice the volume of a single chamber (multiplied by the number of rotors for a multi-rotor engine).

Practical Wankel Engines

The first Wankel engine was tested at the N.S.U. motorcar factory in West Germany on February

1st, 1957. Many experimental engines, and a major change from a rotating housing and rotor to a stationary housing combined with an eccentrically mounted rotor, were made before the first Wankel-powered automobile was offered for sale in 1964. The most serious early problem concerned rotor-tip seals, in particular their short life and the chatter marks that they made on the housing. Much development work has gradually improved the life of the tip seals. Other problems related to poor combustion chamber shape, low speed of flame propagation, poor performance of the recessed sparking plug (necessary to clear the rotor tips) and cooling.

A recess was formed in the rotor flank to improve the shape of the combustion chamber, but the latter still remains longer and thinner than that found in reciprocating engines. This required a change in the R/e ratio to return the compression ratio to a satisfactory value. Combustion problems in the Wankel engine were fundamentally different from those of the reciprocating engine because, in addition to the different combustion chamber shape, the rotor does not slow down and come to rest during the combustion process. Far from it, the gas velocities are many times the typical flame propagation speeds of conventional engines. For this reason, the flame can only travel downstream of rotor rotation. In order to alleviate this condition, some engines use twin sparking plugs, the additional one being timed to spark later than the primary one, to ignite the end gas. The high gas velocity combined with slow combustion results in higher heat losses, combined with lower peak gas temperatures, than those encountered in reciprocating engines. This results in a lower thermal efficiency and, combined with the effect of a lower rate of expansion, less emission of oxides of nitrogen. In contrast, emissions of unburnt hydrocarbons are higher.

Figure 2 shows an engine with peripheral ports, but engines have been built with ports in he sidet walls of the housing, and also with combinations of both. Peripheral ports are rapidly opened and closed by the passing rotor tip seals. Side ports are opened and closed more gradually by the rotor flank and its seals.

Most Wankel engines utilize water-cooling, but some air-cooled versions have been built; similarly while most are of the spark-ignition type using petrol as a fuel, some are of the compression-ignition type using diesel fuel, one example of the latter having a Wankel-type compressor/expander to achieve a suitably high overall compression ratio combined with a reasonable combustion chamber shape.

The primary advantages of the Wankel engine are derived from its small specific size, low specific weight, smooth output torque and the ease with which it can be perfectly balanced (reciprocating engines have unbalanced secondary forces). It has fewer moving parts than conventional engines and therefore may be cheaper to produce, although this cannot be categorically stated at present. Against these advantages lie development problems that are gradually being overcome, and some fundamental problems. In the former category lie the problems of obtaining satisfactory seal and housing life, spark plug life (the recessed sparking plug of the Wankel engine is not cooled by fresh intake charge), and cooling the combustion area of the housing. In the latter category lie the low efficiency due to poor combustion chamber shape, slow combustion rate and high heat losses.

Bibliography

ANSDALE, R. F. (1968) *The Wankel RC Engine*, London: Iliffe Books.
FELLER, F. (1970/71) *Proc. I. Mech. E.* **185**, 13–71.
FROEDE, G. W. (1961) *Trans. S.A.E.* **69**, 179–193.
HURLEY, R. T. and BENTELE, M. (1961) *Trans. S.A.E.* **69**, 194–203 and 641–649.
KENICHI, Y. and TAKASHI, K. (1970) *Trans. S.A.E.* **79**, 216–228.
NORBYE, J. P. (1972) *The Wankel Engine—Design, Development, Applications*, Bailey Bros. & Swinfen Ltd.
WANKEL, F. (1965) *Rotary Piston Machines; Classification of Design Principles for Engines, Pumps and Compressors*, London: Iliffe Books.

J. A. BARNES and N. WATSON

X

X-RAY INTERFEROMETRY. This is concerned with the measurement of phase shifts in the X-ray region of the electromagnetic spectrum, which ranges from 10 pm to 1000 pm in wavelength. In each region of the spectrum the principles and design of interferometers are strongly conditioned by the properties of available radiation sources and by the nature of the interaction between the radiation and materials. Consequently, X-ray interferometers are quite different in operating principle, design and application from, for example, optical or radio interferometers.

The only intense line sources of X-rays available are the spontaneously emitted line spectra of atoms. Their line widths are determined almost entirely by radiation damping (see, for example, Compton and Allison, 1935) so that their line widths $\delta\lambda/\lambda$ are typically 3×10^{-4} and are essentially independent of the X-ray wavelength. Thus, the coherence length for X-rays is about 0.2 μm. Mössbauer lines are much narrower but present sources are too weak for X-ray interferometry.

The refractive index n for X-rays is given by

$$1 - n = \frac{\lambda^2 e^2 N}{2\pi mc^2}(Z + \Delta f' + i\Delta f'')$$

where N is the number of atoms per cm^3, $(Z + \Delta f' + i\Delta f'')$ is the atomic scattering factor and the other symbols have their usual meanings. For solids $(1 - n)$ is a number between 10^{-4} and 10^{-7} in the wavelength range quoted above. Consequently, the specular reflection coefficient at normal incidence is in the range from 10^{-4} to 10^{-7}. On the other hand, the surface finish of X-ray optical components need only be accurate to $\lambda(1 - n)^{-1}$, a length of the order of tens of micrometres. A 1 mm diameter sphere would have a focal length in excess of 10 m and a numerical aperture of less than 10^{-3}. Clearly conventional optical techniques cannot be applied to make X-ray interferometers.

Over very narrow angular ranges, Bragg reflecting crystals are strong reflectors of X-rays and their use as beam amplitude splitters, mirrors and phase sensitive detectors has lead to the successful operation of many different X-ray interferometers.

There are two different cases to consider: the Laue case, in which the incident and diffracted beams are at different surfaces of the crystal and the Bragg case, in which the incident and diffracted beams are formed at the same surface of the crystal, as indicated in Fig. 1a and 1b respectively. According to the Laue equation, using the notation given in these figures,

$$\mathbf{K_0} + \mathbf{h} = \mathbf{K_h}$$

there are two equivalent waves in the crystal during Bragg reflection. Their wave vectors are $\mathbf{K_0}$ and $\mathbf{K_h}$ ($|\mathbf{K_0}| \simeq |\mathbf{K_h}| \simeq \lambda^{-1}$) and they differ in wave vector by $\mathbf{h}$, the reciprocal lattice vector for the acting Bragg reflection: we have two plane waves whose standing wave pattern fits onto the Bragg planes. Thus, the two waves are phase locked together by being locked to the crystal lattice whose spacing d is equal to $|\mathbf{h}|^{-1}$. The Bragg reflecting crystal can be used as a beam amplitude splitter or as a phase maintaining mirror. In the Laue case, with thick crystals, (Fig. 1a) the *position* of the standing wave pattern in the crystal lattice does not depend on the angle of incidence. With thin crystals the position of the standing wavefield depends both on the angle of incidence and on the crystal thickness. In the Bragg case (Fig. 1b) the *position* of the standing wavefield in the crystal lattice changes as the angle of incidence is varied. Details of these variations can be found in general review articles (Bonse, 1969; Hart and Bonse, 1970; Hart, 1971).

A representative X-ray interferometer layout is sketched in Fig. 2. Four "Laue case" crystals are used as the necessary optical components viz. the beam splitter S, two transmission mirrors M and a phase-sensitive detector or analyser A. Interferometers using "Bragg case" components as well as mixtures of Laue case and Bragg case components have also been successfully operated (Bonse and Hart, 1966, 1968). In Fig. 2 the disposition of the crystals is such that two beams are brought to interference at the entrance surface of the analyser crystal where, in the region near P, a two-beam interference pattern is set up. If all of the crystals are perfectly aligned this pattern is simply an image of their atomic planes—the interference pattern formed by two coherent plane waves. In the exit beams of the interferometer, O and H, we observe the moiré pattern formed by the superposition of the lattice image at P onto the Bragg planes of the analyser crystal A. It is obvious that no pattern could with a spacing $\Lambda_R = d/\Phi$ while a small change of lattice spacing $\delta d/d$ produces moiré fringes normal to **h** with a spacing $\Lambda_D = d^2/\delta d$. As a check that this is indeed the mode of operation of the interferometers both of these simple cases have been quantitatively demonstrated (Bonse and Hart, 1966). With existing X-ray interferometers fringe spacings up to 2 cm can be measured so that lattice rotations $\Phi \simeq 10^{-8}$ rad or dilatations $\delta d/d \simeq 10^{-8}$ can be directly measured.

If the analyser wafer of a perfectly aligned X-ray interferometer is translated parallel to **h** then one moiré fringe is recorded in the exit beams when the wafer moves a distance d whatever X-ray wavelength is used! If the translation is simultaneously measured with an optical interferometer, one can make a direct measurement of the crystal lattice spacing without needing to know the X-ray wavelength (Bonse and Hart, 1965; Bonse and te Kaat, 1968; Hart, 1968a; Deslattes, 1969). In this way, the crystal lattice spacing can be used as a secondary standard of length.

Inhomogeneous strain too produces moiré patterns in the output beams of an X-ray interferometer, which can be photographed. This is a useful technique for the characterization of defects in good single crystals. Local variations in lattice spacing and orientation can be measured as well as dislocations and planar defects such as stacking faults and twin lamellae (Bonse and Hart, 1966; Hart, 1968b). Small strains which have been intentionally introduced into crystals, for example by impurity diffusion or by thermal expansion have also been measured by this method (Bonse, Hart and Schwuttke, 1969).

Bibliography

BONSE, U. (1969) *Proc. Vth. International Congress on X-ray Optics and Microanalysis.* Ed. G. Möllenstedt and K. H. Gaukler, Springer-Verlag.

BONSE, U. and HART, M. (1965) *Z. Phys.* **188**, 154.

BONSE, U. and HART, M. (1966) *Z. Phys.* **194**, 1.

BONSE, U. and HART, M. (1968) *Acta Cryst.* A**24**, 240.

BONSE, U., HART, M. and SCHWUTTKE, G. H. (1969) *Phys. stat. sol.* **33**, 361.

BONSE, U. and HELLKÖTTER, H. (1969) *Z. Phys.* **223**, 345.

BONSE, U. and TE KAAT, E. (1968) *Z. Phys.* **214**, 16.

COMPTON, A. H. and ALLISON, S. K. (1935) *X-rays in Theory and Experiment.* New York: Van Nostrand.

CREAGH, D. and HART, M. (1970) *Phys. stat. sol.* **37**, 753.

DESLATTES, R. D. (1969) *Appl. Phys. Letters*, **15**, 386.

HART, M. (1968a) *Brit. J. Appl. Phys.* (*J. Phys. D*) Ser. 2, **1**, 1405.

HART, M. (1968b) *Science Progress (Oxf.)* **56**, 429.

HART, M. (1971) *Reports on Progress in Physics* **34**, 435.

HART, M. and BONSE, U. (1970) *Physics Today*, **23**, 26.

KATO, N. and TANEMURA, S. (1967) *Phys. Rev. Letters*, **19**, 22.

MICHAEL HART.

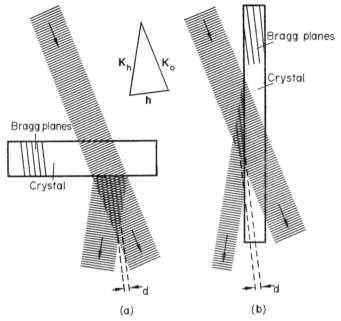

Fig. 1. Waves excited during Bragg reflection. Notice that the standing wavefield, indicated by broken lines, has exactly the same orientation and spacing as the Bragg planes. (a) Laue-case, (b) Bragg case.

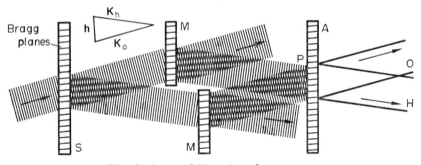

Fig. 2. A typical X-ray interferometer.

be observed unless the whole interferometer was stable to a small fraction of $|\mathbf{h}|^{-1}$; approximately 10 pm. For this reason most of the X-ray interferometers made so far have been cut as monolithic blocks of single crystal.

Applications

While little work has been done in that direction, X-ray interferometers can be used to measure interference fringe visibility from which emission line spectral profiles can be calculated. X-ray atomic scattering factors have been calculated from measurements of the refractive index of various materials (Bonse and Hellkötter, 1969; Creagh and Hart, 1970) and the phase shifts occurring when material is put into one of the interfering beams used as a method of measuring material thickness (Kato and Tanemura, 1967). These same measurements are commonly made too with optical interferometers.

The novel applications of X-ray interferometers stem directly from the moiré-making properties of the Bragg reflecting components. Suppose, for a moment, that one or more of the diffracting crystals is homogeneously or inhomogeneously strained. If the reciprocal vector of the standing wavefield at P is $\mathbf{h} + \delta\mathbf{h}$ and the reciprocal lattice vector of the analyser crystal is $\mathbf{h}$, then the observed moiré pattern will have a reciprocal vector $\delta\mathbf{h}$. For a small rotation Φ, moiré fringes are observed parallel to $\mathbf{h}$

INDEX

Page numbers in parenthesis refer to the first mention of the indexed term in the article referred to. Where no page numbers are given in parenthesis the reference is to the whole article.

Acoustics in the fishing industry 1
Acousto-optic interactions: *see* Advances in oxide materials 4 (7)
Advances in oxide materials 4
Alphanumeric displays: *see* Ferroelectrics, recent applications of 89 (91)
Amino acids, synthesis of: *see* Recent advances in radio and X-ray astronomy 256 (257)
Analysis, electrogravimetric: *see* Electrogravimetric analysis 76
Angiography: *see* Magnetism, medical applications of 153 (156)
Anti-protonic atoms: *see* Mesonic atoms 160 (163)
Applications of reverse osmosis and ultrafiltration 10
Archaeological dating by physical methods 13
Archaeomagnetism: *see* Archaeological dating by physical methods 13 (15)
Asteroid Toro, resonance in orbit of: *see* Resonance in the orbits of the Earth and an asteroid 276 (276)
Astronomy, radio, recent advances in: *see* Recent advances in radio and X-ray astronomy 256
—, **X-ray, recent advances in**: *see* Recent advances in radio and X-ray astronomy 256
Atmosphere, barium cloud experiments in: *see* Barium clouds for experiments in space 18
—, **pollution of the**: *see* Pollution of the air 228 (228)
Atom-probe field-ion microscope: *see* Field-ion microscopy and the atom-probe FIM 96
Atomic frequency standard: *see* Time, atomic 349 (351)
— **time**: *see* Time, atomic 349
Atoms, anti-protonic: *see* Mesonic atoms 160 (163)
—, **detection and identification in mass spectrometry**: *see* Detection and identification of individual atoms and ions in mass spectrometry 62
—, **kaonic**: *see* Mesonic atoms 160 (160)
—, **mesonic, biomedical and industrial uses of**: *see* Mesonic atoms 160 (163)
—, **muonic**: *see* Mesonic atoms 160 (161)
—, **pionic**: *see* Mesonic atoms 160 (162)
—, **sigmic**: *see* Mesonic atoms 160 (163)
—, **xionic**: *see* Mesonic atoms 160 (163)
ATR spectra: *see* Reflection spectroscopy 260 (265)

Autocorrelation function in speech spectrometry: *see* Speech spectrometry 298 (303)
Automation of nondestructive testing: *see* Nondestructive testing, automation of 206

Barium clouds for experiments in space 18
Battery, isotope: *see* Power sources, isotopic 230 (233)
—, **seawater**: *see* Biochemical fuel cells 29 (32)
Bayes criterion: *see* Signal detection, statistical theory of 281 (282)
Beam-foil spectroscopy 25
Big bang model: *see* Recent advances in radio and X-ray astronomy 256 (259)
Biochemical applications of reverse osmosis: *see* Applications of reverse osmosis and ultrafiltration 10 (12)
Biochemical fuel cells 29
Biological applications of spectroscopy: *see* Spectroscopy, biological applications of 288
Biology, lasers in: *see* Lasers in biology and medicine 142
Biomedicine, mesonic atoms in: *see* Mesonic atoms 160 (163)
Black holes in space 32
— **holes**: *see* Black holes in space 32; Neutron stars 203 (203); Recent advances in radio and X-ray astronomy 256 (256)
Blocking patterns from crystals 35
Blood flow, measurement of: *see* Magnetism, medical applications of 153 (155)
Bone, dating of: *see* Archaeological dating by physical methods 13 (17)
Bootstrap sweep generator: *see* Sweep circuits 320 (321)
Bronchography: *see* Magnetism, medical applications of 153 (156)

Cables, development of: *see* Insulation of electric power cables 132 (132)
—, **insulation of**: *see* Insulation of electric power cables 132

Cables, future development of: *see* Insulation of electric power cables 132
—, **high voltage direct current:** *see* Insulation of electric power cables 132 (134)
—, **materials for:** *see* Insulation of electric power cables 132
Calculation of thermodynamic properties from spectroscopic data 38
Carbon fibre 39
Carbon, formation of: *see* Carbon fibre 39 (40)
Catheters, magnetic: *see* Magnetism, medical applications of 153 (154)
Cement, fibre reinforcement of: *see* Fibre reinforcement of cement and concrete 93
Cerenkov counters: *see* Counting technology, recent advances in 51 (53)
Character recognition by machine: *see* Pattern recognition 222 (222)
Charge-transfer complexes 43
Chemical dating: *see* Archaeological dating by physical methods 13 (17)
Circular dichroism 45
Climatology: *see* Paleo-geophysics 220 (220)
Clock paradox, Einstein: *see* Time, atomic 349 (351)
Colour, four-colour problem: *see* Four-colour problem 98
Colour transparencies, copying of 48
Combinational net: *see* Digital learning computers 65 (67)
Communication systems, optical: *see* Optical communication systems 209
Computers, digital learning: *see* Digital learning computers 65
Concrete, fibre reinforcement of: *see* Fibre reinforcement of cement and concrete 93
Conformations, biological: *see* Spectroscopy, biological applications of 288 (288)
Contouring by holography: *see* Holography, engineering applications of 123 (126)
Control, myoelectric: *see* Myoelectric control 192
Cosmic rays, high-energy heavy ions in: *see* High-energy heavy ions 115 (115)
Counters, Cerenkov: *see* Counting technology, recent advances in 51 (53)
—, **Geiger:** *see* Counting technology, recent advances in 51 (51)
—, **proportional:** *see* Counting technology, recent advances in 51 (51)
—, **scintillation:** *see* Counting technology, recent advances in 51 (51)
Counting technology, recent advances in 51
Cramer-Rao bound: *see* Signal detection, statistical theory of 281 (287)
Cranium, laser irradiation of: *see* Lasers in biology and medicine 142 (143)
Critical opalescence: *see* Critical point 54 (56)
— point 54

Critical solution temperature: *see* Critical point 54 (56)
Cryostatic stabilization: *see* Superconductivity, applications of 310 (312)
Crystal structure determination, direct methods of 57
Crystals, blocking patterns from: *see* Blocking patterns from crystals 35
—, **liquid:** *see* Magnetism, medical applications of 153 (154)
—, **orientation of:** *see* Blocking patterns from crystals 35 (37)
Czochralski technique: *see* Advances in oxide materials 4 (4)

Dairy industry, reverse osmosis applications in: *see* Reverse osmosis and ultrafiltration, applications of 10 (12)
Data processing, optical: *see* Optical data processing 213
Dating, archaeological: *see* Archaeological dating by physical methods 13; Thermoluminescent dating 335
—, **thermoluminescent:** *see* Thermoluminescent dating 335
Delay lines: *see* Advances in oxide materials 4 (7)
Delphi method of technological forecasting: *see* Technological forecasting 327 (330)
Detection and identification of individual atoms and ions in mass spectrometry 62
Dielectric loss: *see* Insulation of electric power cables 133 (133)
Digital learning computers 65
Display systems in speech spectrometry: *see* Speech spectrometry 298 (303)
Doppler unit for ultrasonics: *see* Ultrasonics in medical diagnosis 353 (355)
Dosimeter, colour radiation: *see* Radiation dosimeter, colour 248
Dounreay fast reactor: *see* Void formation in fast reactor materials 358 (358)
Dybbuks: *see* Tachyons 325 (327)
Dye dosimeters: *see* Radiation dosimeter, colour 248 (249)
Dynamo theory of Stewart: *see* Barium clouds for experiments in space 18 (21)

Earth, magnetic field of: *see* Palaeogeophysics 220 (220)
—, **resonance in orbit of:** *see* Resonance in the orbits of the Earth and an asteroid 276
—, **rotation of:** *see* Palaeogeophysics 220 (221)
—, **thermal history of:** *see* Palaeogeophysics 220 (221)

Effluent treatment: *see* Applications of reverse osmosis and ultrafiltration 10 (13)
Einstein clock paradox: *see* Time atomic 349 (351)
— theory of relativity: *see* Geometrodynamics 106 (106); Tachyons 325 (325)
Electret transducers 68
Electric field, measurement of: *see* Barium clouds for experiments in space 18 (21)
Electric motors, fractional horsepower: *see* Fractional horsepower motors 99
—, hysteresis: *see* Fractional horsepower motors 99 (104)
—, induction: *see* Fractional horsepower motors 99 (100)
—, reluctance: *see* Fractional horsepower motors 99 (103)
—, single phase commutator or universal: *see* Fractional horsepower motors 99 (102)
—, using transverse flux 72
Electrogravimetric analysis 76
Electrogyro: *see* Regenerative flywheels 270 (271)
— locomotives: *see* Regenerative flywheels 270 (272)
Electrojet, equatorial: *see* Barium clouds for experiments in space 18 (21)
Electromagnetic flowmeters: *see* Magnetism, medical applications of 153 (155)
— gun 81
Electronic absorption spectra: *see* Spectroscopy, biological applications of 288 (288)
— emission spectra: *see* Spectroscopy, biological applications of 288 (290)
— equipment, modular: *see* Modular electronic equipment 174
Electron microscope, mirror: *see* Mirror electron microscope 173
— multipliers, windowless: *see* Detection and identification of individual atoms and ions in mass spectrometry 62 (62)
Electro-optic effect: *see* Ferroelectrics, recent applications of 89 (91)
— lens: *see* Ferroelectrics, recent applications of 89 (91)
Electro-optics: *see* Ferroelectrics, recent applications of 89 (90)
Electrophoretic paint recovery: *see* Applications of reverse osmosis and ultrafiltration 10 (13)
Elementary particles, Regge model of: *see* Regge model of elementary particles 273
Elements, super-heavy: *see* Super-heavy elements beyond the known periodic table 316
Engine, Wankel: *see* Wankel engine 365
Enthalpimetry 84
—, direct injection: *see* Enthalpimetry 84 (85)
Enzyme systems in fuel cells: *see* Biochemical fuel cells 29 (31)
Epitaxial films, study of: *see* Blocking patterns from crystals 35 (37)

Equatorial electrojet: *see* Barium clouds for experiments in space 18 (21)
Ergosphere: *see* Black holes in space 32 (34)
Estimation problem in signal detection: *see* Signal detection, statistical theory of 281 (287)
Eye, laser irradiation of: *see* Lasers in biology and medicine 142 (142)

Feedback net: *see* Digital learning computers 65 (67)
FE-PC (ferroelectric-photoconductive) sandwich: *see* Ferroelectrics, recent applications of 89 (91)
Ferroelectrics, recent applications of 89
Fibre attenuation: *see* Optical communication systems 209 (211)
— bandwidth, single and multimode: *see* Optical communication systems 209 (211)
—, carbon: *see* Carbon fibre 39
— manufacture: *see* Optical communication systems 209 (211)
—, optical: *see* Optical communication systems 209 (211)
— reinforcement of cement and concrete 93
Field-ion microscopy and the atom-probe FIM 96
Films, colour, copying of: *see* Colour transparencies, copying of 48
—, thick, technology of: *see* Thick film technology 344
Filter, electrical: *see* Ferroelectrics, recent applications of 89 (90)
—, optical: *see* Optical data processing 213 (216)
Fishing industry, acoustics in: *see* Acoustics in the fishing industry 1
Fission-track dating: *see* Archaeological dating by physical methods 13 (16)
Flash spectroscopy: *see* Spectroscopy, biological applications of 288 (289)
Flowmeters, electromagnetic: *see* Magnetism, medical applications of 153 (155)
Flywheels, regenerative: *see* Regenerative flywheels 270
Food industry, applications of reverse osmosis and ultrafiltration to: *see* Applications of reverse osmosis and ultrafiltration 10 (12)
Forecasting, technological: *see* Technological forecasting 327
Four-colour problem 98
Fourier transform, optical lenses for: *see* Optical processing of radar and sonar signals 215 (216)
— transform spectroscopy: *see* Reflection spectroscopy 260 (264)
Fractional horsepower motors 99
Frequency standard, atomic: *see* Time, atomic 349 (351)
Fuel cells, biochemical: *see* Biochemical fuel cells 29

Galactic X-ray sources: *see* Recent advances in radio and X-ray astronomy 256 (257)
Galaxies, infrared: *see* Infrared galaxies 128
—, **radio emission from:** *see* Recent advances in radio and X-ray astronomy 256 (258)
—, **Seyfert:** *see* Infrared galaxies 128 (128)
—, **X-ray:** *see* Recent advances in radio and X-ray astronomy 256 (259)
GARP (Global atmospheric research programme): *see* Meteorological observations from satellites 164 (164)
Geiger counters: *see* Counting technology, recent advances in 51 (51)
Gemstones: *see* Advances in oxide materials 4 (8)
Generators, superconducting a.c.: *see* Superconducting machines 304 (309)
—, **sweep:** *see* Sweep circuits 320 (320)
Geodesy: *see* Palaeogeophysics 220 (221)
Geoelectricity: *see* Palaeogeophysics 220 (221)
Geometrodynamics 106
Geon: *see* Geometrodynamics 106 (106)
Glass dosimeters: *see* Radiation dosimeter, colour 248 (250)
Glass-fibre reinforced cement: *see* Fibre reinforcement of cement and concrete 93 (93)
Graphite fibre: *see* Carbon fibre 39 (41)
—, **formation of:** *see* Carbon fibre 39 (40)
Gravitational waves: *see* Geometrodynamics 106 (106)
Guided wave systems: *see* Optical communication systems 209 (211)
Gyreactor: *see* Regenerative flywheels 270 (272)
Gyrobus: *see* Regenerative flywheels 270 (271)

Helium dilution refrigeration 109
—, **negative ion of** 109
High-energy heavy ions 115
Himalayan-Tibetan massif, in monsoons: *see* Monsoon meteorology 187 (190)
Hologram, recording and reconstructing of: *see* Holography, engineering applications of 123 (123)
Holographic recording of radar signals: *see* Optical processing of radar and sonar signals 215 (218)
Holography, engineering applications of 123
—, **interference:** *see* Holography, engineering applications of 123 (124)
Homopolar machines, superconducting: *see* Superconducting machines 304 (308)
Hyperchromism: *see* Spectroscopy, biological applications of 288 (289)
Hypochromism: *see* Spectroscopy, biological applications of 288 (289)
Hypothesis testing: *see* Signal detection, statistical theory of 281 (282)

Ideal observer test: *see* Signal detection, statistical theory of 281 (282)
Infrared galaxies 128
— **interferometer spectrometer:** *see* Meteorological observations from satellites 164 (170)
Insulation of cables: *see* Insulation of electric power cables 132
— **of cables, future developments in:** *see* Insulation of electric power cables 132 (135)
— **of electric power cables** 132
— **of high-voltage direct current cables:** *see* Insulation of electric power cables 132 (134)
—, **parameters of:** *see* Insulation of electric power cables 132 (133)
Interferometer for radio star work: *see* Recent advances in radio and X-ray astronomy 256 (258)
Interferometry, X-ray: *see* X-ray interferometry 368
Ion clouds, observation of: *see* Barium clouds for experiments in space 18 (19)
— **implantation** 136
Ionium and uranium dating: *see* Archaeological dating by physical methods 13 (16)
Ionization cells: *see* Power sources, isotopic 230 (232)
Ionosphere, barium cloud experiments in the: *see* Barium clouds for experiments in space 18 (21)
Ions, detection and identification in mass spectrometry: *see* Detection and identification of individual atoms and ions in mass spectrometry 62
—, **high-energy heavy:** *see* High-energy heavy ions 114
Isotope battery: *see* Power sources, isotopic 230 (233)
Isotopic power sources: *see* Power sources, isotopic 230

Kaonic atoms: *see* Mesonic atoms 160 (162)

Lasers, crystalline: *see* Advances in oxide materials 4 (6)
— **in biology and medicine** 142
—, **hazards of:** *see* Radiation, non-ionizing hazards of 251 (253)
Learning computers, digital: *see* Digital learning computers 65
— **elements:** *see* Digital learning computers 65 (65)
Lens, electro-optic: *see* Ferroelectrics, recent applications of 89 (92)
Limb scanning radiometer: *see* Meteorological observations from satellites 164 (170)
Linear motors: *see* Electric motors using transverse flux 72 (72); Electromagnetic gun 81 (81)
—, **tubular type:** *see* Electromagnetic gun 81 (81)
Location and reduction of noise 144

Loran-C navigation system: *see* Time, atomic 349 (351)

Mach bands: *see* Mach effect (visual) 151 (151)
— **effect (visual)** 151
Magnetic catheters: *see* Magnetism, medical applications of 153 (154)
— **dating:** *see* Archaeological dating by physical methods 13 (15)
— **devices in medicine:** *see* Magnetism, medical applications of 153 (154)
— **fields, direct biological influence of:** *see* Magnetism, medical applications of 153 (153)
— **fields, measurement techniques for:** *see* Magnetism, medical applications of 153 (155)
— **fields of planets:** *see* Planets, magnetic fields of 225
— **instruments in medicine:** *see* Magnetism, medical applications of 153 (154)
— **mirrors:** *see* Thermonuclear power—the present position 341 (343)
— **oxides:** *see* Advances in oxide materials 4 (7)
Magnetism, medical applications of 153
Magnetohydrodynamic power generation: *see* Superconductivity, applications of 310 (315)
Magnetosphere, barium cloud experiments in the: *see* Barium clouds for experiments in space 18 (22)
Magnets, superconducting: *see* Superconductivity, applications of 310 (314)
Mass spectrometer, atom-probe: *see* Detection of individual atoms and ions in mass spectrometry 62 (64)
— **spectrometry, detection and identification of individual atoms and ions in:** *see* Detection of individual atoms and ions in mass spectrometry 62
Medical applications of myoelectric control: *see* Myoelectric control 192 (197)
— **diagnosis, ultrasonics in:** *see* Ultrasonics in medical diagnosis 353
Medicine, lasers in: *see* Lasers in biology and medicine 142
—, **magnetism in:** *see* Medicine, medical applications of 153
Mesonic atoms 160
— **atoms, biomedical and industrial uses of:** *see* Mesonic atoms 160 (163)
Metarelativity: *see* Tachyons 325 (325)
Meteorological observations from satellites 164
Microscopy, field ion: *see* Field ion microscopy and the atom-probe FIM 96
Microwave phonons 171
Miller sweep generators: *see* Sweep circuits 320 (322)
Mirror electron microscope 173
Modular electronic equipment 174
Modulation spectroscopy 176

Molecules, muonic: *see* Mesonic atoms 160 (162)
Monsoon meteorology 187
Monsoons, definition and extent of: *see* Monsoon meteorology 187 (187)
—, **role of Himalayan-Tibetan massif in:** *see* Monsoon meteorology 187 (190)
—, **seasonable transitions of:** *see* Monsoon meteorology 187 (191)
MOST (metal oxide semiconductor transistor): *see* Ion implantation 136 (140)
Motors, electric: *see* Electric motors using transverse flux 72; Fractional horsepower motors 99; Electromagnetic gun 81 (81)
—, **linear:** *see* Electric motors using transverse flux 72 (72); Electromagnetic gun 81 (81)
—, **linear, tubular type:** *see* Electromagnetic gun 81 (81)
Muonic atoms: *see* Mesonic atoms 160 (161)
— **molecules:** *see* Mesonic atoms 160 (162)
Muonium, chemical interactions of: *see* Mesonic atoms 160 (162)
Myoelectric control 192
—, **medical applications of:** *see* Myoelectric control 192 (197)
—, **modes of:** *see* Myoelectric control 192 (196)
— **system, block diagram of:** *see* Myoelectric control 192 (193)
Myoelectric signals: *see* Myoelectric control 192 (193)

Navigation, Loran-C system: *see* Time, atomic 349 (351)
Neuromuscular system: *see* Myoelectric control 192 (194)
Neutral clouds, observation of: *see* Barium clouds for experiments in space 18 (19)
Neutron radiography—the present position 198
—, **detection methods in:** *see* Neutron radiography—the present position 198 (201)
—, **neutron sources for:** *see* Neutron radiography—the present position 198 (199)
Neutron stars 203
Neutrons, chemical binding of 205
Neyman-Pearson criterion: *see* Signal detection, statistical theory of 281 (283)
Nimbus-E microwave spectrometer: *see* Meteorological observations from satellites 164 (170)
Noise, physical description of: *see* Location and reduction of noise 144 (144)
—, **location and reduction of:** *see* Location and reduction of noise 144
Nondestructive testing, automation of 206
—, **use of holography in:** *see* Holography, engineering applications of 123 (126)
Novae: *see* Recent advances in radio and X-ray astronomy 256 (258)

Nuclear chemistry, high-energy heavy ions in: *see* High-energy heavy ions 114 (118)
— **physics, high-energy heavy ions in:** *see* High-energy heavy ions 114 (117)
— **reactor materials, void formation in:** *see* Void formation in fast reactor materials 358
— **reactors, thermonuclear:** *see* Thermonuclear power—the present position 341

Obsidian, dating of: *see* Archaeological dating by physical methods 13 (17)
Oppenheimer-Volkoff limit: *see* Black holes in space 32 (33)
Optical communication systems 209
— **"constants":** *see* Reflection spectroscopy 260 (260)
— **data processing** 213
— **display systems, ferroelectrics in:** *see* Ferroelectrics, recent applications of 89 (91)
— **fibres:** *see* Optical communication systems 209 (211)
— **filter:** *see* Optical processing of radar and sonar signals 215 (216)
— **processing of radar and sonar signals** 215
Orbits of the Earth and an asteroid, resonance in: *see* Resonance in the orbits of the Earth and an asteroid 276
Osmosis, reverse, applications of: *see* Applications of reverse osmosis and ultrafiltration 10
Oxide materials, advances in: *see* Advances in oxide materials 4
—, **amorphous:** *see* Advances in oxide materials 4 (9)
—, **composite:** *see* Advances in oxide materials 4 (8)
—, **magnetic:** *see* Advances in oxide materials 4 (7)
—, **polycrystalline:** *see* Advances in oxide materials 4 (8)
—, **single crystal:** *see* Advances in oxide materials 4 (4)

Paint, electrophoretic recovery of: *see* Applications of reverse osmosis and ultrafiltration 10 (13)
Palaeogeophysics 220
Paraelectric material: *see* Ferroelectrics, recent applications of 89 (89)
Particles, elementary, Regge model of: *see* Regge model of elementary particles 273
Pattern recognition 222
Perceptrons: *see* Pattern recognition 222 (224)
Phase determination in crystal structure analysis: *see* Crystal structure determination, direct methods of 56 (58)
Phonons, microwave: *see* Microwave phonons 171
Phosphenes: *see* Magnetism, medical applications of 153 (153)

Photoelectric spectra, intensities of: *see* Spectroscopy, photoelectron 293 (297)
Photoelectron spectroscopy: *see* Spectroscopy, photoelectron 293
Photomultipliers: *see* Counting Technology, recent advances in 51 (53)
Piezoelectric effect: *see* Ferroelectrics, recent applications of 89 (90)
Pionic atoms: *see* Mesonic atoms 160 (162)
Planets, magnetic fields of 225
Plastics, use of in dosimeters: *see* Radiation dosimeter, colour 248 (250)
—, **use of in fibre-reinforced concrete:** *see* Fibre reinforcement of cement and concrete 93 (95)
Pollution of the air 228
Potassium-argon dating: *see* Archaeological dating by physical methods 13 (16)
Power source, direct-charge collection: *see* Power sources, isotopic 230 (231)
— **sources, isotopic** 230
— **transformer, transcutaneous:** *see* Magnetism, medical applications of 153 (156)
Predictor coefficients: *see* Speech spectrometry 298 (304)
Pressure modulator radiometer: *see* Meteorological observations from satellites 164 (170)
Prosthetic devices: *see* Magnetism, medical applications of 153 (156)
Proton blocking: *see* Blocking patterns from crystals 35 (36)
Pulsars 235
—, **discovery of:** *see* Black holes in space 32 (34); Pulsars 235 (235)
Pulse height analysis: *see* Counting technology, recent advances in 51 (53)
— **shape discrimination:** *see* Counting technology, recent advances in 51 (53)
Pyroelectric effect: *see* Advances in oxide materials 4 (8); Ferroelectrics, recent applications of 89 (89)

Quasars 238
—, **discovery of:** *see* Black holes in space 32 (34); Quasars 238 (238)
—, **nature of:** *see* Quasars 238 (240)
—, **optical properties of:** *see* Quasars 238 (238)
—, **radio properties of:** *see* Quasars 238 (239)
—, **red shifts of:** *see* Quasars 238 (239)

Radar signals, holographic recording of: *see* Optical processing of radar and sonar signals 215 (218)
—, **optical processing of:** *see* Optical processing of radar and sonar signals 215
Radiation and entry of space vehicles into planetary atmospheres 244
— **and shielding in space** 246

Radiation dosimeter, colour 248
—, **laser, hazards of**: *see* Radiation, non-ionizing, hazards of 251 (253)
—, **non-ionizing, hazards of** 251
—, **nuclear, in space**: *see* Radiation and shielding in space 246 (246)
—, **radiofrequency, hazards of**: *see* Radiation, non-ionizing, hazards of 251 (252)
— **shielding in space**: *see* Radiation and shielding in space 246
—, **ultrasonic, hazards of**: *see* Radiation, non-ionizing, hazards of 251 (254)
Radio astronomy, recent advances in: *see* Recent advances in radio and X-ray astronomy 256
Radiobiology, high-energy heavy ions in: *see* High-energy heavy ions 114 (118)
Radiocarbon dating: *see* Archaeological dating by physical methods 13 (13)
Radio emission from galaxies: *see* Recent advances in radio and X-ray astronomy 256 (258)
Radiography, neutron: *see* Neutron radiography—the present position 198
Radiometer, limb scanning: *see* Meteorological observations from satellites 164 (170)
—, **pressure modulator**: *see* Meteorological observations from satellites 164 (170)
—, **selective chopper**: *see* Meteorological observations from satellites 164 (169)
Radio stars: *see* Recent advances in radio and X-ray astronomy 256 (258)
Radiotherapy, high-energy heavy ions in: *see* High-energy heavy ions 114 (118)
Radio waves, pulsating sources of: *see* Pulsars 235
Raman scattering in modulation spectroscopy: *see* Modulation spectroscopy 176 (186)
Recent advances in radio and X-ray astronomy 256
Red shift from astronomical bodies: *see* Quasars 238 (238)
Reflection spectroscopy 260
Refrigeration, helium dilution: *see* Helium dilution refrigeration 109
Regenerative flywheels 270
Regge model of elementary particles 273
— **poles**: *see* Regge model of elementary particles 273 (274)
— **trajectories**: *see* Regge model of elementary particles 273 (274)
Remanent magnetism, natural: *see* Palaeogeophysics 220 (220)
Resonance in orbits of the Earth and an asteroid 276
Reticuloendothelial system: *see* Magnetism, medical applications of 153 (156)
Reverse osmosis, applications of: *see* Applications of reverse osmosis and ultrafiltration 10

Satellite infrared spectrometer: *see* Meteorological observations from satellites 164 (169)

Satellites, meteorological observations from: *see* Meteorological observations from satellites 164
Scene analysis by machine: *see* Pattern recognition 222 (224)
Schwarzchild radius: *see* Black holes in space 32 (33)
Scintillators, crystalline: *see* Counting technology, recent advances in 51 (51)
Screen printing: *see* Thick film technology 344 (348)
Seawater battery: *see* Biochemical fuel cells 29 (32)
—, **treatment of**: *see* Applications of reverse osmosis and ultrafiltration 10 (12)
Seismology: *see* Palaeogeophysics 220 (221)
Selective chopper radiometer: *see* Meteorological observations from satellites 164 (169)
Semiconductor detectors: *see* Counting technology, recent advances in 51 (52)
Semiconductors, substrates for: *see* Advances in oxide materials 4 (7)
Sewage treatment: *see* Applications of reverse osmosis and ultrafiltration 10 (13)
Seyfert galaxies: *see* Infrared galaxies 128 (128); Recent advances in radio and X-ray astronomy 256 (259)
Sigmic atoms: *see* Mesonic atoms 160 (163)
Signal detection, statistical theory of 281
Skin, electromagnetic transmission through: *see* Magnetism, medical applications of 153 (156)
—, **laser irradiation of**: *see* Lasers in biology and medicine 142 (142)
SLAM (Stored logic adaptable microcircuit): *see* Digital learning computers 65 (65)
Solvent perturbation techniques: *see* Spectroscopy, biological applications of 288 (289)
Sonar in the fishing industry: *see* Acoustics in the fishing industry 1 (3)
— **signals, optical processing of**: *see* Optical processing of radio and sonar signals 215
Space, radiation and shielding in: *see* Radiation and shielding in space 246
Space-time singularity: *see* Black holes in space 32 (33)
Space vehicles, re-entry into Earth's atmosphere: *see* Radiation and entry of space vehicles into planetary atmospheres 244 (244)
Spectra, electronic absorption: *see* Spectroscopy, biological applications of 288 (288)
—, **electronic emission**: *see* Spectroscopy, biological applications of 288 (290)
—, **photoelectron, intensities of**: *see* Spectroscopy, photoelectron 293 (297)
Spectrometer, infrared interferometer: *see* Meteorological observations from satellites 164 (170)
—, **Nimbus-E microwave**: *see* Meteorological observations from satellites 164 (170)
—, **satellite infrared**: *see* Meteorological observations from satellites 164 (170)
Spectrometry, Fourier transform: *see* Reflection spectroscopy 260 (264)

Spectrometry, speech: *see* Speech spectrometry 298
Spectroscopic data, calculation of thermodynamic properties from: *see* Calculation of thermodynamic properties from spectroscopic data 38
Spectroscopy, beam-foil: *see* Beam-foil spectroscopy 25
—, **biological applications of** 288
—, **flash**: *see* Spectroscopy, biological applications of 288 (289)
—, **modulation**: *see* Modulation spectroscopy 176; Spectroscopy, biological applications of 288 (289)
—, **photoelectron** 293
—, **reflection**: *see* Reflection spectroscopy 260
—, **thermal desorption**: *see* Thermal desorption spectroscopy 333
—, **wavelength-derivative**: *see* Modulation spectroscopy 176 (180)
—, **wavelength-modulation**: *see* Modulation spectroscopy 176 (180)
Speech recognition by machine: *see* Pattern recognition 222 (223)
— **spectrometry** 298
Sq current system: *see* Barium clouds for experiments in space 18 (21)
Stars, infrared: *see* Recent advances in radio and X-ray astronomy 256 (257)
—, **neutron**: *see* Black holes in space 32 (33); Neutron stars 203; Pulsars 235 (235); Recent advances in radio and X-ray astronomy 256 (256)
—, **radio and X-ray**: *see* Recent advances in radio and X-ray astronomy 256 (258)
Statistical theory of signal detection: *see* Signal detection, statistical theory of 281
Steel fibre reinforced concrete: *see* Fibre reinforcement of cement and concrete 93 (94)
Stellarator: *see* Thermonuclear power—the present position 341 (342)
Superconducting a.c. generators: *see* Superconducting machines 304 (309)
— **homopolar machines**: *see* Superconducting machines 304 (308)
— **machines** 304
— **magnets**: *see* Superconductivity, applications of 310 (314)
Superconductivity, applications of 310
Superconductors: *see* Superconducting machines 304 (304); Superconductivity, applications of 310 (310)
—, **filamentary**: *see* Superconductivity, applications of 310 (313)
— **for electrical machines**: *see* Superconducting machines 304 (306)
—, **types I and II**: *see* Superconductivity, applications of 310 (310)
Super-heavy elements beyond the known periodic table 316
Supernova explosions: *see* Recent advances in radio and X-ray astronomy 256 (257)
Surface layers, thin, study of: *see* Blocking patterns from crystals 35 (37)
Surgery, applications of ultrasonics in: *see* Medical ultrasonics 157 (159)
Sweep circuits 320
— **generators**: *see* Sweep circuits 320 (320)
Symmetry analysis in modulation spectroscopy: *see* Modulation spectroscopy 176 (185)

Tachyons 325
Technological forecasting 327
Thermal desorption spectroscopy 333
Thermionic converters: *see* Power sources, isotopic 230 (232)
Thermodynamic properties from spectroscopic data: *see* Calculation of thermodynamic properties from spectroscopic data 38
Thermoelectric generators: *see* Power sources, isotopic 230 (233)
Thermoluminescence, measurement of: *see* Thermoluminescent dating 335 (336)
Thermoluminescent dating 335
Thermonuclear power—the present position 341
Thick film technology 344
Time, atomic 349
Timebase circuits: *see* Sweep circuits 320 (320)
Tokamak: *see* Thermonuclear power—the present position 341 (342)
Transition radiation: *see* Counting technology, recent advances in 51 (53)

Ultrafiltration, applications of: *see* Applications of reverse osmosis and ultrafiltration 10
Ultrasonic radiation, hazards of: *see* Radiation, non-ionizing, hazards of 251 (254)
Ultrasonics in medical diagnosis 354
— **in medicine**: *see* Medical ultrasonics 157; Ultrasonics in medical diagnosis 353
—, **biological effects of**: *see* Medical ultrasonics 157 (158)
—, **diagnostic applications of**: *see* Ultrasonics in medical diagnosis 353
—, **surgical applications of**: *see* Medical ultrasonics 157 (159)
—, **therapeutic applications of**: *see* Medical ultrasonics 157 (159)
Ultraviolet photoelectron spectroscopy: *see* Spectroscopy, photoelectron 293 (295)
Universal time, coordinated: *see* Time, atomic 349 (349)
Uranium and ionium dating: *see* Archaeological dating by physical methods 13 (16)

Variable function logic elements: *see* Digital learning computers   65 (67)
Vibration, holographic study of: *see* Holography, engineering applications of   123 (125)
Void formation in fast-reactor materials   358

Wankel engine   365
Water, brackish, demineralization of: *see* Applications of reverse osmosis and ultrafiltration   10 (10)
—, high-purity, production of: *see* Applications of reverse osmosis and ultrafiltration   10 (11)
Watson-Crick helical structure: *see* Spectroscopy, biological applications of   288 (289)
Wavelength derivative spectroscopy: *see* Modulation spectroscopy   176 (180)
— modulation spectroscopy: *see* Modulation spectroscopy   176 (180)
White dwarf: *see* Neutron stars   203 (203)

White hole: *see* Black holes in space   32 (35)
Windowless electron multipliers: *see* Detection and identification of individual atoms and ions in mass spectrometry   62 (62)

Xionic atoms: *see* Mesonic atoms   160 (163)
X-ray astronomy, recent advances in: *see* Recent advances in radio and X-ray astronomy   256
— galaxies: *see* Recent advances in radio and X-ray astronomy   256 (259)
— interferometry   368
— photoelectron spectroscopy: *see* Spectroscopy, photoelectron   293 (294)
— sources, galactic: *see* Recent advances in radio and X-ray astronomy   256 (257)
— stars: *see* Recent advances in radio and X-ray astronomy   256 (258)
— structure analysis, direct methods of: *see* Crystal structure determination, direct methods of   56

ENCYCLOPAEDIC DICTIONARY OF PHYSICS

SCOPE OF THE DICTIONARY

For convenience in planning, and to provide a framework on which the Dictionary could be erected, physics and its related subjects have been divided into upwards of sixty sections. The sections are listed below, but, as the Dictionary is arranged alphabetically, they do not appear as sections in the completed work.

- Acoustics
- Astronomy
- Astrophysics
- Atomic and molecular beams
- Atomic and nuclear structure
- Biophysics
- Cathode rays
- Chemical analysis
- Chemical reactions, phenomena and processes
- Chemical substances
- Colloids
- Cosmic rays
- Counters and discharge tubes
- Crystallography
- Dielectrics
- Elasticity and strength of materials
- Electrical conduction and currents
- Electrical discharges
- Electrical measurements
- Electrochemistry
- Electromagnetism and electrodynamics
- Electrostatics
- Engineering metrology
- General mechanics
- Geodesy
- Geomagnetism
- Geophysics
- Heat
- Hospital and medical physics
- Industrial processes
- Ionization
- Isotopes
- Laboratory apparatus
- Low-temperature physics
- Magnetic effects
- Magnetism
- Mathematics
- Mechanics of fluids
- Mechanics of gases
- Mechanics of solids
- Mesons
- Meteorology
- Molecular structure
- Molecular theory of gases
- Molecular theory of liquids
- Neutron physics
- Nuclear reactions
- Optics
- Particle accelerators
- Phase equilibria
- Photochemistry and radiation chemistry
- Photography
- Physical metallurgy
- Physical metrology
- Positive rays
- Radar
- Radiation
- Radioactivity
- Reactor physics
- Rheology
- Solid-state theory
- Spectra
- Structure of solids
- Thermionics
- Thermodynamics
- Vacuum Physics
- X rays